RHEOLOGY

Volume 2: Fluids

Edited by

Giovanni Astarita
Giuseppe Marrucci
Luigi Nicolais

University of Naples
Naples, Italy

PLENUM PRESS · NEW YORK AND LONDON

Library of Congress Cataloging in Publication Data

International Congress on Rheology, 8th, Naples, 1980.
Rheology.

Proceedings of the International Congress on Rheology; 8th, 1980)
Includes indexes.
1. Rheology – Congresses. 2. Polymers and polymerization – Congresses. 3. Fluid dyanmics – Congresses. 4. Suspensions (Chemistry) – Congresses. I. Astarita, Giovanni. II. Marrucci, G. III. Nicolais, Luigi. IV. Title. V. Series: International Congress on Rheology. Proceedings; 8th, 1980.
QC189.I52 8th, 1980 [QC189.5.A1] 531′.11s [531′.11]
ISBN 0-306-40466-4 (v. 2) 80-16929

Proceedings of the Eighth International Congress of Rheology, held in Naples, Italy, September 1–5, 1980, published in three parts of which this is Volume 2.

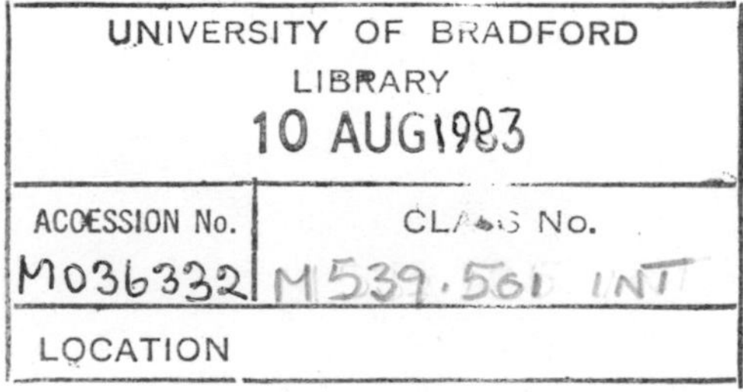

A Division of Plenum Publishing Corporation
227 West 17th Street, New York, N.Y. 10011

Printed in the United States of America

RHEOLOGY

Volume 2: Fluids

VIII INTERNATIONAL CONGRESS ON RHEOLOGY

Naples, September 1–5, 1980

HONORARY COMMITTEE

PRESIDENT

Prof. J. Kubat, *President, International Committee on Rheology*

MEMBERS

Dr. G. Ajroldi, *Past President, Italian Society of Rheology*
Prof. U. L. Businaro, *Director of Research, FIAT*
Dr. E. Cernia, *Director, Assoreni*
Prof. C. Ciliberto, *Vice-President, C.N.R.*
Prof. G. Cuomo, *Rector, University of Naples*
Dr. A. Del Piero, *Director, Tourism Bureau, Town of Naples*
Dr. D. Deuringer, *Director RAI, Radio-Television Network, Naples*
Prof. F. Gasparini, *Dean, Engineering School, University of Naples*
Porf. L. Malatesta, *President, Chemistry Committee, C.N.R.*
Prof. L. Massimilla, *Past Dean, Engineering School, University of Naples*
Prof. A. B. Metzner, *Fletcher Brown Professor, University of Delaware*
Prof. N. Polese, *President, University Social Services, Naples*
Prof. M. Silvestri, *President, Technical Committee, C.N.R.*
Prof. N. W. Tschoegl, *Secretary, International Committee on Rheology*
Sen. M. Valenzi, *Mayor, Town of Naples*
Prof. A. Valvassori, *Director, Istituto Donegani*

ORGANIZING COMMITTEE

Prof. G. Astarita, *President*
Prof. G. Marrucci
Prof. L. Nicolais, *Secretary*

ACKNOWLEDGMENTS

Support from the following Institutions in gratefully acknowledged:

Alitalia, Linee Aeree Italiane, Rappresentanza di Napoli
Assoreni
Azienda Autonoma di Soggiorno, Cura e Turismo di Napoli
Azienda Autonoma di Soggiorno, Cura e Turismo di Sorrento
Centro Ricerche FIAT, S.p.A.
Comitato per la Chimica del Consiglio Nazionale delle Ricerche
Comitato Tecnologico del Consiglio Nazionale delle Ricerche
Istituto Donegani S.p.A.
Opera Universitaria, Napoli
RAI, Radiotelevisione Italiana, Sede Regionale per la Campania
Società Italiana di Reologia
U.S. Air Force
Università di Napoli

CONTENTS OF THE VOLUMES

PREFACE

At the VIIth International Congress on Rheology, which was held in Göteborg in 1976, Proceedings were for the first time printed in advance and distributed to all participants at the time of the Congress. Although of course we Italians would be foolish to even try to emulate our Swedish friends as far as efficiency of organization is concerned, we decided at the very beginning that, as far as the Proceedings were concerned, the VIIIth International Congress on Rheology in Naples would follow the standards of timeliness set by the Swedish Society of Rheology. This book is the result we have obtained. We wish to acknowledge the cooperation of Plenum Press in producing it within the very tight time schedule available.

Every four years, the International Congress on Rheology represents the focal point where all rheologists meet, and the state of the art is brought up to date for everybody interested; the Proceedings represent the written record of these milestones of scientific progress in rheology. We have tried to make use of the traditions of having invited lectures, and of leaving to the organizing committee the freedom to choose the lecturers as they see fit, in order to collect a group of invited lectures which gives as broad as possible a landscape of the state of the art in every relevant area of rheology. The seventeen invited lectures are collected in the first volume of the proceedings. We wish to express our thanks, for agreeing to prepare these lectures on subjects suggested by ourselves, and for the effort to do so in the scholarly and elegant way that the reader will appreciate, to all the invited lectures: R.B.Bird, D.V.Boger, B.D.Coleman, J.M.Dealy, P.De Gennes, C.D.Denson, H.Janeschitz-Kriegl, A.Y.Malkin, R.A. Mashelkar, S.Onogi, C.J.S.Petrie, R.F.Schwarzl, J.Silberberg, K.Te Nijenhuis, C.A.Truesdell, K.Walters, K.Wichterle.

As for the organization of the Congress itself, at the time of writing it is still in the future, and we can only hope that it will work out smoothly. If it does, a great deal of merit will be due to the people who have agreed to act as Chairmen of the individual sessions, and we wish to acknowledge here their help: J.J.Benbow, B.Bernstein, H.C.Booij, B.Caswell, Y.Chen, M.Crochet, P.K.Currie, M.M.Denn, A.T.Di Benedetto, H.Giesekus, J.C.Halpin, A.Hoffmann, Y.Ivanov, L.P.B.Janssen, T.E.R.Jones, W.M.Jones, H.Kambe, J.L.Kardos, E.A.Kearsley, J.Klein, K.Kirschke, S.L.Koh, J.Kubat, R.F.Landel, R.L.Laurence, G.L.Leal, C.Marco, J.Meissner, B.Mena, A.B.Metzner, S.Middleman, Y.F.Missirlis, S.L.Passman, S.T.T.Peng, J.R.A.Pearson, R.S.Porter, P.Quemada, A.Ram, C.K.Rha, W.R.Schowalter, J.C.Seferis, C.L.Sieglaff, S.S.Sternstein, R.I. Tanner, N.Tschoegl, J.Vlachopoulos, J.L.White, C.Wolff, L.J.Zapas.

The contributed papers have been grouped in eleven subject areas: theory; fluid dynamics; rheometry; polymer solutions; polymer melts; suspensions; polymer processing; rubber; polymeric solids; biorheology; miscellaneous. Of these, the first one (theory) has been included in the first volume together with the invited lectures; the next five, which all deal with fluid-like materials, have been included in the second volume, and the last five have benn included in the third volume. Categorizations such as these invariably have a degree of arbitrariness, and borderline cases where a paper could equally well have been included in two different categories do exist; we hope the subject index is detailed enough to guide the reader to any paper which may be placed in a category unexpected from the reader's viewpoint.

Rheology is not synonymous with Polymer Science, yet sometimes it almost seems to be: papers dealing with polymeric materials represent the great majority of the content of this book. Regretting that not enough work is being done on the rheology of non-polymeric materials is an exercise in futility; yet this does seem an appropriate time for reiterating this often repeated consideration.

We would like to have a long list of people whose help in organizing the Congress we would need to acknowledge here. Unfortunately, there are no entries to such a list, with the exception of young coworkers and students who have helped before the time of writing, and will help after it. To these we extend our sincere and warmest thanks; their unselfishness is further confirmed by our inability to report their names. With this exception, we have organized the technical part of the Congress singlehandedly, and we state this not because we are proud of it, but only as a partial excuse for any mishaps that may, and unfortunately will, take place.

We regret that only the abstract of some papers appear in the Proceedings. The mail service being what it is, some papers did not reach us in time for inclusion in the Proceedings; others reached us in time, but were not prepared in the recommended form. Also, some abstracts reached us so late that there was no time left for preparation of the final paper.

At the very end of the third volume, we have collected whatever information (title, abstract, or complete paper) we could on contributed papers the very existence of which became known to us after we had prepared the Table of Contents, Author Index and Subject Index. Again, we apologize for this.

Finally, we want to express our most sincere wishes of success to whoever will be in charge of organizing the IXth International Congress in 1984. Based on our own experience, and in view of the Orwellian overtones of the date, we cannot avoid being pleased at the thought that, whoever it is, it will not be us.

Naples, 1st March 1980

Gianni Astarita
Giuseppe Marrucci
Luigi Nicolais

CONTENTS

VOLUME 2 - FLUIDS

NOTE: Papers identified by the ° sign were not received in time for inclusion in this book, and only the abstract is included.

SUSPENSIONS

FLUID DYNAMICS

THE FLOW OF DILUTE POLYMER SOLUTIONS AROUND CYLINDERS: CHARACTERISTIC LENGTH OF THE FLUID

Jean-Michel Piau

Institut de Mecanique
Domaine Universitaire
38041 Grenoble Cedex

INTRODUCTION

Turbulent drag reduction using dilute polymer solutions has been studied for a long time. However a proper adimensional correlation is still needed. The mechanical parameters of the polymer solutions have still to be identified from experiments and measured, probably by focusing on laminar flows.

In particular, elongational flows such as those around circular cylinders can exhibit many qualitative phenomena. The general pattern of cylinder heat transfer and drag in dilute polymer solutions has been described (1, 2, 3) : a transition occurs as the speed reaches a certain magnitude. Flow visualisation studies (3, 4) showed that the flow can be substantially stagnant around cylinders, which explains why transfers can be so reduced.

New experimental results will be presented. They confirm the existence of a large stagnant zone and they introduce the measurement of a length scale characteristic of the fluid.

To explain the phenomenons and the length scale obtained, a thread trapping model for the flow around cylinders will be discussed and checked experimentaly.

The consequences of the present results and model bear on the formulation of constitutive equations as well as on the correlation of turbulent drag reduction.

EXPERIMENTAL RESULTS

Velocity measurements around cylinders 0,2 mm in diameter have been performed with a specially designed laser Doppler anemometer (5) : the dimension of the probe is about 6 μm. It can be seen figure 1, from the component of velocity along the symetry axis, that the cylinder is really surrounded by a wide zone within which the velocity is very low or zero.

A time scale t_c, characteristic of the fluid can be measured using flat ended total head pressure tubes as in (6) . The time t_c is taken exactly as ϕ_e/U where U = velocity of flow, ϕ_e external diameter of the tubes, at onset of the pressure defect, whereas the maximum velocity gradient is $8U/\pi\phi_e$ in such a flow.

The onset velocity u_c of anormalous heat transfer about submerged cylinders can be measured by towing hot film sensors in the same fluids, and using the same experimental set up as for total head pressure tubes.

A length scale characteristic of the fluid $l_c = u_c t_c$: 5 fits the experimental data which disagree with the shear wave model. This length scale is very near the fully elongated length of macromolecules, due to the numerical factor 5 chosen.

THREAD TRAPPING MODEL

Former studies by Fabula (8) , Metzner and Astarita (9), suggested that Pitot tubes and hot wires were coated by stationary and rigid fluid. Their point of view has been reproduced in several papers on Pitot tubes, and contradicted for hot wires by James and Acosta (1) . Our own view is that Fabula (8) and Metzner and Astarita (9) , are wrong for flows around total head tubes and not wrong for the hot wires.

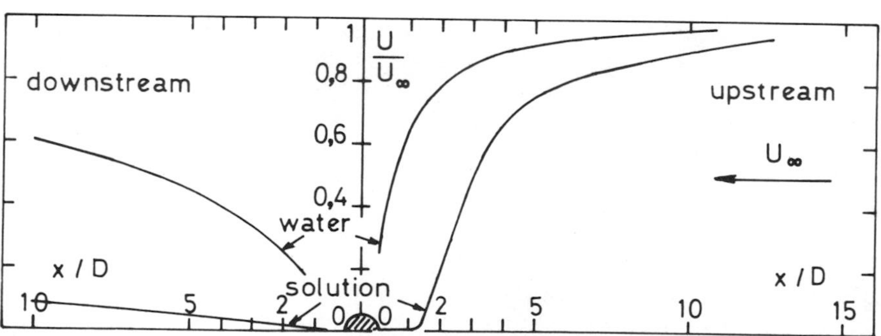

Fig. 1. Velocity profiles for water and for 500 ppm Polyox WSR 301 in water. Reynolds number 5,5.

No screening can be involved for Pitot tubes, as Piau (4),(6) showed. By simultaneously measuring the total pressure error and visualizing the flow with aluminium particles, one can see that the current lines enter the tube just as the pressure error appears, and then enter even deeper, along many diameters, as velocity and pressure errors increase.

On the other hand, the trapping of the long molecular chains by the small wires will be discussed. The strongly elongated macromolecules are caught by the cylinder and more or less kept along its surface. In this scheme the velocity of the center of mass of molecules may differ essentially from that of the fluid and a model for homogeneous monophasic continua may not be suitable. There also appears a concentration gradient with a higher concentration near the wall, in the area where the visualization shows a slow flow.

This model is an adaptation of a visual observation we made of drag reducing asbestos fibers solutions flowing about cylinders : the general flow pattern looks like that obtained with polymers, furthermore it is easy to check that the cylinders are coated by a glue of trapped fibers.

The transition occuring at the onset of the phenomenon is explained by a transition in the extension of the molecules and may be by a molecule wall interaction : the order of this interaction changes abruptly when the molecules are fully extended, as can be expected from the number of molecule-wall contacts.

The phenomenon will then appear when the following conditions are fulfilled :

- there exists a high velocity gradient ahead of the cylinder in an area large enough for the molecules to extend
- the molecules stay in this area long enough to be fully extended
- the molecules leave this area for the wall of the cylinder in a time short enough to keep elongated.

The first condition which is actually decisive for the flow around a cylinder will be further developped, using an approximate analysis in newtonian fluid and critical velocity gradients as measured with total head pressure tubes.

APPROXIMATE CRITERION FOR THE ELONGATION OF MACROMOLECULES NEAR THE STAGNATION POINT OF CYLINDERS

In order to be fully extended, the maximum scale of the molecule must be smaller than the scale of the strain field. In order to evaluate the latter, the elongational flow around a rigid cylinder, more exactly around a prolate spheroid is studied.

Following Happel, Brenner (9) , let us consider prolate spheroidal coordinates (ξ , η , ϕ) obtained by rotating confocal elliptic coordinates about their major axis defined as :

$$\rho = c \text{ sh } \xi \sin \eta \qquad 0 \leq \xi \leq \infty \qquad \tau = \text{ch } \xi \qquad 1 \leq \tau < \infty$$

$$0 \leq \eta \leq \pi \qquad \zeta = \cos \eta \qquad -1 \leq \zeta \leq +1$$

$$z = c \text{ ch } \xi \cos \eta \qquad 0 \leq \phi < 2\pi$$

where $\tau = \tau_0$ is a prolate spheroid the major and minor semi axes of which are $c\,\tau_0$ and $c\sqrt{\tau_0^2 - 1}$.

The spheroid is assumed to remain at rest while fluid streams past it with velocity potential $\varphi = -a z \rho^2/2$ at infinity. The normal and tangential velocities vanish on the spheroid.

For creeping flow, the solution for the stream function is :

$$\psi = -\frac{ac^3}{2} \zeta \tau (\tau^2 - 1)(1 - \zeta^2)\, f(\tau)$$

where we have set :

$$f(\tau) = 1 - \frac{4}{6\tau_0 + (1 - 3\tau_0^2)\,\text{Log}\,\frac{\tau_0 + 1}{\tau_0 - 1}} \left(\frac{\tau_0^2}{\tau} + \frac{1 - 3\tau_0^2}{4}\,\text{Log}\,\frac{\tau + 1}{\tau - 1} + \frac{1 - \tau_0^2}{2}\,\frac{\tau}{1 - \tau^2} \right)$$

We are interested in spheroids that resemble long thin rods. The reduced velocity for this limiting case is

$$\left(\frac{V_\eta}{V_{\eta\infty}} \right)_{\xi_0 \ll 1} \sim 1 - \frac{2 + 3\xi_0^2}{3 - \text{Log}\,4 + 2\,\text{Log}\,\xi_0} \left(\frac{3\tau}{3\tau^2 - 1} - \text{Log}\sqrt{\frac{\tau + 1}{\tau - 1}} \right)$$

To evaluate the disturbed area we can calculate the value $\tau = \tau_\infty$ where we get the reduced velocity 0,99, for several appropriate values of ξ_0.

The largest perturbed area is obtained when the molecule is fully elongated at length 1. This area is a prolate spheroid the major and minor axis of which are about 1,25 1 and 0,74 1 for $\xi_0 = 10^{-4}$. As the exact values depend on ξ_0 we will use 1 and 0,7 1 below, for simplicity.

When a molecule reaches the stagnation point, it stays for a long time in an elongational flow. As the flow is not uniform and the macromolecule response is not exactly known, an approximate criterion will be used : macromolecules fully extend if one eigen value of the velocity gradient exceeds the limiting value t_c^{-1} over an ellipsoidal area the length and thickness of which are 1 and 0,7 1. As the radius R_0 of the cylinders is never much smaller than 1, the only condition to fulfill below will be a velocity gradient t_c^{-1} at radius R_0 + 0,7 1 in the inviscid fluid approximation, and a velocity gradient t_c^{-1} at the tip of the ellipsoids in the viscous fluid approximation.

It is possible to calculate the positive eigenvalues of the velocity gradient and to deduce this criterion at low Reynolds number and at high Reynolds number. As nearly the same criterion is obtained in both conditions, one can expect it to be valid also at moderate Reynolds numbers. The result is $u_c t_c \geqslant 5\ l$ where u_c is the uniform velocity at infinity, very similar to the experimental correlation quoted above.

VISUALISATION TEST

To test further the thread trapping hypothesis, the flow field around a flat plate, spheres and needles has been visualized using aluminium particles added to the fluid. Photographs were taken of water and of a solution of Polyox WSR 301 flowing towards a sharpened blade 0,1 mm thick and 3 mm long, part of a shaving blade. The flow is shown in figure 2a, b, c to be quite similar to that around cylinders, provided their diameter is about the same as the thickness of the plate. This flow is highly dependent on the leading edge phenomena and looks as if the plate was placed in the wake of a thin wire.

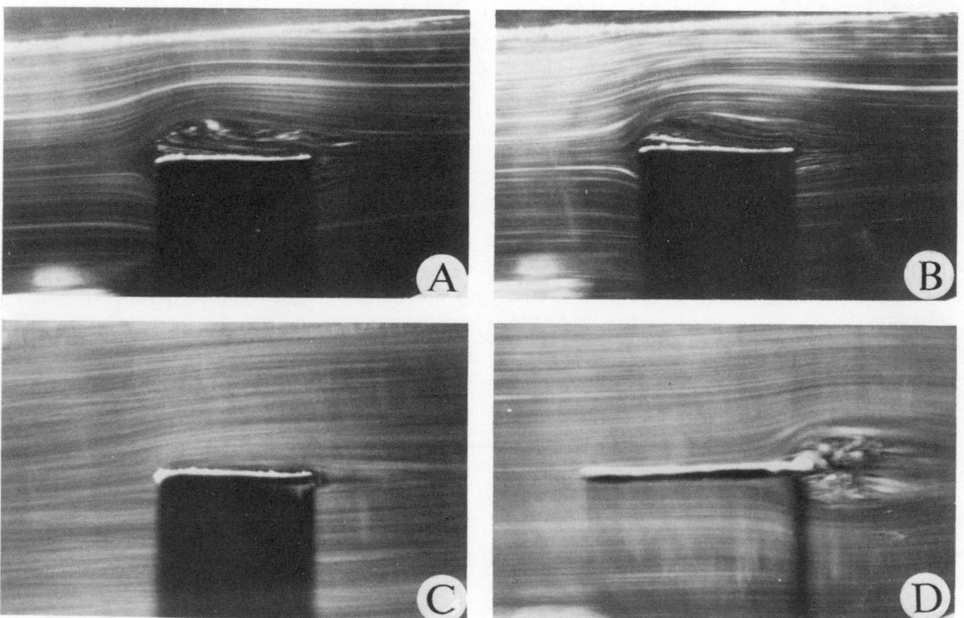

Fig. 2. Flow visualizations around a flat plate : Polyox WSR 301 concentration 400 ppm (a) velocity 0,34 m/s b) velocity 0,074 m/s) water (c) velocity 0,33 m/s)
Flow visualization around a cylinder facing the flow : Polyox WSR 301 concentration 400 ppm (d) velocity 0,13 m/s).

Figure 2(d) shows the flow around a wire 0,2 mm in diameter, bend at right angle and fixed in such a way that one of its ends 4 mm long is parallel to the flow. No disturbance such as those preceeding can be noticed around the tip of this wire. Neither did we notice any disturbance around spheres and needles.

This does not mean that the molecules cannot be stretched around spheres. Moreover, we know from unpublished results (order of velocities : 0,1 m/s, order of Reynolds numbers 20, weight concentration of polyoxethylene in water 100 10^{-6}) that spheres can lead to higher drag coefficients for polymer solutions than from newtonian fluids as well as cylinders do. The point here, is that spheres and needles pointing towards the flow do not trap any molecule as the cylinders and the flat plate do. It is the reason why conical hot film probes work better than cylindrical probes in polymer solutions.

CONCLUSION

The constitutive equations used to describe polymer solution flows will depend on the class of flow studied. Most constitutive equations are valid only for weak flows where the time scale of the flow is larger than the time scale of the fluid (or the molecules). For strong flows the fluid may not have a fading memory as Tanner (11) showed. Moreover, it will not be a Noll simple fluid at all when l_c is involved. Noll simple fluid model is valid for weak flows when the significant length scale $\emptyset$ of the flow is much larger than the coiled molecules diameter, and for strong flows in the limit $l_c \ll \emptyset$.

REFERENCES

(1) D.F. JAMES, A.J. ACOSTA 1970 - J. Fluid Mech. 42, 269.
(2) D.F. JAMES, O.P. GUPTA 1971 - Chemical Engineering Progress 67, 62.
(3) J.M. PIAU 1974 - C.R. Acad. Sci. Paris, B 278, 493.
(4) J.M. PIAU 1974 - Film "Ecoulements de solutions diluées de polymères" SFRS, 96 bd Raspail, Paris.
(5) A. KONIUTA, P.M. ADLER, J.M. PIAU 1980 - J. Non-Newt. Fluid Mech. (to appear).
(6) J.M. PIAU 1974 - Proceedings Colloque CNRS, Paris, n° 223, p 225.
(7) J.M. PIAU - A characteristic length for dilute drag reducing polymer solutions. to be published.
(8) A.G. FABULA 1966 - PhD Thesis, The penssylvania State University.
(9) A.B. METZNER, G. ASTARITA 1967, A.I.Ch.E.J. 13, 550.
(10) J. HAPPEL, H. BRENNER 1965, Low Reynolds number hydrodynamics, Prentice-Hall.
(11) R.I. TANNER 1975 - Trans. Soc. Rheology 19, 557.

SHEAR-THINNING EFFECTS IN CREEPING FLOW ABOUT A SPHERE

R. P. Chhabra, C. Tiu, P. H. T. Uhlherr

Department of Chemical Engineering
Monash University
Clayton, Victoria 3168, Australia

INTRODUCTION

Shear-thinning inelastic fluid behaviour is often described by the power law because of the simplicity of this model. Frequently also zero shear viscosity is difficult to measure and so a more realistic fluid model cannot be used. Despite the severe limitations of the power law in describing regions of a flow field where the shear rate approaches zero, it continues to be applied to the problem of creeping sphere motion. It has not yet been determined over what proportion of the sphere surface this fluid model breaks down. Presumably if the areas of the sphere surface about the front and rear stagnation points and the volume of fluid far from the sphere do not contribute significantly to the total drag of a sphere, then a power law description of the flow field may yield an acceptable result for the drag.

THEORETICAL SOLUTIONS

Many authors have presented approximate solutions for the creeping motion of a sphere through a power law fluid[1-7] and a good review of the area has been given by Acharya et al[1]. All solutions rely on different stream functions arbitrarily chosen only to satisfy the equation of continuity and the required boundary conditions and to contain a dependence on the power law index n. Most authors also applied a variational principle which allowed them to replace an a priori dependence of the stream function on n by an arbitrary parameter, for which the dependence on n was to be determined. In this case, because of the minimisation of energy dissipation rate, the resulting solutions are upper bounds on the drag. Seven solutions are compared in Fig. 1. This figure shows a plot of X(n) where

$$C_D = (24/Re)X \qquad (1)$$

and Re is the usual power law Reynolds number given by $d^n V^{2-n} \rho / K$. Only Tomita[2,3] and Acharya et al.[1] did not employ the variational principle and their solutions are not upper bounds on the drag. In addition, the latter authors introduced further approximations to allow them to give a closed form solution. One of the solutions included in Fig. 1 was obtained by the authors using a stream function first proposed by Ziegenhagen[8]. This choice was based only on the fact that this stream function had not previously been used with the power law; it offers no particular advantages over other stream functions. The solution of Nakano and Tien[6] was obtained for flow about Newtonian fluid spheres and the result shown in the figure is that for an infinite viscosity ratio of the internal to the external phase. Wasserman and Slattery[5], in addition to supplying an upper bound for the drag, also calculated a lower bound based on another variational principle employing a trial stress profile. This lower bound was corrected by Mohan[7].

Fig. 1 shows the strong dependence of the drag correction factor on the choice of stream function. Since none of the solutions are rigorous, nor indeed can possibly be so because of the limitations inherent in the power law, the choice of the best result must be based on experiment.

EXPERIMENTAL RESULTS

Sphere fall tests were carried out in twelve shear-thinning polymer solutions having $0.40 \leq n \leq 0.95$ as well as in three Newtonian fluids. The solutions were well described by the power law over a wide range of shear rate, encompassing the ranges of average shear rate (calculated as 2V/d) generated by all the spheres used. All the solutions were sensibly inelastic, in that normal force was not measurable under steady shear with an R16/19 Weissenberg rheogoniometer. In two of the solutions, normal force just became measurable at shear rates above 180 and 280 s^{-1}, which are very much higher than the shear rates generated in these fluids by any of the spheres used. Certainly in all cases the Weissenberg number was much less than 10^{-3}, and in view of recent results obtained with purely elastic fluids[9], elasticity can be assumed to be absent from the results under discussion here. Such uncertainty concerning the presence of elastic effects will always be encountered with aqueous polymer solutions.

Terminal velocity under gravity was measured for 20 different spheres ($1.59 \leq d \leq 12.69$ mm; $1190 \leq \rho \leq 16600$ kgm^{-3}). The data were corrected for wall effects by carrying out the measurements in five or more cylinders having different internal diameters, and extrapo-

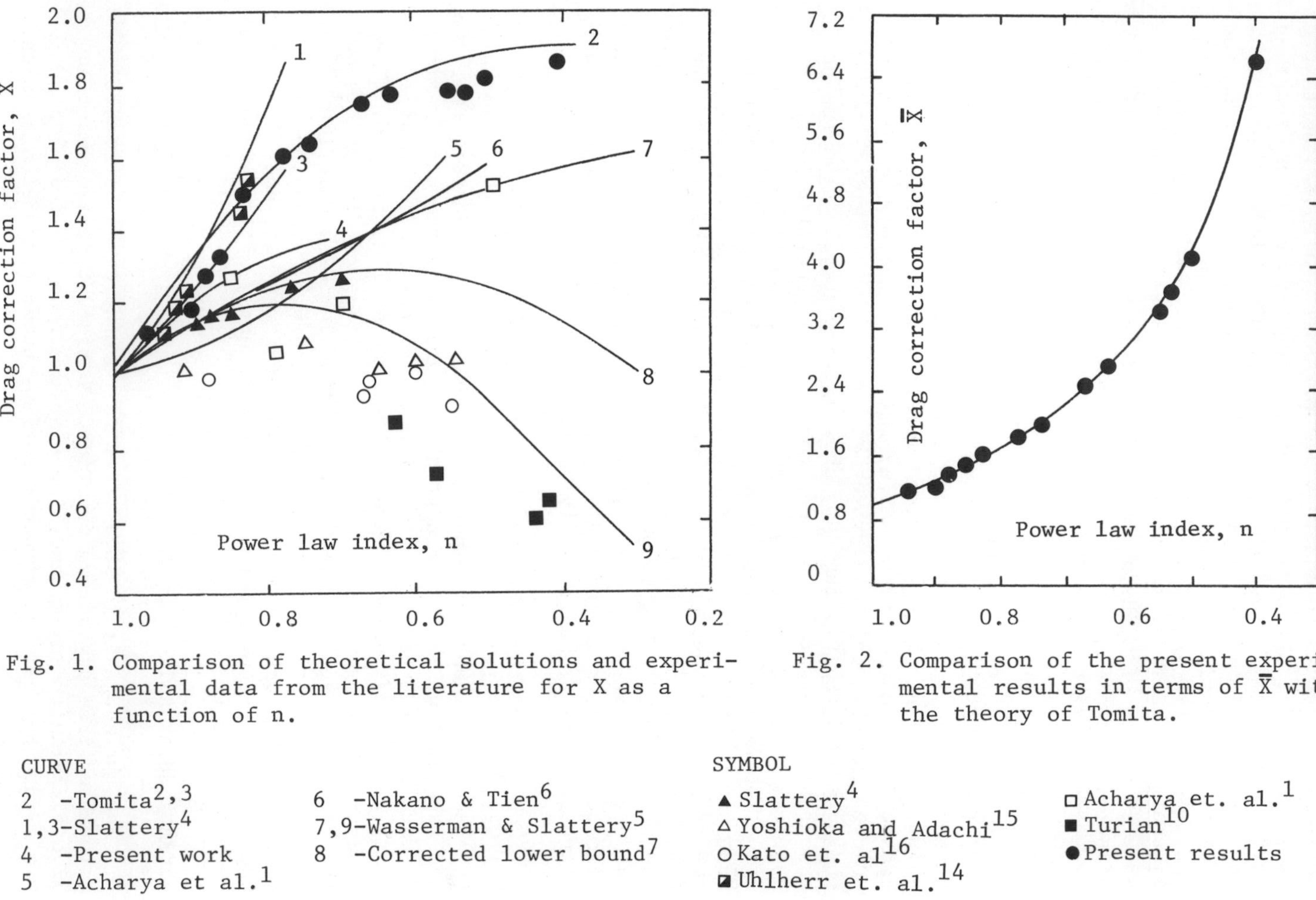

Fig. 1. Comparison of theoretical solutions and experimental data from the literature for X as a function of n.

Fig. 2. Comparison of the present experimental results in terms of $\overline{X}$ with the theory of Tomita.

CURVE

2 -Tomita[2,3]
1,3-Slattery[4]
4 -Present work
5 -Acharya et al.[1]
6 -Nakano & Tien[6]
7,9-Wasserman & Slattery[5]
8 -Corrected lower bound[7]

SYMBOL

▲ Slattery[4]
△ Yoshioka and Adachi[15]
○ Kato et. al[16]
◪ Uhlherr et. al.[14]
□ Acharya et. al.[1]
■ Turian[10]
● Present results

lating the velocity to a zero value of sphere to tube diameter ratio[10,11]. All the usual precautions required by such experiments were observed[9,11-13] and the results showed excellent reproducibility and accuracy; in the case of Newtonian fluids, deviation from the Stokes drag coefficient was less than ± 2%. The results of the measurements are included in Fig. 1. It must be observed that a single point in this figure represents an entire curve of C_D(Re,n). Some 70 values of C_D(Re,n) were obtained, corrected for wall effects. These in turn are based on nearly 1000 terminal velocity measurements in bounded fluids.

Also shown in Fig. 1 are experimental results reported in the literature. A large amount of scatter is evident and no entirely satisfactory explanation for this can be given. Most of the earlier authors did not characterise the elastic properties of their solutions. That these solutions were, in fact, more or less elastic can be inferred from reported information of polymer species and concentrations. Only Acharya et al.[1] and Uhlherr et al.[14] fully characterised their solutions, showing that normal force was not measurable even at the highest shear rates generated by the falling spheres. For the results of other authors, elastic effects certainly cannot be excluded as the cause of the scatter. It has recently been shown that the effect of fluid elasticity is much greater in the presence of shear-thinning than it is for constant viscosity fluids[12]. A value of drag as small as 10% of the Newtonian value may be observed in highly shear-thinning highly elastic fluids. The different treatment of wall effects by different authors may also be reflected in the scatter of the results, although the errors introduced are probably fairly small.

The picture that emerges of creeping sphere motion through inelastic power law fluids is entirely unsatisfactory - from both the theoretical and the experimental aspect. We consider that the new experimental results reported here are among the most reliable to date from the point of view of completeness of fluid characterisation, terminal velocity measurement and wall correction. They indicate that the first approximation solution of Slattery[4] shows the best agreement for $0.8 \leq n \leq 1.0$ and that Tomita's solution[2,3] gives the best description of the drag for $0.4 \leq n \leq 0.8$. The authors had previously[14] reached the same conclusion for $n > 0.8$, and in fact these earlier results were quite independently well reproduced by the present work.

AVERAGE SHEAR RATE

The surface average shear rate for a sphere can be readily calculated from any stream function. Different stream functions give different results, and Stokes stream function leads to $\dot{\gamma} = 2V/d$. Not all the stream functions under discussion give this result when n = 1; those of Tomita and of Wasserman and Slattery do not. These solutions therefore cannot be expected to accurately describe the flow field in the Newtonian case. That they lead to X = 1 at n = 1 is perhaps due to a compensation through the pressure drag term for deviation from the Stokes friction drag. Measurements of total drag tend to be insensitive to effects in the flow field and measurements of velocity distributions would be preferable. These are, however, difficult to obtain close to the sphere surface, where the most important changes are expected. Sigli and Coutanceau[17] have reported visual observations of the flow field about a sphere. They were concerned mainly with the effects of elasticity on the flow field relatively far from the sphere surface and no results are reported with which the validity of a stream function could be directly tested. The stream functions used by various authors are summarised in Table 1, which also includes the surface average shear rate calculated from each and the value of this shear rate for n = 1. The arbitrary parameter σ included in most stream functions depends on n and is given in the original papers. For Ziegenhagen's stream function, $\sigma(n)$ determined by the authors is included at the foot of the table.

The power law Reynolds number $d^n V^{2-n} \rho / K$ arises in Equn 1 through non-dimensionalising the variables. This Reynolds number, like that based on zero shear viscosity, is independent of shear rate. It may be preferable to use a Reynolds number based on an average apparent viscosity which could conceivably better reflect the influence of shear-thinning viscosity on the drag. This Reynolds number $\overline{Re}$ is give by $dV\rho/\overline{\eta}$, where $\overline{\eta}$ is obtained at the surface average shear rate. A drag correction factor can be defined empirically as

$$C_D = (24/\overline{Re})\overline{X} \tag{2}$$

Equn 2 is tantamount to introducing a correction factor directly into the equation for drag force in the Stokes law regime for a Newtonian fluid

$$F_D = 3\pi dV\overline{\eta}\overline{X} \tag{3}$$

The correction factors X and $\overline{X}$ are related by a shear dependent multiplying factor, the value of which is a function only of n, and which can be calculated for each stream function. The multiplying factor $\overline{X}/X$ is included in Table 1 for all the stream functions listed. In each case, the factor reduces to unity for n = 1.

The main disadvantage of this entirely empirical approach is that theoretical solutions can no longer be compared on a single plot of $\overline{X}(n)$ since each solution is now based on a different Reynolds number; solutions can only be compared in pairs. Similarly,

TABLE 1 : STREAM FUNCTIONS USED BY VARIOUS INVESTIGATORS

Author	Stream function ($\xi=r/R$)	Surface average shear rate		Multiplying factor $X/\bar{X}$
		general	n=1	
Stokes	$VR^2\sin^2\theta[-1/4\xi+3\xi/4-\xi^2/2]$		$2V/d$	1
Slattery[4] (0th approxn.)	Same as Stokes	$2V/d$	$2V/d$	2^{1-n}
(1st approxn.)	$VR^2\sin^2\theta\{[-1/4\xi+ 3\xi/4-\xi^2/2]$ $-\sigma[1/\xi^3-2/\xi+\xi]\}$	$(1+4\sigma)2V/d$	$2V/d$	$[2(1+4\sigma)]^{1-n}$
Ziegenhagen[8] (see footnote)	$VR^2\sin^2\theta\{[-1/4\xi+3\xi/4-\xi^2/2]$ $+\sigma[1-\xi/2-1/2\xi]$	$1.33(1.5+\sigma)$ V/d	$2V/d$	$[1.33(1.5+\sigma)]^{1-n}$
Acharya et al.[1]	$VR^2\sin^2\theta\{[-1/4\xi+3\xi/4 - [9n(n-1)/4(2n+1)]$ $[\xi/2-1/2\xi-\xi\ln\xi]\}$	$2V/d$ $2V$	$2V/d$	2^{1-n}
Tomita[2,3]	$VR^2\sin^2\theta[\xi/\sqrt{2} -(1/\sqrt{2\xi})^{1-n}]^2$	$1.33V/n^2d$	$1.33\ V/d$	$(n^2/1.33)^{1-n}$
Wasserman and Slattery[5]	$VR^2\sin^2\theta[\xi/2-(1/\sqrt{2\xi})^{\sigma}]^2$	$1.33\sigma^2V/d$	$1.736\ V/d$	$(1.33\sigma^2)^{1-n}$
Nakano and Tien[6]	$VR^2\sin^2\theta[(\sigma-2)/_{(\sigma+1)\xi} - \xi^2/2 + 3\xi^{\sigma}/_{(\sigma+1)}]$	$2(2-\sigma)V/d$	$2V/d$	$[2(2-\sigma)]^{1-n}$

Parameter σ determined by the authors for Zeigenhagen's stream function :

n	1.00	0.95	0.90	0.85	0.80	0.75	0.70	0.65
σ	0	0.391	0.916	1.320	1.470	1.500	1.500	1.500

experimental results can be compared with only one theoretical solution at a time in terms of its particular $\bar{X}(n)$ definition. This comparison was carried out for the present experimental results and each stream function tabulated and it was found that slight improvements in agreement could be obtained. Maximum deviations were reduced from as much as 30-40% to 20-30% in most cases. However, for the case of Tomita's solution the procedure produced excellent agreement with the experiments over the whole range of n from 1.0 to 0.4. This is shown in Fig. 2. This result may well be fortuitous in view of the shortcomings of Tomita's solution mentioned above, and it is not suggested that this solution gives the best description of the flow field. However, it is suggested that it gives the best macroscopic description of sphere drag in inelastic power law fluids for $0.4 \leq n \leq 1.0$.

The fact that all but one of the solutions employing variational principles and resulting in upper bounds for the drag give values of X smaller than Tomita's has yet to be explained. It may simply be due to the inadequacy of the power law in describing the flow field so that no theoretical solution based on this model can ever rigorously predict sphere drag; agreement with experiment must always be fortuitous. Nevertheless, once one solution has been identified that adequately describes the results of a large number of macroscopic experiments, this solution should be useful for predictions over the range for which it has been tested. If Tomita's solution is used in this way, it must be remembered that the presence of fluid elasticity will influence sphere drag and so produce deviations from any predictions using the solution. The deviations appear to be always towards smaller drag.

CONCLUSIONS

1. If the power law must be used for the description of creeping sphere motion and the power law Reynolds number is used to calculate the drag coefficients, Tomita's theoretical solution gives the best value of total drag for $0.4 \leq n \leq 0.8$ and Slattery's first approximation solution is best for $0.8 \leq n \leq 1.0$, provided that fluid elasticity is absent.

2. An improvement in the prediction of drag using Tomita's solution is observed when the power law Reynolds number is arbitrarily replaced by a Reynolds number based on a surface average apparent viscosity. Excellent agreement with experiment is then obtained for $0.4 \leq n \leq 1.0$. The average viscosity must be determined at the surface average shear rate of $1.33V/n^2d$.

3. The scatter observed in experimental drag results with aqueous polymer solutions and reported in the literature is probably due to the effect of fluid elasticity which tends to reduce drag to an

extent dependant on the shear rate. Excellent reproducibility is obtained with chemically different solutions if elasticity is at a sufficiently low level.

REFERENCES

1. A. Acharya, R. A. Mashelkar and J. J. Ulbrecht, Rheo. Acta, 15:454 (1976).
2. Y. Tomita, Bull. J. S.M.E., 2:469 (1959).
3. G. C. Wallick, J. G. Savins and D. R. Arterburn, Phys. Fluids, 5:367 (1962).
4. J. C. Slattery, A.I.Ch.E.J., 8:663 (1962).
5. M. L. Wasserman and J. C. Slattery, A.I.Ch.E.J., 10:383 (1964).
6. Y. Nakano and C. Tien, A.I.Ch.E.J., 14:145 (1968).
7. V. Mohan, "Fall of liquid drops in non-Newtonian media", Ph.D. Dissertation, I.I.T. Madras (1974).
8. A. J. Ziegenhagen, Appl. Sci. Res., 14A:43 (1964).
9. R. P. Chhabra, P. H. T. Uhlherr and D. V. Boger, J. Non-Newtn. Fluid Mech., 6:187 (1979/80).
10. R. M. Turian, A.I.Ch.E.J., 13:999 (1967).
11. R. P. Chhabra, C. Tiu and P. H. T. Uhlherr, Proc. 6th Aust. Conf. Hydraulics Fluid Mech., Adelaide, 435 (1977).
12. R. P. Chhabra and P. H. T. Uhlherr, Rheo. Acta, in press.
13. R. P. Chhabra and P. H. T. Uhlherr, Rheo. Acta, 18:593 (1979).
14. P. H. T. Uhlherr, T. N. Le and C. Tiu, Can. J. Chem. Eng., 54:497 (1976).
15. N. Yoshioka and K. Adachi, J. Chem. Eng. Japan, 6:134 (1973).
16. H. Kato, M. Tachibana and K. Oikawa, Bull. J.S.M.E., 15:1556 (1972).
17. D. Sigli and M. Coutanceau, J. Non-Newtn. Fluid Mech., 3:107 (1977/78).

ACKNOWLEDGEMENT

The authors acknowledge financial support from the Harold Armstrong Memorial Fund for upgrading a Weissenberg Rheogoniometer. R. P. Chhabra is grateful for support from a Monash Graduate Scholarship.

MEASUREMENTS OF VELOCITY FIELDS AROUND OBJECTS

MOVING IN NON-NEWTONIAN LIQUIDS

O. Hassager, C. Bisgaard, and K. Østergaard

Instituttet for Kemiteknik
Danmarks Tekniske Højskole
Lyngby, Danmark

INTRODUCTION

Gas bubbles and solid objects moving under the influence of gravity in a non-Newtonian liquid are known to exhibit several interesting phenomena. For example Astarita and Apuzzo[1] found that the terminal rise velocity vs. volume curve for single isolated air bubbles may exhibit a discontinuity at a certain "critical" volume, a phenomenon observed later by others[2-4]. Also Riddle, Narvaez and Bird[5] have investigated the interaction between two identical spheres falling along their line of centers in a non-Newtonian liquid. They found that for small initial separations the two spheres eventually converge, while for large initial separations the two spheres eventually diverge. In the present study we have measured the axial component on the axis of gas bubbles and solid spheres moving in a non-Newtonian liquid. The most striking feature of the observed flow fields is the occurrence of a region in the wake behind the objects where the liquid velocity (referred to an observer at rest with respect to the liquid far from the objects) is in the direction opposite to that in which the object is moving. We have called this phenomenon "negative wake".

EXPERIMENTAL EQUIPMENT

The experimental apparatus consists of a vertical 4.5 -inch diameter glass cylinder filled to a depth of 2 m with test fluid. Gas bubbles are injected by a syringe near the bottom of the tube, go through a centering device and proceed axially up through the test section. Solid spheres are released near the top on the axis of the cylinder and proceed down through the test section.

In the test section two photocells are used to measure the velocity of the object. A laser Doppler anemometer is used to measure the axial component of the liquid velocity at a fixed point on the cylinder axis as a function of time. Having crossed the "measuring volume" one of the laser beams impinges on a third photocell. This photocell is used to register the exact time at which the object enters into and exits from the measuring volume and to keep the LDA frequency tracker in "hold" during the passage of the object. The photocells and the LDA output are sampled and stored by a microprocessor with sampling rates up to 20 kH_z possible.

EXPERIMENTAL RESULTS

The constant velocity of the moving objects allows us to transform the experimentally obtained velocity vs. time records into velocity vs. space records that show the axial component of liquid velocity on the axis of the objects. Figure 1 shows the velocity on the axis of a gas bubble rising in a viscous Newtonian liquid. The position and velocity of the bubble is indicated in Figure 1 by the width and height of the rectangle. Figure 1 shows that while some fluid is pushed ahead in front of the bubble, more fluid is pulled along behind the bubble, thereby consituting the "wake" of the bubble.

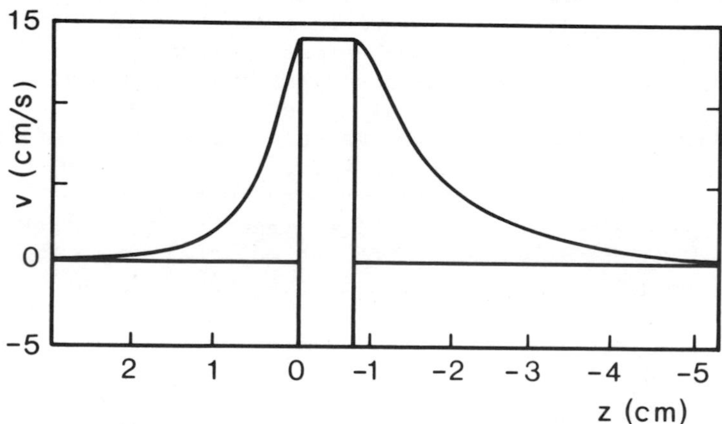

Fig. 1. Axial component of liquid velocity on the axis of an air bubble of volume 400 mm^3 rising in a viscous Newtonian liquid (a 99% glycerol-1% water mixture at 295 K). The abscissa, z is the distance along the axis of the bubble.

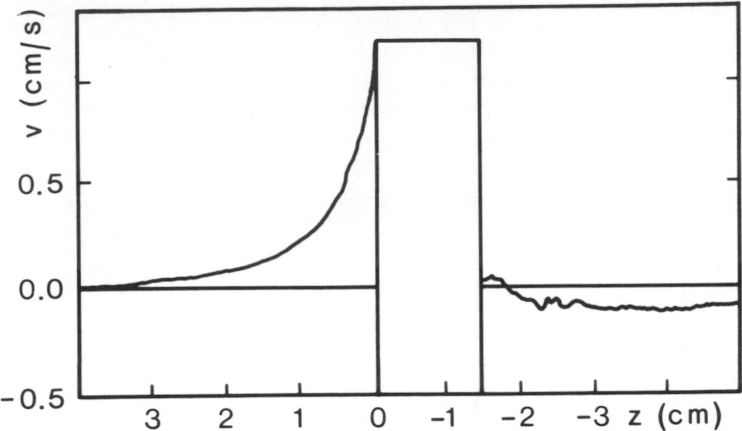

Fig. 2. Representation similar to Fig. 1 for an air bubble rising in a 1% solution of polyacrylamide in glycerol.

Consider now in Figure 2 the velocity on the axis of a gas bubble rising a non-Newtonian test fluid. The figure clearly shows the negative wake in the liquid behind the bubble. The magnitude of the negative wake depends on the size of the bubble, being more pronounced for the larger bubbles. The smallest bubble for which a velocity signal could be obtained had a volume of 250 mm^3 and this also showed a negative wake. Figure 3 shows the axial velocity on the axis of a steel sphere falling through the same test fluid. We see that there is a division between two regions: Close to the sphere the liquid velocity on the axis is in the same direction as the falling sphere, but further away from the sphere there is a region of negative wake. The division between the two regions occurs at about 5.5 sphere diameters behind the sphere for this particular experiment. The velocity in the negative wake decays only very slowly to zero, being still observable 25 sphere diameters behind the sphere.

DISCUSSION

The present experiments provide information that first of all improves our knowledge about the motion of isolated objects in non-Newtonian liquids. Second the information should be useful in connection with the interaction between objects moving along their line of centers. Thus in studies of bubble coalescence in viscous Newtonian fluids[6-7], the wake behind a leading bubble has been

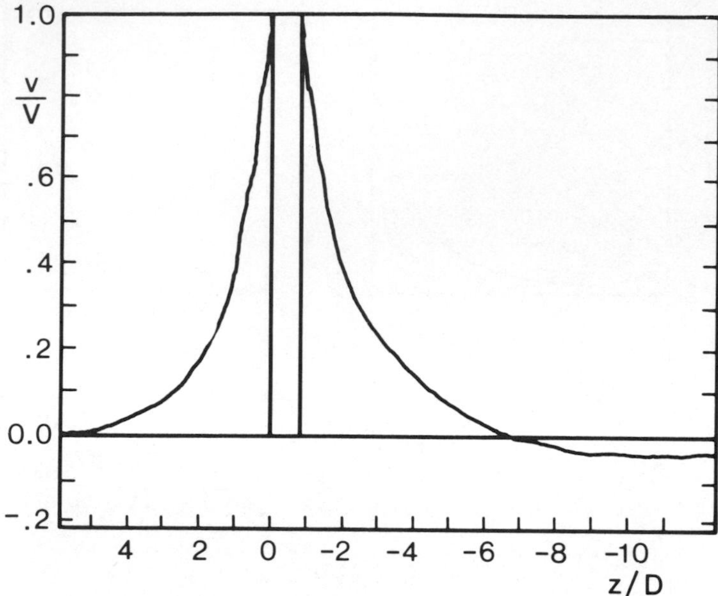

Fig. 3. Representation as in Fig. 2 for a steel sphere falling in the test solution. Here v/V is liquid velocity divided by sphere velocity and z/D is axial distance divided by the sphere diameter.

tied to the increased speed with which a trailing bubble catches up with a leading bubble so that coalescence can take place. If we accept this argument we would make the following conjectures from the present experiments: Two identical gas bubbles rising along their line of centers in the polyacrylamide solution should not coalesce but in fact diverge. Two identical solid spheres falling along their line of centers should converge if the initial separation is smaller than a certain critical distance and diverge if the separation is larger than the critical distance. The latter conclusion is exactly the experimental observation of Riddle, Narvaez and Bird.

ACKNOWLEDGEMENT

The authors gratefully acknowledge the financial assistance provided by STVF. The DISA Company has generously placed a 55L35 frequency tracker at our disposal. Finally we wish to thank Mr. B. Johnson for his invaluable aid with electronics and programming.

REFERENCES

1. G. Astarita and G. Apuzzo, AIChE Journal, 11:815 (1965).
2. P.H. Calderbank, D.S.L. Johnson, and J. Louden, J.Chem. Eng.Sci., 25:235 (1970).
3. L.G. Leal, J. Skoog, and A. Acrivos, Can.J.Chem.Eng., 49:569 (1971)
4. A. Acharya, R.A. Mashelkar, and J. Ulbrecht, J.Chem.Eng. Sci., 32:863 (1977).
5. M.J. Riddle, C. Narvaez, and R.B. Bird, J. Non-Newt. Fluid. Mech., 2:23 (1977).
6. J.R. Crabtree and J. Bridgwater, J.Chem.Eng.Sci., 26:839 (1971).
7. N. de Nevers and J. Wu, AIChE Journal, 17:182 (1971).

THE EFFECT OF ASSIMMETRY ON RAPID BUBBLE GROWTH AND COLLAPSE IN NON-NEWTONIAN FLUIDS

S.K.Hara, W.R.Schowalter

Department of Chemical Engineering
Princeton University, U.S.A.

(Abstract)

There are numerous practical and fundamental motivations for study of time dependent behavior of bubble growth and collapse in non-Newtonian liquids. The former are associated with the subject of cavitation. It is well known that the presence of small amounts of dissolved polymer in an aqueous solvent can have a disproportionately large inhibiting effect on cavitation through reduction of the cavitation number

$$\sigma = \frac{p_s - p_v}{\frac{1}{2}\rho v_o^2}$$

where p_s is the local free-stream static pressure, p_V is the liquid vapor pressure, ρ is the liquid density, and v_o is the free-stream velocity. Ellis et al. observed reductions of σ up to 70% when 300 ppm of Guar gum was dissolved in water. These observations were made in complex flow fields, such as flow past a bluff body placed in a water tunnel.

It is widely held that the large reduction of cavitation number is caused by the elasticity present when the solvent contains dissolved polymer. However, when analyses and experiments have been performed on single bubbles expanding or contracting with spherical symmetry in an infinite fluid otherwise at rest, the results have been disappointing. One finds only a modest change in bubble dynamics because of the presence of viscoelastic fluids. This result has led workers in the field to suspect that the presence of shear fields has an important effect on bubble dynamics. That is an entirely reasonable expectation since the rheology, and hence the response to forcing by bubble pressure, will be strongly dependent on the boundary conditions far from the bubble.

We have studied the predictions for bubble dynamics in a variety of configurations which correspond to boundary conditions realistic enough to provide insight into cavitation under conditions of practical interest but simple enough to retain mathematical tractability. The results serve as a guide for new experiments in cavitating liquids. Our results also permit important comparison to prior work in spherically symmetric bubble dynamics.

A TWO-DIMENSIONAL ASYMMETRIC FLOW OF A VISCOELASTIC FLUID IN A T-GEOMETRY

D.J. Paddon
School of Mathematics
University of Bristol
U.K. BS8 1TW

H. Holstein
Dept. Computer Science
U.C.W. Aberystwyth
U.K. SY23 3BZ

INTRODUCTION

The flow of rheologically complex fluids in channels and pipes has long been the subject of theoretical and practical investigations. Advances in computer technology and in numerical methods have allowed the consideration of more complicated flow situations and more realistic equations of state. The interest towards problems which are industrially relevant has focused attention on flows associated with abrupt changes in geometry. Numerical treatments of such flows have appeared in the literature[1-7] since 1976.

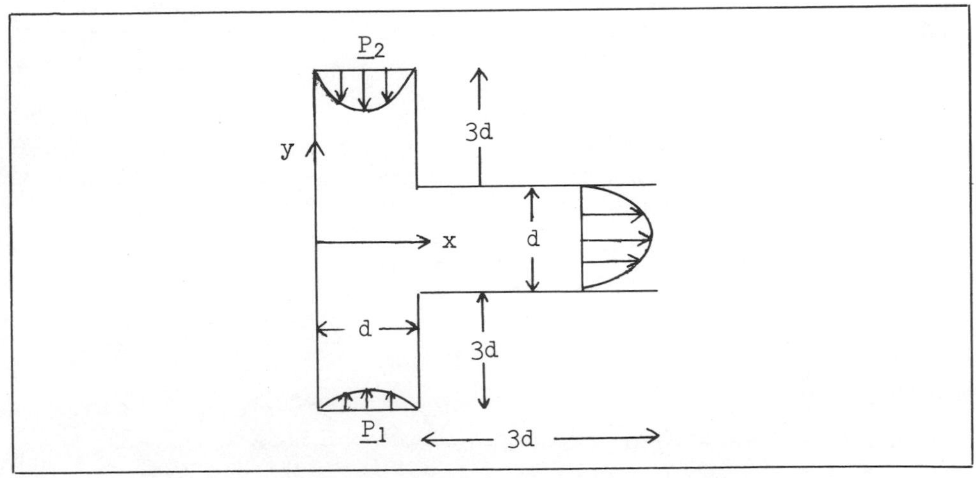

Fig. 1. Schematic view of flow domain.

Most of these studies involve flow fields that contain some degree of symmetry, and therefore require the evaluation of only one half of the full flow domain. In the present work, we investigate an asymmetric flow induced by two unequal pressure gradients $\underline{P}_1$ and $\underline{P}_2$. These are applied at the inlets of a two-dimensional T-formation, as shown in figure 1.

The T-formation is an interesting flow geometry. For a given fluid, there is a variety of flow regimes according to the chosen ratio of the imposed inlet pressure gradients. A study of this geometry may also be regarded as a preliminary investigation of a mixing problem, involving two dissimilar fluids.

The asymmetric nature of the problem naturally requires a large computational domain. This in turn makes demands on the computation time and store. Under these circumstances, if accuracy is not to be sacrificed, numerical efficiency is of the utmost importance.

THEORETICAL CONSIDERATIONS

It has been argued that differential equations of state must be implicit in the stresses, when a flow involving abrupt changes in geometry is modelled.[8] In this investigation we consider the flow of a Maxwell fluid. Its equation of state is defined by

$$p^{ik} = -pg^{ik} + p'^{ik} \tag{1}$$

$$p'^{ik} + \lambda\, \mathfrak{d}p'^{ik}/\mathfrak{d}t = 2\eta_0 e_{(1)}^{ik} \tag{2}$$

where p^{ik}, g^{ik} and $e_{(1)}^{ik}$ are controvariant tensors of the stress, the co-ordinate metric and the first rate of strain respectively, $\mathfrak{d}/\mathfrak{d}t$ is the convected time derivative for a contravariant tensor, and p is the isotropic pressure. The fluid is characterised by a constant viscosity η_0 and a relaxation time λ.

We express all tensor qualities in physical components relative to a two-dimensional Cartesian x-y co-ordinate system. In terms of the stream function ϕ and vorticity ω, related through

$$\frac{\partial^2\phi}{\partial x^2} + \frac{\partial^2\phi}{\partial y^2} = -\omega \tag{3}$$

the stress equation of motion is

$$\rho\left(\frac{\partial\phi}{\partial x}\frac{\partial\omega}{\partial y} - \frac{\partial\phi}{\partial x}\frac{\partial\omega}{\partial y}\right) = \frac{\partial^2 p'_{xx}}{\partial x\partial y} + \frac{\partial^2 p'_{xx}}{\partial y^2} - \frac{\partial^2 p'_{xy}}{\partial x^2} - \frac{\partial^2 p'_{yy}}{\partial x\partial y} \tag{4}$$

The implicit nature of the equation of state (2) prevents the direct evaluation of the stress components for substitution into the stress equation of motion. We are forced to look upon the implicit equation of state as a set of differential relationships for the stress components, to be solved in parallel with the equation of motion.

The numerical solution of the coupled system is possible only if the equation of motion contains second order derivatives in the vorticity.[9] This is achieved by decomposing the stress components into Newtonian and non-Newtonian contributions, given by

$$p'^{ik} = \bar{p}^{ik} + 2\eta_0 e^{(1)ik} \tag{5}$$

Substitution of (5) into (3) gives the modified equation of motion:

$$\eta_0\left(\frac{\partial^2\omega}{\partial x^2} + \frac{\partial^2\omega}{\partial y^2}\right) + \rho\left(\frac{\partial\phi}{\partial x}\frac{\partial\omega}{\partial y} - \frac{\partial\phi}{\partial y}\frac{\partial\omega}{\partial x}\right)$$

$$= \frac{\partial^2\bar{p}_{xx}}{\partial x\partial y} + \frac{\partial^2\bar{p}_{xy}}{\partial y^2} - \frac{\partial^2\bar{p}_{xy}}{\partial x^2} - \frac{\partial^2\bar{p}_{yy}}{\partial y^2} \tag{6}$$

Although we can also transform (2) into an equation involving $\bar{p}^{ik}$, it is easier to solve (2) for p'^{ik} and use (5) before substituting the stresses into (6).

The formulation of the problem is completed by a specification of the boundary conditions. These are derived from the no-slip velocity condition on the walls of the geometry, and the imposition of fully developed flow with specified pressure gradients at the inlet boundaries.

NUMERICAL CONSIDERATIONS

We discretise the flow domain by imposing a square mesh of side h, and with respect to this mesh, replace the differential equations by finite difference equations. We solve these equations using the successive overrelaxation method. Although in principle this is not the most efficient way of handling large systems of linear equations, we have found that the modest convergence rate of this method is effective in controlling some of the numerical instabilities inherent in the non-linear coupling of the equations.

The successive overrelaxation iteration matrices should be diagonally dominant, if favourable convergence properties are to be maintained. Diagonal dominance is usually retained in the presence of first order derivatives by their replacement with first order accurate upwind differences.[10] Here we propose the use of a weighted finite difference

$$\left.\frac{\partial\omega}{\partial x}\right|_i \simeq (\tfrac{1}{2} + \alpha)\,\frac{\omega_{i+1} - \omega_i}{h} + (\tfrac{1}{2} - \alpha)\,\frac{\omega_i - \omega_{i-1}}{h}$$

where the optimum weight α is determined by minimising the truncation error $\alpha h\omega_i'' + O(h^2)$ while maintaining diagonal dominance.

Finally, the boundary conditions need to be formulated for ϕ, ω and the stresses. The variation of these variables across the inlets follows from the imposed inlet and the no-slip wall conditions. These also determine the constant ϕ values along the walls. The no-slip condition reduces (2) to algebraic relations, from which the wall stress components are easily found.

Wall vorticity values were obtained from the Woods method. Careful consideration was given to the approximation of vorticity at the corners of the geometry. A variety of corner strategies was considered. At this stage of our work we cannot favour with certainty any one method. Results presented here were based on the discontinuous vorticity method of Stevenson.[11] The corner stresses were evaluated from the appropriate discontinuous values of the corner vorticities.

Conditions at the outflow boundary were obtained through extrapolation of the developing interior solution.[12]

Referring to figure 1, a grid size of d/20 and entry/exit lengths of 3d were used. Numerical experiments showed that changes in the flow field were not obtained when the mesh was refined or longer entry/exit lengths were applied.

RESULTS

A selection of flows of a Maxwell fluid is shown in figures 2, 3 and 4 for three combinations of inlet pressure gradients. In each of these diagrams the equivalent Newtonian flow is represented by broken contours. The channel width d is taken as 1cm, and the fluid density ρ as 1 g cm^{-3}.

Examination of the diagrams reveals only small variations between Newtonian and non-Newtonian streamlines. The vorticity contours show greater differences, but are qualitatively similar. However, marked differences appear in the shear stress results.

Significant differences in the Newtonian and non-Newtonian flows are confined to the regions influenced by the corners. When the dividing streamline falls in this region, as in figure 3, this line is locally displaced, although the position of the stagnation point

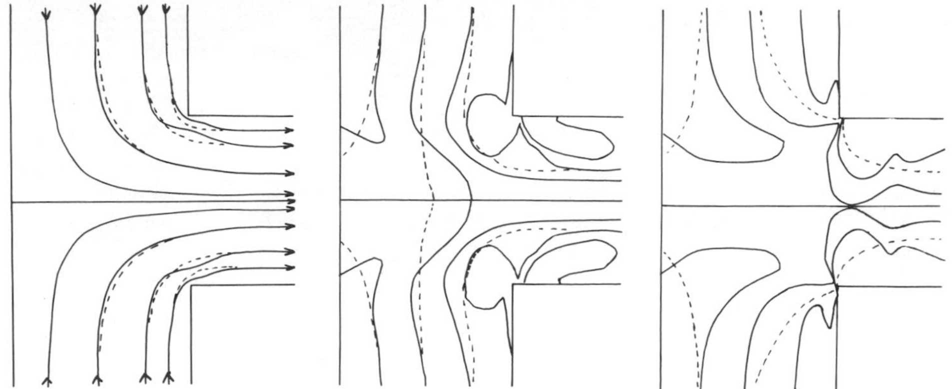

Fig. 2 Contours for $\underline{P}_1 = 50$, $\underline{P}_2 = -50$ dynes cm^{-3}, $\lambda = 0.15\ s^{-1}$, $\eta_0 = 5$ poise.

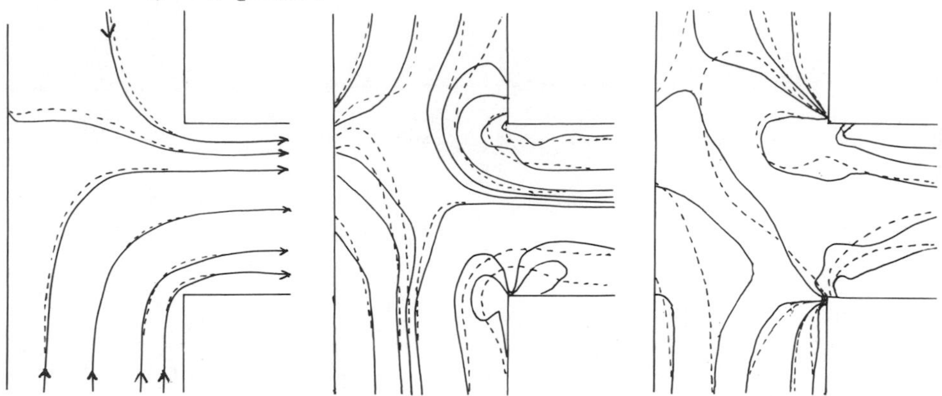

Fig. 3 Contours for $\underline{P}_1 = 50$, $\underline{P}_2 = -5$ dynes cm^{-3}, $\lambda = 0.15\ s^{-1}$, $\eta_0 = 5$ poise .

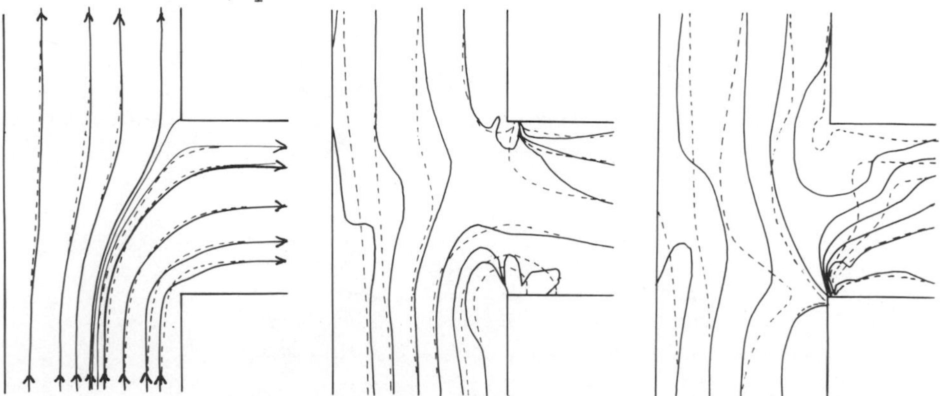

Fig. 4 Contours for $\underline{P}_1 = 100$, $\underline{P}_2 = 50$ dynes cm^{-3} $\lambda = 0.15\ s^{-1}$, $\eta_0 = 5$ poise .

is unaffected. At a distance of two channel widths from the corner, however, the Newtonian and non-Newtonian contours become indistinguishable.

The use of weighted finite differences, as presented in the previous section, showed an improvement of 15% in convergence rates when compared to our preliminary work using standard upwind differences. We also believe the method achieved an increase from order 2 to order 3.5 of the Weissenberg number in corner region, beyong which numerical instabilities became uncontrollable.

CONCLUSION

We have demonstrated that the implicit Maxwell model can give flow fields significantly different in the regions influenced by the corners. We do not have, however, quantitative methods available to judge the accuracy of the numerical solution.

The research is continuing. We are examining our vorticity boundary approximation and corner treatment, and hope to evaluate the work relative to experimental results.

We are grateful for useful discussion with Professor K. Walters and Dr. R.S. Jones.

REFERENCES

1. M.G.N. Perera and K. Walters, J. Non-Newtonian Fluid Mech. 2 (1977) 49.
2. M.G.N. Perera and K. Walters, J. Non-Newtonian Fluid Mech. 2 (1977) 191.
3. M. Kavahara and N. Takeuche, Computers and Fluids, 5 (1977) 33.
4. T.B. Gatski and J.L. Lumley, J. Comp. Phys. 27 (1978) 42.
5. T.B. Gatski and J.L. Lumley, J. Fluid Mech. 86 (1978) 623.
6. A.R. Davies, K. Walters and M.F. Webster, J. Non-Newtonian Fluid Mech. 6 (1979).
7. M.J. Crochet and M. Bezy, J. Non-Newtonian Fluid Mech. 5 (1979) 201.
8. K. Walters, J. Non-Newtonian Fluid Mech. 5 (1979) 113.
9. L.G. Leal, J. Non-Newtonian Fluid Mech. 5 (1979) 33.
10. D. Greenspan, Univ. Wisconsin, Tech. Report 20(1968).
11. J.F. Stevenson, ASME Paper No. 72-APM-SSS.
12. E.O. Macagno and T.K. Hung, J. Fluid Mech. 28 (1967) 43.

EXTRUSION FLOW BETWEEN PARALLEL PLATES

H. Holstein

Dept. Computer Science
U.C.W. Aberystwyth
U.K. SY23 3BZ

D.J. Paddon

School of Mathematics
University of Bristol
U.K. BS8 1TW

INTRODUCTION

Parallel pipe flow and non-viscometric flow between parallel plates are of industrial relevance, and have received attention theoretically[1] and experimentally.[2] A flow which combines both characteristics, such as injection moulding, is equally relevant, though the complex transition from pipe to radial flow is much less understood. In this paper we calculate the flow field using numerical methods. The flow domain is shown in fig. 1.

A circular pipe of diameter d opens out at right angles into a pair of parallel plates with separation d/2. Fully developed flow enters the pipe at AD and leaves the plates over a circumferential area of radius EF. The flow is taken as axially symmetric about the centre line DE, and steady. Thus all flow variables are functions of r and z, and only the half-domain DABCFE need be considered. The boundaries and the centreline of the geometry are streamlines, and there is a stagnation point at the centre of the end disc.

THE FLUID MODEL

The case for implicit differential models has been argued for complex flows subjected to abrupt changes in geometry.[3] Accordingly, computations have been carried out for an upper convected Maxwell model. This model is characterised by a single relaxation time λ_1 and a constant viscosity η_0. In view of the existence of non-Newtonian fluids with viscosities that are largely shear-rate indepen-

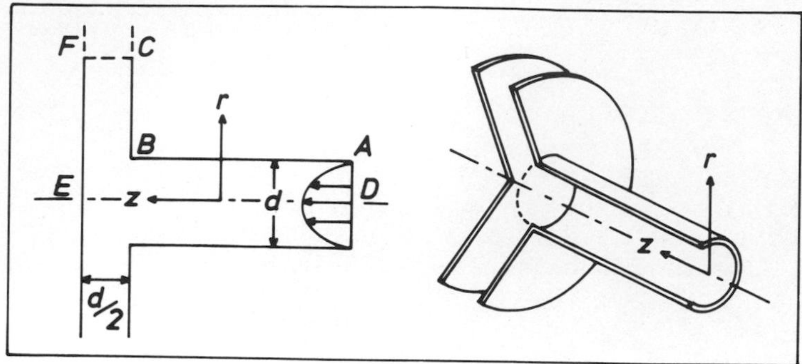

Fig. 1. Sectional and schematic views of the extrusion geometry.

dent,[4] the use of a constant viscosity model is justified. Furthermore, predictions about any non-Newtonian behaviour can be ascribed to the single parameter λ_1.

BASIC EQUATIONS

Let $\underline{u} = (u, 0, w)$ be the velocity vector
p be the isotropic pressure
$\underline{\underline{\tau}}'$ be the extra stress tensor with four independent components τ'_{rr}, τ'_{rz}, τ'_{zz}, $\tau'_{\theta\theta}$.

The equation of continuity, stress equation of motion and upper convected Maxwell equation are respectively

$$\nabla.\underline{u} = 0 \tag{1}$$

$$\rho\underline{u}.\nabla\underline{u} = -\nabla\rho + \nabla.\underline{\underline{\tau}}' \tag{2}$$

$$\underline{\underline{\tau}}' + \lambda_1 \left(\underline{u}.\nabla\underline{\underline{\tau}}' + \nabla\underline{u}^{\dagger}.\underline{\underline{\tau}}' + \underline{\underline{\tau}}'.\nabla\underline{u} \right) = \eta_0 \left(\nabla\underline{u} + \nabla\underline{u}^{\dagger} \right) \tag{3}$$

These equations must be solved subject to conditions for $\underline{u}$ and $\underline{\underline{\tau}}'$ on the boundary.

Across the inlet AD, $\underline{u}$ and $\underline{\underline{\tau}}'$ are determined by the imposed pressure gradient and the condition of fully developed flow. The no-slip condition $\underline{u} = \underline{0}$ holds on ABC and EF. On the centreline, u and $\partial w/\partial r$ are zero from symmetry.

The stress components are obtained from (3), which reduces to simple algebraic relationships along ABC and EF. Along the centreline, (3) reduces to a set of first order ordinary differential equations, which may be solved for the stress components using known inlet values at D. The integrations need only be performed for τ'_{rr} and τ'_{zz}, since those for τ'_{rr} and $\tau'_{\theta\theta}$ are identical, and τ'_{rz} is zero by symmetry.

The boundary CF is treated by a numerical method, outlined below. This boundary is assumed to be sufficiently far from the plate exit to be unaffected by free surface conditions.

NUMERICAL METHOD

We define the modified stress tensor $\overline{\underline{\underline{\tau}}}$ by

$$\underline{\underline{\tau}}' = \overline{\underline{\underline{\tau}}} + \eta_0 (\nabla\underline{u} + \nabla\underline{u}^{\dagger}). \tag{4}$$

Substituting (4) into (2) gives the modified equation of motion

$$\rho\underline{u}.\nabla\underline{u} = -\nabla p + \eta_0\nabla^2\underline{u} + \nabla.\overline{\underline{\underline{\tau}}}. \tag{5}$$

This substitution is necessary if the elliptic character of the system (1) - (3) is to be exploited by numerical method.[5] Equations (1), (5), (3) and (4), with (1) and (5) recast in the vorticity/stream function formulation, are expressed in finite difference form over a regular rectangular mesh. The resulting equations, linear except in their coupling, are solved for the stress and kinematic variables by successive overrelaxation.[6] Diagonal dominance of the iteration matrices is ensured by using upwind finite differences where necessary.

The vorticity along the walls and particularly at the corner (B) must be given careful consideration. We found, as previously reported by Roache,[7] that the Woods formulation of boundary vorticity gave reliable results. The specification of vorticity in the corner region is a more contentious issue. We took the vorticity as discontinuous, using limited forms of the vorticity (and stresses) along AB and CB, as appropriate.

The outflow at CF was allowed to evolve from the numerical solution in the neighbourhood of CF using numerical extrapolation. This method was preferred to an imposed flow field based on some independent calculation.

Calculations were carried out with an inlet length AB of 4 inlet radii, an outlet length EF of 5 inlet radii, and with r and z grid lengths of 1/16 of the inlet radius. Computational experiments with higher length ratios or with finer grids did not add significant detail to the results.

RESULTS

Numerical computations were performed for a geometry with an inlet diameter of 1 cm and a fluid density of $\rho = 1$ g cm^{-3}.

The pressure gradient $\bar{P}$ ($= -\partial p/\partial z$) at the inlet and the viscosity η_0 determine the volumetric flow rate. Contour diagrams for a non-inertial flow (Re=0.0625) are shown in fig. 2. The Weissenberg number based on the inlet shear rate is 0.75. Fig. 3 shows an inertial flow (Re=9.375) with an inlet Weissenberg number of 1.12. The equivalent Newtonian flows are represented by broken line contours.

In the neighbourhood of the corner, the streamlines of the elastic flow are displaced towards the corner. The effect of elasticity on the vorticity contours is noticeably greater than found for the streamlines. The shear stresses are qualitatively different. Changes between the Newtonian and non-Newtonian flows rapidly diminish with increasing distance from the corner, and at a distance of one pipe diameter the differences are insignificant.

Pressure differences were calculated by integrating equation (5) along streamlines. For non-inertial flow, it was found that the Maxwell fluid required an increased pressure drop over the equivalent Newtonian fluid, to maintain the same flow rate. These findings are in broad agreement with Co and Bird.[1] It may be noted that the kinematic similarity of the Newtonian and non-Newtonian flows implies that changes in the pressure fields are derived mainly from changes in the stress fields.

When the flow was inertial, fig. 3, no significant differences in pressure drop were found. Comparison with the analysis of Co and Bird is not valid, as their results were derived for creeping flow.

CONCLUSIONS

The calculations showed that high shear rates were obtained in the neighbourhood of the corner flow, and this justified the use of an implicit rheological model. Away from the influence of the corner, it would appear that a simple explicit rheological model would suffice.

The results indicate that the pressure drop and the flow characteristics would differ significantly from Newtonian behaviour at large Weissenberg numbers.

In common with other researchers,[8] we found the numerical technique to be limited by a maximum Weissenberg number (of order 2.5), beyond which convergence could not be achieved. Therefore, we are not able to simulate flows in which the corner has a significant

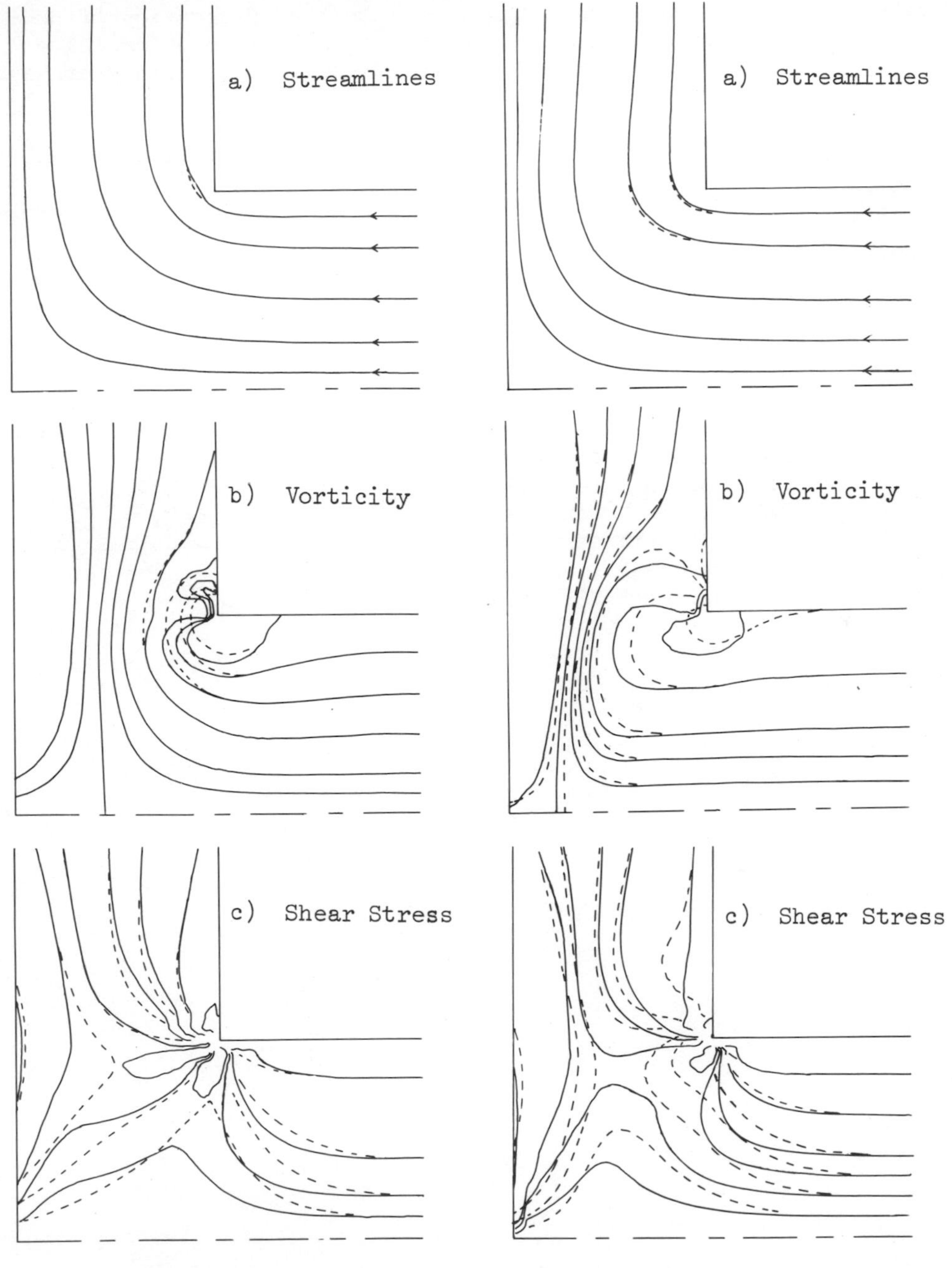

Fig. 2 $\overline{P}$ = 100 dyn.cm^{-3}
λ_1 = .15 s^{-1}
η_0 = 5 poise

Fig. 3 $\overline{P}$ = 1.5 dyn.cm^{-3}
λ_1 = .15 s^{-1}
η_0 = .05 poise

effect. Further research is required into the application of numerical methods applied to the solution of implicit rheological equations of state, in order to extend these studies to higher Weissenberg number flows.

We are grateful for useful discussion with Professor K. Walters and Dr. R.S. Jones.

REFERENCES

1. A. Co and R.B. Bird, Appl. Sci. Res. 33 (1977) 385.
2. L.R. Smith, J. Rheol. 22(6) (1978) 571.
3. M.G.N. Perera and K. Walters, J. Non-Newtonian Fluid Mech. 2 (1977) 49.
4. D.V. Boger, J. Non-Newtonian Fluid Mech. 3 (1977/8) 87.
5. L.G. Leal, J. Non-Newtonian Fluid Mech. 5 (1979) 33.
6. D.J. Paddon, Doctoral Thesis, Univ. Wales, (1979).
7. P.J. Roache, "Computational Fluid Dynamamics", Publ. Hermosa (1972).
8. A.R. Davies, K. Walters and M.F. Webster, J. Non-Newtonian Fluid Mech. 5 (1979).

THE INFLUENCE OF CHAIN STIFFNESS ON THE FLOW BEHAVIOR OF POLYMERS IN THE ENTRANCE OF A CAPILLARY

D.G. Baird

College of Engineering, Virginia Polytechnic Institute
U.S.A.

(Abstract)

A glass capillary attached to an Instron Rheometer has been used to study the flow behavior of a flexible, a semi-rigid, and a rigid chain polymer in a capillary. The flexible chain polymer exhibits the "wine-glass" pattern in the entrance region and undergoes a flow instability in the entrance associated with the onset of elastic fracture at a wall shear stress (τ_w) of 10^4 N/m^2. The semi-rigid polymer does not exhibit the "wineglass" entrance flow pattern. However, a flow instability arises in the capillary at a wall shear stress of 10^2 to 10^3 N/m^2. The rigid chain polymer exhibits behavior in the capillary entrance similar to that expected for an inelastic fluid even though large normal stresses are measured. No flow instabilities were observed for values of τ_w up to 2×10^4 N/m^2. Die swell decreased with increasing chain stiffness.

AN EXPERIMENTAL INVESTIGATION OF FLOW IN THE DIE LAND REGION OF A CAPILLARY RHEOMETER

A.V.Ramamurthy, J.C.H. McAdam

CSIRO, Division of Mineral Engineering, Clayton, Victoria, Australia 3i68

(Abstract)

A Laser Doppler Anemometer and a high speed data logging system are used to study the details of flow in the die land region. An abrupt entry cylindrical glass die, 6 mm I.D. and 210 mm long, is used. A 9.7 poise Newtonian Silicone Oil and a highly elastic polymer solution are investigated under low Reynolds number flow conditions ($N_{Re} < 1$).

The results indicate that, for the Newtonian fluid, the axial velocity distribution is fully developed at 1 mm from the die entry plane and is independent of the Reynolds number within the range studied. Some unusual effects are observed for the elastic fluid. In particular, it is found that a form of vena contracta is formed immediately downstream of the entry which leads to acceleration-deceleration effects. At higher flow rates, the axial velocity distribution becomes asymmetric followed by the development of instability. Under unstable flow conditions, the velocity at the wall of the die is non-zero indicating a slip flow, in addition.

THE ENTRY AND EXIT FLOW INTO A SLIT

W. Philippoff

New Jersey Institute of Technology

323 High Street, Newark, N.J. 07012, U.S.A.

INTRODUCTION

The stress-distribution at the entry and exit of a cylindrical capillary is as yet not completely known, even if technologically important phenomena, such as the so-called "melt-fracture" occur in or near them. Here the results of both a theoretical and experimental investigation (using birefringence) are described, resulting in some quantitative relations for a slit and Newtonian liquids, as a first approximation to capillary flow.

THEORY

Starting from the well-known Jeffery-Hamel flow between converging planes the influence of the entry-angle α on the overall stress-distribution at some distance from the apex was calculated.[1] It could be experimentally verified.[2] But this flow into a mathematical "point sink" cannot be experimentally realized; also at the apex infinitely large stresses appear.

Therefore, to approximate realizable conditions the expansion of the theory to a slit of finite breadth is necessary. For the infinitely short slit (a hole in a mathematical plane) Morse and Feshbach[3] gave a solution for the stream-function, velocity and pressure, not the stresses. Solutions for the same quantities are given by Weissberg[4] for the capillary, Foerste[5] for the slit.

The same problem has been treated recently by the finite difference method[6,7], the results are very similar to the ones here.

The stresses are second spatial derivatives of the stream-function, therefore are very sensitive to a change in the shape of the streamlines. In elliptical coordinates μ, ϑ introduced in[3] the stream-function is: $\psi = Q/\pi(\vartheta - \frac{1}{2}\sin 2\vartheta) \neq f(\mu)$, evaluated in Fig. 1, for a slit of total breadth 2.

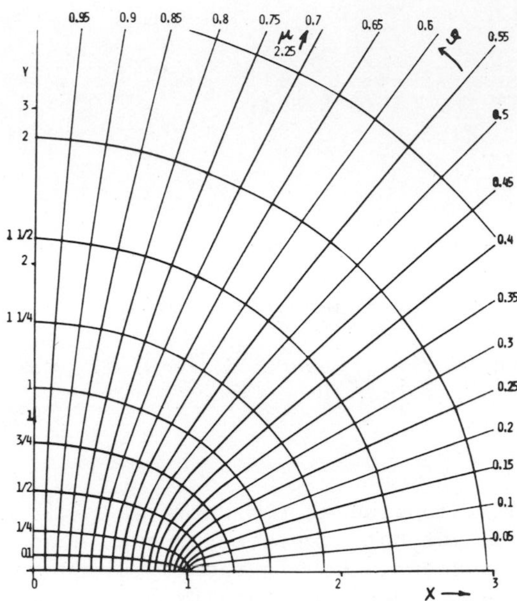

Fig. 1 Flow-lines for an infinitely short slit.

With cartesian coordinates $x = \cosh\mu\cos\vartheta$ and $y = \sinh\mu\sin\vartheta$ and the components of velocity $u = (-2Q/\pi).\sin^2\vartheta(\sinh^2\mu + \sin^2\vartheta)^{-\frac{1}{2}}$ and $v = 0$, the stress-components are: $\sigma_{\mu\mu} = 2\eta(1/h)\partial u/\partial\mu$, $\sigma_{\vartheta\vartheta} = 2\eta\,.\,(u/h^2)\partial h/\partial\mu$ and $\sigma_{\mu\vartheta} = \eta\frac{\partial}{\partial\vartheta}(u/h)$ with Q the flow per unit length of the slit and B = 2 in sec and the scale-factor $h = (\sinh^2\mu + \sin^2\vartheta)^{\frac{1}{2}}$, the result for the difference in the principal stresses, that according to the stress-optical law is proportional to the observed birefringence (isochromatics) is:

$$\sigma_1-\sigma_2 = ((\sigma_{\mu\mu}-\sigma_{\vartheta\vartheta})^2 + 4\sigma_{\mu\vartheta}^2)^{\frac{1}{2}} = (8Q\eta/\pi)\sin\vartheta.\sinh\mu(\sinh^2\mu + \sin^2\vartheta)^{-3/2}$$

For a numerical evaluation a parameter $A = \pi(\sigma_1 - \sigma_2)/8Q\eta$ is introduced, leading in cartesian coordinates to:

$$x_{1,2} = (1 - y^2 \pm ((y/A)^{4/3} - 4y^2)^{\frac{1}{2}})^{\frac{1}{2}}$$

The stress-field, meaning the function y= f(x) for constant A is shown in Fig. 2, suitable to compare with birefringence results.

From Fig. 2 one sees that in the entry-plane the stress is constant = 0, that there is a singularity at the corner where the stress becomes infinite and a maximum of stress on the axis at y= 0.707 and A = 0.3849. These conditions are different from the ones for the Jeffery-Hamel flow: the singularity at the apex is split into two at the corners. The axial stress is a pure tension at the entry, a compression at the exit. With $p_o = \frac{1}{2}$ (applied pressure), the pressure from [3] is shown in Fig. 3, from which follows that there is a pressure drop ($\Delta p/p$) near the axis, followed by a line of 0 drop at $\sim 45°$ and a pressure build-up at larger angles, tending to infinity at the corner, where the stresses in Fig. 2 equally tend to infinity. The pressure at the entry-plane is $\Delta p/p_o = -1 + x/(x^2 - 1)^{-\frac{1}{2}} = \text{cotanh}\mu - 1$, with no solution below x = 1. There the pressure is constant = p_o.

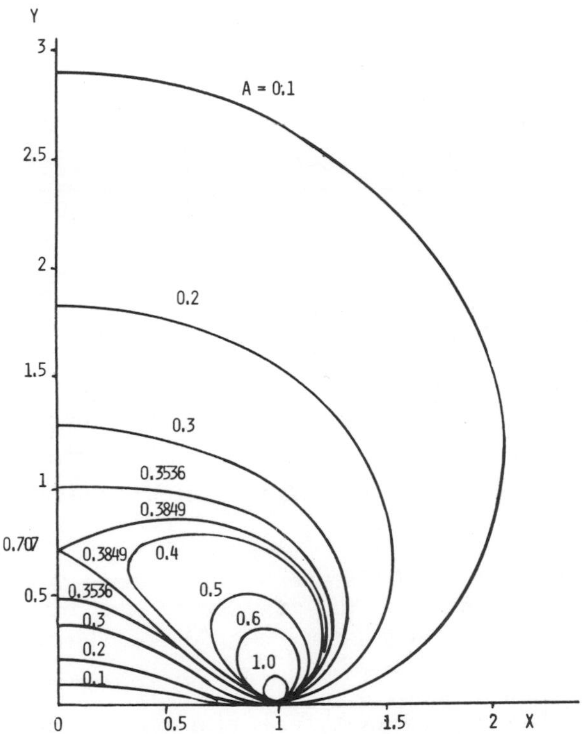

Fig. 2 Stress-distribution for an infinitely short slit. Half a "bun"

For the slit of finite length, the stream-function is modified as in[4], using the one for the slit: $\psi = Q/4(\cos^3\vartheta + 2 - 3\cos\vartheta)$ at the entry-plane, assuming that the slit-flow is established there [4,5], and letting it decrease with $(1 + 7.5\mu^2)^{-\frac{1}{2}}$ as in [4]. This amounts to a small change in the shape of streamline, at the most by 8%. The calculation becomes very complicated, as $v \neq 0$. The result is, that the stresses shown in Fig. 2 must be multiplied by a correction factor, which is a closed analytical expression, but can only be evaluated by a large computer. The results are shown in Fig. 4.

Qualitatively Fig. 2 and 4 are similar, the differences are in the value of the maximum axial stress A = 0.5093 at y = 0.5210

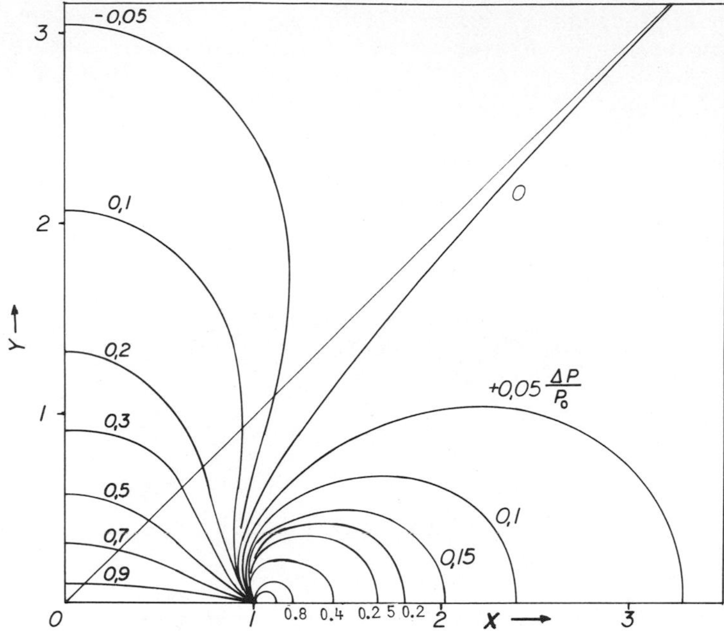

Fig. 3 The pressure-distribution $\Delta p/p_o$ at the entry to a slit.

and the curves between A = 0.51 and 0.9 that have a peculiar curvature near the entry-plane. This is due to the introduced boundary condition that the stream-function at the entry-plane must be the one for the slit.

$\sigma_1 - \sigma_2 = 2\,\sigma_w$ at the wall, calculated from the stream-function for the slit is A = $3\pi/8$ = 1.1781 (or $Q=2\sigma_w/3\eta$ for a slit with B = 2). Therefore the tensile stress at the axis is 0.5093/1.1781 = 0.4323 of $\sigma_1 - \sigma_2$ at the wall or 0.8646 of σ_w the shear stress at the wall usually used in Rheology.

A further relation is the ratio of the breadth of the "bun" to its height. For the "point sink" it is a constant = $\sqrt[4]{64/27}$ = 1.241, for the finite slit it depends on A and has a limit at A= 0.5093 and a minimum between A = 0.51 and 0.45.

EXPERIMENTAL RESULTS

The optical experiments were performed on an Epoxy-resin with a viscosity in the range used of about 10^8 poises, the experimental arrangement and the method of evaluation have been previously described[2]. From numerous experiments with a slit of length/breadth of 4:1 the axial stress-distribution with different driving pressures is shown in Fig. 5. The curve drawn is the one calculated.

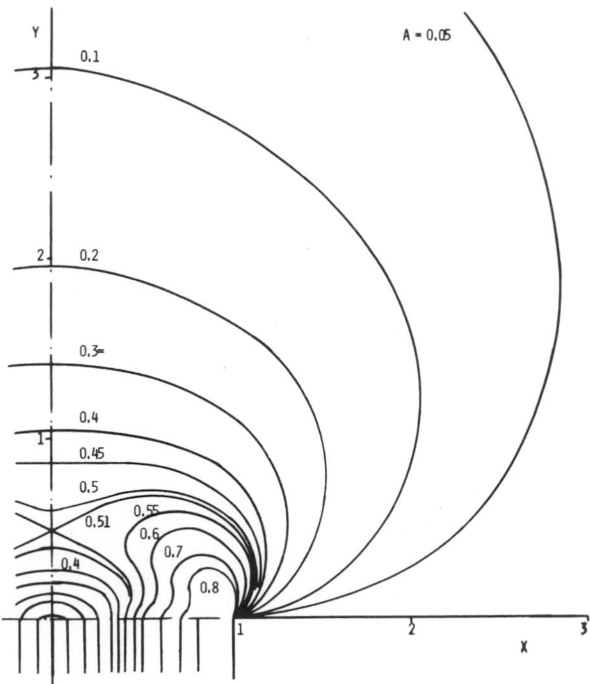

Fig. 4. Stress-distribution for a slit of finite length.

The predicted maxium at y = 0.521 is indeed there, the points lie closely on the calculated curve beyond the maximum. However the 0-stress is not at y = 0, but at y = -0.87, meaning that the slit-flow is present only at distances from the entry-plane larger than 0.435 B, valid also for the submerged exit, not at the entry-plane as assumed. For the infinitely short slit the experimental values at different pressures followed the theoretical curve in the whole range. The check of the values for different pressures (σ_w in the 10^6 dynes/cm^2 range) proves the Newtonian behavior of the material used. Counting the number of observed fringes, estimating to ¼ fringe, the average gave a ratio of 0.40 ± 0.02 instead of 0.432 at a point $y = 0.50 \pm 0.04$ instead of 0.521. The direction of $\sigma_1 - \sigma_2$ followed the theory closely, not being normal to the isochromatic. Inside a triangle determined by the points (0,0), (0,$\pm$ 1) and (-0.87, 0) the direction of $\sigma_1 - \sigma_2$

is axial, outside this area in the slit it is at 45° to the axis as expected. The transition is quite sharp. This shows that the assumption of an established slit flow at the entry-plane is untenable, but no other limit could be found. The axial stresses at the exit into air are smaller than for the submerged exit, the ratio is 0.87. A curvature of the entry corner with radii of 0.405 and 0.920 B was investigated. The position of the maximum stress was shifted towards or even below the entry-plane, but its distance from the 0-fringe was unchanged. The stress was also unchanged. The singularity at the corner disappeared for different α (flat entry $\alpha = 90°$, slit $\alpha = 0°$) the ratio of $\sigma_1 - \sigma_2$ to σ_W

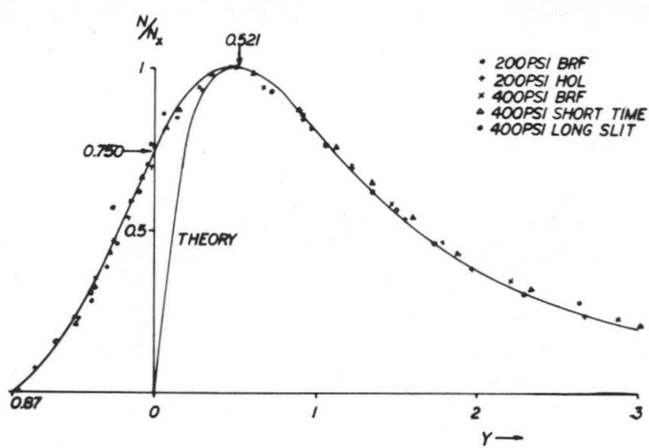

Fig. 5 Experimental axial stress-distribution for the 4:1 slit.

changes smoothly from 0.4 to 0, a reduction to 0.2 occurs at 15°. If this stress is the origin of the "melt fracture" a curvature of the entry will not change it, but as is known empirically, a smaller α will decrease it.

This work was supported by NSF-Grant ENG-76-08862.

1. W. Philippoff, The Analysis of the Entrance Flow into a Slit Part I. Trans. Soc. Rheol. 20:423 (1976)
2. R. Racano, T. Mitslal, L. Buteau, G. Lei and W. Philippoff, The Analysis of the Entrance Flow into a Slit. Part II Experimental. Trans. Soc. Rheol. 23:39 (1979)
3. P. M. Morse and H. Feshbach, Methods of theoretical Physics. McGraw Hill, N.Y. 1953, Vol. II P. 1197.
4. H. L. Weissberg, End Correction for Slow Viscous Flow through long tubes, Physics of Fluids 5:1033 (1962)
5. J. Foerste, Die ebene laminare Stroemung in einem Halbraum bei kleiner Reynolds-Zahl. Z.A.M.M. 43:353 (1963)
6. R. E. Nickell, R. I. Tanner and B. Caswell. The solution of viscous incompressible jet and free surface flows using finite-element methods. J. Fluid Mech. 65 189 (1974)
7. M. Viriyayuthakorn and B. Caswell, Finite element Simulation of Viscoelastic Flow. Report NSF Eng7800722 (1979)

ON NON-NEWTONIAN FLOW THROUGH A SLIT PLATE

Kitaro Adachi and Naoya Yoshioka

Department of Chemical Engineering
Kyoto University
Kyoto, 606, JAPAN

INTRODUCTION

The flow through a two-dimensional orifice, i.e., a slit plate, in a two parallel plate channel involves two flow geometries of conversion and diversion. Both of them are widespread and important in the polymer processing of extrusion and injection mouldings. Therefore, there has been much discussion. Giesekus[1,2] paid attention to the remarkable difference in the flow pattern between the converging flow and the diverging flow although the corresponding, creeping flow of a Newtonian fluid has a symmetric flow pattern on both the sides of the orifice plane. That asymmetry is one of the characteristic features of viscoelastic behaviors, as is also seen in the flow past a sphere.[3] Most existing studies have been directed to the converging, entry flow in connection with the elongational flow, but the diverging, exit flow is also important. Thus, it would be interesting to correlate the pressure drop with the related flow pattern in each flow region, or, at least, to divide the additional pressure drop due to the existence of the orifice plate into the entrance and exit portions, if possible. Further points to be made clear are the difference in the pressure drop between the 2-dimensional, plane flow through a slit plate and the 3-dimensional, axisymmetric flow through an orifice, the critical condition for the onset of the steady oscillating flow, its period, and the change in the pressure drop with the onset of viscoelastic flow disturbances. The present work will be more or less concerned with these points.

AN EXPERIMENTAL STUDY ON FLOW PATTERNS

The flow visualization experiments were performed in a vertical channel of a 10×15 cm^2 rectangular cross-section and 230 cm height. It was equipped with a head tank of a 30×30 cm^2 cross-section at the top and a bulb controlling the flow rate at the bottom. The duct in the test section had a slit plate of 1 cm thickness with a slit of 15 cm in length and six different width (h_0=0.043, 0.065, 0.108, 0.211, 0.494 and 1.00 cm), which is illustrated in Fig. 1. Sodium polyacrylate aqueous solutions of 0.3, 0.5, and 0.75 wt% were used. The flow was visualized by fine aluminium dust.

One example of the photographs, which were taken at the vertical center plane, is shown in Fig. 2. The flow patterns are similar to those reported by others. There is an elongational flow with secondary vortices in front of the slit plate and a radial flow behind it. The vortex length data are presented in Fig. 3. Here, the characteristic quantities are defined at the slit and designated by the

$$\varepsilon=h_0/H=A_0/A \quad X=L/H \quad W_s=\lambda\bar{u}_0/h_0 \quad R_e=h_0\bar{u}_0\rho/\eta(\dot{\gamma}_w)$$
$$\lambda=(\tau_{11}-\tau_{22})/2\tau\dot{\gamma}\Big|_{\dot{\gamma}=\dot{\gamma}_w} \quad \dot{\gamma}_w=(2\bar{u}_0/h_0)(2n+1)/n \quad Q=W\bar{u}_0h_0=W\bar{U}H \tag{1}$$

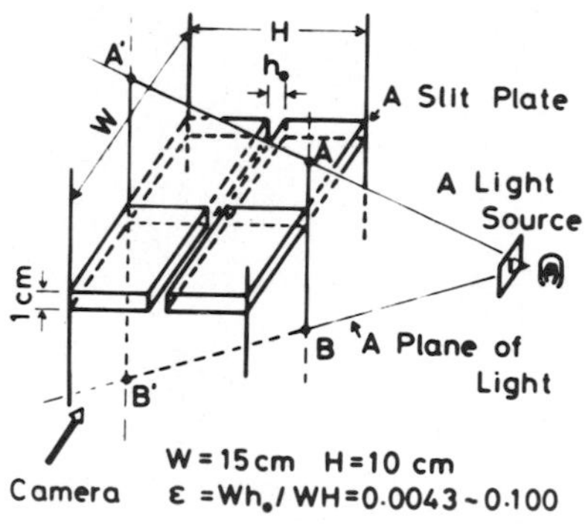

Fig. 1. Test Section with a Slit Plate.

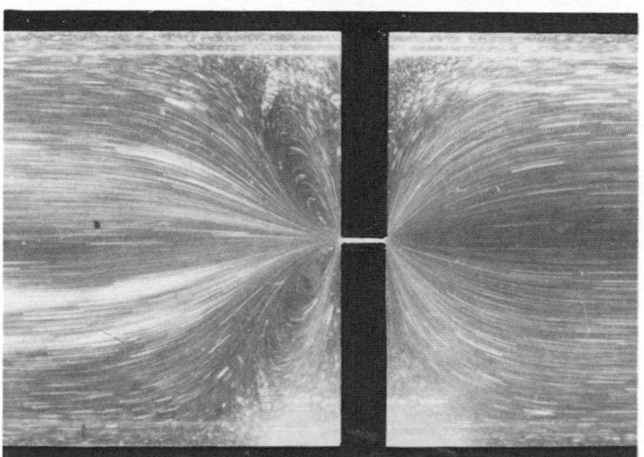

Fig. 2. Steam Pattern of Stable Flow

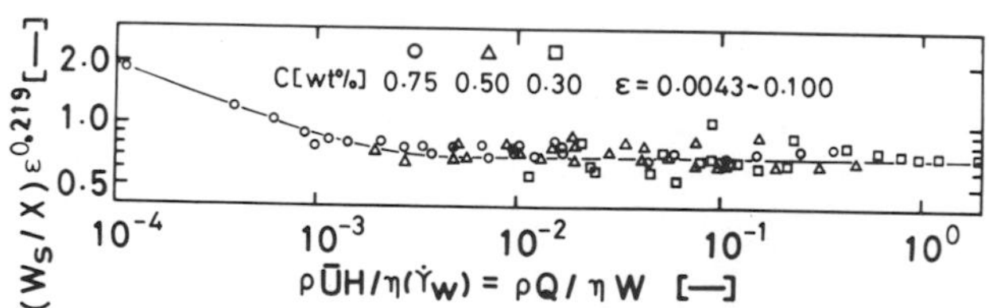

Fig. 3. Vortex Length

subscript " 0 ". Nguyen and Boger[4] have presented a relation, $(W_s/X)\varepsilon^{-1/2}$=const., for the tubular entry flow, but some of their data have a clear trend that W_s/X decreases as the contraction ratio ε increases. Thus, the present result, $(W_s/X)\varepsilon^{0.2186}$=const., seems to be reasonable. The rising curve at small Reynolds number indicates that the effect of the reservoir wall supporting the vortices begins to vanish. There is a critical value of the stress ratio, as is shown in Fig. 4, beyond which the flow becomes unstable. The value of 9.7 is almost two times larger than that of 5 given by Cable and Boger.[5] However, the present result seems to be quite consistent with the critical values for melt fracture reported by many others,[6] if we take into account that polymer solutions behave more like low density, branched polymer melts than high density polymer melts. At the critical state, W_s also takes a constant value of 0.811±0.100. The main flow oscillates from side to side with a long, regular period. The period becomes shorter as the flow-rate increases. As is presented in Fig. 5, $T \propto 1/\bar{u}_0^{0.5}$ and the value of the proportional constant changes with the concentration of the solution. In the case of tubular entry flow, Cable and Boger's data[5] give $T \propto 1/\bar{u}_0^{0.820}$ for four different polymer solutions, and den Otter's data[7] show $T \propto 1/\bar{u}_0^{1.15}$ for branched PDMS liquid and $T \propto 1/\bar{u}_0^{1.28}$ for LDP melt. The value of the power index seems to change depending upon whether the flow is plane or tubular. It is not clear how the rheological properties have an effect on the period of oscillation.

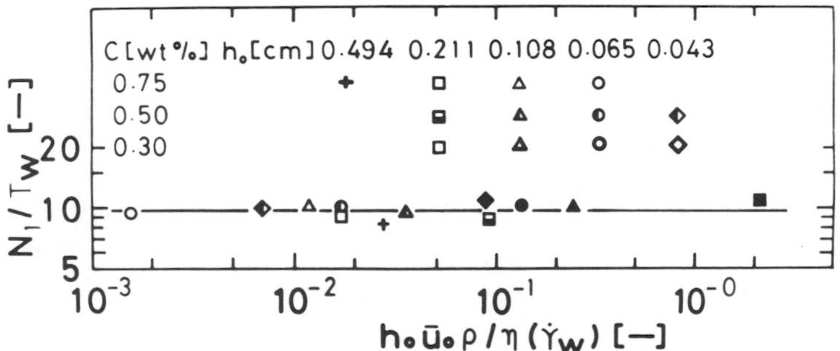

Fig. 4. Critical Stress Ratio for the Onset of Steady Oscillation

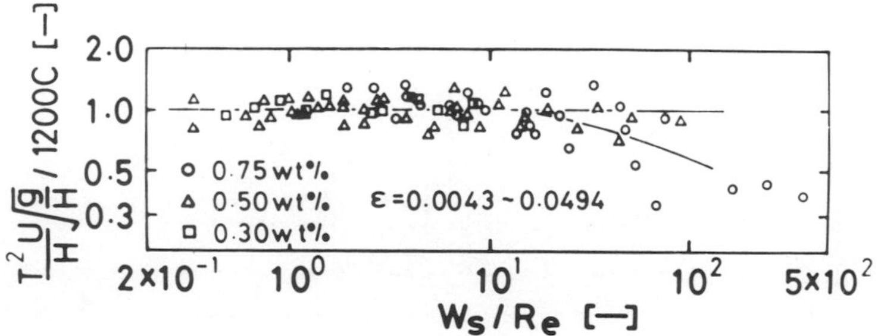

Fig. 5. Periodic Time of Steady Oscillation

A STUDY ON PRESSURE DROPS

The same duct as that for the flow visualization experiments was set horizontally for the pressure drop measurements. The fluid thrusts normal to the duct wall were measured through the two pairs of 5^{ϕ} holes which were drilled in the surface of the wall, 1 cm and 17 cm, upstream and downstream from the slit plate. The pressure difference between the two holes 1 cm upstream and downstream was almost the same as that between another pair of holes under any flow condition studied. This pair of holes more remote from the slit was used so that the hole-pressure errors[8] could be cancelled by the two similar flow states at their holes. It must be noted that large periodic osillations as well as small fluctuations of streaming did not yield an striking change in the flow rate-pressure drop relationship. In what follows, the pressure drop for stable flow will be considered.

The measured pressure difference consists of the entry loss, the friction loss at the slit wall and the exit loss. The pressure drop at the slit may be estimated by assuming a fully developed flow in a parallel plate channel. The rest may be divided into the entry and the exit loss by considering the macroscopic momentum balance for a domain which is illustrated in Fig. 6. From the equations of motion and continuity, the following relation can be derived without any approximation if $\tau_{xy}=0$ on the center line (y=0).

$$\left[\int_0^{h/2}(-p+\tau_{xx}-\rho u^2)\,dy\right]_{x(D)}^{x(A)}=\int_{x(A)}^{x(D)}\left[(p-\tau_{xx})\frac{dh}{dx}+\tau_{xy}\right]_{h/2}dx \tag{2}$$

The entry line AA' and the exit line BB' are so remote from the slit that the flow can not be perturbed by the existence of the slit plate . No difference in the measured thrust force difference between the two pairs of holes suggests the following approximation upstream

$$\int_{h_e/2}^{y(A')}(-p+\tau_{xx})\Big|_{x(A)}dy\simeq-\int_{x(A)}^{x(B)}(-p+\tau_{xx})(x,h)\frac{dh}{dx}dx \tag{3}$$

and the analogous approximation downstream. Then, finally, we get

$$\Delta p_{total}=-(2/h_e)\left[\int_{x_A}^{x_B}+\int_{x_B}^{x_C}+\int_{x_C}^{x_D}\right]\tau_{xy}(x,h)\,dx=\Delta p_{entry}+\Delta p_{slit}+\Delta p_{exit} \tag{4}$$

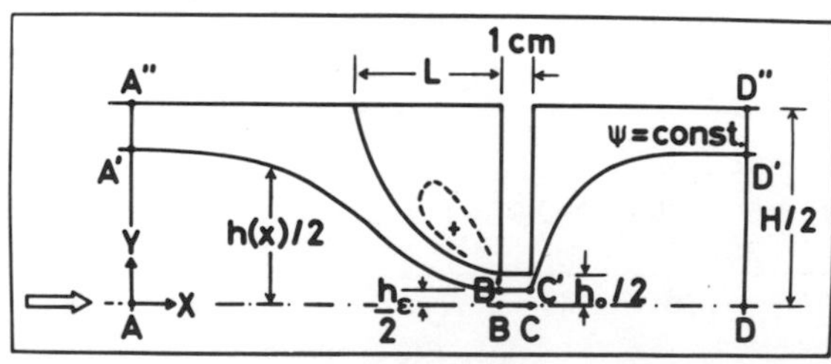

Fig. 6. A Domain for Calculation of Macroscopic Momentum Balance

Table 1

$\dot{\gamma}_w$	Δp^{elast}_{conv}	Δp^{elast}_{div}	$\Delta p_{div}/\Delta p_{conv}$	Δp^{visc}_{slit}	Δp^{visc}_{div}	$\Delta p^{meas}_{total}/\Delta p^{calc}_{total}$
1.34	4130	1290	0.312	3930	618	0.810
2.13	4250	1520	0.358	4520	710	0.917

The shear stress on the streamline y=h(x) may be roughly estimated using the steady shear viscosity data. The strain-rate tensors are calculated from differentiation of the stream function. To do so, three adjacent streamlines are selected on the photograph (Fig. 2). The three constant values of the stream function can be determined from the velocity distributions measured at the entry and the exit boundary. The estimated values of Δp_{total} are not discouraging at all, as is shown in Table 1. The ratio $\Delta p_{exit}/\Delta p_{entry}$ is about 0.335 at middle flow-rates. The exit pressure drops are about two times larger than the viscous exit losses, so that the exit flow is not a viscous flow although it is not elongational. The entry pressure drop can also be estimated using Cogswell's theory[9] if the angle of flow convergence is known. The predicted total pressure drops, (1.335 Δp_{entry}+Δp_{slit}), were plotted in Fig. 7 and compared with experimental data. It was rather difficult to determine the flow convergence angle. A more accurate relationship between the angles and the flow conditions is needed for a better agreement.

Black, Denn and Hsiao[10] obtained a negative, elastic pressure loss for two dimensional converging flow by using a convected Maxwell model. Tanner's theorem[11] makes it easy to calculate the pressure loss for the plane, creeping flow of the second-order fluid on the basis of the corresponding Newtonain flow solution.

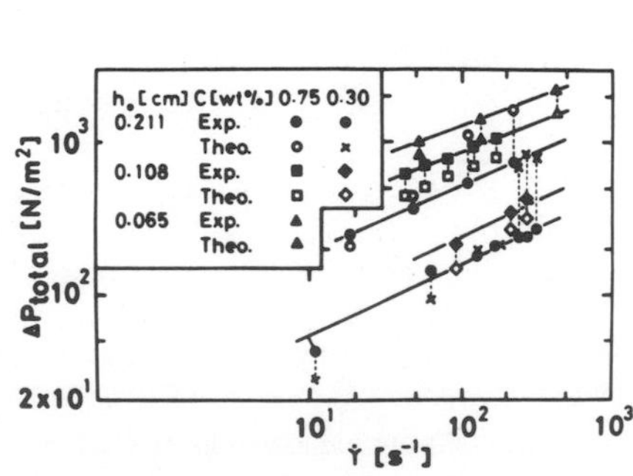

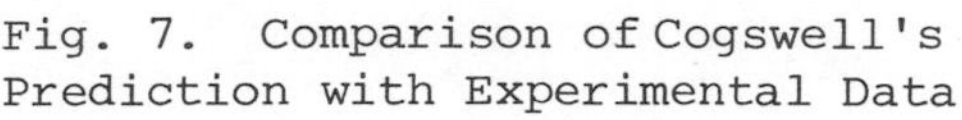
Fig. 7. Comparison of Cogswell's Prediction with Experimental Data

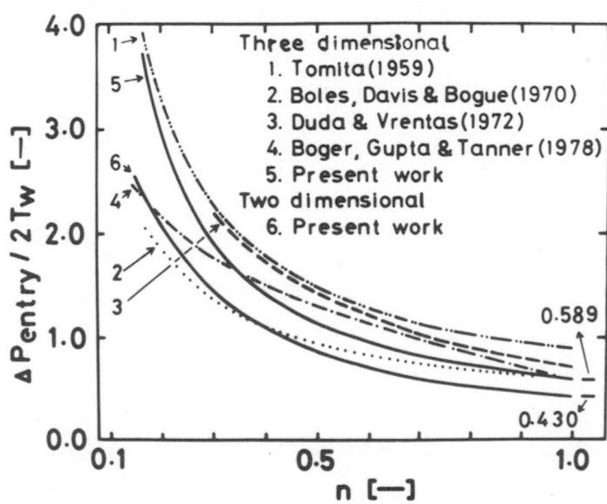

Fig. 8. Entry Losses for Inelastic Power-Law Fluids

$$\Delta p_{total} = -\int_{x(A)}^{x(D)} \frac{\partial \tau_{xy}(x,0)}{y} dx \tag{5}$$

$$\tau_{xy}(x,y) = \mu\left(\frac{\partial v}{\partial x}+\frac{\partial u}{\partial y}\right) + \beta\left[\left(u\frac{\partial}{\partial x}+v\frac{\partial}{\partial y}\right)\left(\frac{\partial v}{\partial x}+\frac{\partial u}{\partial y}\right)+2\left(\frac{\partial u}{\partial x}\frac{\partial u}{\partial y}+\frac{\partial v}{\partial x}\frac{\partial v}{\partial y}\right)\right] \tag{6}$$

The Newtonian symmetrical flow on both sides of the slit yields a relation, $\Delta p_{entry} = -\Delta p_{exit}$, for the additional, elastic pressure loss. Thus, Wissler's theoretical prediction[12], that the elastic term is proportional to W_s^2, seems to be questionable. The present experimental results are expressed by

$$Y \equiv \left(\Delta p_{total}^{meas} - \Delta p_{slit}^{visc}\right)/\Delta p_{div}^{visc}\Big|_{\dot{\gamma}_w(slit)} = 1.00 + 0.0025\ \varepsilon^{-1} W_s^{1.4} \tag{7}$$

which is much lower than that for an orifice. It is natural since the maximum stretch-rate is smaller for Newtonian, creeping flow through a slit than for the corresponding flow through an orifice at the same value of ε. In Eq.(7), Δp_{div}^{visc} is given in Fig. 8. That was computed using the eliptic coordinates by the variational method.

REFERENCES

1. V. H. Giesekus, Nicht-Lineare Effekte beim Strömen Viscoelastischer Flüssigkeiten durch Schlitz- und Lochdüsen, Rheol. Acta , 7:127 (1968).
2. V. H. Giesekus, Vershiedene Phänomene in Strömungen Viscoelastischer Flüssigkeiten durch Düsen, Rheol. Acta, 8:411 (1969).
3. K. Adachi, N. Yoshioka, and K. Sakai, An Investigation of Non-Newtonian Flow past a Sphere, J. Non-Newtonian Fluid Mech., 3:107 (1977/1978).
4. H. Nguyen and D. V. Boger, The Kinematics and Stability of Die Entry Flows, J. Non-Newtonian Fluid Mech., 5:353 (1979).
5. P. J. Cable and D. V. Boger, A Comprehensive Experimental Investigation of Tubular Entry Flow of Viscoelastic Fluids, AIChE Journal, 25:152 (1979).
6. J. Vlachopoulos, Die Swell and Meltfracture: Effects of Molecular Weight Distribution, Rheol. Acta, 13:223 (1974)
7. J. L. den Otter, Mechanisms of Melt Fracture, Plastics and Polymers, 38:155 (1970).
8. J. M. Broadbent, A. Kaye, A. S. Lodge, and D. G. Vale, Possible Systematic Error in the Measurement of Normal Stress Differences in Polymer Solutions in Steady Shear Flow, Nature, 217 :55 (1968).
9. F. N. Cogswell, Converging Flow of Polymer Melts in Extrusion Dies, Polym. Eng. Sci., 12:64 (1972).
10. J. R. Black and M. M. Denn, Converging Flow of a Viscoelastic Liquid, J. Non-Newtonian Fluid Mech., 1:83 (1976).
11. R. I. Tanner, Plane Creeping Flows of Incompressible Second-Order Fluids, 9:1246 (1966).
12. E. H. Wissler, Viscoelastic Effects in the Flow of Non-Newtonian Fluids through a Porous Medium, I & EC Fund., 10:411 (1971).

ELASTIC EFFECTS IN DIE ENTRY FLOWS

M. J. Crochet and M. Bézy

Université Catholique de Louvain
B1348 Louvain-la-Neuve, Belgium

INTRODUCTION

The kinematics of die entry flow is a complex phenomenon in which inertia, shear-thinning and elasticity play a significant role. A clear understanding of the effect of these three features upon the flow field on the basis of experimental results (e.g. Cable and Boger, 1978) is difficult because of the practical impossibility of separating these effects in general. The discovery by Boger and Nguyen (1978) of a viscous fluid which is highly elastic at low shear rates while not shear thinning and optically clear led to a series of experimental observations of die inlet flow at very low Reynolds numbers where the sole parameter is the elasticity of the flow. Dramatic results have been reported by Nguyen and Boger (1979). At low flow rates, a corner vortex upstream of the small tube entrance remains stationary; when the flow rate (or the elastic character of the flow) increases, the vortex grows and leads eventually to various modes of unstable flow. All test fluids (glucose syrup-Separan solutions) presented a domain of shear rate associated with second-order viscometric behavior. Nguyen and Boger (1979) made the observation that the initiation of the vortex growth regime occurs at shear rates just beyond the domain of second-order behavior.

Much progress has been made over the last few years on the theoretical analysis of viscoelastic flow by means of numerical techniques. Die entry flow has been studied by Crochet and Bézy (1979) and by Viriyayuthakorn and Caswell (1980); Perera and Walters (1977) considered the case of abrupt contractions in plane flow. In the present paper, we will first examine the flow of an upper convected Maxwell fluid through a 4:1 contraction, and pursue the calculation up to the highest flow rate allowing convergence of the numerical calculation; we will find that the corner vortex grows only slightly. In order to test the validity of Nguyen and Boger's conjecture on the inception of the vortex growth regime, we will calculate the flow through the same die of a White-Metzner fluid with viscometric functions identical to those of a test fluid used by Nguyen and Boger (1979).

ENTRY FLOW OF A MAXWELL FLUID

First, we consider the motion of an upper-convected Maxwell fluid which has the following constitutive relations,

$$\underset{\approx}{\sigma} = -p\underset{\approx}{I} + \underset{\approx}{T} \ , \quad \underset{\approx}{T} + \lambda\overset{\nabla}{\underset{\approx}{T}} = 2\mu\underset{\approx}{D} \ , \tag{1}$$

where p denotes the pressure, $\underset{\approx}{\sigma}$ is the Cauchy stress tensor, $\underset{\approx}{T}$ the extra-stress tensor and $\underset{\approx}{D}$ the rate of deformation tensor; μ is the (constant) shear viscosity and λ is a (constant) relaxation time, while $\overset{\nabla}{\underset{\approx}{T}}$ is the upper-convected stress time derivative. The momentum equations in the absence of body forces and inertia terms and the equation of conservation of mass are

$$-\nabla p + \nabla\cdot\underset{\approx}{T} = \underset{\sim}{0} \ , \quad \nabla\cdot\underset{\sim}{v} = 0 \ . \tag{2}$$

The numerical method used here is a finite element mixed method with 7 unknown fields (2 velocity components, pressure, and 4 stress components). We use Galerkin's procedure for discretizing equations (1) and (2), and the non-linear problem is solved by Newton-Raphson iterations; 4 to 5 iterations are sufficient when the iterative procedure converges. More details on the numerical method may be found in Crochet and Keunings (1980). The velocity components are given everywhere on the boundary except on the axis of symmetry where we impose a vanishing axial contact force. The finite element grid used for the calculations is shown on Fig. 1; we have selected long entrance and exit lengths in order to obtain fully developed Poiseuille flow in the inlet and outlet sections.

The calculations are performed with a vanishing Reynolds number. The non-dimensional quantity selected for characterizing the elastic character of the flow is the product $\lambda\dot{\gamma}$ of the relaxation time and the wall shear rate in the fully developed downstream flow; this quantity equals eight times the Weissenberg number defined by Nguyen and Boger (1979). We were able to obtain convergence up to a value of $\lambda\dot{\gamma} = 3.5$; the Newton-Raphson iterative procedure does not converge for higher values of $\lambda\dot{\gamma}$. It is worth noting that the test fluids used by Boger and Nguyen exhibit a quadratic normal stress difference up to some value of $\lambda\dot{\gamma}$, depending upon the concentration of Separan and the glucose solution; a value of 3.5 for $\lambda\dot{\gamma}$ corresponds to the range of second order behavior of a 0.2% Separan-glucose solution.

The stream function has been normalized and takes the value 0 on the boundary and 1 on the axis of symmetry. Figure 2 shows the streamlines for Newtonian flow and for viscoelastic flow with $\lambda\dot{\gamma} = 3.5$; the increment between streamlines is 0.1 in the main flow and 0.001 in the secondary vortex. We have also indicated the value of the ratio X = L/D (used by Nguyen and Boger) where L is the distance to the corner and D is the upstream diameter. It is found that the vortex size increases slightly with the elastic character of the flow; its intensity however is multipled by 5. Such a behavior has been observed by Nguyen and Boger (1979) when the first normal stress difference remains a quadratic function of $\dot{\gamma}$. It should be noticed that a small vortex is growing right at the die entrance; it is unfortunate that the numerical method does not converge for higher values of $\lambda\dot{\gamma}$. It is suspected that this divergence is due to the high elongational stresses appearing near the die entrance, and which are due to the peculiar behavior of the upper convected Maxwell fluid. Fig. 3 shows contour lines of T_{zz}/T^{W}_{rz} , where T^{W}_{rz} is the value of the wall shear stress in the downstream fully developed Poiseuille flow. Our results for the present problem are in good agreement with those obtained by Viriyayuthakorn and Caswell (1980) who developed a finite element method for constitutive relations of the integral type.

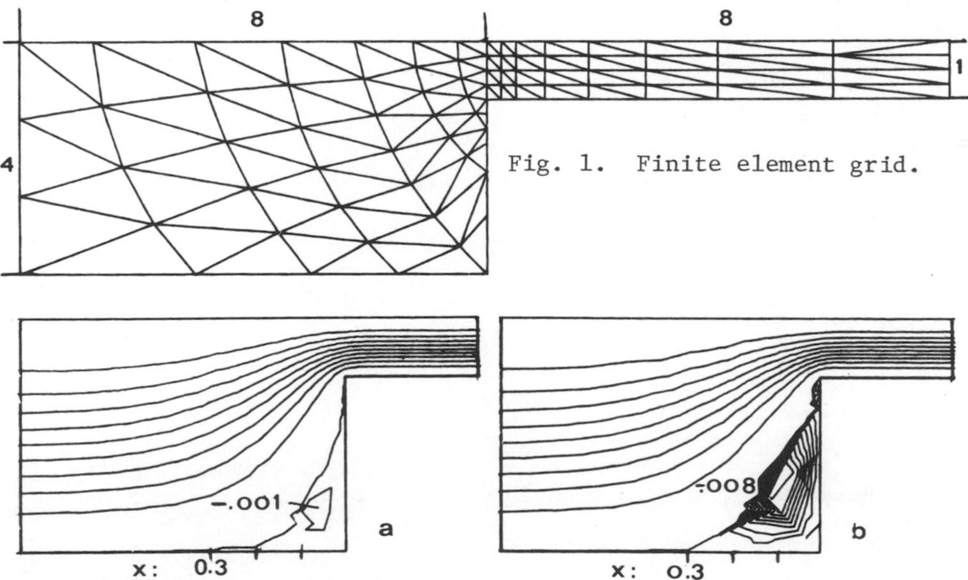

Fig. 1. Finite element grid.

Fig. 2. Streamlines for $\lambda\dot{\gamma} = 0$ (a) and $\lambda\dot{\gamma} = 3.5$ (b).

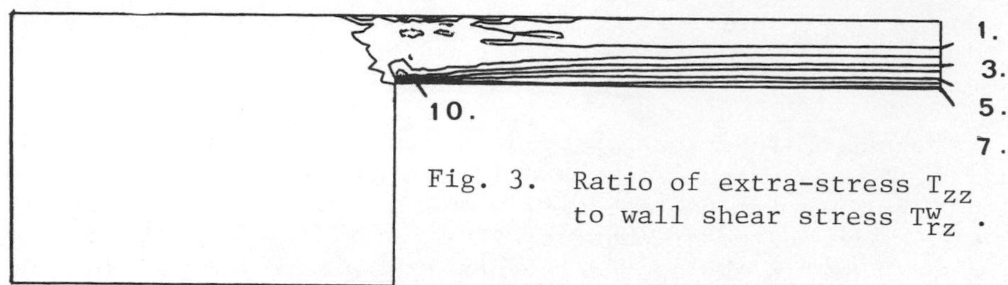

Fig. 3. Ratio of extra-stress T_{zz} to wall shear stress T^w_{rz}.

ENTRY FLOW OF A WHITE-METZNER FLUID

In order to calculate if departing from second order viscometric behavior at some critical value of $\lambda\dot{\gamma}$ leads to vortex growth in die-entry flow, we must select a test fluid such that this critical point lies well below 3.5, which is the limit of convergence of the numerical method. Fluid E3, cited by Nguyen and Boger (1979), consisting of 98.5% glucose syrup, 1.5% water and 0.03% Separan, satisfies these criteria, since the first normal stress difference departs from quadratic behavior when $\lambda\dot{\gamma} = 0.5$; Fig. 4a gives a plot of the shear stress T_{12} and the first normal stress difference N_1 against the shear rate $\dot{\gamma}$. The fluid is practically not shear-thinning, with a power law index of 0.97; we will assume that the shear viscosity maintains the constant value of 21.15 Nsm^{-2}.

Instead of an upper convected Maxwell model, we will now consider the following constitutive relation,

$$\underset{\approx}{T} + \lambda(II_D)\overset{\nabla}{\underset{\approx}{T}} = 2\mu\underset{\approx}{D} \tag{3}$$

which is a modified form of the White-Metzner (1963) constitutive relation. On the basis of the data provided in Fig. 4a, we may now calculate the value of λ as a function of the shear rate,

$$\lambda(\dot{\gamma}) = N_1/2T_{12}\dot{\gamma} , \tag{4}$$

which is plotted in Fig. 4b. An appropriate analytical representation of $\lambda(\dot{\gamma})$ is given by

$$\lambda(\dot{\gamma}) = \lambda_o[1 + (K\dot{\gamma})^2]^m , \tag{5}$$

with $\lambda_o = 0.112s$, $K = 0.090s$ and $m = -0.722$. The theoretical curve plotted on Fig. 4b shows excellent agreement with experimental data when $\dot{\gamma} < 20s^{-1}$; this corresponds to the range of our calculations.

It is interesting to note that the product $\dot{\gamma}\lambda(\dot{\gamma})$, where the function $\lambda(\dot{\gamma})$ is given by (5), reaches a maximum when $\dot{\gamma} = 17s^{-1}$,

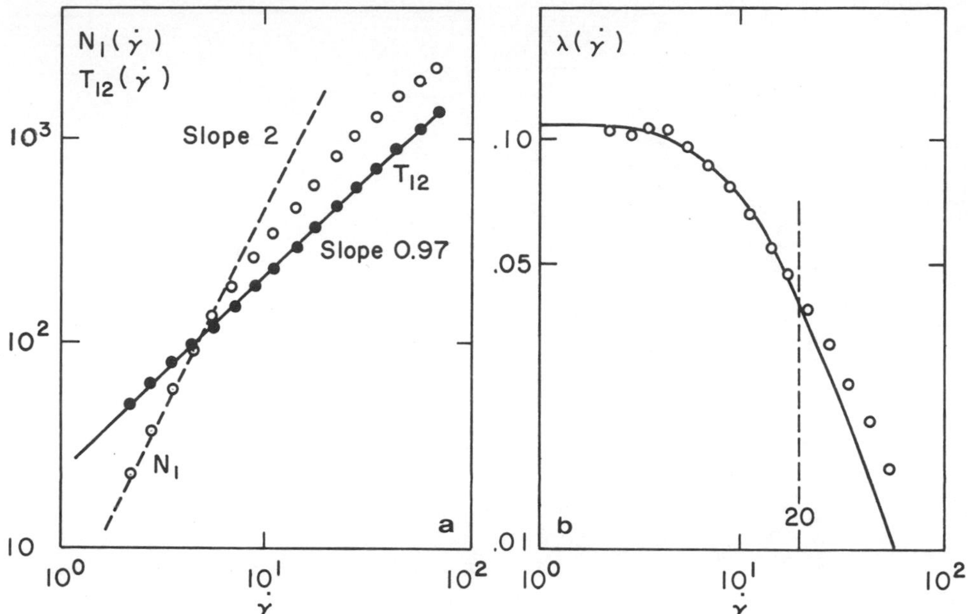

Fig. 4. Shear stress and first normal stress difference (a) and relaxation time (b) of the test fluid.

and is not a meaningful non-dimensional parameter. Let V be the mean velocity in the downstream pipe and R be its radius; we may define the non-dimensional quantities $\underset{\approx}{T}^*$, $\underset{\approx}{D}^*$ and II_{D*} as follows

$$\underset{\approx}{T} = \mu V/R\underset{\approx}{T}^*, \quad \underset{\approx}{D} = V/R\underset{\approx}{D}^*, \quad II_{D*} = (V/R)^2 II_D \ . \tag{6}$$

The constitutive relation (3) now becomes

$$\underset{\approx}{T}^* + \frac{\delta}{\lambda_o}\lambda[\delta^2/\lambda_o^2 \ II_{D*}] \ \overset{\nabla}{\underset{\approx}{T}^*} = 2\underset{\approx}{D}^* \ , \tag{7}$$

where $\delta = \lambda_o V/R$ is our new non-dimensional parameter.

The calculation has been pursued up to a value of $\delta = 0.5$; in the experiments carried by Nguyen and Boger (1979) with fluid E3, it was found that the length of the vortex should be about four times its size in the Newtonian flow, i.e. 0.8 times the diameter of the upstream pipe. Fig. 5a shows the streamlines for the case $\delta = 0.5$; it is quite evident that the experimental results are not represented by a numerical simulation based on the constitutive relation (3). The extra-stress on the downstream wall and at the corner decreases sharply when λ is a decreasing function of $\dot{\gamma}$; this may be observed on Fig. 5b, where we have plotted contour lines of T_{zz}/T_{rz}^W .

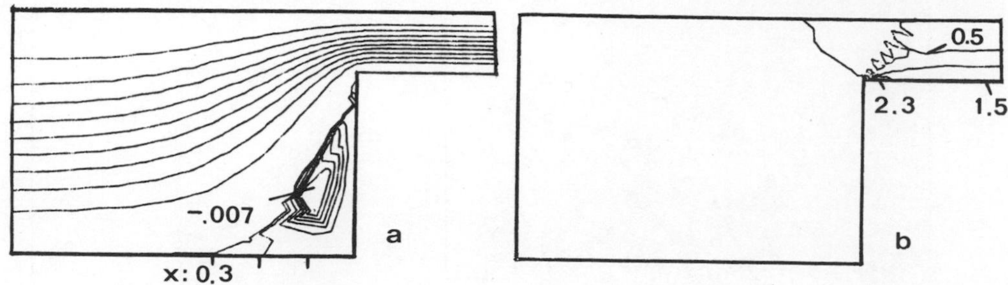

Fig. 6. Streamlines for $\delta = 0.5$ (a) and ratio of extra-stress T_{zz} to wall shear stress T^{w}_{rz} (b).

CONCLUSIONS

The mixed finite element method is able to calculate the die entry flow of an upper convected Maxwell fluid for values of $\lambda\dot{\gamma}$ up to 3.5. A particular model has been selected for reproducing the viscometric properties of a test fluid beyond its domain of second-order behavior. However, the use of such a model in numerical calculations does not simulate the vortex growth regime revealed by experiments on die entry flow.

REFERENCES

Boger, D. V., and Nguyen, H., 1978, "A Model viscoelastic fluid," Polymer Engineering and Science, 18:1037.

Cable, P. J., and Boger, D. V., 1978, "A comprehensive investigation of tubular entry flow of viscoelastic fluids," AIChE Journal, 24:869.

Crochet, M. J., and Bézy, M., 1979, "Numerical solution for the flow of viscoelastic fluids," J. Non-Newtonian Fluid Mech., 5:201.

Crochet, M. J., and Keunings, R., 1980, "Die swell of a Maxwell fluid: numerical prediction," J. Non-Newtonian Fluid Mech., 7:to be published.

Nguyen, N., and Boger, D. V., 1979, "The kinematics and stability of die entry flows,' J. Non-Newtonian Fluid Mech., 5:353.

Perera, M. G. N., and Walters, K., 1977, "Long range memory effects in flows involving abrupt changes in geometry," J. Non-Newtonian Fluid Mech., 2:49.

Viriyayuthakorn, M., and Caswell, B., 1980, "Finite element simulation of viscoelastic flow," J. Non-Newtonian Fluid Mech., 7:to be published.

White, J. L., and Metzner, A. B., 1963, "Development of consitutive equations for polymeric melts and solutions," J. Appl. Polym. Sci., 7:1867.

ON HOLE PRESSURE ERROR FOR VISCOELASTIC FLUIDS

Kitaro Adachi, Kazuhiro Kawai and Naoya Yoshioka

Department of Chemical Engineering
Kyoto University
Kyoto, 606, JAPAN

INTRODUCTION

It is a common practice to measure the fluid thrust normal to a solid wall through the small hole which is drilled in the surface of the wall. For viscoelastic flow, however, this experimental technique is not a common practice since some disturbance of the primary flow occures due to the hole, thus yielding in principle an error in measuring the undisturbed pressure. Therefore, there has been much discussion[1,2] on this hole pressure error.

Assuming that the flow is so slow that the second-order fluid model is valid, Tanner and Pipkin[3] showed that for plane, Poiseuille flow as well as Couette flow, of a viscoelastic fluid past a narrow and deep slot the hole pressure error, P_h, is expressed by

$$P_h=0.5\beta\dot{\gamma}_w^{\,2} \qquad (1)$$

where β is a fluid model constant and $\dot{\gamma}_w$ the undisturbed shear rate at the slot mouth. This theoretical prediction appears to be in good agreement with experimental data. However, they did not study the disturbed flow pattern which yields the hole pressure error although the following boundary conditions were used: the slot is deep enough so there is negligible motion at the bottom of it, and the slot is narrow enough so the primary flow is not disturbed except very near the slot mouth. On the other hand, O'Brien[4] computed flow patterns of plane, creeping, Newtonian flows past a slot of a various size. As a result, the following was concluded: it depends upon the channel and the slot size as well as upon the whole velocity profile approaching the slot how much the main flow is disturbed.

From the two works described above, the present authors inferred that Eq.(1) will be valid for any type of creeping flow only in the limit case when $d/h \rightarrow 0$ and $d/l \rightarrow 0$, where h is the gap between the parallel channel walls, d the slot width, and l the slot depth. Generally there will be a threshold, although it may not be clear, at which d/h and d/l begin to have an effect on the hole pressure error. The critical values of d/h and d/l will vary with the flow state. The plane, creeping flow of the second-order fluid past a slot was calculated mainly for the Couette flow by a direct finite difference method to study the magnitude of disturbance due to the existence of a hole and to make it clear how the hole pressure error depends upon d/h and d/l.

The secon-order fluid model is often regarded as too simple to describe the nonlinear viscoelastic phenomena. Therefore, it may be interesting to estimate a stress field with a more realistic constitutive equation when the flow field is given. Wagner's[5] integral constitutive equation with a strain-dependent memory function was used to calculate the distributions of shear and normal stresses for the non-viscometric, relevant flow problem.

CALCULATION OF FLOWS OVER A SLOT AND HOLE-PRESSURE ERRORS FOR THE SECOND-ORDER FLUID

The flow situation under consideration is shown in Fig. 1. The equation of motion for plane flows of the second-order fluid can be expressed in terms of the stream function, ψ, and vorticity, ω.[6]

$$\Delta\omega - R_e\left(u\frac{\partial\omega}{\partial x}+v\frac{\partial\omega}{\partial y}\right) - W_s\left(u\frac{\partial\Delta\omega}{\partial x}+v\frac{\partial\Delta\omega}{\partial y}\right)=0 \qquad \Delta\psi=-\omega \qquad \left(R_e=\frac{\rho Uh}{\mu} \quad W_s=-\frac{\beta U}{\mu h}\right) \quad (2)$$

This equation was solved numerically by a finite difference method

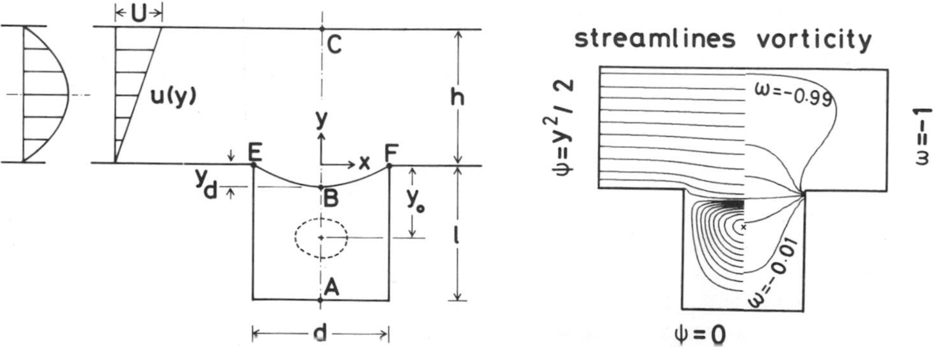

Fig. 1. Schematic Representation of the Flow Situation (left)

Fig. 2. Streamlines and Equivorticity Lines for l/h=d/l-1.0 (right)

at Re=0.0. Fig. 2 shows the flow patterns (left) and the vorticity patterns (right) of Newtonian fluids for d/h=d/l=1.0. Accuracy of the calculation seems to be satisfactory when the results are compared with O'Brien's ones.[4] It was difficult to obtain a solution when $W_s>0.1$ although not impossible. In several cases, moreover, there was no difference in the flow patterns and the stress distributions between the two solutions at $W_s=0.0$ and 0.1. Accordingly the corresponding Newtonian solutions were used to calculate the stress fields for the second-order fluid. Such an approximation is also justified by Tanner's theorem.[3] The hole pressure error is estimated by the following formula which is derived from the stress momentum equation without the inertia term:

$$\bar{P}_h/N_1=\frac{h}{d}\int_{x(E)}^{x(F)} P_h(x)/N_1 dx \qquad P_h(x)/N_1=-\int_{-1/h}^{1}\frac{\partial}{\partial x}\left(\frac{\tau_{xy}}{N_1}\right)dy \qquad (N_1=-2\beta\dot{\gamma}_w^2)$$

$$\tau_{xy}(x)/N_1=\frac{1}{2W_s}\left(\frac{\partial v}{\partial x}+\frac{\partial u}{\partial y}\right)-\frac{1}{2}\left[\left(u\frac{\partial}{\partial x}+v\frac{\partial}{\partial y}\right)\left(\frac{\partial v}{\partial x}+\frac{\partial u}{\partial y}\right)+2\left(\frac{\partial u}{\partial x}\frac{\partial u}{\partial y}+\frac{\partial v}{\partial x}\frac{\partial v}{\partial y}\right)\right] \qquad (3)$$

Some computed results are presented in Table 1. Fig. 3 (right) sug-

Table 1. Results of Computation

d/h	l/d	$\Delta x/h$	$-\omega_o h/U$	y_o/d	y_d/d	$\bar{P}_h/N_1$	$P_h(AC)/N_1$	$P_h(DC)/N_1$
1/1	1/1	1/20	0.1738	0.30	0.0799	-0.0673	-0.0238	-0.0105
1/2	1/5	1/80	——	——	——	-0.1641	-0.1394	——
1/2	1/2	1/80	0.1580	0.25	0.127	-0.1682	-0.1562	-0.1474
1/3	1/2	1/60	0.1624	0.25	0.144	-0.2121	-0.2060	-0.1941
1/3	1/1	1/60	0.1838	0.30	0.0886	-0.2149	-0.2100	-0.1710
1/4	4/1	1/80	0.1846	0.30	0.0883	-0.2252	-0.2219	-0.1816
1/8	1/1	1/80	0.1877	0.30	0.120	-0.2473	-0.2413	-0.2155
1/2*	2/1	1/20	0.2155**	0.40	0.197	-0.2387	-0.2337	-0.2105
1/4*	4/1	1/40	0.2297**	0.40	0.162	-0.2523	-0.2470	-0.2218

* Poiseuille flow **$-\omega_o h/2U_{max}$

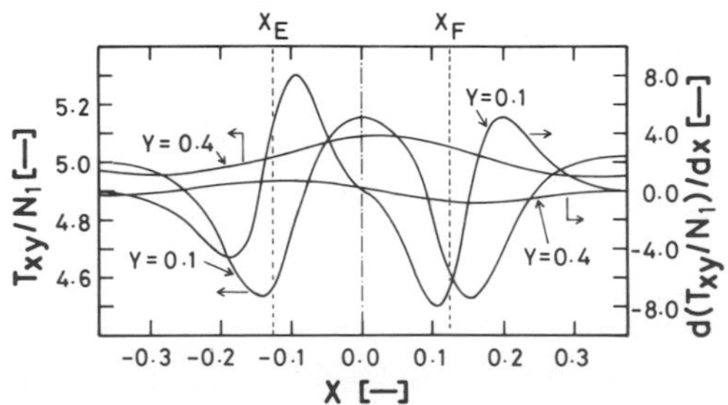

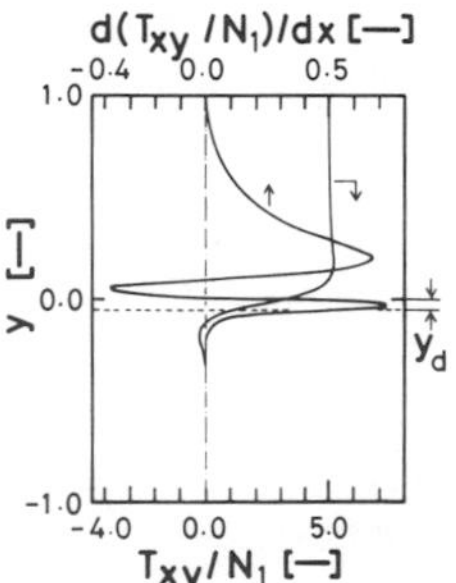

Fig. 3. Shear Stress Distributions on the Slot Center Line (right) and on Two Horizontal Lines (left) for d/h=d/l=1/4

gests that the secondary vortex flow in a hole does not contribute to the hole pressure error $\bar{P}_h$. This idea is supported by the fact that $P_h(AC)$ is nealy equal to $P_h(BC)$ which is calculated on the center line above the dividing line. The value of $\bar{P}_h$ varies considerably with changing d/h when d/h>1/8. However, it approaches to the theoretical value[3] -1/4 when $d/h \leq 1/8$. The slot depth of $l/d \geq 1/2$ has no effect on $\bar{P}_h$. Rejecting Han and Kim's experimental work, Higashitani and Lodge[8] insisted that there is no difference in $\bar{P}_h$ between the Couette flow and the Poiseuille flow. But the present work shows the clear existence of the difference, as has been expected from O'Brien's work as far as the hole pressure error is assumed to be caused by the disturbance of flow due to a hole.

THE STRESS CALCULATION BY USE OF WAGNER'S INTEGRAL CONSTITUTIVE EQUATION WITH A STRAIN-DEPENDENT MEMORY FUNCTION

For very slow flow there is no significant changes in the streamlines between the viscoelastic fluid and the Newtonian fluid, but the stress field of the viscoelastic flow seems to exhibit an appreciable change from that of the Newtonian flow. Then, it may be natural to try to estimate the viscoelastic stress field by substituting the Newtonian velocity field into a more realistic constitutive equation. To this end Wagner's constitutive equation was used:

$$\tau^{kl}(p)=\int_{-\infty}^{t}\mu(t-t';I_{\bar{c}^{-1}},\ II_{\bar{c}^{-1}})(\bar{c}^{-1kl}(t,t')-g^{kl}(p))dt' \quad (4)$$

$$\bar{c}^{-1kl}(t,t')=g^{\alpha\beta}(P,t')\frac{\partial x^k}{\partial X^\alpha}\frac{\partial x^l}{\partial X^\beta} \quad (5)$$

where μ is the strain-dependent memory function proposed by Wagner[5], and $\bar{c}^{-1kl}$ is the Finger relative strain tensor. The material point, which occupied the position $P(X^\alpha)$ at time t', is assumed to move to the position $p(x^k)$ at time t. The deformation gradient tensor $\partial x^k/\partial X^\alpha$ may be calculated by introducing the Protean coordinate system[9]:

$$x^1=X^1+\int_{t'}^{t}v^1(\xi)d\xi \qquad x^2=X^2 \qquad x^3=X^3 \qquad (x^1=x \quad x^2=\psi \quad x^3=z) \quad (6)$$

$$\frac{\partial x^1}{\partial X^1}=1+\int_{t'}^{t}\frac{\partial v^1}{\partial x^k}\frac{\partial x^k}{\partial X^1}d\xi \qquad \frac{\partial x^1}{\partial X^2}=-\int_{t'}^{t}\frac{\partial v^1}{\partial x^k}\frac{\partial x^k}{\partial X^2}d\xi \qquad \frac{\partial x^2}{\partial X^1}=0 \qquad \frac{\partial x^2}{\partial X^2}=1 \quad (7)$$

Here, it is not easy to solve even numerically the above integral equation of $\partial x^k/\partial X^\alpha$. From physical consideration, however, the following approximate solution can be obtained:

$$\frac{\partial x^1}{\partial X^1}=(x(\psi,\ t+dt)-x(\psi,\ t))/(x(\psi,\ t'+dt)-x(\psi,\ t'))=u(p)/u(P) \quad (8)$$

$$\frac{\partial x^1}{\partial X^2}=(x(\psi+d\psi,\ t''')-x(\psi,\ t))/d\psi \qquad (t'''=t-t'+t'') \quad (9)$$

where ψ=const is the streamline and one of the coordinate lines too.

As is shown in Fig. 4, the material points P,R and p are located at the grid points, but generally Q, r and q are not there. Here, the vales of u,v,y and t at the grid points in the whole computation space should have been stored in the computer. Thus, the position $x_r(\equiv x(\psi+d\psi,t'''))$ is the only one unknown. It can be found by interpolation, for the time t''' is known and then the two grid points, nearest the point r and between which it is located on the streamline $\psi+d\psi$=const., can be recognized by the computer. The metric tensors g^{kl} for the Protean coordinate system was given by Sagendorph and Leigh.[9] The time integral on the streamline was done tracing back as far as the one integral increment in the interval $[x_P-\Delta x\ x_P]$ was more than 10^{-3} % of the total amount in the interval $[x_P,x_p]$. Computation was made using the boundary conditions even if the point P or r was located outside the computation space in which the numerical solution of the Newtonian flow was stored. The stress field for the steady, Couette flow of the LDPE melt[5] at 150°C and in the shear-rate of 1.0 s^{-1} over the slot (d/h=d/l=1.0) was computed after the used program was checked for the Couette flow without a slot. The

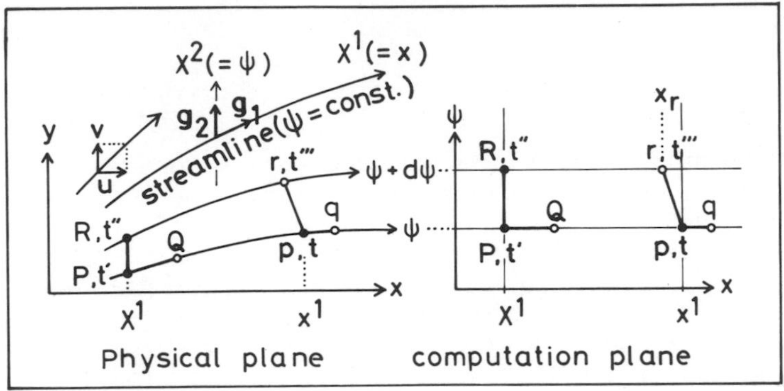

Fig. 4. Calculation of the Strain History by Use of the Protean Coordinate System

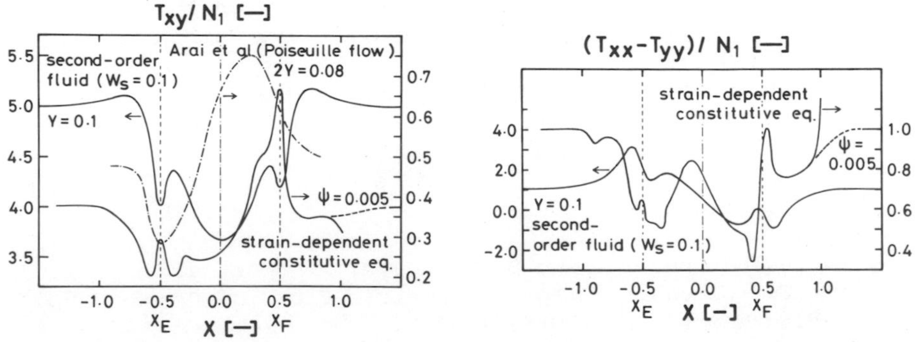

Fig. 5. Stress Distributions Obtained from Computations and Measurements for Viscoelastic Fluids

distributions of the shear stress and the primary normal stress difference on the streamline, $\psi/Uh=0.005$, are compared with those for the second-order fluid ($W_s=0.1$) and Arai and Hatta's[10] experimental work in Fig. 5. Their data was obtained for the Poiseuille flow of a HDPE melt at 160°C with the slot of d/h=0.4 and d/l=2.0. The theoretical prediction based on the integral equation is quite encouraging and seems to be more real than that for the second-order fluid.

CONCLUSIONS

According to the present work, the following conslusions may be deduced: (1) The hole diameter-channel height ratio, d/h, is the most important factor for the hole pressure error. (2) There is a considerable difference in the hole pressure error between the Couette flow and the Poiseuille flow. (3) The integral constitutive equation with a strain-dependent memory function is useful even for practical problems.

REFERENCES

1. J. M. Broadbent, A. Kaye, A. S. Lodge, and D. G. Vale, Possible Systematic Error in the Measurement of Normal Stress Differences in Polymer Solutions in Steady Shear Flow, Nature, 217 :55 (1968).
2. K. Higashitani and W. G. Pritchard, A Kinematic Calculation of Intrinsic Errors in Pressure Measurements Made with Holes, Trans. Soc. Rheol., 16:687 (1972).
3. R. I. Tanner and A. C. Pipkin, Intrinsic Errors in Pressure-Hole Measurements, Trans. Soc. Rheol., 13:471 (1969).
4. V. O'Brien, Closed Streamlines Associated with Channel Flow over a Cavity, Phys. Fluids, 15:2089 (1972).
5. M. H. Wagner, Zur Netzwerktheorie von Polymer-Schmelzen, Rheol. Acta, 18:93 (1979).
6. M. J. Crochet and G. Pilate, Numerical Study on the Flow of a Fluid of Second Grade in a Square Cavity, Comp. and Fluids, 3:283 (1975).
7. C. D. Han and K. U. Kim, Measurements of Pressure-Hole Error in Flow of Viscoelastic Polymeric Solutions: Effects of Hole Size and Solution Concentration, Trans. Soc. Rheol., 17:151 (1973).
8. K. Higashitani and A. S. Lodge, Hole Pressure Error Measurements in Pressure-Generated Shear Flow, Trans. Soc. Rheol., 19:302 (1975).
9. F. E. Sagendorph, IV. and D. C. Leigh, A Numerical Solution Method for Two-Dimensional Non-Viscometric Flows of Fluids Exhibiting Fading Memory, PB 200269 (1971).
10. T. Arai and H. Hatta, Estimation of Hole Pressure Error by Flow Birefringence Technique, JSME 858th Kōen Ronbunshū, No.750-8: 105 (1975).

ELASTIC GEL BIREFRINGENCE METHOD AS APPLIED TO THE STRESS ANALYSIS OF HOLE PRESSURE ERROR

Teikichi Arai

Department of Mechanical Engineering, Faculty of Engineering, Keio University
14-1, Hiyoshi 3 chome, Kohoku-ku, Yokohama, 223 Japan

INTRODUCTION

From the comparison between birefringence data of flowing polmmer melt through a duct and of the jelly solidified in the same duct exhibited when deformed under static load, I presented a paper in 1975 informing the following concept: The same distributions of the relative deviatoric stress intensity and direction as a polymer melt shows will be generated by a deformation of an elastic gel in case its selection is appropriate with respect to its shear modulus, birefringence sensitivity and the magnitude of the applied load.[1] Although the theoretical verification of its rationality was regarded as difficult, the anticipation on the interrelations between birefringence responses of arbitrary liquid and dolis was expected to afford, when experimentally proved valid, a very useful analogical method of stress analysis of flowing liquid and provide a lot of data useful for the industrial purposes.[2] Hence I called this method "Elastic gel birefringence method" or more briefly "Gel birefringence method" and have continued experimental clarification without finding any practical contradiction. The present paper is concerned with its extended application accompanied by relevant simultaneous pressure determinations to giving insight into the hole pressure error[3] on which inconsistent interpretations have been reported in literature.

BASIC RELATIONS BETWEEN STRESS AND BIREFRINGENCE

For the two-dimensional stress analysis,[4] the following equations from (1) to (3) were used for the simple shear deformation of elastic gel near a slot explanatory shown by Fig. 1:

$$n = N_{fr}/B = C_\sigma(\sigma_1 - \sigma_2) \tag{1}$$

$$\tau p = (\sigma_1 - \sigma_2)/2 \tag{2}$$

$$\tau p = \sigma_{yx}/\sin 2\chi \tag{3}$$

In these equations, n is the fringe order per unit optical path, N_{fr} observed fringe order , C_σ stress optical coefficient, B the duct width which is identical with the optical path, $(\sigma_1 - \sigma_2)$ the principal stress difference, τ_p principal shear stress usually called maximum shear stress, σ_{yx} shear stress exerted to x direction on the plane normal to y axis, and χ the extinction angle. It must be notified here that since the spacial direction of the polarised light transmission can statistically be regarded as paralled to the direction of principal stess, χ proves identical with the angle made by the stess direction with any of the reference axis used for the determination of χ.

When G denotes the shear modulus of the gel, h_s the clearance between the parallel plates, and s_x the displacement of the upper parallel plate to the direction of shear, shear stress between the plates may be expressed by

$$\sigma_{yx} = Gs_x/h_s = G\gamma_{ss} \tag{4}$$

where γ_{ss} is called nominal shear strain of the gel between the parallel plates. The application of shear difference integration method to the calculation of hole pressure error was given by the

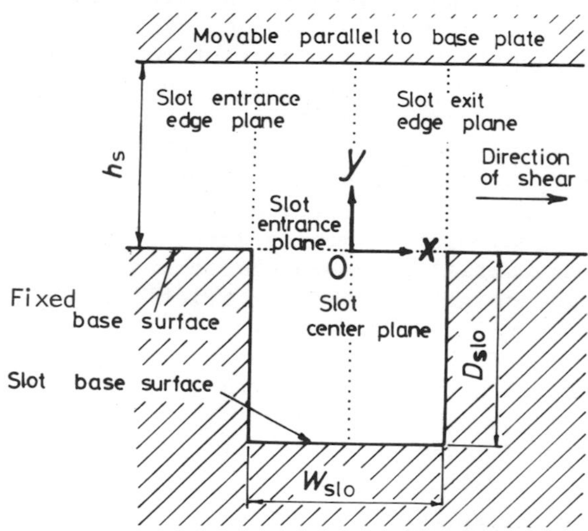

Fig. 1 Explanatory diagram for the simple shear deformation about a slot entrance with indications of the Cartesian coordinates x and y, z being paralled with the incident polarised light axis.

following equations from (5) to (9) by neglecting the effect of the gravitational load[5]:

$$\frac{\partial \sigma_{xy}}{\partial x} + \frac{\partial P_{yy}}{\partial y} = 0 \tag{5}$$

$$(p_{yy})_{bot} - (p_{yy})_{w} = \int_{y_{bot}}^{y_w} \frac{\partial \sigma_{xy}}{\partial x}\, dy \tag{6}$$

$$P_0 \simeq P_w = -(p_{yy})_{bot} \tag{7}$$

$$\Delta P_{slo} = P_0 - P_{bot} = -(p_{yy})_w + (p_{yy})_{bot} \tag{8}$$

$$\langle \Delta P_{slo} \rangle = \frac{1}{W_{slo}} \int_{y_{bot}}^{y_w} \left\{ (\sigma_{xy})_{exit} - (\sigma_{xy})_{ent} \right\} dy \tag{9}$$

In these equations, P is the pressure, p the total stress, σ the deviatoric stress, the subscripts bot and w mean the values on the slot bottom and on the movable opposite parallel wall surface, respectively. P_0 and P_{slo} are the reference wall pressure presumable on the wall provided without slot and the one on the slot bottom, respectively. ΔP_{slo} and $\langle \Delta P_{slo} \rangle$ are the sectional hole pressure error on the slot bottom and its mean value integrated over the slot width, respectively. The subscript ent and exit mean the values on the slot entrance and exit edge planes, respectively.

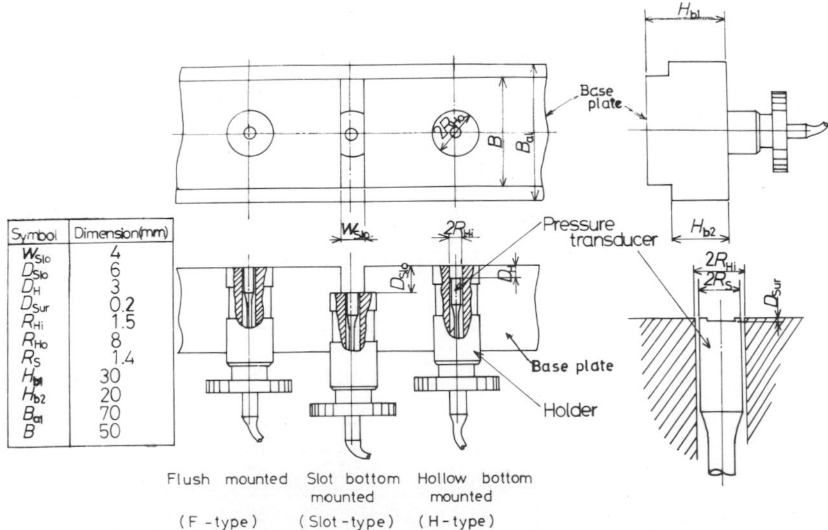

Symbol	Dimension(mm)
W_{Slo}	4
D_{Slo}	6
D_H	3
D_{Sur}	0.2
R_{Hi}	1.5
R_{Ho}	8
R_S	1.4
H_{b1}	30
H_{b2}	20
B_{a1}	70
B	50

Fig. 2 Explanatory diagram for the three setting types of micro-pressure transducers on the base wall.

EXPERIMENTAL

The optical system used was an improved one of the previous work . Used duct model was constructed conveniently with black and transparent polymethyl methacrylate resin plates. Jellies of different gelatin (Type M-159, supplied by Miyagi Chem. Co.) concentration c of 2, 4 and 15 wt.% were used as the test specimens after solidified in the duct model from their aqueous solutions, respectively for every run of test. Pressures were measured by mico-pressure transducers(Type PML-C, produced by Kyowa Electric Co., measurable pressure range: 0-500 g/cm^2) under three different setting conditions shown by Fig. 2. The effect of the minute hollow depth of 0.2 mm for the protection of diaphragm was neglected for the pressure reading.

RESULTS AND DISCUSSION

Fig. 3 shows the change of hole pressure error P_e of (P_{slo}-P_F) and (P_H-P_F) with γ_{ss} for the deformation of jelly of c=4 wt.%. Therein, P_H is the measured pressure with H type, and P_F with F type. It must be notified that within the observed range of γ_{ss}, (P_H-P_F) was larger than (P_{slo}-P_F) and at certain critical points of γ_{ss}, these values changed their sign from positive to negative, respectively. Since it may be accepted that P_e with small strain in gel method would correspond to low shear late flow of liquid, the positive value would correspond to the slow flow of Newtonian liquid. This tendency was found in Baird's paper[6] as P_e vs shear stress relation for glycerine. On the other hand, maximum values of p_e found in Tanner and Pipkin's paper[7] for a polymer solution would correspond to the points where $dPe/d\gamma ss$ =0. Fig. 4(a) and (b) show corresponding isochromatics and isoclinics at γ_{ss}=0.63, respectively. In these figures, we could find slight but clear asymmetries with respect to the center plane, which became large with γ_{ss} regardless of the specimen composition. Fig. 5(a) is ·corresponding superposed map of isoclinics, from which the map of principal stresses was obtained as shown by

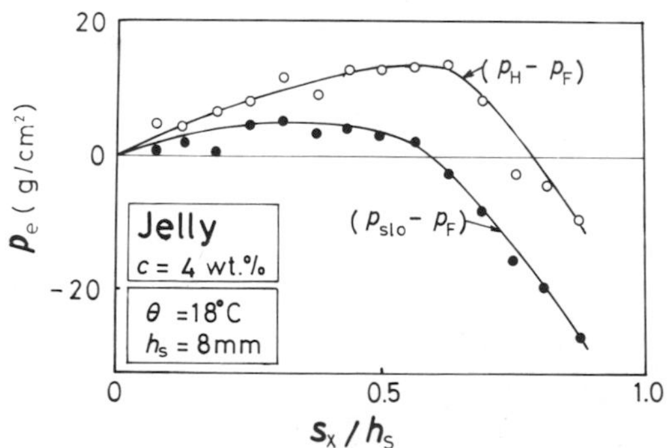

Fig. 3 Change of hole pressure error P_e with γ_{ss}.

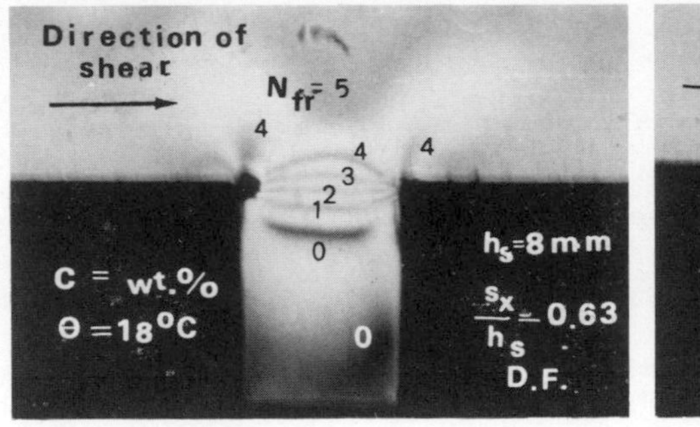

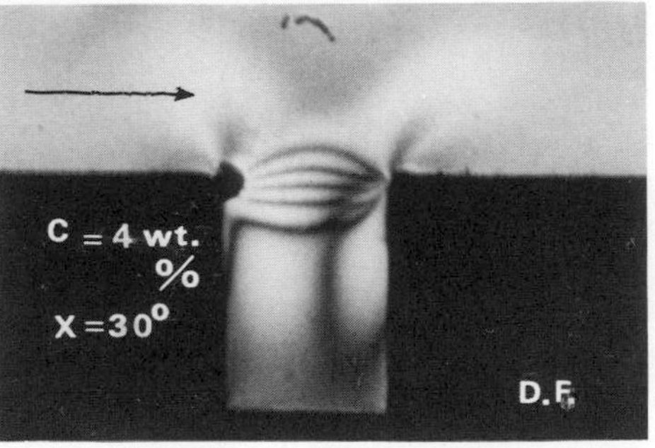

(a) (b)

Fig. 4 Isochromatic fringe patterns of jelly of c=4 wt.% at γ_{ss}= 0.63 (a), and corresponding isoclinics for χ=30° (b).

Fig. 5(b). In this figure, solid lines show the extensional stress and broken ones the compressive, C_{01} an isotropic singular point, and zero fringe order isochromatic line is the one transposed from the location in Fig. 4(b). It must be notified with Fig. 5(a) that at the entrance and exit edges, isoclinics of different extinction angle crossed together but isochromatics didn't show the maximum values there. Therefore, they couldn't be regarded as the stress concentration points but would well be defined as the stress discontinuity points or, more specifically, conversion points of stress direction. This birefringence response was a sharp contrast to the former observation at the entrance edges for the flow into a parallel slit[8]. Moreover, judging from the informations given by photoelastic stress analysis[4], the zero fringe order isochromatic line in Fig. 5(b) must correspond to the boundary streamline between primary and secondary flows, the existence of the latter flow having been well demonstrated

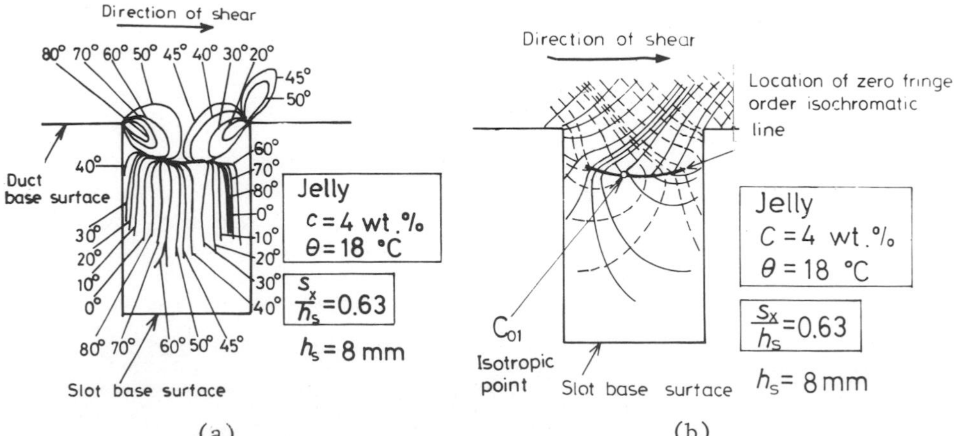

(a) (b)

Fig. 5 Maps of superposed isoclinics (a), and corresponding principal stresses (b).

by experiments for the pressure flow of polymer solutions and melts[9] as well as by computation based upon Stokes' assumption for both simple shear and pressure flows of Newtonian liquid.[10] In supporting the conjecture on the zero order isochromatic line, C_{01} in Fig. 5(b) was regarded to correspond to the rigorous geometrical site of the initial diverging point of stress on the streamline. $\langle \Delta P_{slo} \rangle$ for Fig. (5) was calculated as 2.6 g/cm^2 by applying shear difference integration method of eq.(5) to (9) using the observed C_σ value of 55 cm/kg and σ_{xy} values calculable by eqs. (1) to (3). In comparison, the observed value of (P_{slo}-P_F) obtained by interpolation on the curve in Fig. 3 was 4 g/cm^2.

Finally, I would like to express my sincere thanks to undergraduate students Mrs. H. Taira, A. Tatsumi, and K. Nakazawa for their collaborations in carrying out this work. I am also grateful to Prof. D.C. Bogue of the University of Tennessee who gave me the valuable advice of the rational terminology "Elastic gel birefringence method."

REFERENCES

1. T. Arai , H. Hatta and T. Ikeda, Birefringences of Rubber-like Solids in Slit Duct as Compared with Those of Polymer Melts Flowing through the Same Ducts, in Proc. 7th. Int. Congr. Rheology, Gothenburg (1977).
2. T. Arai and H. Hatta, Analogical Stress Analysis of Liquid Flow by the Birefringence Determination of Jelly under Load, in "Flow Visualization", T. Asanuma, ed., Hemisphere Pub.,Washington(1979).
3. J. M. Broadbent, A. Kaye, A. S. Lodge and D. C. Vale, Possible Systematic Error in the Measurement of Normal Stress Differences in Polymer solutions in Steady Shear Flow, Nature, 217:55 (1968).
4. R. B. Heywood, "Photoelasticity for Designers", Pergamon Press, Oxford (1969).
5. T. Arai and H. Hatta, Hole Pressure Error, Streamline and Flow-birefringence, paper submitted to J. Soc. Rheol. Jpn.
6. D. G. Baird, Fluid Elasticity Measurements from Hole Pressure Error Measurement Data, J. Appl. Polym. Sci., 20:3155 (1976).
7. R. I. Tanner and A. C. Pipkin, Intrinsic Errors in Hole Pressure Measurements, Trans. Soc. Rheol., 13:471 (1969).
8. T. Arai and H. Asano, Analysis of Flow of Molten Polyethylene through a parallel slit by the Birefringence Method, Kobunshi Kagaku (in Japanese), 18:292 (1972).
9 T. H. Hou, P. P. Tong and L. DeVargas, On the Origin of the Hole Pressure, Rheol. Acta, 16:554 (1977).
10. V. O'Brien, Closed Streamlines Associated with Channel Flow over a Cavity, Phy. Fluids, 15:2089 (1972).

ON A PULSATING FLOW OF POLYMERIC FLUIDS††

N. Phan-Thien

Department of Mechanical Engineering
The University of Newcastle
NEW SOUTH WALES 2308
AUSTRALIA

INTRODUCTION

We consider here the flow of a non-Newtonian liquid in a straight circular pipe under a fluctuating pressure gradient of the form

$$\frac{\partial P}{\partial z} = P_o(1 + \varepsilon n(t)), \quad \varepsilon << 1 , \tag{1}$$

where P is the pressure, P_o, the steady state pressure gradient, n(t), a stationary random function and ε is a perturbation parameter.

This problem has received considerable attention in the literature (Barnes et al, 1971; Davies et al, 1978; Phan-Thien, 1978). These authors (except Phan-Thien, who allowed n(t) to be a random function) considered a sinusoidal pressure gradient noise and used perturbation techniques to show that an increase in the volumetric flow rate is possible and that

(i) the flow enhancement is an order $0(\varepsilon^2)$ effect,
(ii) the fluid has to be shear-thinning in order to exhibit any positive flow enhancement.

However, with regard to the frequency of the fluctuating part of the pressure gradient, only the Maxwell model (Phan-Thien, 1978) describes the flow field realistically: it predicts an increase in the mean flow rate with increasing frequency of fluctuation, a fact observed by Barnes et al (1971). This has been shown (Phan-Thien, 1980) to be a general feature of complex-shear-thinning† non-Newtonian fluids which are describable by the generalized Maxwell

† By "complex-shear-thinning" we mean that the modulus of the complex viscosity $|\eta^*(\omega)|$ decreases with increasing frequency ω.

†† Work supported by the Australian Research Grant Committee.

integral constitutive equation. Phan-Thien's (1980) study points to a dynamic mechanism in which the frequency-dependent flow enhancement directly relates to oscillatory shear property of the fluid.

In this talk, I will attempt to explore further in this direction. A discussion of the pressure gradient noises is now in order.

PRESSURE GRADIENT NOISES

We assume that the pressure gradient noise is stationary in the weak sense. That is, its mean and its correlation function is invariant under a time-shift transformation. Without loss of generality, we let

$$\langle n(t) \rangle = 0 , \tag{2}$$

where the angular brackets denote an ensemble average quantity. Thus P_o is the mean pressure gradient. Secondly, we assume that the correlation function

$$R(\tau) = \langle n(t)\, n(t+\tau) \rangle \tag{3}$$

tends to zero fast enough as $|\tau| \to \infty$ so that it can be represented by a Fourier integral

$$R(\tau) = \int_{-\infty}^{\infty} e^{i\lambda\tau}\Omega(\lambda)d\lambda , \quad i^2 = -1 , \tag{4}$$

where $\Omega(\lambda)$ is the spectral density of $R(\tau)$. Under this assumption, $n(t)$ can be represented by (spectral representation, Yaglom, 1965):

$$n(t) = \int_{-\infty}^{\infty} e^{i\lambda t}\, dZ(\lambda) , \tag{5}$$

where $dZ(\lambda)$ is a unique random function with the following properties (Yaglom, 1965):

$$\langle dZ(\lambda) \rangle = 0 \tag{6}$$

$$\langle dZ(\lambda_1)\, dZ_c(\lambda_2) \rangle = \delta_{12}\Omega(\lambda_1)\, d\lambda_1 , \tag{7}$$

where the subscript c denotes a complex conjugate quantity and δ_{12} is the Kronecker delta.

ANALYSIS

If the rotational Reynolds number ($\frac{\rho a^2\omega}{\eta}$, η/ρ: fluid kinematic viscosity, a: pipe radius, ω: a representative frequency of fluctuation) is small, inertial effects may be neglected and the shear stress is given by

$$\tau = \frac{1}{2}rP_o(1 + \varepsilon n(t)). \tag{8}$$

This equation, together with a constitutive law for the liquid, determine the local shear rate $\dot{\gamma}(r,t)$ as a stochastic quantity from which the mean flow rate can be determined:

$$\langle Q \rangle = -\pi \int_0^a \langle \dot{\gamma}(r,t) \rangle \, r^2 dr \tag{9}$$

The percentage flow enhancement, I, can be defined as $100(\langle Q \rangle - Q_o)/Q_o$, where Q_o is the steady flow rate, viz.

$$I = 100 \int_0^a \langle \dot{\gamma} - \gamma_o \rangle r^2 dr \Big/ \int_0^a \gamma_o \, r^2 dr \, , \tag{10}$$

where γ_o is the steady shear rate corresponding to P_o.

To complete the formulation we nominate the following constitutive equations for the extra stress tensor:

$$\underset{\sim}{\tau} = \int_{-\infty}^{t} G(t-t') \, f(\Pi_{\underset{\sim}{D}}) \underset{\sim}{E}(t,t') \underset{\sim}{D}(t') \, \underset{\sim}{E}(t,t')^T \, dt' \tag{11a}$$

and

$$\underset{\sim}{\tau} = \eta \underset{\sim}{A}_1 + (\nu^1 + \nu^2) \underset{\sim}{A}_1^2 - \frac{1}{2} \nu^1 \underset{\sim}{A}_2 \tag{11b}$$

In case (a), G(s) is the relaxation spectrum, $f(\Pi_D)$ is an arbitrary function of the second invariant of $\underset{\sim}{D} = 1/2(\underset{\sim}{L} + \underset{\sim}{L}^T)$, $\underset{\sim}{L}$ is the velocity gradient tensor and $\underset{\sim}{E}(t_1 t')$ satisfies $\partial/\partial t \, \underset{\sim}{E}(t,t') = \underset{\sim}{L}(t) \, \underset{\sim}{E}(t,t')$, $\underset{\sim}{E}(t,t') = \underset{\sim}{1}$.

In case (b), η, ν^1, ν^2 are the viscosity and normal stress coefficient functions and $\underset{\sim}{A}_1$, $\underset{\sim}{A}_2$ are the first two Rivlin-Ericksen tensors.

Equation (11a) is the generalized Maxwell model considered by Phan-Thien (1980), which behaves quite decently in a superposed oscillatory shear flow†. Equation (11b) is the well-known Criminale-Ericksen-Filbrey constitutive equation (Criminale et al, 1958) which is completely general in a steady viscometric flow.

In a steady shear flow with shear rate $\dot{\gamma}$, both models predict the shear stress to be

$$\tau = \eta(\dot{\gamma}^2)\gamma = \eta_o \Gamma(\dot{\gamma}) \, , \tag{12}$$

where η_o is the zero-shear-rate viscosity.

---

† Ultrasonic frequency is being excluded here. This is not a stringent requirement since the smallness of the rotational Reynolds number demands that ω be small. Consequently fast-varying pressure gradient noises are not considered.

In an oscillatory shear flow of frequency ω, model (a) predicts a complex viscosity of

$$\eta^*(\omega) = \int_0^\infty G(s)e^{-i\omega s}ds \tag{13}$$

whereas (b) requires that

$$\eta^*(\omega) = \eta_o - \frac{1}{2}\omega\nu_o^1 i , \tag{14}$$

where $\nu_o^1 = \nu^1(\dot{\gamma}=0)$ is the zero-shear-rate first normal stress coefficient.

Since the analysis in both cases are similar, we record here only case (a) to save space. The results for case (b) will be indicated for comparison.

We seek a perturbation solution of the form

$$\dot{\gamma} = \gamma_o + \varepsilon\gamma_1 + \varepsilon^2\gamma_2 + 0(\varepsilon^2) , \tag{15}$$

from which the shear stress is given by

$$\tau = \tau_o + \varepsilon\tau_1 + \varepsilon^2\tau_2 + 0(\varepsilon^3) , \tag{16}$$

A comparison between (16) and (8) gives

$$\tau_o = \frac{1}{2}rP_o = \eta_o \, \Gamma_o \tag{17}$$

$$\tau_1 = \eta_o \, \Gamma_o \, n(t) = \Gamma_o' \int_0^\infty G(s) \, \gamma_1(t-s) \, ds \tag{18}$$

$$\tau_2 = 0 = \Gamma_o \int_0^\infty G(s)\gamma_2(t-s)ds + \frac{\Gamma_o''}{2} \int_0^\infty G(s) \, \gamma_1^2(t-s)ds \tag{19}$$

where we have used the shorthand notation

$$\Gamma_o = \Gamma(\dot{\gamma})\Big|_{\dot{\gamma}=\gamma_o} , \quad \Gamma_o' = \frac{d\Gamma(\dot{\gamma})}{d\dot{\gamma}}\Big|_{\dot{\gamma}=\gamma_o} , \text{ etc.} \tag{20}$$

As the operation on the right-hand side of (18) is linear, $\gamma_1(t)$ is also a stationary process with zero mean. In fact, in terms of the spectral representation of $\gamma_1(t)$:

$$\gamma(t) = \int_{-\infty}^\infty e^{i\lambda t} \, dZ_1(\lambda) \tag{21}$$

we have, by virtue of its uniqueness,

$$\Gamma_o' \int_o^\infty G(s)e^{-i\lambda s}\, ds\, dZ_1(\lambda) = \eta_o\, \Gamma_o\, dZ(\lambda) ,$$

that is

$$dZ_1(\lambda) = \frac{\eta_o \Gamma_o}{\Gamma_o' \eta^*(\lambda)}\, dZ(\lambda) \tag{22}$$

Note that $\langle dZ_1(\lambda)\rangle = 0$ and the spectral density of $\gamma_1(t)$, $\Omega_1(\lambda)$ is given by

$$\langle dZ_1(\lambda_1) dZ_{1c}(\lambda_2)\rangle = \delta_{12}\, \Omega_1(\lambda_1) d\lambda_1 ,$$

that is,

$$\Omega_1(\lambda) = \frac{\eta_o^2 \Gamma_o^2}{\Gamma_o'^{\,2} |\eta^*(\lambda)|^2}\, \Omega(\lambda). \tag{23}$$

By the same argument, the spectral representation of $\gamma_2(t)$ is

$$\gamma_2(t) = \int_{-\infty}^\infty e^{i\lambda t} dZ_2(\lambda) , \tag{24}$$

where

$$dZ_2(\lambda) = -\frac{\Gamma_o''}{2\Gamma_o' \eta^*(\lambda)} \int_{-\infty}^\infty \int_o^\infty G(s)e^{-i(\lambda-\lambda')s} e^{-i\lambda' t} dZ_1(\lambda) dZ_{1c}(\lambda') \tag{25}$$

Since

$$\langle dZ_2(\lambda)\rangle = -\frac{\eta_o''}{2\Gamma_o' \eta^*(\lambda)} \int_o^\infty G(s)e^{-i\lambda s} \Omega_1(\lambda)\, ds\, d\lambda ,$$

$$= -\frac{\Gamma_o''}{2\Gamma_o' \eta^*(\lambda)}\, \Omega_1(\lambda)\, d\lambda\, \eta^*(\lambda)$$

the mean of $\gamma_2(r,t)$ is given by,

$$\langle\gamma_2\rangle = -\frac{\eta_o^2 \Gamma_o^2 \Gamma_o''}{2(\Gamma_o')^3} \int_{-\infty}^\infty \frac{\Omega(\lambda)}{|\eta^*(\lambda)|^2}\, d\lambda \tag{26}$$

Thus, the percentage flow enhancement is

$$I = -50\varepsilon^2\eta_o^2 \frac{\int_o^a \frac{\eta_o^2\Gamma_o''}{(\Gamma_o')^3} r^2 dr}{\int_o^a \gamma_o r^2 dr} \int_{-\infty}^{\infty} \frac{\Omega(\lambda)}{|\eta^*(\lambda)|^2} d\lambda \qquad (27)$$

The corresponding formula for case (b), after some straightforward manipulation, is

$$I = -\frac{50\varepsilon^2\eta_o^2}{\int_o^a \gamma_o r^2 dr} \int_o^r dr \int_{-\infty}^{\infty} r^2 \frac{\eta_o^2\Gamma_o'\Gamma_o' - \frac{1}{2}\nu_o\nu_o'\lambda^2}{|\eta_o\Gamma_o' - \frac{1}{2} i\nu_o\lambda|^4} \Gamma_o^2\Omega(\lambda)d\lambda \qquad (28)$$

where we have dropped the superscript "1" on ν for simplicity.

DISCUSSIONS

At low frequency of oscillations, the spectral density of n(t) is virtually a delta function situated at $\lambda=0$ from which both models predict the same inelastic flow enhancement of

$$J = -50\varepsilon^2 \frac{\int_o^a \frac{\Gamma_o^2\Gamma_o''}{(\Gamma_o')^3} r^2 dr}{\int_o^a \gamma_o r^2 dr} \langle n(t)^2 \rangle \qquad (29)$$

In (29) we have used the fact that $\langle n^2(t) \rangle = \int_{-\infty}^{\infty} \Omega(\lambda)d\lambda$; the error in this approximation is of the order $O(\int_{-\infty}^{\infty} \lambda^2\Omega(\lambda)d\lambda)$. Note that due to the shear-thinning property of the fluid, J is always positive indicating an increase in the flow rate.

The role of the dynamic mechanism of the flow enhancement can be exposed more clearly by assuming a one-frequency pressure gradient noise:

$$n(t) = \sin\omega t \ , \quad \Omega(\lambda) = \frac{1}{4} (\delta(\lambda-\omega) + \delta(\lambda+\omega)) \qquad (30)$$

from which (a) gives

$$I = J \frac{\eta_o^2}{|\eta^*(\omega)|^2} \qquad (31)$$

whereas (b) predicts

$$I = -\frac{25\varepsilon^2\eta_o^2}{\int_o^a \gamma_o r^2 dr} \int_o^a \frac{\eta_o^2\Gamma_o'\Gamma_o'' - \frac{1}{2}\nu_o\nu_o'\omega^2}{|\eta_o\Gamma_o' - \frac{1}{2} i\omega\nu_o|^4} \Gamma_o^2 r^2 dr \qquad (32)$$

(32) may be simplified further if both ν_o and Γ_o' vary slowly with γ_o. In that case, the flow enhancement for case (b) becomes

$$I \simeq J \frac{\eta_o^2}{|\eta^*(\omega)|^2} \qquad (33)$$

While (31) predicts that the flow enhancement should increase with increasing frequency ω, (33) (or (32)) predicts the opposite trend due to equation (14). However, both agree that the dynamic flow enhancement mechanism directly links with the oscillatory shear behaviour of the fluid.

It is hoped that experimental data and predictions of strain-dependent constitutive equations will be available by the opening time of the Congress.

REFERENCES

Barnes, H.A., Townsend, P. and Walters, K., 1971, Rheol. Acta, 10:517.

Davies, J.M., Bhumiratana, S. and Bird, R.B., 1978, J. non-Newt. Fluid Mech., 3:237.

Phan-Thien, N., 1978, J. non-Newt. Fluid Mech., 4:167.

Phan-Thien, N., 1980, Rheol. Acta, in press.

Yaglom, A.M., 1965, "An Introduction to the theory of stationary Random functions", Translated and edited by R.A. Silverman, Prentice-Hall, Inc., New Jersey.

ABOUT A POSSIBLE CAUSE OF VISCOELASTIC TURBULENCE

U. Akbay, E. Becker, S. Krozer and S. Sponagel

Institut für Mechanik
Technische Hochschule
D-6100 Darmstadt

INTRODUCTION

Viscoelastic turbulence is an instability which can be observed in extrusion flows. Fig. 1 shows schematically an extruder. Flow instabilities can arise, in principle, in the entrance region upstream of the capillary, in the capillary or at the exit. In this note the stability of fully developed channel flow in a plane capillary is investigated. The spatial variation of shear velocity causes a complication of the theory which is avoided by first studying the stability of plane Couette-flow between parallel plates. Then it is shown that the same mechanism for instability, found here, is also responsible for instability of plane channel flow.

COUETTE FLOW

Because the undisturbed basic flow is viscometric it is suggestive to use, in the framework of a linearized analysis, the ideas of nearly viscometric flow theory [1]. For further simplification only disturbances within the flow plane are studied, so that the disturbed flow field remains a plane one. Finally, the memory

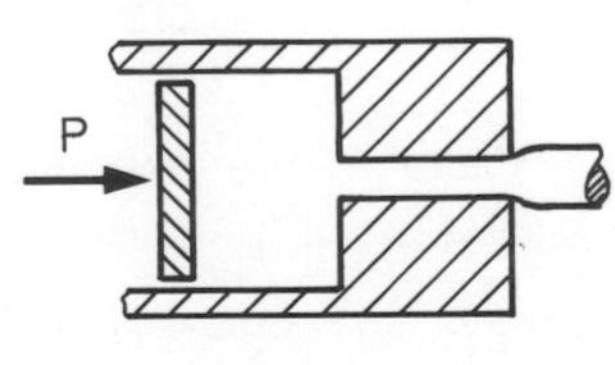

Fig. 1

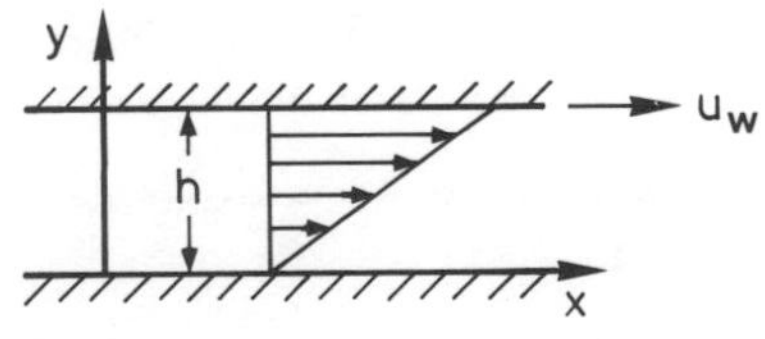

Fig. 2

time of the fluid which characterizes the stress reaction to the flow disturbance is assumed to be small [2]. In that case the disturbance part of stress, τ_{ik}', is connected with the disturbance part of the velocity field, v_i', by the following constitutive equation:

$$\tau_{ik}' = A_{ik12}\left(\frac{\partial v_1'}{\partial x_2} + \frac{\partial v_2'}{\partial x_1}\right) + A_{ik22}\frac{\partial v_2'}{\partial x_2} \tag{1}$$

The compatibility conditions of nearly viscometric flow theory establish relations between the coefficients A_{ijkl} and the viscometric functions of the fluid. For example:

$$A_{1222} = \kappa N/2 \tag{2}$$

Here, $N(\kappa)$ is the first normal stress coefficient, and κ denotes the shear velocity ($= U_w/h$) of the basic Couette-flow.

The streamfunction of the disturbance part of the velocity field is written as

$$\psi(x_1, x_2, t) = \phi(x_2)\exp(i\alpha(x_1 - ct)) \tag{3}$$

By inserting (3) into the momentum balance equation and by linearizing one obtains the following equation of Orr-Sommerfeld-type:

$$\begin{aligned} iRe\alpha(y+c)(\phi''-\alpha^2\phi) &= \phi^{IV}+2\alpha^2(1-\varepsilon)\phi''+\alpha^4\phi \\ &\quad + i\Gamma(\phi''+\alpha^2\phi)' \end{aligned} \tag{4}$$

Here, $\varepsilon = \eta/\eta_d$ denotes the ratio of viscosity $\eta = \tau/\kappa$ to differential viscosity $\eta_d = d\tau/d\kappa$; $\tau = \tau_{12}$ is the shear stress in viscometric flow. The Reynolds number is defined by $Re = U_w h/\eta_d$, and the parameter Γ has the following meaning:

$$\Gamma = \frac{\kappa}{\eta d}\,\frac{d(\kappa N)}{d\kappa}\,\alpha \tag{5}$$

The amplitude function ϕ has to satisfy the following boundary conditions (noslip condition at both walls):

$$\phi(o) = \phi'(o) = \phi(1) = \phi'(1) = 0 \tag{4^+}$$

(Note that velocities have been nondimensionalized with U_w and lengths with h).

Now, the parameter $\lambda := i\alpha Re\,c$ is defined, and λ and ϕ are expanded in powers of αRe:

$$\lambda = \lambda_o + \lambda_1\,\alpha Re + \ldots \qquad \phi = \phi_o + \phi_1\,\alpha Re + \ldots \tag{6}$$

For the lowest order terms eq. (4) reduces to:

$$\lambda_0(\phi_0''-\alpha^2\phi_0)=\phi_0^{IV}+2\alpha^2(1-\varepsilon)\phi_0''+\alpha^4\phi_0+i\Gamma(\phi_0''+\alpha^2\phi')\qquad(7)$$

Γ is a real-valued parameter. In the following way it can be shown that λ_0 is also real-valued: We multiply eq. (7) by $\bar{\phi}_0$ (the conjugate complex of ϕ_0); in the same way the equation for $\bar{\phi}_0$, derived from eq. (7), is multiplied by ϕ_0. Subtracting from each other the equations thus obtained and integrating the resulting equation between the limits 0 and 1, lead

$$-(\lambda_0-\bar{\lambda}_0)\int_0^1[\phi_0''\bar{\phi}_0''+\alpha^2\phi_0\bar{\phi}_0]dx_2=i\Gamma\int_0^1(\phi_0'\bar{\phi}_0'+\alpha^2\phi_0\bar{\phi}_0)'dx_2\qquad(8)$$

ϕ_0 and $\bar{\phi}_0$ both satisfy the homogeneous boundary conditions (4^+); hence the right side of (8) is zero. Because the integral on the left side is positive, one can conclude:

$$\text{Im}(\lambda_0)=0\qquad(9)$$

The stability limit is given by $\text{Re}(\lambda_0)=0$. Therefore, at the stability limit eq. (7) reduces to

$$0=\phi_0^{IV}+2\alpha^2(1-\varepsilon)\phi_0''+\alpha^4\phi_0+i\Gamma(\phi_0''+\alpha^2\phi_0)'\qquad(10)$$

Because it is not clear a priori that eq. (10) has eigen-solutions, we first study this equation in the limit for small values of $\alpha^2 \ll 1$. The expansion $\phi_0=\phi_{00}+\alpha^2\phi_{01}+\ldots$, $\Gamma=\Gamma_0+\alpha^2\Gamma_1+\ldots$ leads to

$$\phi_{00}^{IV}+i\Gamma_0\phi_{00}'''=0\qquad(11)$$

By integrating this equation one obtains

$$(2+i\Gamma_0)/(2-i\Gamma_0)=\exp(i\Gamma_0)\qquad(12)$$

The smallest root of (12) is $\Gamma_0=8.986$. Taking account of the α^2-term by applying Fredholm's alternative theorem [3] to the equation for ϕ_{01} one derives the following approximate result:

$$\frac{\kappa}{\eta}\frac{d(\kappa N)}{d\kappa}\ \overset{<}{\underset{>}{=}}\ 4.6\,\varepsilon^{-1}(\varepsilon-0.24)\ \begin{cases}<\ \text{stable}\\=\ \text{indifferent}\\>\ \text{instable}\end{cases}\qquad(13)$$

In fig. (3) the results of exact integration of (7), displayed as fully drawn curves, are compared with the approximate results (dotted lines). They agree well in the range $1<\varepsilon<50$.

Channel flow

Transferring the ideas explained above from Couette-flow to plane channel flow leads to the following equation:

$$\eta_o \lambda_o (\phi_o'' - \alpha^2 \phi_o) = L(\eta_d L\phi_o) - 4\alpha^2 (\eta\phi_o)' + i\alpha \left(\frac{d(\kappa^2 N)}{d\kappa} L \phi_o\right)' - i\alpha L(\kappa N\phi_o') \quad (14)$$

ϕ has to satisfy the following no-slip boundary conditions (see fig. 4):

$$\phi_o(-1/2) = \phi_o'(-1/2) = \phi_o(+1/2) = \phi_o'(+1/2) = 0 \quad (14^+)$$

The following notation has been used: κ: = du/dy, ': = d/dy, $L\phi$: = $\phi'' + \alpha^2\phi$; η_o is the value of η for y = o. To make matters as simple as possible without loosing essential features of the problem we take $\eta = \eta_d$ = const and N = const. Thereby eq. (14) is reduced to

$$\lambda_o(\phi_o''-\alpha^2\phi_o) = L^2\phi_o - 4\alpha^2\phi_o'' + i\Gamma\{(2\kappa L\phi_o)' - L(\kappa\phi_o')\} \quad (15)$$

Here $\Gamma = \alpha\kappa_m \frac{N}{\eta}$ with κ_m denoting the maximum shear velocity on the channel wall ($y= \frac{1}{2}$).

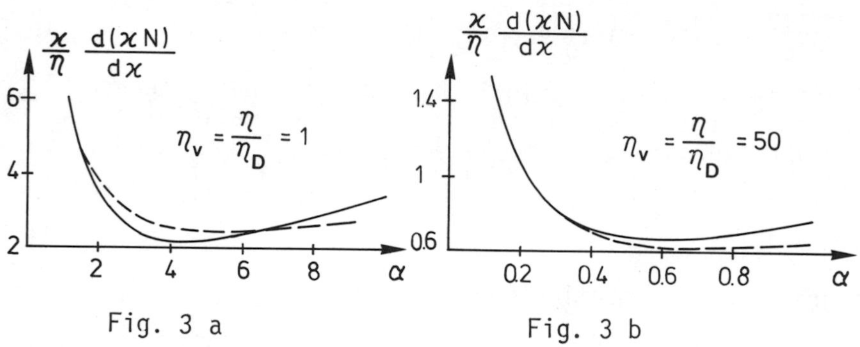

Fig. 3 a

Fig. 3 b

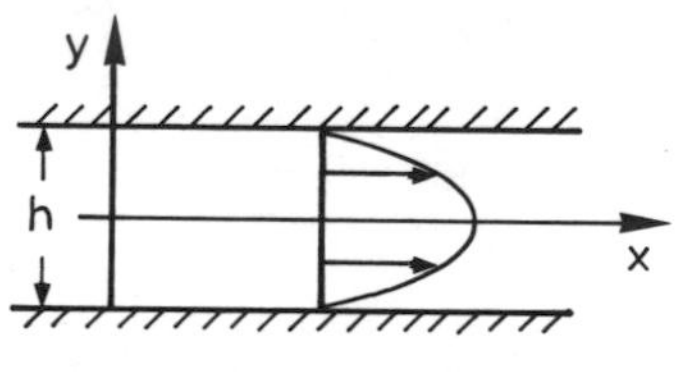

Fig. 4

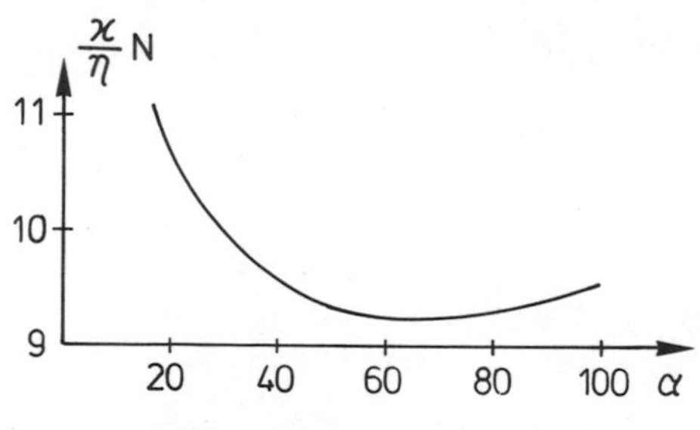

Fig. 5

The eigenvalue λ_o of (14) is now complex. In order to show this and to calculate the imaginary part of λ_o we use the same procedure as explained above for Couette-flow. Instead of eq. (8) we now obtain

$$-(\lambda_o-\bar{\lambda}_o)\int_{-0.5}^{+0.5}(\phi_o'\bar{\phi}_o'+\alpha^2\phi_o\bar{\phi}_o)dy=3i\Gamma\int_{-0.5}^{+0.5}y(\bar{\phi}_o'L\phi+\phi_o L\bar{\phi}_o')dy \quad (16)$$

In deriving (16) we have used the fact that L is a self-adjoint operator. The right side of (16) is transformed as follows:

$$\int_{-0.5}^{+0.5}y(\bar{\phi}_o'L\phi_o+\phi_o L\bar{\phi}_o')dy = \int_{-0.5}^{+0.5}[\bar{\phi}_o'(\phi_o''+\alpha^2\phi_o)+\phi_o(\bar{\phi}_o''+\alpha^2\bar{\phi}_o)]y\,dy$$
$$= \int_{-0.5}^{+0.5}y[(\phi_o'\bar{\phi}_o')+\alpha^2(\phi_o\bar{\phi}_o)]'dy=-\int_{-0.5}^{+0.5}[\phi_o'\bar{\phi}_o'+\alpha^2\bar{\phi}_o]dy \quad (17)$$

By inserting this result into (16) one obtains Im (λ_o) = 1,5 Γ. The stability limit is again given by Re(λ_o) = 0. Therefore, at the stability limit, eq. (15) reduces to

$$0 = \phi_o^{IV} - 2\alpha^2\phi_o''+\alpha^4\phi_o-i\Gamma\ ((y\phi_o'''+1,5\phi_o'')+\alpha^2(y\phi_o'+0,5\phi_o)) \quad (18)$$

Fig. 5 shows the stability limit as determined by numerical integration of (18). Obviously, channel flow is more stable than Couette-flow for the same value of the shear velocity at the wall. This means that one can realize higher shear velocities at the wall of a plane channel than in plane Couette-flow without encountering instability. Furthermore, fig. 5 shows that the wavelength of the most unstable disturbance is a smaller fraction of the channel width for channel flow than for Couette-flow.

References

1. A. C. Pipkin and R. D. Owen, Phys. Fluids 10 (1967), 836 - 843
2. E. Becker, will be published in Advances in Applied Mechanics 20, (1980)
3. M. M. Denn, Stability of reaction and transport processes, Prentice Hall, Englewood Cliffs N. J. (1975)

FREQUENCY ANALYSIS OF ELECTRICAL FLUCTUATIONS - A NEW METHOD TO STUDY FLOW INSTABILITIES IN CAPILLARY FLOW

K. Hedman, C. Klason, J. Kubat

Chalmers University of Technology, Gothenburg, Sweden

(Abstract)

Results of measurements of electrical thermal noise (fluctuations in the frequency range 0.01 - 1000 Hz) on elastic, shear thinning polymer solutions of polyethylene oxide, polyacrylamide and maltose/polyacrylamide during capillary flow are reported. The flow of these solutions gave rise to a thermal noise exhibiting a 1/f-frequency distribution, increasing in intensity with the flow rate. At a critical flow rate, when flow instabilities were observed, the noise showed a pulsating character. The corresponding frequency spectra contained a number of sharp peaks, all being multiples of a fundamental frequency of the order of 1Hz, indicating a very regular nature of the flow instabilities.

Solutions with comparable shear thinning properties but lacking elasticity did not produce pulsations in the frequency range used.

The observed effects were compared with the results of frequency analysis of the pressure fluctuations observed during extrusion of polymer melts through dies (melt fracture and shark skin phenomena).

INSTABILITY OF JETS OF NON-NEWTONIAN FLUIDS

P. Schümmer and K. H. Tebel

Institut für Verfahrenstechnik RWTH Aachen
Turmstraße 46
D-5100 Aachen

INTRODUCTION

The instability and the break-up of liquids is gaining more and more importance in several cases of technical application e.g. for the production of powder out of a melt, for the spray drying and for the atomization of fuel and cooling fluid. Very often an optimal additional treatment and further processing require the production of monodispersed droplets, i.e. the initial disturbance has to be constant and well known. So it happens for instance at the reprocessing of fuel kernels[1] and at the production of liquid-filled capsules[2]. By using additives like long chain polymer molecules to increase the stability of the solutions they receive non-Newtonian characteristics and hinder the occurance of separate droplets because of ligaments in between. A lot of experimental and theoretical work has been done concerning the break-up of Newtonian[3] and non-Newtonian[4,5] fluids. In the case of Newtonian systems a linearized stability analysis describes very well the droplet production of a steady lequid jet. The extension to non-Newtonian systems sometimes disagrees with the experimental observations. This paper is restricted to the discription of the ligaments and their formation as it is shown in Fig. 1.

The experiments show that it seems to be possible to approximate the shape of the ligament by a cylinder of the radius R within an average range of time. This flow situation looks like the kinetics of a spinning process where a fibre is stretched continuously in axial direction.

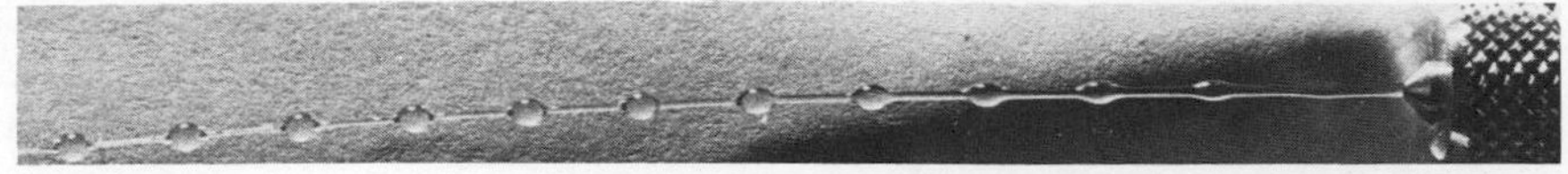

Fig. 1. Breakup of a viscoelastic jet

THEORY

Fig. 2 shows the result of this description. A model containing two ideal geometries: cylinders and spheres. The flow in the ligaments is unterstood as an uniaxial, homogenous, transient and inertialess elongational flow. The behaviour of the liquid can be described by a Jeffreys-law with a contravariant time derivative.

$$\underline{\underline{S}}^* + t_1 \overset{\Delta}{\underline{\underline{S}}}{}^* = 2 \eta_o \left(\underline{\underline{D}} + t_1 \eta_s / \eta_o \cdot \overset{\Delta}{\underline{\underline{D}}}\right) \tag{1}$$

$$\overset{\Delta}{\underline{\underline{S}}}{}^* = \dot{\underline{\underline{S}}}{}^* - (\nabla \underline{v})^T \underline{\underline{S}}^* - \underline{\underline{S}}^* (\nabla \underline{v}) \tag{2}$$

If the polymer macro-molecules are understood as elastic dumbbells in a Newtonian carrier liquid, one will get this material equation from an exact statistic. These assumptions lead to the following system of equations:

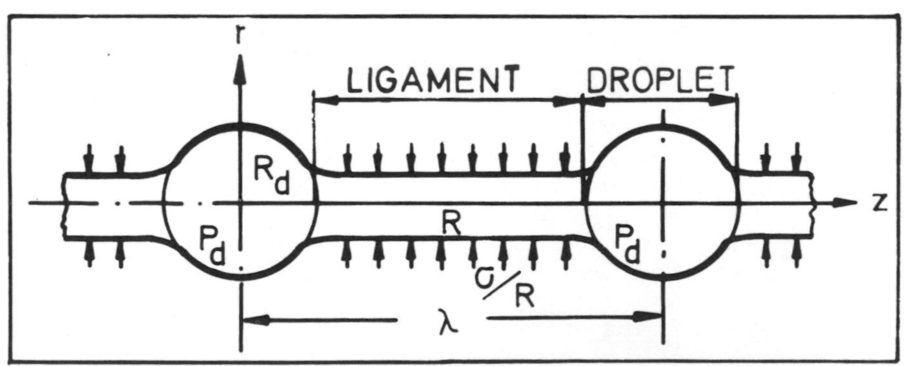

Fig. 2. Model of the droplet production

$$S^{*}_{zz}-S^{*}_{rr}=\sigma/R-p_d \tag{3}$$

$$S^{*}_{zz}+t_1(\dot{S}^{*}_{zz}-2\dot{\varepsilon}S^{*}_{zz})=2\eta_o\dot{\varepsilon}+2\eta_s t_1\ddot{\varepsilon}-4\eta_s t_1\dot{\varepsilon}^2 \tag{4}$$

$$S^{*}_{rr}+t_1(\dot{S}^{*}_{rr}-\dot{\varepsilon}S^{*}_{rr})=\eta_o\dot{\varepsilon}-\eta_s t_1\ddot{\varepsilon}-\eta_s t_1\dot{\varepsilon}^2 \tag{5}$$

$$\dot{R}/R=-\dot{\varepsilon}/2 \tag{6}$$

The pressure $p_d = 2\sigma/R_d$ is negligible in the range of $R_d \gg R$. It is possible to appraise the value of p_d by discussing the conservation of mass. Hence it follows that p_d can be taken as a constant counter pressure for $R_d > R$.

The solution of an elongational flow for a Newtonian fluid leads to a time constant or system time t_s:

$$t_s = 6\,\eta_o R_o/\sigma \tag{7}$$

Using η_o, R_o and σ as single parameters or in combination like equation (7) the dimensionless form with the following abbreviations can be introduced:

$$R^+=R/R_o \quad \eta_s^{\,+}=\eta_s/\eta_o \quad S^{*\,+}_{ii}=S^{*}_{ii}/(\sigma/R_o) \quad p_d^+=p_d/(\sigma/R_o) \tag{8}$$

$$t^+=t/t_s \quad t_1^+=t_1/t_s \quad \dot{\varepsilon}^+=t_s\dot{\varepsilon} \quad \ddot{\varepsilon}^+=t_s^2\ddot{\varepsilon} \tag{9}$$

The condition of inertialess, i.e. the compatibility of equation (6) with the balance of momentum is only satisfied if the values of the following expressions are negligible:

$$R^+(-\ddot{\varepsilon}^+ + \dot{\varepsilon}^{+2}/2)/(144\ Oh^2) \ll 1 \tag{10}$$

$$R^+(\ddot{\varepsilon}^+ + \dot{\varepsilon}^{+2})\ /(144\ Oh^2) \ll 1 \tag{11}$$

where Oh is the Ohnesorge number

$$Oh = \eta_o/(2\rho\sigma R_o) \tag{12}$$

and λ^+ is the dimensionless wavelength

$$\lambda^+ = \lambda/R_o \tag{13}$$

R^+ and λ^+ are used as the upper values for the dimensionless coordinates r^+ and z^+.

After having combined the equations (3 ÷ 6) an ordinary differential equation of the third order in R^+ is obtained:

$$\eta_s^+ t_1^{+2} R^{+2} \dot{R}^+ \dddot{R}^+ - \eta_s^+ t_1^{+2} R^{+2} \ddot{R}^{+2} + \eta_s^+ t_1^+ R^{+2} \dot{R}^+ \ddot{R}^+ +$$

$$- 9\eta_s^+ t_1^{+2} \dot{R}^{+4} + \eta_s^+ t_1^+ R^+ \dot{R}^{+3} + \eta_s^+ t_1^{+2} R^+ \dot{R}^{+2} \ddot{R}^+ +$$

$$- t_1^+ R^{+2} \ddot{R}^+ - 9 t_1^{+2} \dot{R}^{+3} + t_1^+ R^+ \dot{R}^{+2} + R^{+2} \dot{R}^+ + R^{+2} \dot{R}^{+2} +$$

$$+ 8 p_d^+ t_1^{+2} R^+ \dot{R}^{+3} + p_d^+ t_1^+ R^{+3} \ddot{R}^+ - 3 p_d^+ t_1^+ R^{+2} \dot{R}^{+2} - p_d^+ R^{+3} \dot{R}^+ = 0 \quad (14)$$

The equivalent equations for a contravariant Maxwell-fluid and a Newtonian fluid respectively with $\eta_s^+ = 0$ and $t_1^+ = 0$ respectively can easily be obtained.

A Runge-Kutta fourth order process modified due to Gill[6] is used for the numerical solution of the differential equation (14). It has a local error of approximation of the order $O(h^5)$ where h is equal to the time step Δt. A time step control is integrated on the basis of a equation to approximate the overall error of approximation[7].

The number of initial conditions determined by experiment concerning $R_o, \dot{R}_o, \ddot{R}_o$ depends on the kind of material law, i.e. the order of the differential equation. A pure numerical discussion shows that it is admissible to take the Maxwell-law instead of the Jeffreys-law if η_s^+ is less then 0.1. That means that the determination of the initial condition $\ddot{R}_o$ is not necessary. Furthermore there is a great sensitivity of the solution against very small relaxation times.

EXPERIMENTAL

Samples of aqueous solutions of polyvinylalcohol were used to study the formation of ligaments. The experimental observations were done by using an electromagnetic vibration system LING 403. The polymer solutions issue from a shortlength, cylindrical nozzle into the surrounding air. The nozzle vibrates under reproduceable conditions in flow direction. All optic measurements were done by means of a 35 mm NIKKOMAT camerabody combined with a 105 mm f/4.5 lens, a middle ring to double the focal length and an electron flasch unit METZ 45 CT. A synchronized stroboscope was used for visual observations. The negatives were measured at 125 X magnification with a LEITZ microscope. Using a screwmicrometerocular the accuracy was better than 25 µm. The surface tension was measured by means of a LAUDA tensimeter according to the ring method due to du Noüy. Measuring at low deformation rates in order to avoid additional stresses because of the stretching motion was of great importance. The aqueous solutions of polyvinylalcohol show a constant shear viscosity in the lower concentration range[9]. The shear viscosity was measured with the RHEOTRON using a couette system with a narrow gap. At a constant rate of flow there are well defined conditions for every combination of disturbance amplitude and disturbance frequency. The distance between two droplets is constant over the whole jet length. There is a suitable range to begin with the evaluation according to the model theory. The experimental data were used to estimate the initial conditions. The numerical solution was used to find the best fit with a trial and error method with variation of the parameter t_I^+ on a RADIO SHACK TRS-80 micro computer. It was possible to value the quality of the numerical approximation immediately on a X-Y recorder.

RESULTS

Fig. 3 shows a typical result for a 5% aqueous polyvinylalcohol solution at a frequency of 396 s^{-1} and an average velocity of 2.85 m/s. The nozzle has both a diameter and a length of 1 mm. The amplitude of the disturbance is 69 µm. With the initial slope of -0.175 and a time constant of 3.09 ms the solutions of the differential equation (14) for a Maxwell-law, i.e. $\eta_s^+ = 0$, leads to a best fit of the experimental data in case the relaxation time has the value 10.8 ms.

CONCLUSIONS

It seems to be possible to describe the flow in the ligaments as a surface tension driven elongational flow. Besides the principal results a new method occurs for measuring low relaxation times by using the so far undesired effect of increased stability during the break-up of viscoelastic fluids occurs.

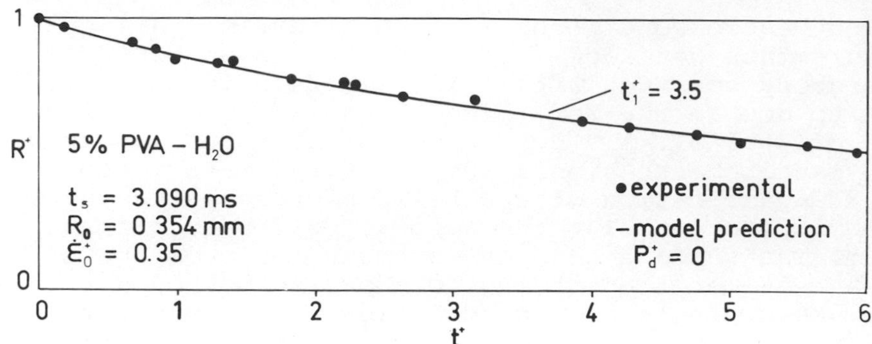

Fig. 3. Ligament stability of a aqueous polyvinyl-alcohol solution

REFERENCES

1. M.J. McCarthy and N.A. Molloy, Review of Stability of Liquid Jets and the Influence of Nozzle Design, Chem. Eng. J. 7,1 (1974)
2. H. Ringel, "Kontinuierliche Herstellung von THO_2- und (TH, U)O_2-Kernen für Hochtemperaturreaktoren", Kernforschungsanlage Jülich, Dissertation JÜL-1258, (1975)
3. R.D. Dietert, Application of vibration to the encapsulation process, Int. Lab., 11 (1979)
4. M. Goldin, "Breakup of a Laminar Capillary Jet of a Non-Newtonian Fluid", The City University of New York, Dissertation, (1970)
5. M. Gordon, J. Yerushalmi, and R. Shinnar, Instability of Jets of Non-Newtonian Fluids, Trans. Soc. Rheol 17:2, 303 (1973)
6. A Ralston and H.S. Wilf, "Mathematical Models for Digital Computers", Vol. 1, 117 (1965)
7. G. Jordan-Engeln and F. Reutter, "Formelsammlung zur Numerischen Mathematik mit Fortran IV Programmen", BI-Hochschultaschenbuch Band 106, 193, (1974)
8. R.T. Balmer, A Surface Tension Driven Extensional Flow, Trans. Soc. Rheol. 19:3, 407 (1975)
9. B. Roberts and T.K. O'Brien, The Rheology of Poly (Vinyl Alcohol) in Concentrated Aqueous Solutions, Proc. VII Int. Cong. Rheol. Gothenburg, (1976)

EXPERIMENTAL STUDIES OF HEAT TRANSFER IN VISCOELASTIC FLOW THROUGH PIPES. APPLICATIONS TO SOLAR ENERGY COLLECTORS

B. Mena, F. Avila & M. Sen

Instituto de Investigaciones en Materiales
and Faculty of Engineering
National University of Mexico
Apdo. Post. 70-360, México 20, D.F.

INTRODUCTION

It is well known that for viscoelastic fluids, rectilinear flow through pipes of non-circular cross-section is not possible in general and some secondary flows appear in the cross-section of the pipe. These secondary flows have a strong effect upon the flow characteristics but do not influence appreciably the rate of flow through the pipe. (1). Nevertheless it has been shown that the secondary flows improve substantially the heat transfer properties along the pipe when compared to pipes of equivalent circular cross-section (2,3,4,5)

This heat transfer improvement suggests numerous applications never before examined. One such application is the object of this paper, namely, a solar energy collector of the thermo-syphon type which uses a non-Newtonian liquid, as opposed to water, as the heat conducting element. Such a collector could be used as a closed loop heater for various industrial and home applications.

EXPERIMENTAL APPARATUS

The experimental arrangement is depicted schematically in Figure 1. It is a closed loop thermosyphon with a heating section whose temperature is monitored by a hot water pump

(P) which circulates hot water through a jacket that surrounds the pipe. The initial experiments where at a constant wall temperature as opposed to a constant heat flux, but this second case was examined in another experimental arrangement. A cooling section is provided using cold tap water inside a similar jacket.

Two thermo couples (T) located at the beginning and at the end of the heating section respectively provide the necessary readings for entrance and exit temperatures. The pressure drop between the lower and upper parts of the thermosyphon is measured by a sensitive pressure transducer (P). Finally, the flow rate is detected via a flow-meter (F) of the floating-ball type. The pipe which forms the closed-loop has a square cross-section of 1 cm^2 and the whole apparatus is constructed of aluminium.

As mentioned above, a second experimental apparatus, similar to the one just described, was used for the case of constant heat flux. The basic difference being that the hot-water jacket and the pump, were substituted by a heating element surrounding the pipe and suitably insulated by a jacket.

PRELIMINARY RESULTS

Some preliminary results have been obtained for the constant heat flux thermo-syphon. Two non-Newtonian solutions were tested, namely aqueous solutions of polyacylamide Separan AP-30 of 0.05% and 0.25% p.p.m.w. The measurements obtained for these solutions were then compared to similar measurements for distilled water. The results are shown in Table 1 for two different values of the heat flux input (Qr). An overall efficiency of the system may be calculated as follows:

$$\text{EFFICIENCY (\%)} = \frac{m\ Cp/\Delta T}{Qr}$$

where m is the mass flow rate, Cp is the heat capacity of the fluid and ΔT is the temperature difference between the inlet and the outlet.

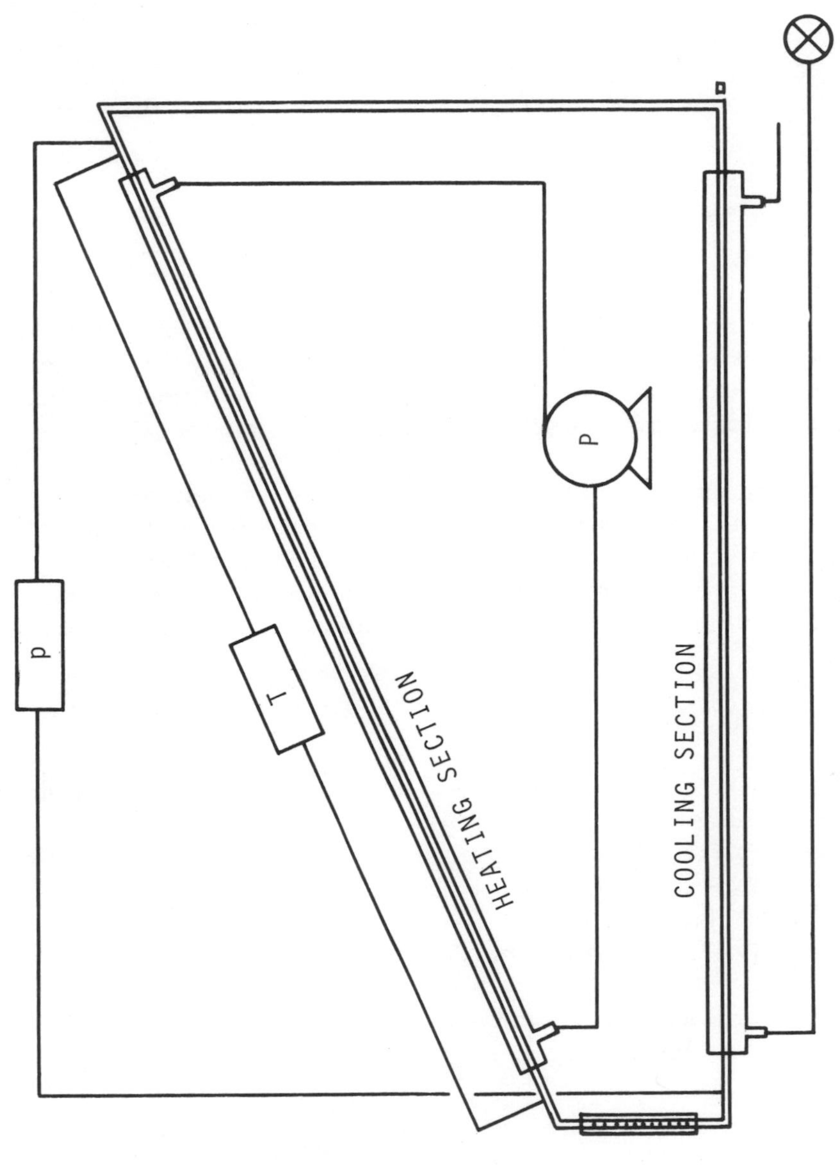

SCHEMATIC DESCRIPTION OF THE THERMO-SYPHON APPARATUS

TABLE 1. Thermo-syphon efficiencies for various fluids at different heat fluxes.

	HEAT FLUX (Qr) 9.7 Watts	HEAT FLUX (Qr) 53.5 Watts
FLUID	EFFICIENCY %	EFFICIENCY %
Water	55	55
0.05% Soln.	45	68
0.25% Soln.	76	87

It may be observed from the above results that the heat transfer properties of the system are enhanced with the non-Newtonian solutions. This is due not only to a higher viscosity value but also to the presence of secondary flows (5).

Experiments are at present being performed for the case of constant wall temperature and for a wide range of non-Newtonian solutions, in particular for Toluene-sulphonic acid and Cetil-Trimethyl ammonium aqueous solutions which are known to exhibit large elastic properties at very low viscosities.

REFERENCES

1) Dodson, A.G., Townsend, P. and Walters, K., Computers and Fluids 2, 317 (1974).
2) Oliver, D.R., Trans. Instn. Chem. Engrs. 47, 18 (1969).
3) Oliver, D.R. and Karim, R.B., Can J. Chem. Eng. 49 236 (1971).
4) Oliver, D.R. and Ashgar, S.M., Trans. Instn. Chem. Engrs. 54, 218 (1976).
5) Mena, B. Best, G., Bautista, P. and Sanchez, T., Rheol. Acta 17, 454 (1978).

NATURAL CONVECTION OF A REINER-RIVLIN-FLUID IN A RECTANGULAR ENCLOSURE

K. J. Röpke and P. Schümmer

Institut für Verfahrenstechnik RWTH Aachen
Turmstraße 46
D-5100 Aachen

INTRODUCTION

The principle mechanism of laminar natural convection in rectangular enclosures has been investigated by several scientists. But nearly all papers published are dealing with Newtonian fluids and constant physical properties. Regarding technical fluids however the influence of non-Newtonian fluid behaviour and variable properties is of more interest.

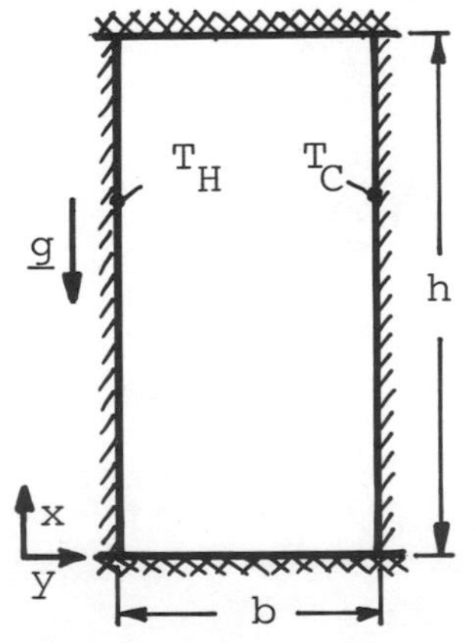

$t < o : T = T_o$ in the whole system

$t > o : T = T_H$ at $y = o$

$T = T_C$ at $y = b$

$\partial T / \partial x = 0$ at $x=o$ and $x=h$

$v_x = v_y = o$ on the walls

Fig. 1 Physical model

For the theoretical and experimental studies an enclosed cavity is chosen in which the motion is generated by a temperature gradient normal to the direction of the body force. The heating of such an enclosure caused by heat flux and/or by a suitable temperature level can be with/without heat transfer, i.e. the sidewalls have got the same/differential boundary conditions. In this paper the latter case is discussed (s. fig. 1).

THEORY

The governing differential equation describing this problem generally are those of state (1), energy (2) and motion (3) (assumptions about the incompressibilty, heat conduction and inner energy are already regarded)

$$\nabla \cdot \underline{v} = 0 \tag{1}$$

$$\varrho c_v \frac{DT}{Dt} = \nabla \cdot (\lambda \nabla T) + \underline{\underline{S}}^{*} : \nabla \underline{v} \tag{2}$$

$$\varrho \frac{D\underline{v}}{Dt} = \nabla \cdot \underline{\underline{S}} + \varrho \underline{g} \tag{3}$$

It is assumed that the deviation from the state of equilibrium is sufficiently small ($\Delta T/T_{medium} \ll 1$) and that just small changes in velocity exist. So the dissipation term can be neglected and the Boussinesq approximation can be used (the density variation is only considered in the bouyance term and described by the coefficient of thermal expansion)
Discussing the influence of non-Newtonian fluid behaviour on the natural convection we need an equation of state that is

- material objective
- "handy" in the mathematical sense
- describing the fluid behaviour in regard to the physical problem

Prelimanary examinations had to clarify which rheological parametres have got a dominant influence on flow

- numerical calculations for natural convection at a vertical plate of a "contravariant" Maxwell fluid have shown that the elasticity has just an influence in the very beginning of motion

- it is proved that the parameters considered by equations of the type

$$\underline{\underline{S}}^* = 2\,\eta(II_D)\,\underline{\underline{D}} \qquad (4)$$

are able to describe the plane flow field in the rheological point of view up to the third order
- the streak lines observed while heating a non-Newtonian fluid filled cavity do not show any flow anomaly differing from those of Newtonian fluids

Therefore a special Reiner-Rivlin-Fluid was chosen for theoretical investigations

$$\underline{\underline{S}}^* = \frac{2\,\eta_o}{1 + t_o^2\left(1 - \frac{1}{\chi^2}\right) Sp(\underline{\underline{D}}^2)}\,\underline{\underline{D}} \qquad (5)$$

Besides the rheological influence already discussed the influence of variable physical properties is to be examined. The variation of the thermal diffusivity $\alpha = \lambda / \rho c_v$ and the thermal expansion coefficient in the temperature region of interest is less than 1%. Just the dynamic viscosity varies about 10%. The temperature dependence of the dilute polymere solutions used in the experiments can be described by a modified Arrhenius term[1]

$$\eta = \eta_o \; e^{(E_s/R(1/T - 1/T_o)} \qquad (6)$$

All the other physical properties are considered to be constant.

Looking at the equation of momentum one sees that the stress term is influenced directly by the temperature dependence of $\eta_o(T(x,y))$ as well as indirectly by the gradient $\nabla\cdot(\eta_o(T(x,y))$. Latter is relevant just in the very beginning of motion. Therefore only the direct dependence is considered in the numerical study.

By suitable transformation of (1), (2), (3), (5) and (6)

- elimination of the isotropic pressure by applying the rotation on the equation of motion
- introduction of the stream function ψ and the vorticity function ω
- introduction of dimensionless units (characterized by capital letters) in an analog way to the Newtonian case

one gets the following equations (7)- (15)

$$\frac{\partial^2\Theta}{\partial X} + \frac{\partial^2\Theta}{\partial Y} - Pr\,U\frac{\partial\Theta}{\partial X} - Pr\,V\frac{\partial\Theta}{\partial Y} - Pr\frac{\partial\Theta}{\partial\tau} = 0$$

$$U\frac{\partial\Omega}{\partial X} - V\frac{\partial\Omega}{\partial Y} - Pr\frac{\partial\Omega}{\partial\tau} =$$

$$\frac{\partial}{\partial X\partial Y}(S^*_{YY} - S^*_{XX}) + \frac{\partial^2 S_{YX}}{\partial X^2} - \frac{\partial^2 S_{YX}}{\partial Y^2} - Gr\frac{\partial\Theta}{\partial Y}$$

$$\frac{\partial^2\Psi}{\partial X^2} + \frac{\partial^2\Psi}{\partial Y} = -\Omega$$

$$S^*_{XX} = \delta\frac{\partial V}{\partial Y} \qquad S^*_{YY} = \delta\frac{\partial U}{\partial X} \qquad S_{YX} = 0.5\,\delta\left[\frac{\partial U}{\partial Y} + \frac{\partial V}{\partial X}\right]$$

$$\delta = \frac{2}{1 + t^{*2}Sp(D)} \qquad \frac{Pr}{Pr_0} = \vartheta^{*\left(1 - \frac{1}{2(\Theta + \Theta_0)}\right)} \qquad \frac{Gr}{Gr_0} = \vartheta_*^{\left(2 - \frac{1}{(\Theta + \Theta_0)}\right)}$$

Numerical and experimental investigations

This set of differential equations is numerically solved by a computer program based on the Peaceman-Rachford algorithm[2]. Unfortunately there are still some instabilties for large Grashof numbers Gr at the moment.

Besides the numerical investigations experiments with dilute aqueous solutions of polyvinylalcohol and polyacrylamid are done. The temperature distribution is measured by a holographic Mach-Zehnder-interferometer. Figure 2 shows two characteristical slopes: "1" for water and "2" for 0.3 w% polyvinylalcohol-water-solution with a Gr-ratio $Gr_1/Gr_2 = 1/4$.

A velocity control is not possible yet. It is planned to install a special back-scattering-laser-doppler-anemometer.

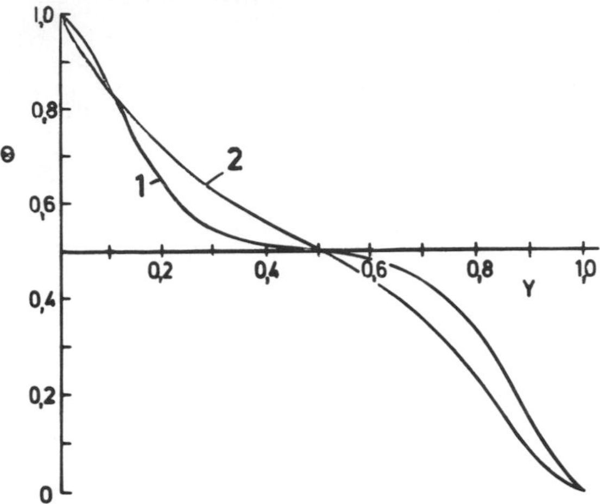

Fig. 2 temperature distribution at X = 1/2

RESULTS

Discussing the heating with heat transfer the Nusselt number is defined as

$$Nu = \frac{1}{T_H - T_C} \quad \frac{h}{b} \int_0^h \left. \frac{\partial T}{\partial y} \right|_{x=o} \quad dx$$

The Nusselt number found theoretically and experimentally will be shown as a function of Grashof number Gr, Prandtl number Pr, aspect ratio h/b, dimensionless temperature coefficient ϑ^* and rheological coefficient t^*. Especially the order of influence of the last two parameters will be compared.

REFERENCES

1. G. Berg, "Turbulenzentstehung in der Rohrströmung viskoelastischer Flüssigkeiten", Dissertation RWTH-Aachen (1979)
2. M.E. Newell, F.W. Schmidt, Heat Transfer by Laminar Natural Convection Within Rectangular Enclosures, Heat Transf., 92, 159 (1970)

HEAT EFFECT AND TEMPERATURE RISE

FOR CIRCULAR TUBE LAMINAR FLOW

Akira Ooiwa and Kimio Kurase

National Research Laboratory of Metrology
1-4, 1-Chome, Umezono, Sakura-Mura, Niihari-Gun
Ibaraki, 305 Japan

INTRODUCTION

Viscous heat effect for laminar flow in circular tube has been investigated by various authors[1.-4.], and solved analytically under the condition that the tube walls are kept at the constant temperature, or that the tube walls are thermally insulated. Therefore, those results can be applied to the system where the thermal conductivity of the tube is relatively large or small enough. But when the thermal conductivity of the tube is comparable with that of fluid and the thickness of the tube is comparable with the tube bore, their results are not effective. So authors investigate the viscous heat effect and the temperature rise in circular tube laminar flow when the heat drain passing across the tube walls is not negligible.

THEORETICAL ANALYSIS

For the purpose of mathematical development, we have the circular tube illustrated in Fig.1. and the following assumptions;

(1) The fluid is Newtonian.
(2) The flow in the tube is steady both dynamically and thermally.
(3) The end effect is negligible.
(4) There is no slip of fluid on tube wall.
(5) The fluid velocity has only the axial component.
(6) External forces, such as gravity, are neglected.
(7) Viscosity and thermal conductivity of fluid are constant and uniform.
(8) The temperature of the inflow fluid and that of the external tube wall are same and constant.

(9) The axial component of the heat conduction is negligible as compared with the radial component of the heat conduction.
(10) In case of liquid flow, the fluid is incompressible, and in case of gaseous flow, the fluid is assumed to be ideal gas.

From the assumptions (1)-(7), the flow in the tube is considered to be Poiseuille flow.

Except radiation, the temperature of material is determined generally by following three factors; 1) heat generation and absorption, 2) heat conduction, and 3) heat transfer by convection. Three factors are unified and the rate of temeprature change dT/dt is given as follows.

$$\frac{\partial T}{\partial t} = (\sigma + \kappa \nabla^2 T)/c - u \cdot \nabla T \tag{1}$$

where σ is heat generation per unit volume and time, κ thermal conductivity, c volumetric heat capacity, u velocity of fluid. The eq.(1) is applied to the tube illustrated in Fig.1 and is solved under the condition of steady state ($\partial T/\partial t = 0$) and the assumptions.

Viscous heat effect for liquid flow

From the definition of viscosity, the heat generation of the fluid per unit volume and time is expressed as follows.

$$\sigma = (\text{SHEAR STRESS}) \times (\text{VELOCITY GRADIENT}) \tag{2}$$

For Poiseuille flow,

$$\sigma(r) = \frac{1}{4\eta}\left(\frac{dp}{dz}\right)^2 \cdot r^2 \tag{3}$$

The average temperature T_a of the fluid over the cross section of the tube is defined as

$$T_a(z) = T_0 + \Delta T = \frac{1}{q}\int_0^a T(r,z) \cdot u(r) \cdot 2\pi r dr. \tag{4}$$

where q is the volumetric flow rate, T_0 temperature at the tube entrance. The average temperature rise ΔT is analyzed as $\Delta T = \Delta T_c + \Delta T_f$, where ΔT_c is temerature difference between the tube walls, ΔT_f temperature difference between T_a and that of the inner tube wall.

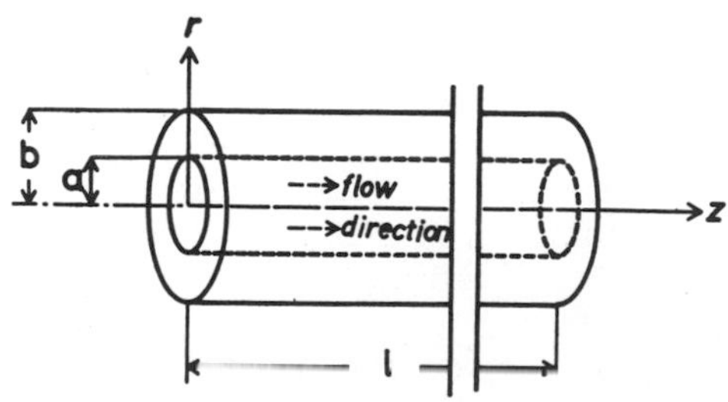

Fig.1. Dimensions and cylindrical coordinates of a circular tube.

If the length of the tube is long enough, the fluid temperature at the exit end of tube will reach that of equilibrium state, in which the viscous heat generation and the heat drain across the tube walls are in equilibrium. But, as the tube length is generally not so long, the temperature of fluid about the exit end of tube is still in the rising process. The rising rate of the fluid temperature is considered to be proportional to the difference between the viscous heat generation Q and heat drain L passing across the tube walls per unit length of the tube. Consequently, the heat drain L increases to reach the level of Q. This process is expressed by the following equation.

$$\frac{1}{c}(Q-L) - q\frac{\partial T_a}{\partial Z} = 0 \tag{5}$$

The above eq.(5) appears to be the integrated form of the eq.(1). The heat generation Q and the heat drain L per unit length of the tube is expressed as follows.

$$Q = \frac{\pi a^4}{8\eta}\left(\frac{dp}{dz}\right)^2 = -\frac{d}{dz}(pq) \tag{6}$$

$$L = \frac{2\pi\kappa_c}{\ln(b/a)}\Delta T_c \tag{7}$$

κ_c is κ of the tube.

In order to solve the eq.(5), we must detérmine the relation between the temperature difference ΔT_c and the average temperature T_a, i.e. the relation between ΔT_c and ΔT_f. In the process of temperature rise, the relation $\Delta T_f \propto \Delta T_c^2$ is generally found in the results of the numerical solution, explained in the next section. Using this relation, the eq.(5) is solved under the boundary condition (T=0 at Z=0).

$$\Delta T = \Delta T_c(1+\alpha\cdot\Delta T_c) \tag{8}$$

where

$$\alpha = \frac{1}{\Delta T_{ec}}\ \frac{5\kappa_c}{24\kappa_f\ln(b/a)}$$

$$\Delta T_c = \Delta T_{ec}\left(1-\exp\left[-\frac{Z}{Z_{c1}}-\frac{\Delta T_c}{\Delta T_0}\right]\right) \tag{9}$$

$$Z_{c1} = -\frac{c\,a^4}{16\eta}\frac{dp}{dz}\left(\frac{\ln(b/a)}{\kappa_c}+\frac{5}{12\kappa_f}\right)$$

$$\Delta T_0 = \frac{a^4}{16\eta}\left(\frac{dp}{dz}\right)^2\cdot\left(\frac{12\kappa_f\cdot\ln(b/a)}{5\kappa_c}\right)\cdot\left(\frac{\ln(b/a)}{\kappa_c}+\frac{5}{12\kappa_f}\right)$$

κ_f is κ of the fluid.

where ΔT_{ec} is ΔT_c in equilibrium state, $\Delta T_{ec} = Q\cdot\ln(b/a)/(2\pi\kappa_c)$.
The eq.(8) is not easy to be calculated except by means of computer, because ΔT_c is implicitly defined in the eq.(9).

In order to simplify the equation of fluid temperature rise, we suppose that the relation between ΔT_c and ΔT_f is expressed as $\Delta T_c \propto \Delta T_f$

This condition is available in the equilibrium state and the error from the real temperature distribution increases with the state goes far from the equilibrium. Under this condition the temperature rise is given as follows.

$$\Delta T = \Delta T_e (1 - \exp[-\frac{z}{z_{02}}]) \quad (10)$$

where

$$z_{02} = -\frac{c a^4}{16\eta}\frac{dp}{dz}\left(\frac{\ln(b/a)}{\kappa_c} + \frac{5}{24\kappa_f}\right) \quad (11)$$

where ΔT_e is ΔT in equilibrium state, $\Delta T_e = \Delta T_{ec} + 5Q/(48\kappa_f)$.
In order to get the temperature rise ΔT_l at the exit end of tube, we can replace z and dp/dz with tube length l and pressure drop per length $-\Delta p/l$, respectively. ΔT_l is given by the following form.

$$\Delta T_l = A\cdot(1 - \exp[-\frac{1}{Ac\Delta p}])\,\Delta p^2 \quad (12)$$

where

$$A = \frac{a^4}{16\eta l^2}\left(\frac{\ln(b/a)}{\kappa_c} + \frac{5}{24\kappa_f}\right) \quad (13)$$

<u>Viscous heat effect for gaseous flow</u>

It is necessary to consider the heat absorption due to the bulk expansion as well as the viscous heat generation. From the assumption(10), the following state equation is available.

$$pq = m\cdot R\cdot T \quad (14)$$

where m is mol flow rate. R is gas constant. The work E lost from the fluid per unit length and unit time is expressed by the following equation.

$$E = -\frac{d}{dz}(pq) = -mR\frac{dT}{dz} \quad (15)$$

This equation means that there is no energy exchange between the fluid and the tube, if the temperature T is constant in the axial direction. Therefore, the heat energy generated by viscous resistance is spent in compensation for the reduction of internal energy owing to the bulk expansion of flowing gas in consequence of the decrease of pressure. Through the tube, temperature of flowing gas is thought to be kept constant except near the inlet and outlet region where the fluid is in acceleration and deceleration.

NUMERICAL SOLUTION

The temperature distribution in the tube is calculated by the method of asymptotical approximation. For the efficiency of numerical treatment, dimensions in the tube (r,z) is devided into (100 x 100) quantized parts. The results are shown in Fig.2. The reduced axial distance ζ is defined as $\zeta = z/z_{02}$.

The temperature distribution in the equlibrium state (i.e. $\zeta \gg 1$) is assumed to be equal to that expressed by the analytical reults. So we can examine the propriety of the numerical treatment by comparison between the analytical results and numerical ones. The differences between both are less than 2 %. Since the round error by quantization in numerical treatment is about 2 %, it is concluded that the numerical calculation is performed suitably. From the numerical results, we find that the relation between ΔT_c and ΔT_f is expressed approximately as $\Delta T_f \propto \Delta T_c^2$.

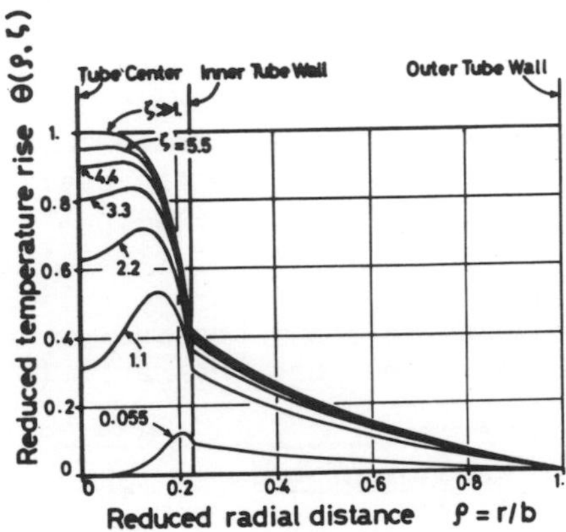

Fig.2. Temperature distribution over the cross section of a tube, as a function of reduced axial distance ζ from the entrance end of the tube, calculated numerically with the iterative method applied to the eq.(2).

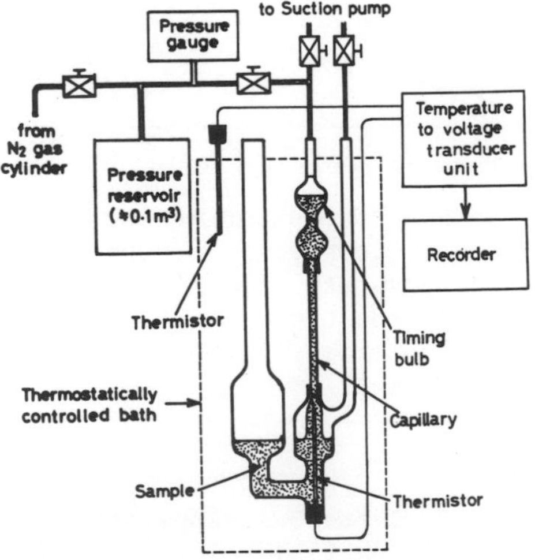

Fig.3. A schematic diagram of a temperature rise measuring aparatus, the main part is a variable-pressure type viscometer.

COMPARISON BETWEEN THE THEORETICAL RESULTS AND THE EXPERIMENTAL ONES

The apparatus used is shown in Fig.3. Viscosity and temperature rise are measured at the same time under temperatures, 20 °C, 30 °C, and 40 °C, and pressures, 0.5-2.5 x 10^5 Pa (0.5-2.5 atm.). Viscosities are measured by the timing bulb method, and temperature rises are measured by thermistors.

From the results (Fig.4), there are differences, less than 11 %, between the curve(1) calculated by the eq.(12) and the curve(2) by the eq.(8). Experimental results are between these two theoretical curves. This differences are realized to be caused by estimation error of $\Delta T_f / \Delta T_c$. Therefore, the difference is negligible, when the pressure drop is small and the temperature rise ΔT_l at the exit end reaches almost to that of equilibrium state ΔT_e, or when the pressure drop is large enough and it is still in the beginning of the process of temperature rising ($\Delta T_l \ll \Delta T_e$). Between the above mentioned two states, the difference has the biggest value that depends on the ratio $\Delta T_{ef} / \Delta T_{ec}$. In this experiment, the ratio $\Delta T_{ef}/ \Delta T_{ec}= (5/24)\kappa_c/(\kappa_f \ln(b/a))$ is 1.02. In order to apply the analytical result, i.e. the eq.(12), it should be certified that the ratio is less than unity.

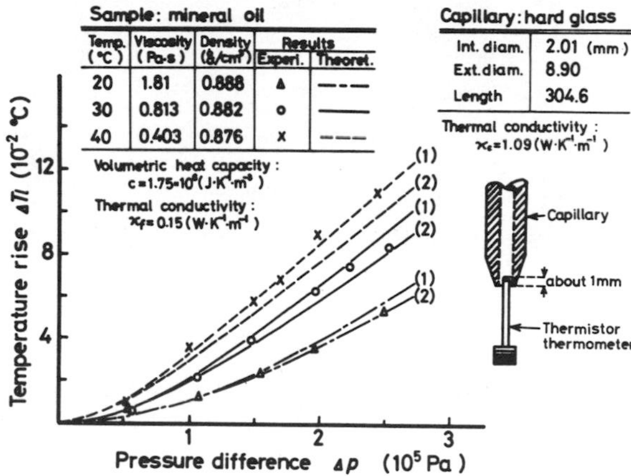

Fig.4. Comparison between the experimental results (marks) and the theoretical ones (curves), representing the relation between the temperature rise at the exit end of the tube ΔT_l and the pressure drop through the tube Δp. The theoretical relations are obtained by the eq.(12) for curve(1) and by the eq.(8) for curve(2).

REFERENCES

1. M. D. Hersey and J. C. Zimmer, J. Appl. Phys., 8, 359 (1937)
2. R. B. Bird, SPE J., 11, 35 (1955)
3. H. L. Toor, Ind. Eng. Chem., 48, 922 (1956)
4. R. E. Gee and J. B. Lyon, Ind. Eng. Chem., 49, 956 (1957)

NUMERICAL SOLUTIONS FOR FLOW OF AN OLDROYD FLUID BETWEEN COAXIAL ROTATING DISKS

R.K.Bhatnagar

State University of Campinas, (Sao Paulo), Brasil

(Abstract)

The paper deals with the problem of flow of anelastico-viscous fluid of Oldroyd type confined between two infinite parallel disks. The motion is treated as steady, and choosing a cylindrical polar coordinate system, the lower disk is taken to rotate with a constant angular velocity Ω while the upper disk to rotate with constant angular velocity $K\Omega$. The non-dimensionalized forms of equation of continuity, three equations of motions and six equations for the stress components are written. This set of ten non-linear partial differential equations is first reduced to set of ordinary non-linear differential equations by the use of certain appropriate transformations.

For the numerical solution of this set of non-linear ordinary differential equations, their finite-difference analogues are written using either backward or forward differences so that the resulting system retains diagonal dominance. In a similar manner, the finite-difference analogues of the bounda-y conditions are also written. The governing system is solved for various values of Reynolds number from 1 to 1000 using the successive over-relaxation method. The elastic constants are also varied to obtain comparative results. The rotation parameter K is chosen to be 0,-1 and 2. This vould cover the cases of one disk being held at rest, both disks rotating in same direction with differents angular velocities and the disks rotating in opposite directions.

The axial, radial and transverse velocities are plotted graphically for large number of cases and spme very interesting results are obtained for an Oldroyd fluid.

MEDIUM AND LARGE DEBORAH NUMBER SQUEEZING FLOWS

D. M. Binding, F. Avila, A. Maldonado and M. Sen

División Profesional and División de Estudios de Posgrado
Facultad de Ingeniería
Universidad Nacional Autónoma de México

INTRODUCTION

In this paper we present the results obtained from an experimental study of 'reverse' squeezing flows at relatively large Deborah numbers. The test fluid is contained between two flat, horizontal, circular plates each of radius a, the lower one being fixed while the upper one is free to move vertically. For time $t <$ 0 the separation of the plates, $h(t)$, is maintained at the constant value h_0. At $t=0$ a vertical force F is applied to the upper plate and we focus attention on the behaviour of $h(t)$ as time increases. Schematic diagrams of the flow geometry and initial conditions are shown in fig. 1.

There have been many works published about this class of flow [1-10] but most, if not all, have been restricted to the particular case when F is negative, i.e. when the plates are forced together under the applied load. (For convenience we call this flow A). For this flow, experimental observations have indicated that for small Deborah numbers the behaviour of $h(t)$ is governed only by the viscosity of the fluid, whereas for sufficiently large Deborah numbers the plates come together more slowly than would be predicted on the basis of an inelastic fluid of the same viscosity [7,8,10]. The effect is more pronounced at short times after the onset of the flow. Because of these observations it is usually concluded that elastic fluids behave as better lubricants than inelastic fluids.

Let us now consider the case when F is positive (flow B),

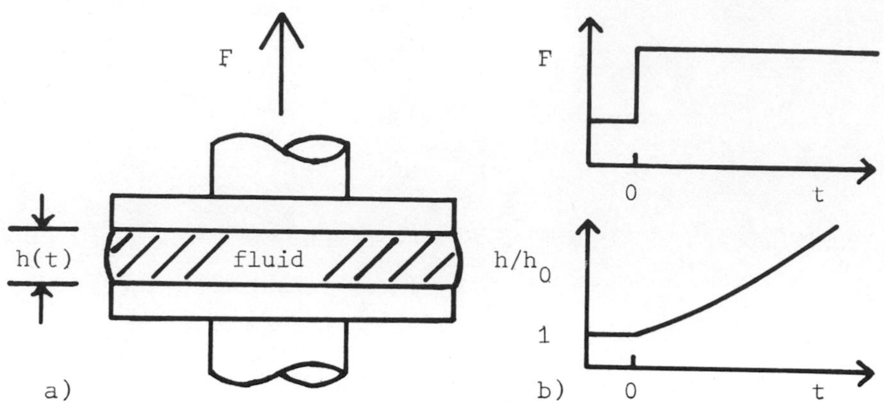

Fig. 1 a) Schematic diagram of flow geometry
b) Schematic diagram of initial conditions

i.e. when the plates are forced to separate. It is not difficult to imagine that the same mechanism giving rise to better lubrication in flow A may cause poorer lubrication in flow B. In other words, the plates might be expected to separate more slowly for an elastic fluid compared with an inelastic fluid with the same viscosity. However, it is now generally accepted that elastic fluids do in fact make better lubricants.

In the present study we concern ourselves with flow B, the motives being the following:

i) It could be argued that flow B is more representative of certain lubrication systems than flow A.

ii) In view of the above discussion there exists the possibility that flow B is essentially different to flow A.

INSTRUMENTATION

To perform the experiments a Weissenberg Rheogoniometer (Sangamo Controls, model R19) was converted following closely the conversions already reported by other authors[10], the principal difference being that a system of two ball bearings supported on an independent frame had to be incorporated in order to apply a force F in the upward vertical direction. The Mooney platen arrangement, instead of the lower flat plate, was used so that the fluid filled the test region throughout the duration of each experiment. Edge effects introduced by this measure were kept to a minimum by using only the volume of fluid essential for maintaining a full test region. Shear stress data were obtained in the usual way, with a cone and plate arrangement.

EXPERIMENTAL RESULTS AND DISCUSSION

The first tests were performed on Newtonian fluids for which the separation h(t) should satisfy[1]:

$$F = \frac{3\pi\eta\dot{h}a^4}{2h^3} \qquad \ldots(1)$$

where η is the constant Newtonian viscosity and $\dot{h}$ is the time derivative of h. Several fluids, such as silicone, glycerine and mixtures of glycerine and water, were tested under loading conditions ranging from 2N to 106N depending on the viscosity and h_0. Two values of h_0, 0.5mm and 0.8mm, were considered. The viscosities ranged from 0.02 Pa.s to 6.0 Pa.s. In all cases good agreement with equation (1) was obtained. Fig. 2 illustrates the results for glycerine (η= 1.69 Pa.s) at a temperature of 17°C for two different loads F.

The non-Newtonian fluids used were 2, 2.5 and 3% concentrations by weight of Separan A.P.30 in water. Typical results from these experiments are displayed in figs.3 and 4 for the 2.5% concentration fluid. The most noticeable feature is the initial rapid increase in the separation h, followed by a more gentle accelerating motion. Although the effect decreases with decreasing F it is interesting to note that even at relatively low values of F the initial rapid increase is still perceivable. This is seen in fig. 5 for the

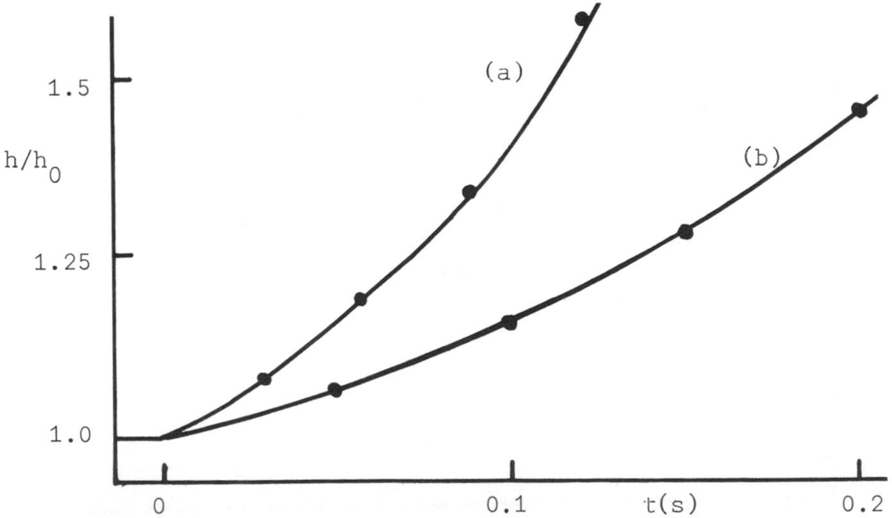

Fig. 2 h(t) traces for glycerine. Temp. 17°C, η =1.69 Pa.s, h_0= 0.8mm. a) F = 63.9N b) F = 32.8N
—— Experimental trace. ● Prediction based on eqn.(1)

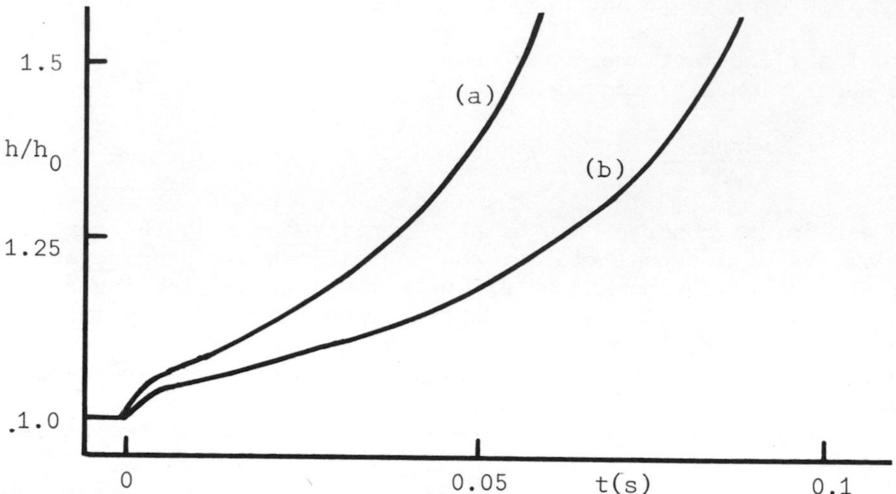

Fig. 3 h(t) traces for 2.5% Separan A.P. 30 in water.Temp. 16°C h_0=0.8mm. (a) F = 42.5N (b) F = 32.8N

same fluid, the value of F being 11.9N.

At this point it is worth comparing the observed behaviour with the theoretical behaviour based on viscosity considerations alone. To do this in the simplest way possible, a 'power-law' model is

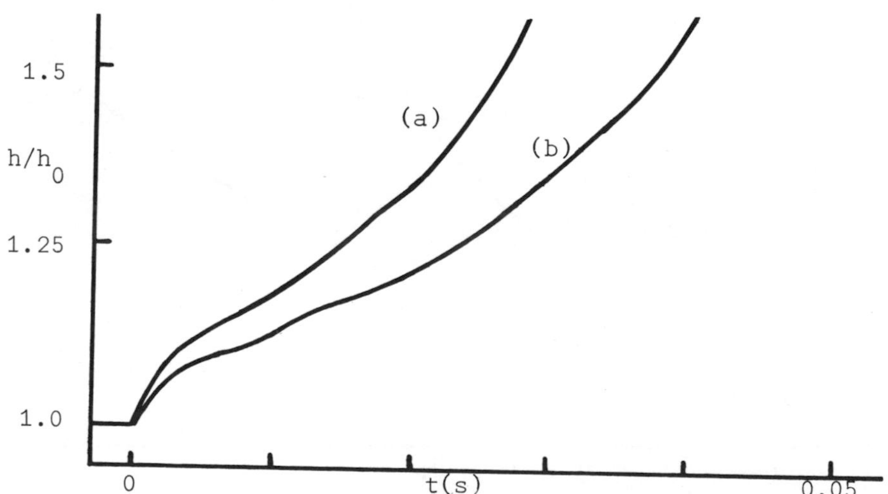

Fig. 4 h(t) traces for 2.5% Separan A.P.30 in water. Temp. 16°C h_0= 0.8mm. (a) F=63.9N. (b) F=53.4N.

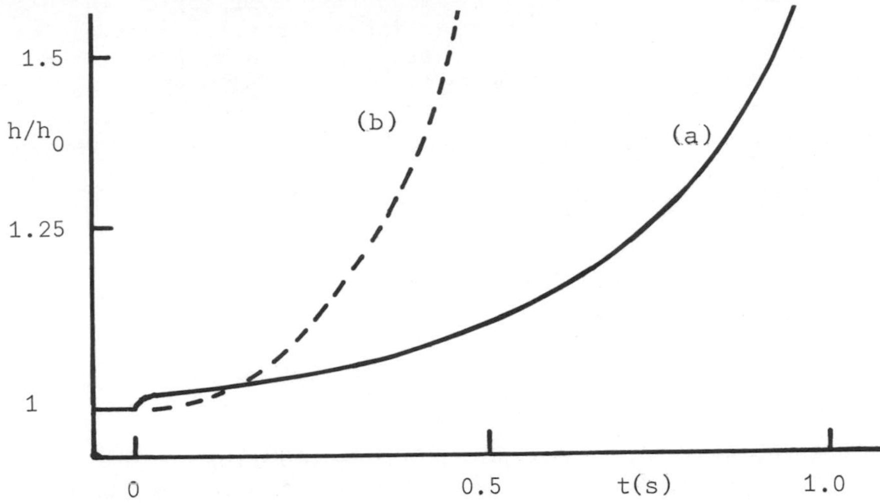

Fig. 5. h(t) traces for 2.5% Separan A.P.30 in water. Temp. 16°C h_0=0.8mm. (a) Experimental trace. (b)Prediction based on eqn. 3.

used to represent the shear stress function:

$$\sigma = m\,\dot{\gamma}^{n}, \qquad \ldots(2)$$

σ being the shear stress, $\dot{\gamma}$ the shear rate and m and n, two constants. For this model, the function h(t) is given by[2]:

$$F = \frac{2\pi m\,(2n+1)^{n}(\dot{h})^{n}\,a^{n+3}}{n^{n}(n+3)\,h^{n+3}} \qquad \ldots(3)$$

Since the 'power-law' model overestimates the viscosity in the low shear rate range, equation (3) would underestimate the rate at which the separation increases, particularly if the load F is small. The comparison of the observed behaviour with the theoretically predicted behaviour, based on equation (3), is shown in fig. 5. It is clear that viscosity effects alone cannot account for the observed slow rate of separation of the plates. The same discrepancy was observed in all the experiments on the Separan-water solutions, all of which had 'power-law' exponents of approximately 0.23.

CONCLUSION

As a result of the experiments described in this paper we conclude that, as far as lubrication is concerned, the initial rapid separation of the plates observed for the elastic fluids studied, implies that these behave as better lubricants. In this context, we refer to dynamic lubrication effects (such as during start-up or sudden load changes) rather than steady situations. From the fundamental point of view, it would appear that there exist basic differences in the mechanics of flows A and B, the latter being, perhaps, dominated more by the effects of extension than the former.

REFERENCES

1. J. Stefan and K. Sitzgber, Akad. Wiss. Math. Natur. Wein, 69:713 (1874).
2. J.R. Scott, Trans. Inst. Rubber Ind., 1:169 (1931).
3. R. I. Tanner, A.S.L.E. Trans., 8:179 (1965).
4. A. B. Metzner, Trans. A.S.M.E., 90F:531 (1968)
5. J. M. Kramer, Appl. Sci. Res. 30:1 (1974)
6. A. B. Metzner, Rheol. Acta, 10:434 (1971)
7. P. Parlato, M. Ch. E. Thesis, Univ. of Delaware, (1969)
8. P.J. Leider and R. B. Bird, Ind. Eng. Chem. Fund., 13:336 (1974).
9. J. M. Davies, Ph D. Thesis, Univ. of Wales, (1974)
10. G. Brindley, J. M. Davies and K. Walters, J. Non-Newtonian Fluid Mech., 1:19 (1976)

OSCILLATORY FLOW IN PIPES OF NON-CIRCULAR CROSS-SECTION

B. Mena & F. Nuñez

Instituto de Investigaciones en Materiales and Faculty of Engineering, National University of Mexico
Apdo. Post. 70-360 México 20, D.F.

INTRODUCTION AND PREVIOUS WORK

The flow of viscoelastic liquids through pipes of circular cross-section has been recently examined for the case in which the pipe oscillates longitudinally about a mean position (1-5). The effects of amplitude and frequency of the oscillations upon the mean flow rate were evaluated and increases of an order of magnitude in the flow rate when compared to purely rectilinear flow were found. A close examination of the problem showed the following conclusions:

a) Experimental

(1) Exceedingly large effects (due to the oscillations of the pipes) were present producing increases in flow rate of up to 20 times. The increases appeared at moderate values of amplitude and frequency of the oscillations and took place while the pressure gradient along the pipe remained virtually constant.

(2) A visual examination of the flow pattern inside the pipe showed that the dominating factor was the shear dependent viscosity and that the elastic properties of the fluid were of secondary importance in the phenomenon.

b) Theoretical

1) Linear viscoelastic models as well as inelastic "power law type" models were unable to predict quantitative agreement (3).

2) Qualitative agreement may be obtained using higher order approximations (4) and in particular using an upper convected Maxwell Model characterized by a viscosity function and one relaxation time (5).

3) Quantitative agreement has not yet been predicted.

In the light of the previous work, a new experimental investigation is being carried out by the authors in order to examine the following points:

a) The effect of longitudinal oscillations upon the flow rate for pipes of non-circular cross-section. i.e. square and rectangular (6).

b) A visual examination of the flow pattern inside the pipes.

c) The effect of transverse oscillations upon the flow rate.

d) The combined effect of transverse and longitudinal vibrations upon the flow rate.

e) The effect of oscillations upon the turbulent flow of Newtonian liquids in pipes (7).

EXPERIMENTAL ARRANGEMENT

The experimental apparatus has been described elsewhere (3) and the only modification is a minor one, namely a motor to provide the transverse vibrations upon the pipe. Therefore, the author does not think that it requires to be described in detail.

PRELIMINARY RESULTS

At the time of press, the experiments are unfortunately at an early stage but similar effects appear to be present for the pipes of non-circular cross-section as were present for the pipes of circular cross-section. Some changes have appeared in the flow pattern which might be due to secondary flows. It is nevertheless expected that an overall evaluation of the topic will be ready in time for the Congress.

REFERENCES

1) Manero, O. and Mena, B., Rheol. Acta 16, 573 (1977)
2) Manero, O., Mena, B. and Valenzuela, R., Rheol. Acta 17, 693 (1978)
3) Mena, B., Manero, O. and Binding, D. M., J.N.N.F.M. 5, 427 (1979)
4) Kazakia J.Y. and Rivlin, R.S., Rheol. Acta 17, 210 (1978)
5) Manero, O. and Walters, K., J.N.N.F.M. in press.
6) Kazakia J.Y. and Rivlin, R.S., J.N.N.F.M. 6, 145 (1979)
7) Rivlin, R.S. (personal communication)

SQUEEZING FLOWS OF VISCOELASTIC LIQUIDS

P. Shirodkar, S.Middleman[°]

University of Massachusetts, U.S.A.
[°] University of California, U.S.A.

(Abstract)

Squeezing flows occur in a variety of industrial applications, including lubrication, printing, and sheet molding. Squeezing flow also occurs in the region between two bubbles growing in a foaming liquid. An experiment designed to study squeezing flow has been designed, and data indicate that the force-time behavior cannot be modeled using purely viscous, non-Newtonian constitutive equations. The force resisting squeezing first rises above the purely viscous level, and then falls below the viscous level as time increases.

Mathematical models of squeezing flows have been examined, and analytical solutions based on lubrication theory have been evaluated through comparison to numerical solutions based on the finite element method. It appears that a model based on lubrication theory, using the known kinematics of a power law viscous fluid, but using Wagner's viscoelastic fluid for the stresses, predicts the behavior of force-time curves in qualitative agreement with the observations for strongly elastic fluids.

DILUTE POLYMER SOLUTION FLOWS THROUGH PERIODICALLY CONSTRICTED TUBES

V.N. Kalashnikov

Institute for Mechanical Problems, The USSR Academy of Sciences, Moscow, USSR

(Abstract)

The results of experiments on the drag of dilute Polyox solution flows in tubes with periodic step change in diameter are presented. Constricted parts of tubes have 0.8, 1.85 and 4.15 mm i.d. Large to small inner diameter ratio for the tube is 2.5. In addition to tubes, transparent periodically constricted ducts with square cross section were tested. Several different concentrations up to 200 ppm were used.

Significant drag increase was observed. For some concentrations and flow rates ratio of Polyox solution friction to Newtonian was found as high as 20. The maximum drag increase asymptotes were noted. It was showen that in some cases pressure drop for Polyox solution flow falls abruptly with increase of flow rate.

Connection of anomalous drag increase with a flow instability (elastic turbulence) was shown by visualisation and laser Doppler measurements in transparent ducts. The results were compared with well known results for polymer solution flows in porous media.

Diagrams of flow type in coordinates Reynolds number-Weissenberg number were introduced. Flow in nonuniform tubes and flows in uniform tubes were analyzed on the basis of these diagrams. The analog between the jump of pressure drop in nonuniform tube flow and the onset of the Toms phenomenon is discussed.

RHEOLOGICAL EFFECTS IN THE DYNAMICS OF LIQUID FILMS AND JETS

V.M. Entov

Institute for Mechanical Problems of the USSR Acad. Sci., Moscow, USSR

(Abstract)

The dynamics of free jets and films of non-Newtonian liquids is investigated. The general dynamical equations are derived under assumption of thinness of the jet or film. The Break-up of capillary jets of nonlinearly-viscous (shear-thickening and shear-thinning) as well as elasto-viscous fluids is studied by analytical and numerical methods. It is shown that the shear-thickening and (or) elastic tension tend to delay the jet disintegration. Some experimental data are reported which conform to the theory presented. The significance of molecular orientation for stabilization of fibers and jets of polymeric liquids is discussed.

The influence of elasticity on the geometry of dynamical equilibrium of a film of elastoviscous liquid is investigated. The possibility of estimation of rheological properties of a fluid by means of experiments with jets and films is analyzed.

THE IMPORTANCE OF RHEOLOGY IN THE DETERMINATION OF THE CARRYING CAPACITY OF OIL-DRILLING FLUIDS

M.A. Lockyer
Engineering Department
Oxford Polytechnic
Oxford OX3 0BP

J.M. Davies & T.E.R. Jones
Mathematics Department
Plymouth Polytechnic
Plymouth PL4 8AA.

INTRODUCTION

The ability of a drilling fluid to convey drill cuttings from a well is not fully understood and this is particularly so in the case of highly deviated wells where difficulties are frequently experienced in cleaning the hole. The cuttings travel with a lower velocity than the drilling fluid and they can accumulate in the well bore. If this is not kept to a minimum then this can either lead to degradation of the cuttings or the drill string may get lodged in the hole.[1] This study forms the first part of a long term project, sponsored by International Drilling Fluids, to understand and predict the carrying capacity of oil drilling muds.

Figure 1 contains a schematic diagram of a typical oil well. The drilling mud is pumped down the inside of the rotating drill string, injected out through nozzles at the drill bit and pumped up through the annulus to carry the drill cuttings to the surface.

Experiments were carried out to determine the flow properties of drilling fluids and it was found that the shear stress/shear rate curves were power law over the range of shear rates considered. None of the samples* exhibited a yield stress and therefore were not Bingham in character. For this project our initial approach has concentrated on the study of the flow of power law fluids past a stationary sphere (Figure 2).

---

*Samples supplied by International Drilling Fluids, St. Austell, England.

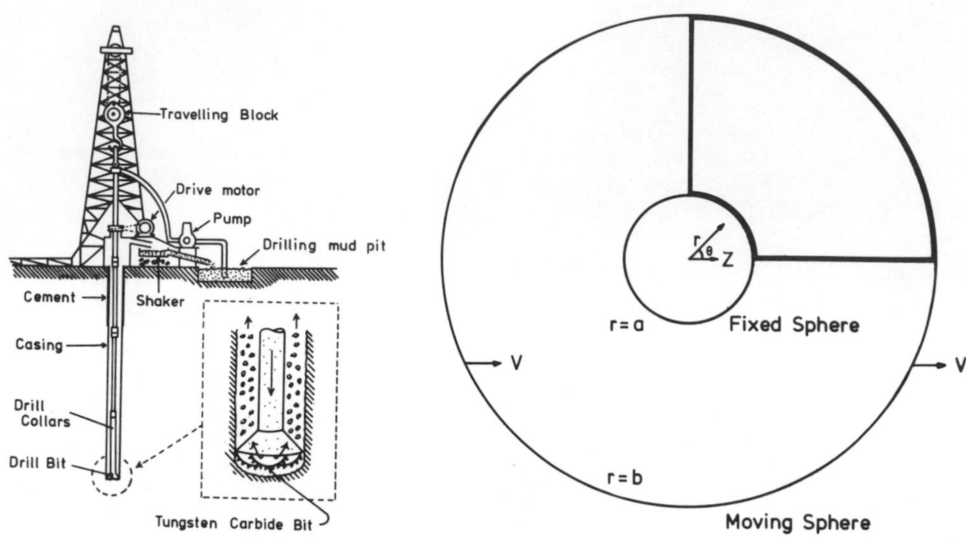

Figure 1. Schematic diagram of an oil well

Figure 2. Sphere with coordinate system

A number of workers have previously investigated theoretically the slow flow of a power law fluid moving past a stationary sphere[2,3,4,5,6]. Ulbrecht[2], for instance, used a perturbation analysis with some further approximations. Slattery[3] used the variational principle to obtain an upper and lower bound for the drag coefficient and Tomita[4] used a stream function which satisfied the boundary conditions to evaluate an approximate value for the drag coefficient. However various approximations have been used by all the authors in order to obtain an analytical solution. Large discrepancies exist between the drag coefficients obtained by these authors as will be shown later in Figure 5. An analysis which does not make any assumptions is therefore required.

Our initial numerical work concentrates on the solution for n near 1 using a pertubation analysis without making any further approximations. The partial differential equation involved now becomes linear.

We follow closely the work of Ulbrecht[2] and we present the corrected form of his analytical solution which was obtained by neglecting certain high order terms in the analysis.

Theoretical Analysis

We assume that we have slow axisymmetric flow, steady state conditions and negligible edge effects. The solution we consider is that of a stationary solid sphere, radius a, within a concentric spherical shell, radius b, filled with the fluid and moving with velocity U (Figure 2). We use a spherical polar coordinate system with the relevant equations of motion,

$$\frac{\partial p}{\partial r} = -\left[\frac{1}{r^2}\frac{\partial}{\partial r}(r^2\tau_{rr}) + \frac{1}{r\sin\theta}\frac{\partial}{\partial\theta}(\tau_{\theta r}\sin\theta) - \frac{(\tau_{\theta\theta}+\tau_{\phi\phi})}{r}\right] \quad (1)$$

$$\frac{\partial p}{\partial \theta} = -\left[\frac{1}{r^3}\frac{\partial}{\partial r}(r^3\tau_{r\theta}) + \frac{1}{r\sin\theta}\frac{\partial}{\partial\theta}(\tau_{\theta\theta}\sin\theta) - \frac{\tau_{\phi\phi}\cot\theta}{r}\right] \quad (2)$$

The power law model is given by

$$\tau_{ik} = -K\,|\dot{\gamma}|^{n-1}\,\dot{\gamma}_{ik} \quad (3)$$

where

$$|\dot{\gamma}| = \sqrt{\tfrac{1}{2}\sum_{i}\sum_{j}\dot{\gamma}_{ij}\,\dot{\gamma}_{ji}} \quad (4)$$

τ_{ik} is the extra stress tensor and $\dot{\gamma}_{ik}$ is the rate of strain tensor.

The velocity components can be expressed in terms of a stream function ψ as

$$V_r = \frac{-1}{r^2\sin\theta}\frac{\partial\psi}{\partial\theta} \quad (5)$$

$$V_\theta = \frac{1}{r\sin\theta}\frac{\partial\psi}{\partial r} \quad (6)$$

By symmetry we need only consider solving the equations over the region outlined in Figure 2 and the boundary conditions are,

$$\psi = \frac{\partial\psi}{\partial r} = \frac{\partial\psi}{\partial\theta} = 0 \quad \text{on } r = a \quad (7)$$

$$\psi = \frac{-Ub^2\sin^2\theta}{2}, \quad \frac{\partial\psi}{\partial r} = -Ub\sin^2\theta, \quad \frac{\partial\psi}{\partial\theta} = -Ub^2\sin\theta\cos\theta \quad \text{on } r = b \quad (8)$$

$$\psi = \frac{\partial\psi}{\partial\theta} = 0 \quad \text{on } \theta = 0 \quad (9)$$

$$\frac{\partial\psi}{\partial\theta} = 0 \quad \text{on } \theta = \frac{\pi}{2} \quad (10)$$

Equations (1) - (6) may be combined to give

$$E^4(\psi) = g(r,\theta,n,d,\psi_r,\psi_\theta,\psi_{rr},\psi_{r\theta},\psi_{\theta\theta},\psi_{rrr}.\psi_{rr\theta},\psi_{r\theta\theta}\ \psi_{\theta\theta\theta},$$
$$\psi_{rrrr},\ \psi_{rrr\theta},\psi_{rr\theta\theta},\psi_{r\theta\theta\theta},\psi_{\theta\theta\theta\theta}) \quad (11)$$

where
$$E^4=\left[\frac{\partial^2}{\partial r^2} + \frac{\sin\theta}{r^2}\ \frac{\partial}{\partial\theta}\ \left(\frac{1}{\sin\theta}\frac{\partial}{\partial\theta}\right)\right]^2 \quad (12)$$

and where g contains non linear ψ terms.
For Newtonian liquids g = 0 to give

$$E^4(\psi) = 0 \quad (13)$$

An exact analytical solution of equation (11), for any $n \neq 1$, subject to the boundary conditions (7) - (10) is not possible. At the present time we are involved in solving the equations numerically. We have, though, carried out the analysis for n near 1, as the ψ terms in the expression for g then become linear.

On neglecting terms of order $(n - 1)^2$ equation (11) reduces to

$$E^4(\psi) = (n-1)\ g_1(r,\theta) \quad (14)$$

where $g_1(r,\theta)$ is a known function of r and θ.
Ulbrecht[2] now makes a further approximation by neglecting the higher order terms in $(\frac{a}{r})$ to obtain (in corrected form)

$$E^4(\psi) = \frac{9(n-1)\ U\ a\ \sin^2\theta}{r^3} \quad (15)$$

The total drag on the solid sphere can be expressed in the form

$$F_D = 12\pi a^2 K\left(\frac{U}{2a}\right)^n\ F(n,d) \quad (16)$$

where $d = \frac{b}{a}$. Ulbrecht[2] solved equation (14) for ψ and obtained

$$F(n) = 3^{\frac{3n-3}{2}} \left\{\frac{33n^5 - 63n^4 - 11n^3 - 97n^2 + 16n}{4n^2\ (n+1)(n+2)(2n+1)}\right\} \quad (17)$$

This result appears to be incorrect as it reduces to (neglecting $(n-1)^2$ terms).

$$F(n) = 1 - 0.602(n-1) \quad (18)$$

whereas the correct form can be shown to be

$$F(n) = 1 - 0.375(n-1) \quad (19)$$

We have plotted F(n) against n for equations (17), (18) and (19) in Figure 3.

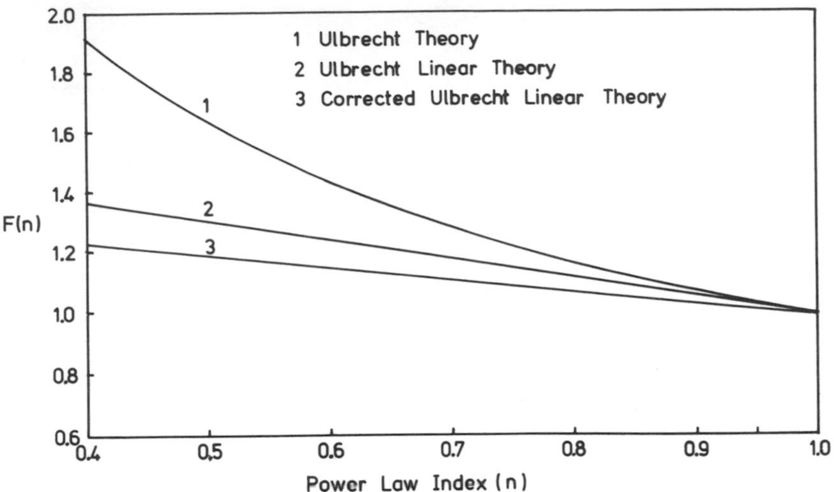

Figure 3. Ulbrecht curves (infinite outer sphere) F(n) against n.

The assumption used by Ulbrecht[2] of neglecting terms of higher order in $(\frac{a}{r})$ is certainly valid for large distances away from the sphere. However as the drag force calculation is carried out over the surface of the sphere given by $r = a$ then the assumption may not be valid. We will be able to check the validity by solving the equations numerically.

Throughout the analysis we have derived the equations for a finite outer sphere. It can be shown that for a Newtonian liquid with $d = 100$ the drag coefficient differs by approximately 2% from the infinite outer sphere result. Therefore a diameter ratio of 100 will be sufficiently large enough for our numerical calculations.

The fourth order partial differential equation (14) was solved by examining the pair of coupled second order differential equations given by

$$E^2 \psi = G, \quad E^2 G = (n-1)\, g_1 \ (r,\theta) \tag{20}$$

The grid mesh was taken over a quarter sphere with linear divisions in the θ direction and exponential divisions in the r direction. Central differences were used at all non boundary points and the resulting matrix was solved using successive over relaxation iterative procedure. A conjugate gradient method of solution was also considered. Our numerical results for the normalised drag coefficient $F(n,d)$ against n are shown in Figure 4 for diameter ratios d = 2, 10, 20, 40 and 100. These results therefore imply that the Ulbrecht[2] assumption is not a valid approximation and leads to an incorrect form for the drag.

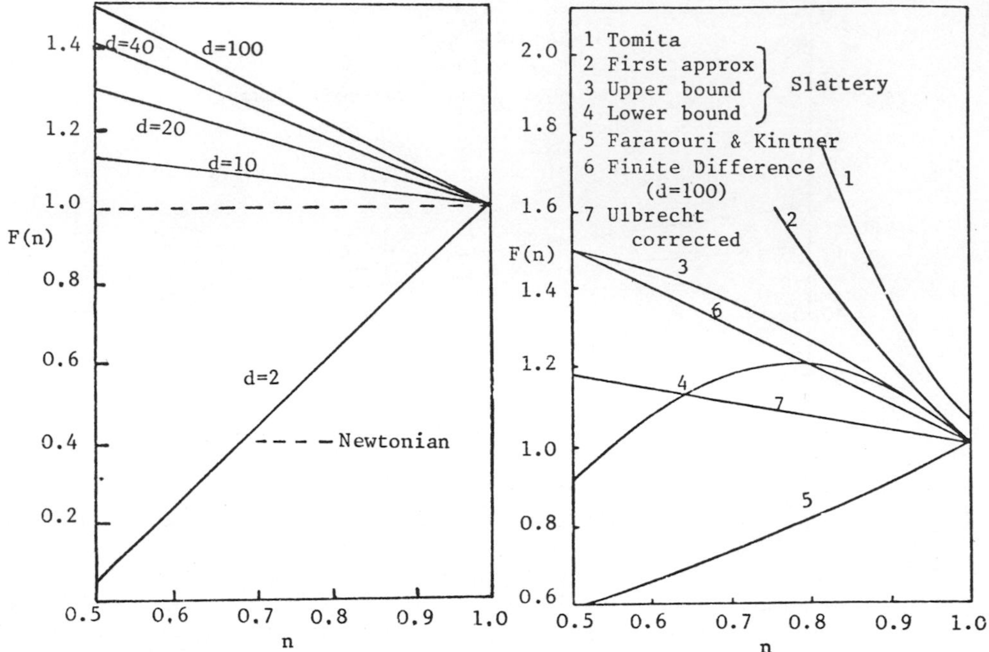

Figure 4. Numerical calculations for F(n) using the n near 1 approximation.

Figure 5. Comparison of our numerical results with other workers.

In Figure 5 we present our numerical result for d = 100 in comparison with other workers[2,3,4,5,6].

In conclusion this work seeks to solve a difficult problem and when completed will give new insight into the flow of a power law liquid past a sphere.

We would like to thank International Drilling Fluids, St. Austell, Cornwall, England and Dr P. Lundie in particular for all the help we have received during the preparation of this work.

References

1. B. Windle, I.D.F. Ltd. Internal report (1979).
2. A. Acharya, R.A. Mashelkar and J. Ulbrecht. Rheol. Acta 15:454 (1976).
3. M.L. Wasserman and J.C. Slattery, A.I.Ch.E. Journal 103:383 (1965).
4. Y. Tomita, Bull. Japan.Soc.Mech.Eng. 2:469 (1959).
5. A. Fararouri and R.C. Kintner, Trans. Soc. Rheol. 5:369 (1961).
6. Y. Nakaro and C. Tien, A.I.Ch.E. Journal 16:56- (1970).

STRUCTURE OF TURBULENCE IN PIPE FLOW OF VISCOELASTIC FLUIDS

P. Schümmer, W. Thielen

Institut für Verfahrenstechnik
D-5100 Aachen
Turmstraße 46

INTRODUCTION

There is a great interest in turbulent flows of polymer solutions, particularly in the phenomenon of "drag reduction". In this work we would like to show some results of our investigations on the turbulent pipe flow and the interpretation of these data.

RHEOLOGY OF THE INVESTIGATED FLUIDS

The rheological behaviour of aqueous polymer solutions depends on the type of polymer and on its concentration. At very low concentrations, the solutions might have constant shear viscosity but they already show a dominant relaxation time ($< 10^{-1}$s). With increasing concentration or molecular mass the shear viscosity becomes shear-dependent.
One way of obtaining relaxation times is to use structural models for the polymer molecule.
The relaxation times calculated with these models are of the order of one hundreth of a second and smaller.
A model of the rheological behaviour of dilute polymer solutions is the one of a suspension of elastic coils, the so-called Rouse-Zimm-Coils, in a Newtonian solvent. We obtain for the total stress of polymer and solvent equation (1).

$$\underline{\underline{\sigma}}(t) = - p\underline{\underline{I}}+2(\eta_s+\eta_p)\underline{\underline{D}}(t)+\eta_p \int_{-\infty}^{t} \bar{\mu}(t-t')\frac{D^2\underline{\underline{G}}}{Dt'^2}\, dt' \qquad (1)$$

where $\underline{\underline{G}}$ is a deformation history:

$$\underline{\underline{G}}(t,t')=(g^{\alpha\beta}(t')\frac{\partial x^i(t)}{\partial x^\alpha(t')}\frac{\partial x^j(t)}{\partial x^\beta(t')} - g^{ij}(t))\, \underline{e}_i\underline{e}_j$$

$\underline{\underline{D}}(t)$ is rate of deformation,

η_s is solvent viscosity,

η_p is additive polymer viscosity,

$$\bar{\mu}(t-t')=\sum_{\nu=1}^{n} t_\nu \frac{e^{-\frac{t-t'}{t_\nu}}}{\sum_{\nu=1}^{n} t_\nu}$$ is a special relaxation function

and t_ν is a discret spectrum of relaxation times describing a coil. A more advanced model takes into account an additional dependence of the function $\bar{\mu}$ on $\underline{\underline{G}}$.

In equation (1) we see that we have in addition to a quasi Newtonian stress an integral part which is especially important in case of dynamic deformation, for example in a turbulent pipe flow.

BALANCE OF MOTION AND MECHANICAL ENERGY

Using eq. (1) in case of stationary turbulent pipe flow the equation of motion gives

$$\tau_w \cdot \frac{r}{R} = \eta \frac{d\bar{v}_z}{dr} - \rho \overline{v_z' v_r'} + \eta_p \overline{\int_{-\infty}^{t} \bar{\mu}(t-t')(\ddot{\underline{\underline{G}}})_{rz}(t,t')dt'} \qquad (2)$$

with $\bar{\underline{v}} = \frac{1}{T} \int_{t-T}^{t} \underline{v}(t')dt'$; $\underline{v} = \bar{\underline{v}} + \underline{v}'$ etc.

It should be remarked that $\eta = (\eta_s + \eta_p)$ is the total shear viscosity of the solution.
For dilute solutions η has a shear independent value as predicted by the coil-model.
By dividing with wall shear stress we obtain

$$\tau^+ = \tau_v^+ + \tau_t^+ + \tau_M^+ \qquad (3)$$

with: $\tau^+ = -\tau_w \cdot \frac{r}{R} / \rho u_*^2$; $\tau_v^+ = \frac{dv^+}{dy^+}$; $\tau_t^+ = \frac{\rho \overline{v_z' v_r'}}{\rho u_*^2}$

$$\tau_M^+ = -\frac{\eta_p}{\rho u_*^2} \overline{\int_{-\infty}^{t} \bar{\mu}(t-t')(\ddot{\underline{\underline{G}}})_{rz}(t,t')dt'} ,$$

$$v^+ = \frac{\bar{v}_z}{u_*} ; \quad y^+ = \frac{(R-r)u_* \rho}{\eta} ; \quad \rho u_*^2 = \tau_w$$

In case of variable shear viscosity a representative value is used.
A further insight into turbulence gives the balance of mechanical energy.

In turbulence there are two ways balancing mechanical energy; in the first way one balances mechanical energy of the time averaged velocity and in the other way the mechanical energy of velocity of fluctuations. This leads first to the following relation:

$$- \overline{v_r' \, v_z'} \frac{d\bar{v}_z}{dr} + \frac{\eta}{\varrho} \left(\frac{d\bar{v}_z}{dr}\right)^2 + \frac{d\bar{v}_z}{dr} \, (\underline{\underline{\sigma}}_M)_{rz} = \frac{d\bar{v}_z}{dr} \frac{r}{2\varrho} \frac{\partial \bar{p}}{\partial z} \qquad (4)$$

with: $\underline{\underline{\sigma}}_M = \eta_p \int_{-\infty}^{t} \bar{\mu}(t-t') \underline{\underline{\ddot{G}}}(t,t')dt'$

Multiplication with $\eta / \varrho u_*^4$ leads to the dimensionless form

$$\underbrace{\frac{dv^+}{dy^+} \cdot \tau_t^+}_{T^+} + \underbrace{\left(\frac{dv^+}{dy^+}\right)^2}_{E^+} + \underbrace{\frac{dv^+}{dy^+} \tau_M^+}_{M^+} = \underbrace{\frac{dv^+}{dy^+} \tau^+}_{Z^+} \qquad (5)$$

The second way leads to

$$\overline{v_r' \, v_z'} \frac{\partial \bar{v}_z}{\partial r} + \varepsilon + \varepsilon_M$$

$$+ \frac{1}{r} \frac{\partial}{\partial r} \left[r \overline{\left(\frac{q^2}{2} + \frac{p'}{\varrho}\right) v_r'} - \frac{\eta}{2\varrho} \frac{\partial \overline{q^2}}{\partial r} - \frac{\eta}{\varrho} \frac{1}{r} \frac{\partial}{\partial r} (r \overline{v_r'^2}) - \frac{1}{\varrho} \left\{ (\overline{\sigma_{M_{rr}} v_r'}) \right.\right.$$

$$\left.\left. + (\overline{\sigma_{M_{r\theta}} v_\theta'}) + (\overline{\sigma_{M_{r\theta}} v_\theta'}) \right\} \right] = 0 \qquad (6)$$

ε : turbulent dissipation
ε_M: turbulent dissipation cause of elastic properties

In total we have three additional energy parts, resulting form elastic fluid proportions.

EXPERIMENTAL INVESTIGATIONS

The theory of dilute polymer solutions in turbulent pipe flow represented here shows the existence of additional terms in equation of motion and balance of energy which exceed the classical theory of turbulence.

We have performed measurements of time averaged velocity and velocity fluctuations in various directions with a Laser Doppler Anemometer to prove these terms in a quantitative way.
Measuring Reynolds-stress, the stress resulting from elastic fluid properties in equation of motion could be obtained as the difference between total shear stress and the sum of viscous stress and Reynolds-stress. Time series analysises have been performed on the velocity signals to get an insight in the small structure of turbulence. Some of the experimental results are shown in the next pictures.

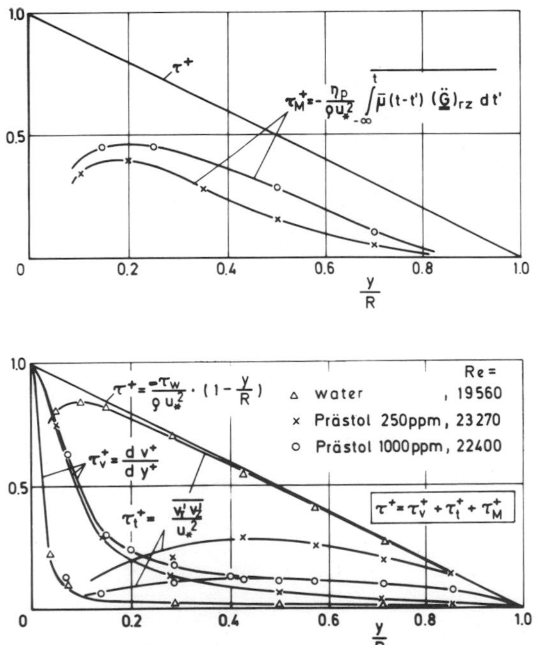

Fig. 1 Stresses in Turbulent Pipe Flow

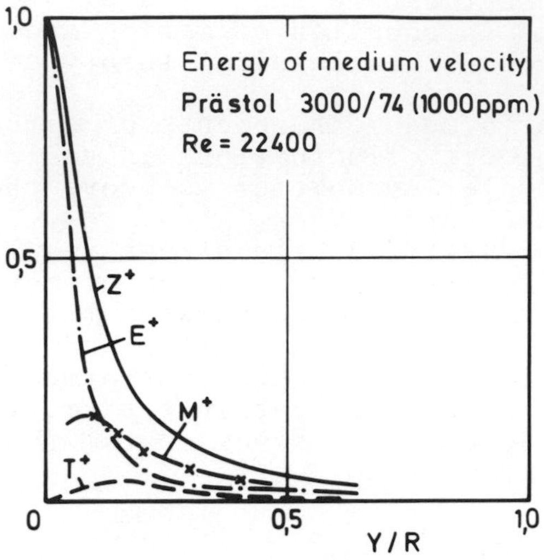

Fig. 2 Mechanical Energy of Time Averaged Velocity Over Tube Radius (from equ. 5)

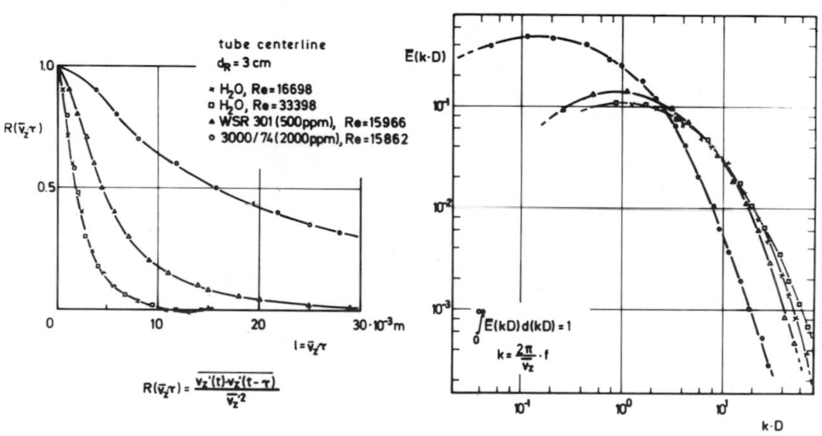

Fig. 3 Correlation Functions and Dimensionless Energy Spectra

In fig. 1 we see that elastic fluid properties play a dominant part in stress balance, rising with increasing viscoelastic fluid properties.
Fig. 2 results from fig. 1.
In fig. 3 we see some corresponding dimensionless energy spectra and correlation functions.

For water we have a constant integral scale of longitudinal velocity fluctions. From correlation functions and energy spectra we see that large scale fluctuations become greater at the cost of small scales which are damped starting with length scales which can be correlated with rheological fluid properties.
We suggest that this damping effect results from turbulent dissipation due to elastic fluid properties obtainend in balance of energy.

VELOCITY FIELD IN AN ELONGATIONAL POLYMER SOLUTION FLOW

Lyazid, A., Scrivener, O., Teitgen, R.

Institut de Mécanique des Fluides - ERA CNRS
2, rue Boussingault, F - 67083 Strasbourg Cedex

INTRODUCTION

The experimental evidence has shown that the presence of polymer molecules in a fluid flow has an influence on its characteristics. It becomes more and more clear that a phenomenon, like friction reduction, can be related either to the energy storage abilities of the molecules or to the existence of an extensional viscosity. But, until now, the conditions of extension of the molecules are poorly understood.

De Gennes[1] has studied theoretically the change in conformation that can be produced by the sharp acceleration in an extensional flow field. His work, and the results of other authors, has shown that the uncoiling of the molecular chains can only occur when the elongational shear stress is high enough and if the macromolecule remains a sufficient time (so called : persistence time) in that shear stress.

The purpose of our investigation is to study the effect of the macromolecules on the velocity field and gradient of an elongational flow so as to try to relate the appearance of extension, as visualized by flow birefringence, to the characteristics of the flow. Observations on the conditions of appearance of localized flow birefringence in cells of different shapes for several polymer-solvant pairs and concentrations obtained for a joined research programme with the University of Metz and reported in previous papers[2,3,4]. Velocity profiles of abnormalous shape and measurement of birefringence parameters are also reported in[5]. In a parallel paper, further measurements of birefringence in similar experimental conditions are reported by Cressely and Hocquart[6].

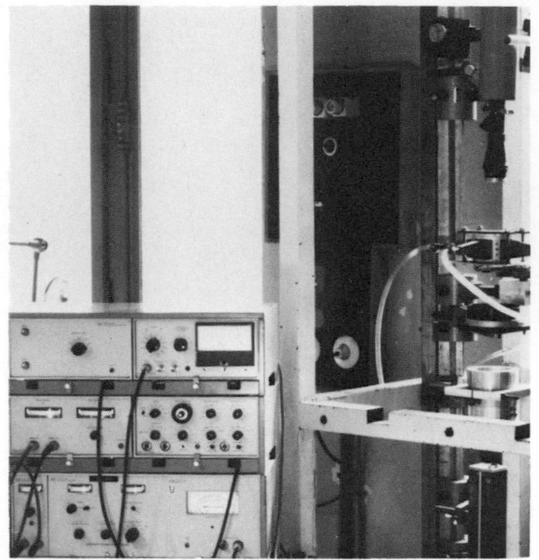

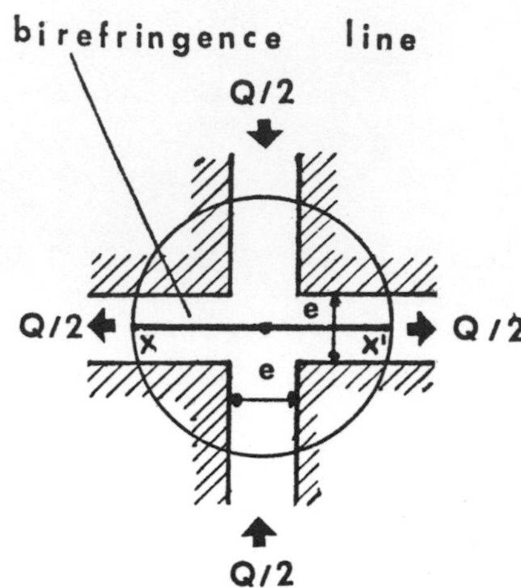

Fig. 1. Experimental set-up and "cross-flow" cell.

In this paper, we will present the concentration and flow regime conditions of appearance of abnormalous velocity profiles in a "cross flow cell", which was designed for the joined research programme with the University of Metz.

EXPERIMENTAL FACILITIES

The cell

The elongational flow is performed in the gap of a cell, termed "cross flow cell". This device consists of a chamber of 30 mm depth which cover and base are equiped with stress free glasses. Four right angle corner pieces are placed in the chamber, leaving flow passages of 2 mm width arranged as a cross (Fig. 1.). At the ends of the flow passages, four slits are provided in the walls of the chamber, each with a profiled inlet. Two opposite slits are used as fluid inlets and the two others as outlets.

Flow in the cell is laminar ; and the pathlines were visualized by particle tracers. At high flow rates appearance of instabilities was observed in the outlet channels.

With polymer solutions a birefringence line was observed on the axis of the outlet channels starting at the zero velocity central point[4]. This line was found to increase in length and width when increasing the flow rate.

Velocity measurement system

The velocities were measured in the gap of the cell by Laser Doppler Anemometry. The system is used in forward scatter fringe mode. The measuring volume is kept fixed and the cell moved along two orthogonal axes by a translation stage. The location of the measuring point in the cell was determined optically by reference to the positions of the corners and the walls. Due to the small width of the gap only the velocity component parallel to the axis was measured. For each profile 20 velocity measurements were taken in the 2 mm gap.

The Doppler Frequency, proportional to the velocity, was measured by use of a DISA Frequency Tracer (55 L 20 or 55 N 20 with more accuracy). The results were computed and plotted with an Hewlett Packard computing system.

Due to finite non negligible dimensions of the measuring volume corrections are generally necessary in the wall region. These were not taken in account in the present work. The difference between computed and measured flow rate was found to be less than 1 per cent.

The measurements were performed with water, glycerol added water (60 % of glycerol in volume) and PEO 301 of 0.1, 0.2 or 0.4 % in concentration.

The fluid had to be seeded by tracer particles and we chose particles of 16 mμ of diameter. The very low quantity needed was insufficient to change the rheological properties of the solutions.

EXPERIMENTAL RESULTS

Water and Glycerol water mixture

In both cases, the distribution of the velocities is closed to a laminar profile in the inlet channels. At the same flow rate the profiles are similar showing that glycerol has no effect on the rheological behaviour of the solution.

In the outlet channels the same shape is found after rearrangement of the flow. As expected, addition of glycerol produces an enlargement of the entrance length due to the higher viscosity.

PEO solutions in the inlet channels

In the inlet channels the velocity profiles the difference with the solvant one is not significant. The shape is closed to the laminar profile even at high polymer concentrations and flow rate (Fig. 2.) showing that the viscoelastic behaviour of the solution

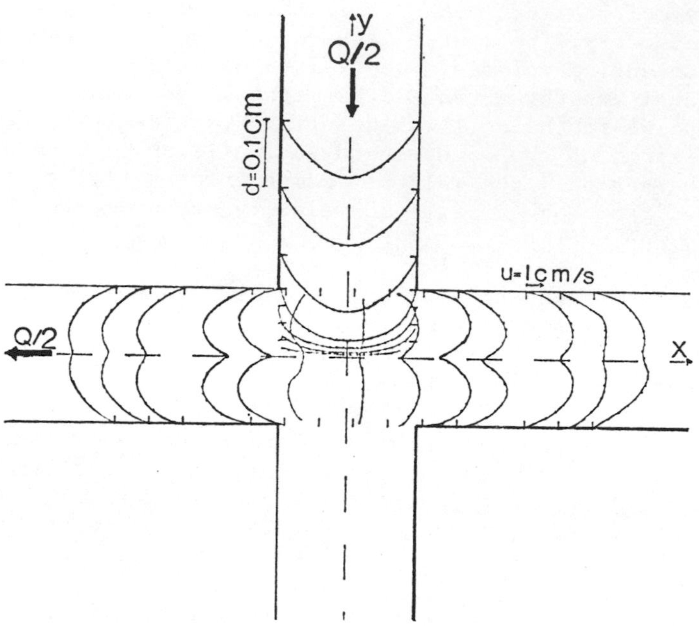

Fig. 2. Velocity profiles in the inlet and outlet channels : PEO WSR 301 ; C = 0.4 % ; Q = 2.63 cm^3/s.

have no significant effect on a laminar flow. In the central part of gap the profiles are gradually flattened until the zero velocity on the xx' axis.

PEO solutions in the outlet channels

The results are different from those generally admitted for the newtonian flow. As shown on the fig. 2 the profiles have an unexpected shape. Even in the center of the gap the profile presents a minimum on the axis and two maxima on both sides.

At the lower concentrations (.1 %) the minimum of velocity is only slightly visible in the center part of the gap and the profile rearranges itself at a short distance from the corners in the channels (fig. 3) . The same results are obtained at higher concentrations but only for low flow rates.

On the other hand, for higher concentrations (.2 and .4 %) the minimum value remains visible in the outlet channels when the flow rate is high enough (fig. 2 and 4) and higher the flow rate is longer the distance needed by the profile to rearrange, probably

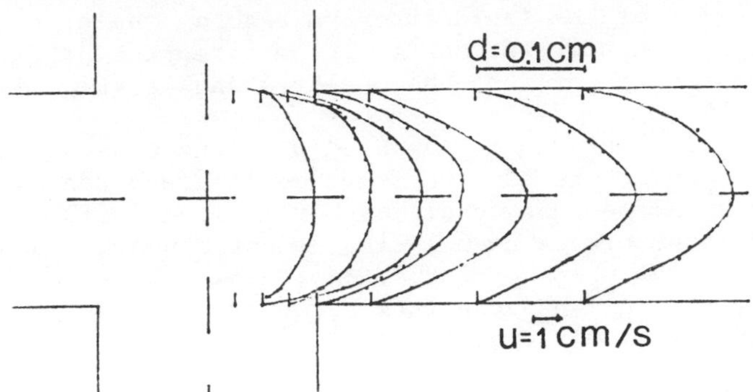

Fig. 3. Velocity profiles in the outlet channel at low concentrations PEO WSR 301 ; C = .1% ; Q = 4.51 cm^3/s.

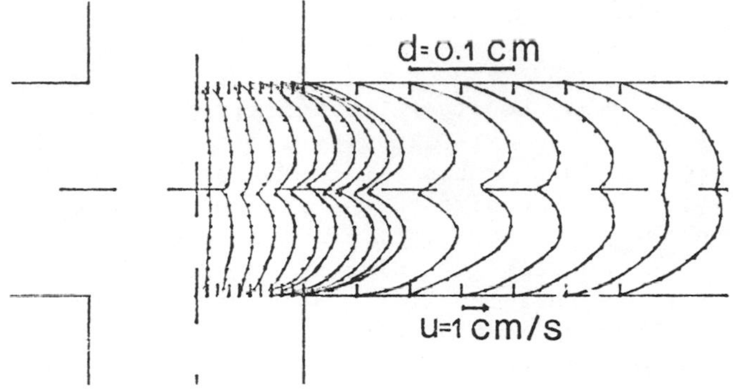

Fig. 4. Velocity profiles in the outlet channels at high concentrations. PEO WSR 301 ; C = .2 % ; Q = 3.36 cm^3/s

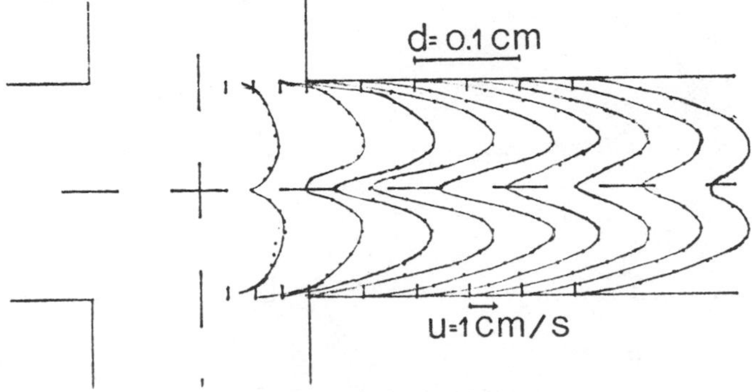

Fig. 5. Velocity profiles in the outlet channels at high concentrations and flow rates.
PEO WSR 301 ; C = .4 % ; Q = 6.02 cm^3/s

due to the memory of the fluid after release of the elongational gradient. At high concentrations and flow rates the profiles are not rearranged at the exit of the outlet channels (Fig. 5).

These results have to be compared with the observations of the conditions of appearance of birefringence. As reported by Cressely and Hocquart[6], in the same conditions of flow and with the same polymer sample the birefringence line do not appear, independently from the flow rate, when concentration is lower than . 1 % . It proves that a strong linkage exists between extension of the macromolecules and the flow field.

Evolution of the velocities at different distances from the axis

From the previous measurements we have plotted the evolution of the longitudinal velocity component along the channels and hence its gradient at different distances from the axis.

It can be seen that (Fig. 6), in the central part of the gap the gradient is roughly constant for different distances from the axis in exception to the axis itself where the gradient is continuously decreasing. In the channels themselves the velocities reach a constant level corresponding to the rearrangement of the profile. This rearrangement comes more from a decrease of the two maxima than of the slight increase of velocity on the axis. It seems to be important to note that the little difference between the velocity gradient in the central part of the gap is not sufficient to explain the location of birefringence on the axis and its vicinity.

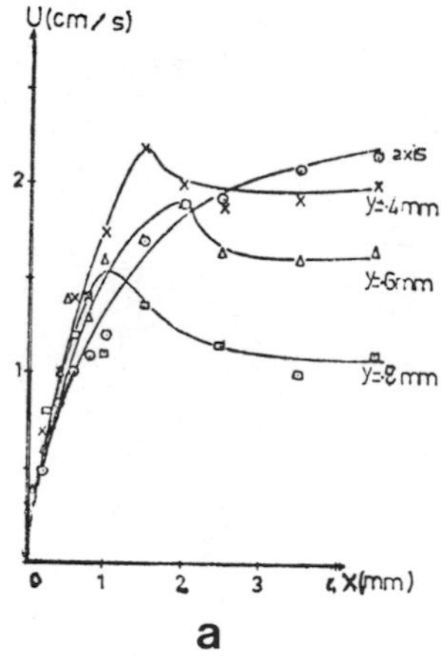

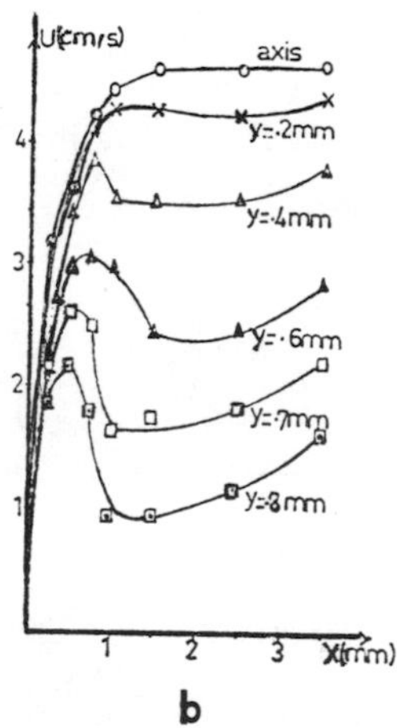

Fig. 6. Evolution of the velocities at different distances from the axis in the outlet channels.
a) PEO WSR 301 C = .4 % Q = 1.75 cm^3/s
b) PEO WSR 301 C = .1 % Q = 3.34 cm^3/s

CONCLUSION

Velocity profile and birefringence measurement in an elongational flow suggest that the condition of high velocity gradient is not enough to explain the extension of the macromolecules. A second parameter must be introduced with the persistence time. A relation certainly exists between the flow field and the birefringence intensity which decrease has to be related to the rearrangement of the profile. A more complete knowledge of the velocity field is necessary before making any assumption on the conditions of extension of the macromolecular chains. The observed shape of the profile and the evolution of the gradient can be attributed either to the viscoelastic properties of the solutions or to the existence of extensional viscosity both linked to the elastic properties of the macromolecule. An attempt was made to compute the flow field[4] by solving by a finite difference method the equations of the movement. But due to the use of a power law model as stress strain relation the experimental results were not confirmed by the computation. A new attempt is presently made in using a 3 parameter Oldroyd model more accurate for taking in account the elastic properties of the solutions. Calculations made by Perera[7] in a T shape cell have show an effect of the elastic properties of the solution on the velocity profiles similar to those we measured in the cross flow cell.

ACKNOWLEDGEMENT

This work was partially sponsored by the CNRS. We would like to thank J.M. Krempff of the IMFS for his help with the computing of the experimental results.

REFERENCES

1. P. G. De Gennes, Coil-stretch transition of dilute flexible polymers, Chem. Phys. 60:5030 (1974).
2. R. Cressely, R. Hocquart and O. Scrivener, Birefringence d'écoulement dans un dispositif à deux rouleaux. Optica Acta 25:559 (1978).
3. R. Cressely, R. Hocquart and O. Scrivener, Lignes de biréfringence localisée dans un dispositif à deux rouleaux tournant en sens contraire. Optica Acta 26:1173 (1979).
4. O. Scrivener, C. Berner, R. Cressely, R. Hocquart, R. Sellin and N. S. Vlachos, Dynamical behaviour of drag-reducing polymer solutions, J. of N.N. Fluid Mech. 5:475 (1979).
5. R. Cressely, R. Hocquart, A. Lyazid and O. Scrivener, Study of the dynamics of macromolecules in extensional flow, Proc. IUTAM Symp. Makromainz, Mainz, (1979).
6. R. Cressely and R. Hocquart, Dynamics of flexible and large macromolecules in elongational flow, Proc. VIII Int. Cong. Rheol., Naples, (1980).

7. M. G. N. Perera and K. Walters, Long range effects in flows involving abrupt changes in geometry, J; of N.N. Fluid Mech. 2:49 (1977).

DETERMINATION OF CHARACTERISTIC FUNCTIONS OF A VISCOELASTIC LIQUID IN A NON-VISCOMETRIC FLOW

J. R. Clermont, P. Le Roy, J. M. Pierrard

Institut de Mécanique de Grenoble
Laboratoire Associé au CNRS
France

INTRODUCTION

The purpose of the present work is the numerical determination of characteristic functions of a viscoelastic fluid by kinematic data. These are obtained from the axisymmetrical flow in a convergent, between two Poiseuille flows, with a velocity field of the form :

$$\vec{V}(\rho, z) = u(\rho, z)\, \vec{e}'_\rho + v(\rho, z)\, \vec{e}'_z \quad (1)$$

in the orthonormal frame $(\vec{e}'_\rho, \vec{e}'_\theta, \vec{e}'_z)$ associated to the cylindrical coordinates $\rho = 1$, $\theta = 2$, $z = 3$. In this frame, the components of the strain-rate tensor $\underset{\sim}{\mathcal{D}}$ are given by the following matrix :

$$\mathcal{D} = \begin{bmatrix} \frac{\partial u}{\partial \rho} & 0 & \frac{1}{2}\left(\frac{\partial v}{\partial \rho} + \frac{\partial u}{\partial z}\right) \\ 0 & \frac{u}{\rho} & 0 \\ \frac{1}{2}\left(\frac{\partial v}{\partial \rho} + \frac{\partial u}{\partial z}\right) & 0 & \frac{\partial v}{\partial z} \end{bmatrix} \quad (2)$$

If B_1, B_2, B_3, denote the three invariants of $\mathcal{D}$, we have : $B_1 = \sum_i \mathcal{D}_{ii} = 0$ (incompressibility), $B_2 = \sum_i \sum_h \mathcal{D}^{ih} \mathcal{D}^{hi}$

$$B_3 = \sum_i \sum_j \sum_k \mathcal{D}^{ih} \mathcal{D}^{hj} \mathcal{D}^{ji} \quad (3)$$

At the wall where the fluid adheres, B_3 is zero, which is a property of viscometric flows. Every where else, $B_3 \neq 0$, and near the axis the flow is elongational. The investigation of such a flow gives us the possibility to obtain information which cannot be available from viscometric experiments generally performed in lab, while the flow in a convergent is closely related to industrial processes.

EXPERIMENTAL DATA OF THE CONVERGING FLOW

Experiments have been performed with a test fluid of 2000 poises (solution of 50 % by weight of PIB in mineral oil). The velocity components u and v were obtained by means of a laser anemometry technique. The pressure drops related to the upstream and downstream fully developed regions were also determined. Details concerning the experimental techniques are given in (1). In the same paper, we presented a data smoothing method for the calculation of the kinematic field, the incompressibility condition being verified.

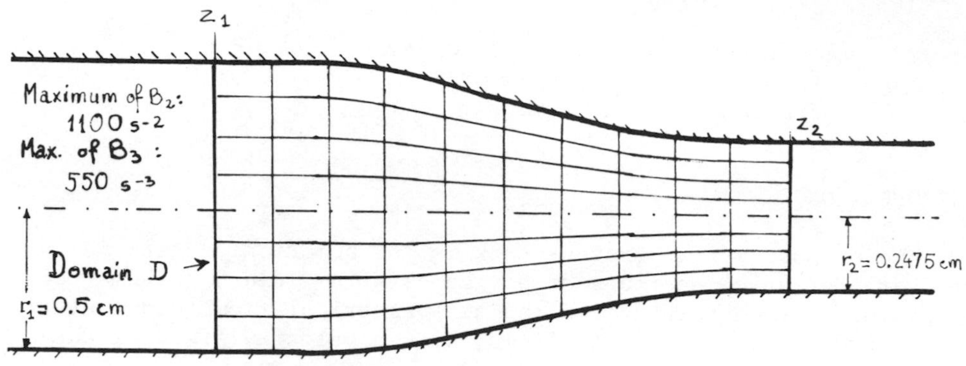

Fig. 1. The converging region.

One of the main results of this previous work was that for a non-viscometric flow of a viscoelastic liquid, the third invariant has to be taken into account. Consequently, any rheological model who wants to describe complex (non-viscometric) flows must involve characteristic functions of B_2 and B_3. This point of view concerns Le Roy-Pierrard model (LR-P).

THE COROTATIONAL LR-P MODEL

In the LR-P model, the deviatoric stress tensor $\underset{\sim}{S}$ is given by the following equation :

$$\underset{\sim}{S} = \underset{\sim}{S}_{(1)} + \underset{\sim}{S}_{(2)}$$

with

$$S^{*ij}_{(1)}(t) = \int_{-\infty}^{t} \Psi_1(t-\tau, B_2, B_3) . E_1(\tau) d\tau \tag{4}$$

$$S^{*ij}_{(2)}(t) = \int_{-\infty}^{t} \Psi_2(t-\tau, B_2, B_3) . E_2(\tau) d\tau$$

S^{*ij} are the components of $\underset{\sim}{S}$ in the corotational frame at time t.

$\underset{\sim}{E}_1$ and $\underset{\sim}{E}_2$* (2) are combinations of kinematic tensors related to

---

* $\underset{\sim}{E}_1 = \underset{\sim}{D}$, $\underset{\sim}{E}_2 = B_3 \underset{\sim}{D} - B_2 \underset{\sim}{D}\underset{\sim}{D} + \frac{1}{3}(B_2)^2 \underset{\sim}{I}$

the flow.

Ψ_1 and Ψ_2 are kinematic functions of the model :

$$\Psi_1(t-\tau, B_2, B_3) = \sum_m a_m \exp[-\lambda_m(t-\tau)].\hat{\Phi}(B_2,B_3) \quad (5)$$

and
$$\Psi_2(t-\tau, B_2, B_3) = \sum_m b_m(B_2)\exp[-\lambda_m(t-\tau)].\hat{\Phi}(B_2,B_3)$$

where a_m, λ_m are positive constants, the functions $b_m(B_2)$ are such that the generally accepted assumption $N_2 = -0.1\ N_1$ is verified, N_1 and N_2 denoting the normal stress functions in a viscometric flow.

The quantities a_m, λ_m , $b_m(B_2)$ are known from viscometric experiments. Although the function $\hat{\Phi}(B_2,B_3)$ is unknown, $\hat{\Phi}(B_2,0)$ was calculated from Poiseuille experiments, on the form (Fig. 2):

$$\hat{\Phi}(B_2,0) \simeq 1 + \alpha_1 (B_2)^{1/2} + \alpha_2 B_2 + \alpha_3 (B_2)^{3/2} \quad (6)$$

for $B_2 \leq 8 \times 10^3\ s^{-2}$

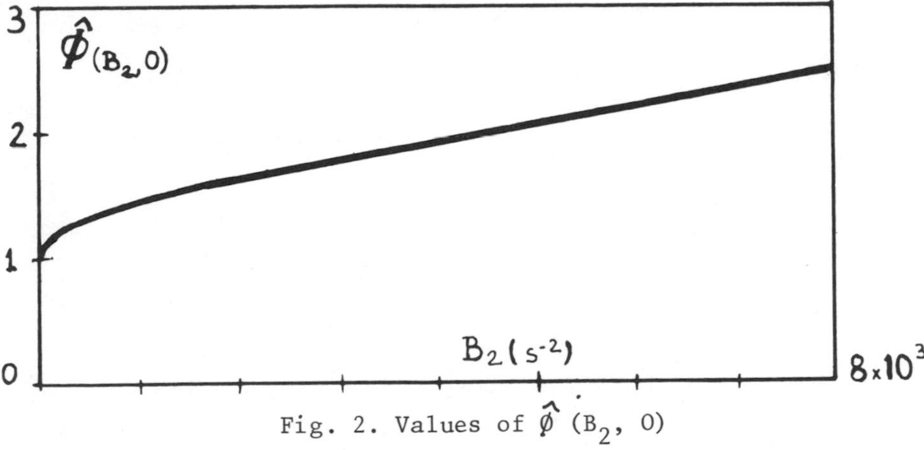

Fig. 2. Values of $\hat{\phi}$ $(B_2, 0)$

DETERMINATION OF THE FUNCTION $\hat{\phi}$ (B_2, B_3)

Position of the problem

To solve the problem of $\hat{\phi}$ determination in non-viscometric flows, we consider the domain D of Fig. 1 between the last upstream Poiseuille section $z = z_1$ and the first downstream Poiseuille section $z = z_2$. The bounds z_1 and z_2 are obtained experimentally from the velocity profiles v and, as precised in the foregoing, the parameters of the model are completely known in the viscometric case (B_3 = 0). Then, using radial stress data of the two Poiseuille regions, the isotropic pressure p and the deviatoric stress tensor $\underset{\sim}{S}$ are calculated for the sections z_1 and z_2 . The total stress tensor is :

$$\underset{\sim}{\sigma} = -p\underset{\sim}{I} + \underset{\sim}{S} \tag{7}$$

In the orthonormal frame ($\vec{e}'_\rho$, $\vec{e}'_\theta$, $\vec{e}'_z$) the components of $\underset{\sim}{S}$ for the LR-P model take the form :

$$S^{kl}_{(t)} = \sum_m \int_{-\infty}^{t} \hat{\Phi}(B_2,B_3)\,c_m(B_2,\tau)\exp[-\lambda_m(t-\tau)]\left\{E^{kl}_{(\tau)}\right\}d\tau \tag{8}$$

where c_m is written in terms of a_m and $b_m(B_2)$. $\{E^{kl}\}$ denotes a set of kinematic functions of the invariants B_2 , B_3 and the quantities involved by the rate of deformation $\underset{\sim}{D}$ and the vorticity tensor $\underset{\sim}{R}$

In the present problem, the unknowns are $\underset{\sim}{S}$ and p , or $\hat{\phi}$ and p and the boundary conditions are the followings :
$\hat{\phi}$ is known at the wall ($B_3 = 0$) and at the sections z_1 and z_2

p is known at $z = z_1$ and $z = z_2$, unknown at the wall.

Linearization procedure

The determination of $\hat{\phi}$ is carried out by writing the dynamical equations. Since $R_e \sim 10^{-3}$, the inertial terms become negligible.

To begin with, the set of functions $\hat{\phi}$ and p is discretized on the irregular grid of the streamlines and sections z = constant. To compute the stress at any point of a streamline (i), we use a Lagrangian description.

Equation (8) implies the calculation of integrals of the type :

$$\begin{aligned} I_1 &= \sum_m \int_{-\infty}^{t} \hat{\Phi}(B_2,B_3)\exp[-\lambda_m(t-\tau)]\,K(\tau)\,d\tau \\ I_2 &= \sum_m \int_{-\infty}^{t} \hat{\Phi}(B_2,B_3)\exp[-\lambda_m(t-\tau)]\,F(\tau)\cos A(t,\tau)\,d\tau \\ I_3 &= \sum_m \int_{-\infty}^{t} \hat{\Phi}(B_2,B_3)\exp[-\lambda_m(t-\tau)]\,G(\tau)\sin A(t,\tau)\,d\tau \end{aligned} \tag{9}$$

$A(t,\tau)$ is the integral of the rotation term which can be expressed as : $A(t,\tau) = a(t) - a(\tau)$.

A linear interpolation of $\hat{\phi}$ between two successive points $j-1$ and j leads to the following equations :

$$\hat{\phi}(\tau) = \hat{\phi}^{(i)}_{j-1} + \frac{\hat{\phi}^{(i)}_j - \hat{\phi}^{(i)}_{j-1}}{t_j - t_{j-1}}(\tau - t_{j-1}) \tag{10}$$

$$\begin{aligned} \int_{t_{j-1}}^{t_j} H(\tau).\hat{\Phi}(\tau)\,d\tau = {} & \hat{\Phi}^{(i)}_{j-1}\left\{\int_{t_{j-1}}^{t_j} H(\tau)\,d\tau - \frac{1}{t_j - t_{j-1}}\int_{t_{j-1}}^{t_j} H(\tau)\cdot(\tau - t_{j-1})\,d\tau\right\} \\ & + \hat{\Phi}^{(i)}_j\left\{\frac{1}{t_j - t_{j-1}}\int_{t_{j-1}}^{t_j} H(\tau).(\tau - t_{j-1})\,d\tau\right\} \end{aligned} \tag{11}$$

and the deviatoric stress $\underset{\sim}{S}$ is written in the linear form :

$$S^{(i)}_j = \sum_k A^{(i)}_{jk}\,\hat{\Phi}^{(i)}_k \tag{12}$$

Since the stress at point j of the streamline (i) is a function of the times τ before the present time t_j, $A^{(i)}_{jk}$ is a triangular matrix.

For the line (i), the time t = 0 corresponds to the position $z = z_1$ of the fluid particle. The functions $\{E^{k\ell}\}$ being expressed into polynomial forms of τ, the matrix $A^{(i)}$ is determined by equations (8), (9) and (11) and the use of finite differences for the derivatives leads to a linear system of $\hat{\phi}^{(i)}_j$ and $p^{(i)}_j$.

Dynamical equations

By the change of variables $(\rho, z) \longmapsto (\rho_0, z_0)$:

$$\left.\begin{aligned} \rho &= f(\rho_0, z_0) \\ z &= z_0 \end{aligned}\right\} \tag{13}$$

where ρ_0 denotes the radial position of the particle at t = 0, the domain $\mathcal{D}$ is mapped into a rectangular domain $\mathcal{D}'$ and the dynamical equations are written as follows :

$$\frac{\partial \Pi}{\partial \rho_0} = \frac{\partial}{\partial \rho_0}(S^{11} - S^{33}) - f'_{\rho_0} \cdot \frac{\partial S^{13}}{\partial \rho_0} + f'_{\rho_0} \cdot \frac{\partial S^{13}}{\partial z_0} + \frac{f'_{\rho_0}}{f}(S^{11} - S^{22}) \tag{14}$$

$$\frac{\partial \Pi}{\partial z_0} = \frac{f'_{z_0}}{f'_{\rho_0}} \frac{\partial}{\partial \rho_0}(S^{11} - S^{33}) + \frac{1 - f'_{z_0}}{f'_{\rho_0}} \cdot \frac{\partial S^{13}}{\partial \rho_0} + f'_{z_0} \cdot \frac{\partial S^{13}}{\partial z_0} + \frac{f'_{z_0}}{f}(S^{11} - S^{22}) + \frac{1}{f} S^{13} \tag{15}$$

where $\Pi = p - S^{33}$

For $\rho_0 = 0$, $f'_{z_0} = 0$, $S^{13} = 0$ and $S^{11} - S^{22} = 0$; equation (14) vanishes and equation (15) takes the form :

$$\frac{\partial \Pi}{\partial z_0} = \frac{1}{f'_{\rho_0}} \cdot \frac{\partial S^{13}}{\partial \rho_0} + \frac{1}{f} S^{13} \tag{16}$$

Delicate points of the resolution are the vicinity of $\rho_0 = 0$ and errors arising from numerical derivation.

For $\rho_0 = 0$ (centre line (i_0)), the kinematic terms are written in terms of ρ_0 such that :

$$\left(\frac{\partial S^{13}}{\partial \rho_0}\right)^{(i_0)}_j \simeq \left(\frac{S^{13}}{\rho_0}\right)^{(i_0)}_j \simeq \sum_k B^{(i_0)}_{jk} \hat{\phi}^{(i_0)}_k \tag{17}$$

On the other hand, in the variable system (ρ_0, z_0) we have :

$$\frac{\partial S^{13}}{\partial z_0} = \left(\frac{1}{v}\right) \cdot \frac{dS^{13}}{dt} \tag{18}$$

The derivative dS^{13}/dt can be expressed in the same form than equation (12). Then, the right-hand side of equations (14) and (15) involves only derivatives of the variable ρ_0 .

The curves $\hat{\phi}_{B_2}(B_3)$ will be presented at the Congress.

REFERENCES

1. J.R. Clermont, P. Le Roy, J.M. Pierrard : "Theoretical and experimental study of a viscoelastic fluid in the converging region of a pipe" : Jnl of NNFM 5, (1979) 387-408.
2. P. Le Roy, J.M. Pierrard : "Fluides viscoélastiques non linéaires satisfaisant à un principe de superposition. Etude théorique et expérimentale" : Rheol. Acta, 12, 449-454 (1973).
3. J.R. Clermont, P. Le Roy, J.M. Pierrard and S. Zahorski : "Ecoulement non viscométrique de fluides non Newtoniens", Symposium Franco-Polonais, Krakow, June 1977, (Problèmes non linéaires de Mécanique), Scientific Editions of Poland, in press.

THE WALL EFFECT IN ORTHOGONAL STAGNATION FLOW

T. Hsu, P. Shirodkar, R. L. Laurence, and H. H. Winter

Department of Chemical Engineering
Goessmann Laboratory
University of Massachusetts
Amherst, MA 01003

1. INTRODUCTION

The kinematic boundary condition has a profound effect on the velocity field in a flow channel. This effect will be studied with respect to an extensional rheometer of planar extensional flow.

The need for the development of an extensional rheometer is most obvious in polymer engineering. The extensional properties of polymeric liquids determine the processing behavior, particularly in shaping operations with free surfaces (film blowing, blow molding), and the end use properties after solidification (orientational phenomena). Extensional rheometry has made great progress (see Petrie[1]), but the extensional properties can still be measured only in a limited region (high viscosity, low extension rates).

The kinematics of extensional flow are very well understood[2,3]. However, it is difficult to achieve this kinematics and simultaneously measure the stress in a flow apparatus. Such would be the requirements of an extensional rheometer. Winter, Macosko and Bennett[4] proposed stagnation flow in dies with lubricated walls as a possible approach. The shape of the die follows a stream surface of steady homogeneous extension. At the wall, a thin lubricant layer reduces the shear stress.

The tangent to a stream surface is not a principal plane of the desired extensional flow. It therefore is questionable whether it will be possible to achieve the necessary stress boundary condition, even with lubricated flow. In what follows, several kinematical boundary conditions and their effect on the flow will be studied. The analysis will concentrate on planar stagnation flow of Newtonian fluids.

2. PROBLEM DESCRIPTION

The geometry of planar stagnation flow is best described using hyperbolic cylindrical coordinates[5] ξ, μ given by

$$x_1 = (\rho + \mu)^{1/2} \quad ; \quad \rho = (x_1{}^2 + x_2{}^2) = (\xi^2 + \mu^2)^{1/2} \tag{1}$$

$$x_2 = (\rho - \mu)^{1/2} \quad ; \quad \nabla^2\psi = 2\rho \left(\frac{\partial^2\psi}{\partial\xi^2} + \frac{\partial^2\psi}{\partial\mu^2}\right) \tag{2}$$

$$x_3 = x_3 \quad ; \quad g_{\mu\mu} = g_{\xi\xi} = (2\rho)^{-1}$$

The walls of the extensional flow rheometer[4] correspond to a surface of constant ξ. If we consider the flow of a Newtonian fluid at low Reynolds number ($N_{Re} << 1$), the flow is described by the biharmonic equation

$$\nabla^4\psi = 0 \quad \text{or equivalently} \tag{3}$$

$$\nabla^2\psi = \rho \left(\frac{\partial^2\psi}{\partial\xi^2} + \frac{\partial^2\psi}{\partial\mu^2}\right) = -w \quad ; \quad \nabla^2 w = \rho \left(\frac{\partial^2 w}{\partial\xi^2} + \frac{\partial^2 w}{\partial\mu^2}\right) = 0 \ , \tag{4}$$

where w is the z-component of vorticity and ψ is the stream function.

Ideal planar stagnation flow is given by a stream function independent of μ (Berker[3]). It can be readily shown that $\psi = \psi(\xi)$ if and only if $w = 0$ everywhere, i.e., the system is vorticity free. The zero slip boundary condition offers, however, a source of vorticity and, therefore, the flow field will not conform to that of ideal planar stagnation flow.

The boundary conditions along the planes of symmetry ($\xi = 0$) and at the wall ($\xi = 1$) are

$$\left.\begin{array}{ll} \text{at } \xi = 0 & w = 0 \\ \xi = 1 & \dfrac{\partial\psi}{\partial\mu} = 0 \end{array}\right\} \text{all } \mu \tag{5}$$

At the entrance and exit, the flow field is prescribed. For example, if we have plug flow at $\mu = \alpha$, then possible solutions are

$$w = 0 \text{ and } w = \frac{A \sin q\xi \sin q(\mu - \alpha)}{\sin q\alpha}. \tag{6}$$

The eigenvalues q and the constant A are to be determined from the conditions at the rheometer boundary (slip, no slip, or lubricating fluid). Then, a solution for ψ can be obtained from a Green's function analysis of equation (4). The results of this analysis will be presented in a separate paper.

3. FINITE ELEMENT SIMULATION

The Finite Element Method (FEM) is chosen because of its success in handling irregularly shaped boundaries. The computer programs developed by Malone[6] were adapted to the problem of planar stagnation flow. The basic discretization unit is a six-node triangular element discussed by Nickell, et al[7]. The grid is shown in fig. 1. Three distinct boundary conditions are investigated using FEM.

Ideal Planar Stagnation Flow

In planar stagnation flow the shear and normal stress distributions along the wall are given by the expressions[4]

$$T_{nn} = \frac{x_2^2}{(x_1^2 + x_2^2)} (T_{11} - T_{22}) + T_{22} \tag{7}$$

$$T_{nt} = \frac{x_1 x_2}{(x_1^2 + x_2^2)} (T_{11} - T_{22}) \tag{8}$$

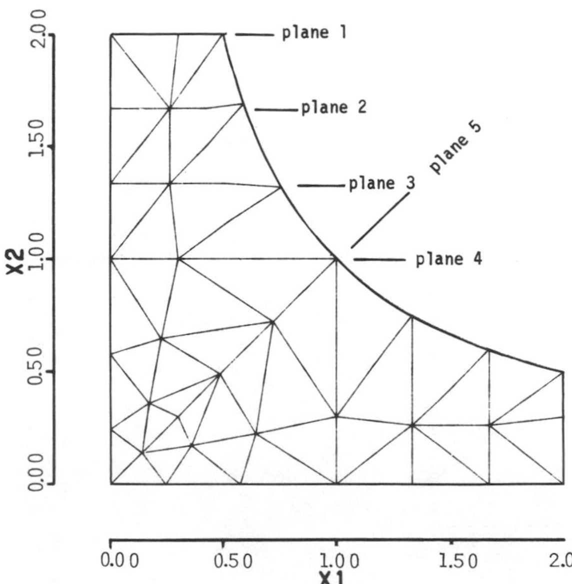

Fig. 1: The finite elements grid for the planar stagnation flow problem. Stress values on grid points in planes 1 to 5 are plotted in the figures below.

However, in the FEM simulation it is easier to utilize the kinematic conditions, namely

$$(\underline{v}) = (\dot{\varepsilon}x_1, -\dot{\varepsilon}x_2, 0), \text{ at wall.} \qquad (9)$$

The results of the numerical computation show excellent agreement with eqs. 7 and 8 (see fig. 2). Note that ideal planar stagnation flow is achieved by a non-zero shear stress distribution at the wall.

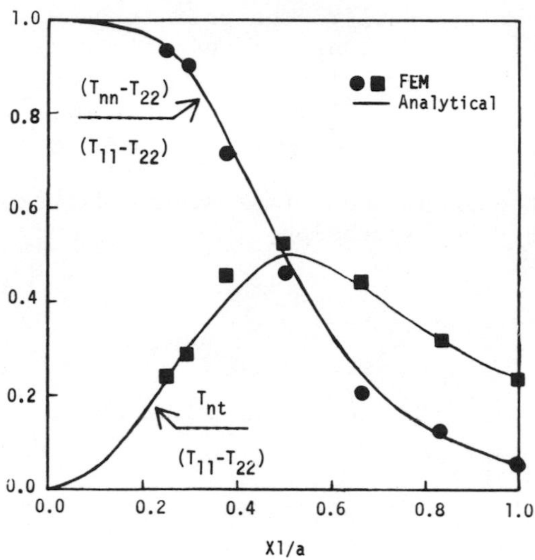

Fig. 2: Ideal planar stagnation flow: Normal and shear stress distribution along the stream surface. The coordinate x_1 is normalized by the die dimension a.

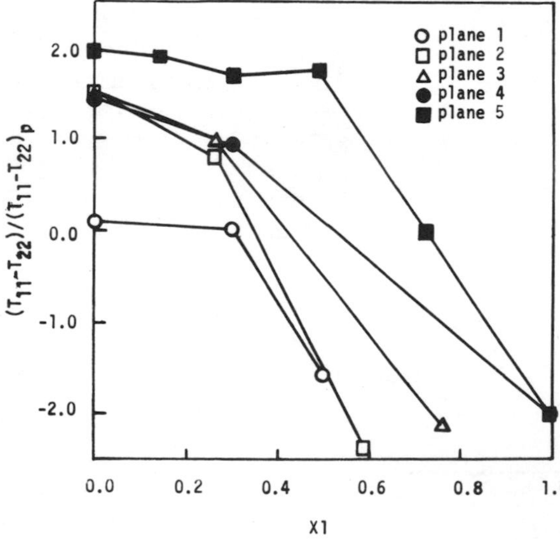

Fig. 3: Stress distribution (normalized with respect to ideal planar stagnation flow)--no slip at the wall.

Zero Slip at the Wall

In the absence of a lubricant, the fluid sticks to the wall. The results of the FEM simulation using the zero slip boundary condition

$$(\underline{v}) = (0,0,0), \text{ at wall} \tag{10}$$

are shown in fig. 3. The normal stress difference (normalized with respect to the planar stagnation case) is plotted as a function of the x_1 coordinate at five different planes (see fig. 1). Because we are dealing with a Newtonian fluid, the stress distribution also reflects the uniformity of the deformation field. The elongation rate at the centerline ($x_1 = 0$) increases as a fluid element approaches the stagnation point, with a region of fairly constant elongation rate in the middle. The local elongation rates near the stagnation point are significantly greater than those expected from ideal stagnation flow. The region of uniform extension rate is very small.

Free Surface

If one assumes that the stream surface is in fact a free surface, then the boundary conditions at the surface are

$$\underline{\underline{T}}\cdot\underline{n} = \underline{0} \quad \text{and} \quad \underline{v}\cdot\underline{n} = 0. \tag{11}$$

From fig. 4 it is obvious that the stress distribution is not uniform. Thus, stagnation flow with a free surface may be unsuitable

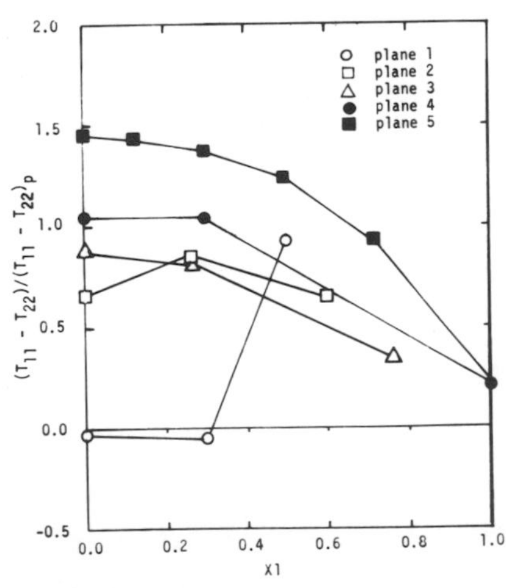

Fig. 4: Stress distribution (normalized with respect to the ideal case)--stream surface is a free surface.

to measure elongational viscosity because of the nonuniformity of extension rates, throughout the flow.

4. CONCLUSIONS

The boundary conditions of ideal planar stagnation flow appear to be impossible to achieve in an experiment. Possible boundary conditions are no-slip, free surface, and lubricated wall. Each of these boundary conditions causes the flow to deviate from ideal planar extension. For no-slip, there is a region near the stagnation point where the extension rate is constant; however, this region is very small. Planar stagnation flow with a free surface, which would resemble a spinning experiment, seems to be governed by non-uniform extension rates. The slip boundary condition currently is under investigation.

5. NOMENCLATURE

A	constant in equation (6)
$g_{\mu\mu}$, $g_{\xi\xi}$	the components of the metric tensor in hyperbolic coordinates
G	the Green's function
n	the direction normal to wall
q	the eigenvalue of the Laplacian operator in equation (4)
t	the direction tangential to wall
T_{ij}	the component of the stress tensor in Cartesian coordinates
$\underline{\underline{T}}$	the stress tensor
w	the vorticity
x_1, x_2, x_3	the Cartesian coordinates; x_3 is the neutral coordinate

Symbols

α	$\mu = \alpha$ at the entrance
δ	delta function
μ, ξ	the hyperbolic cylindrical coordinates
ρ	$= (\xi^2 + \nu^2)$
ψ	stream function

6. REFERENCES

1. C. J. S. Petrie, Elongational Flows, Pitman, London (1979).
2. H. Giesekus, Rheol. Acta. 2:101 (1962).
3. P. R. Berker, in "Handbuch der Physik," Band VIII, 2., pp. 1-384, S. Flügge, ed., Springer, Berlin (1963).
4. H. H. Winter, C. W. Macosko, and K. Bennett, Rheol. Acta 18: 323 (1979).
5. P. Moon and D. E. Spencer, Field Theory Handbook, Springer, Berlin (1961), p. 204.
6. M. F. Malone, Ph.D. Dissertation, Univ. Mass., Amherst (1979).
7. R. E. Nickell, R. I. Tanner, and B. Caswell, J. Fluid Mech. 65:189 (1976).

R H E O M E T R Y

NORMAL STRESS MEASUREMENTS IN LOW-ELASTICITY LIQUIDS

A.S.Lodge, T.H.Hou, P.P.Tong

University of Wisconsin, U.S.A.

(Abstract)

Recent results obtained for lubricating oils containing polymer additives will be described. Pressure distributions measured in a Truncated-Cone & Plate apparatus yield values for $N_3 = N_1 + 2N_2$. The SEISCOR/Lodge Stressmeter, a new slit-die apparatus, yields values for hole pressure P, wall shear stress σ, and wall shear rate s. According to the Higashitani-Pritchard-Baird theory, $N_1 - N_2 = 3 \ dP/d\sigma$. N_1 and N_2 denote the first and second normal stress differences in unidirectional shear flow.

IN-LINE ELASTICITY MEASUREMENT FOR MOLTEN POLYMERS

A.S.Lodge, L. de Vargas

University of Wisconsin, U.S.A.

(Abstract)

A new Melt Stressmeter has been constructed. This is a slit-die apparatus, having two flush-mounted and one slot-mounted pressure transducer, which yields values for hole pressure P, extrapolated exit pressure P_{xt} , wall shear stress σ, and wall shear rate $\dot{s}$. For comparison, values of the first normal stress differente N_1 and shear stress σ were measured using two Weissenberg Rheogoniometers. For two low-density polyethylenes at 150°C, the hole pressure measures elasticity; the exit pressure does not.

MEASUREMENT OF THE FIRST NORMAL STRESS COEFFICIENT OF POLYMER MELTS IN A CIRCUMFERENTIAL FLOW THROUGH AN ANNULUS BY MEANS OF A PRESSURE DIFFERENCE TRANSDUCER.

Kalman Geiger

Institut für Kunststofftechnologie
Universität Stuttgart
Böblinger Str. 70, 7000 Stuttgart 1

1. INTRODUCTION

In a special class of flow, the steady shear flow, the following equations describe the components of the extra-stress tensor $\underline{\tau}$:

$$\begin{aligned} \tau_{21} &= \eta \dot{\gamma} \\ \tau_{11} - \tau_{22} &= \psi_1 \cdot \dot{\gamma}^2 \\ \tau_{22} - \tau_{33} &= \psi_2 \dot{\gamma}^2 \end{aligned} \qquad (1)$$

The viscosity η , the first normal stress coefficient ψ_1, and the second normal stress coefficient ψ_2 are three material constants. The values of these constants depend on the shear rate and the temperature.

The measurement of the first normal stress coefficient is possible in the steady shear flow only, where the streamlines are curved. In a circumferential pressure flow through a curved slit (Fig. 1) the curved streamlines are supported by the inner (in the direction of the radius of curvature) streamlines and in the annulus a normal stress gradient perpendicular to the flow direction is generated. This is a measure for the first normal stress.

The geometry of the curved slit is characterized by the gap $h = r_a - r_i$ and the radius-ratio $\kappa = r_i/r_a$.

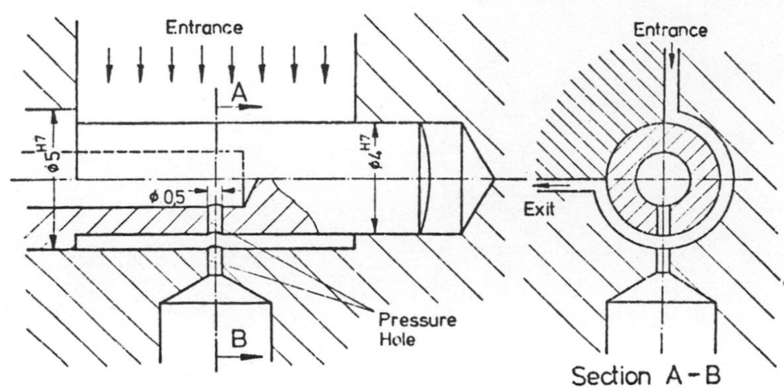

Fig. 1: Annular slit die for circumferential flow

In a curved slit a steady shear flow for high shear rates can be induced ($1 < \dot{\gamma} < 10000\ s^{-1}$, like the capillar rheometer).

The normal stress gradient is a <u>pressure-difference</u> between the inner and outer wall of the curved slit (Fig. 1, pressure-holes). In this paper a calculation will be presented for the pressure-difference /2/ and a new measurement system for measuring of this pressure-difference for various polymer melts.

2. CALCULATIONS OF PRESSURE-DIFFERENCES Δp FOR A LDPE-MELT

The pressure differences induced by steady shear flow in the curved slit between the inner and outer wall can be calculated from the r-component of momentum equation in cylinder coordinates /1/ (Θ - first direction, r - second direction, z - third-direction):

$$\Delta p = \int_{r_i}^{r_a} (\tau_{\theta\theta} - \tau_{rr}) \cdot \frac{dr}{r} \tag{2}$$

The first normal stress N_1 is known for a steady shear flow from (1):

$$\Delta p = \int_{r_i}^{r_a} \psi_1(\dot{\gamma}) \cdot \dot{\gamma}^2 \cdot \frac{dr}{r} \tag{3}$$

The first normal stress coefficient Ψ_1 is known from the relaxation-spectrum /2/ :

$$\Psi_1(\dot{\gamma}) = \sum_{j=1}^{N} \frac{2 \cdot a_j \cdot \lambda_j^3}{(1 + n \cdot \lambda_j \cdot \dot{\gamma})^3} \quad ; \quad N = 8 \tag{4}$$

The shear rate $\dot{\gamma}$ can be calculated with the viscosity function $\eta(\dot{\gamma})$ known from the relaxation-spectrum for the circumferential flow in the curved slit:

$$\eta(\dot{\gamma}) = \sum_{j=1}^{N} \frac{a_j \cdot \lambda_j^2}{(1 + n \cdot \lambda_j \cdot \dot{\gamma})^2} \quad ; \quad N = 8 \tag{5}$$

The pressure differences were calculated for different average velocities in the die $\bar{v}$ with (3), (4) and (5). In Fig. 2 these pressure-differences for different temperature and radius ratio $\kappa = 0{,}8$ and gap with $h = 0{,}5$ mm are shown (e.g. $\Delta p = 1{,}2$ bar at high shear rate and low temperature). Fig. 3 shows the pressure difference Δp for different radius ratio κ. That is the influence of curvature. The decrease of the pressure difference with increasing radius ratio is approximately linear (in the interval $0{,}8 \leq \kappa \leq 1$).

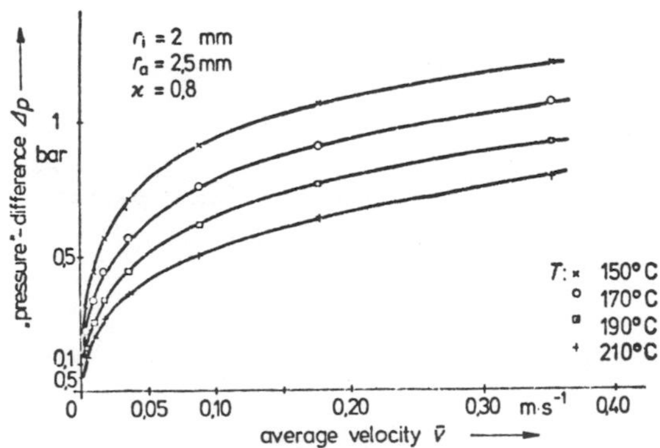

Fig. 2: Calculated "pressure"-difference in a circumferential flow through an annulus Ref. /3/ .

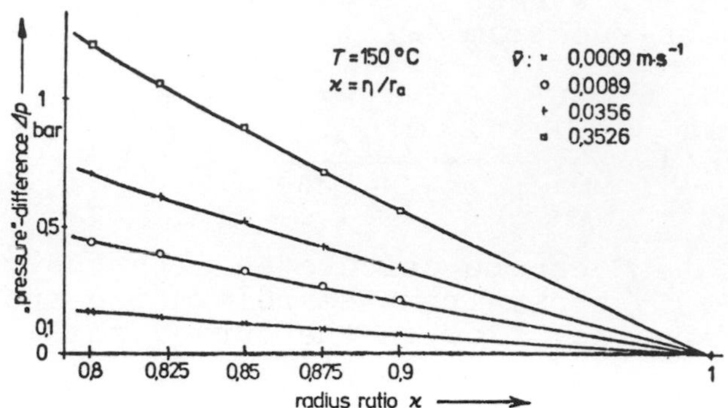

Fig. 3: Calculated "pressure"-difference in a circumferential flow through an annulus for different radius ratio and average velocity. Ref./3/.

3. MEASUREMENT OF THE PRESSURE-DIFFERENCE Δ p:

The calculated pressure-differences in the curved slit die between the inner and outer wall shown in Fig.2 are in the range from 0 to 2 bars at an absolute pressure up to 100 bars (p_i, $p_a \gg \Delta p$). Therefore the problem lies in the measurement of a very low pressure-difference Δ p at a high pressure niveau. It can be solved only with a differential pressure transducer with high sensitivity. Fig. 4 shows the experimental arrangement.

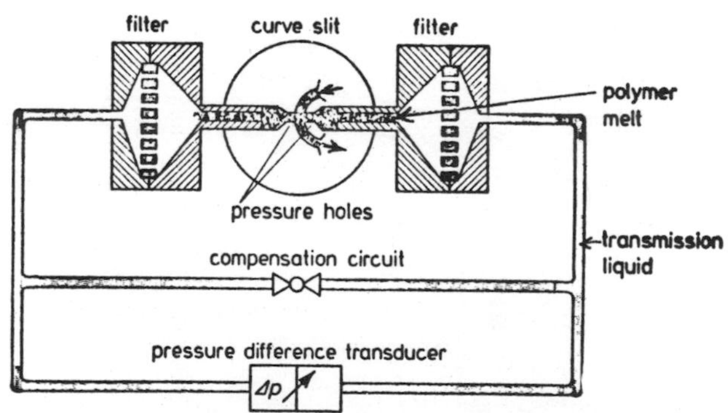

Fig. 4: Arrangement for the pressure-difference measurement in a curved slit die.

4. RESULTS

Fig. 5 shows experimental results for a LDPE-melt (Lupolen 1800 H, BASF) at different temperatures. The behaviour can be described with the Carreau-approximation for the first normal stress coefficient /3/ :

$$\psi_1(\dot{\gamma}) = \frac{a}{(1 + b \cdot \dot{\gamma}^2)^c} \tag{6}$$

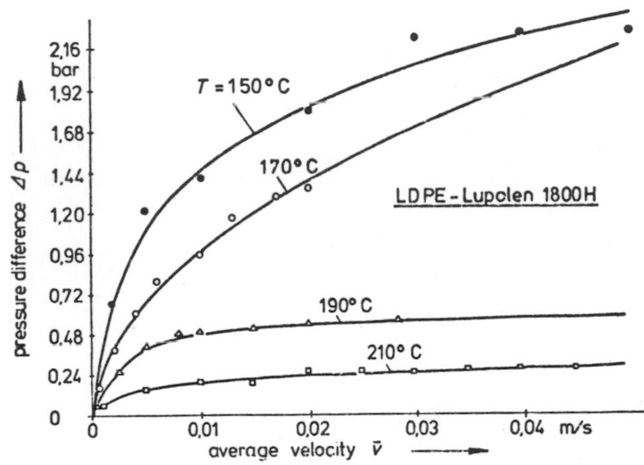

Fig. 5: Measured and calculated pressure-differences in a circumferential flow through an annulus. Ref. /1/

REFERENCES

/1/ Geiger, K. and Winter, H.H., Messung der ersten Normalspannungskoeffizienten von Kunststoffschmelzen im Bogenspalt, Rheol.Acta 17, 264-273 (1978).

/2/ Laun, H.M., Description of the non-linear shear behaviour of a low density polyethylene melt by means of an experimentally determined strain dependent memory function, Rheol.Acta 17, 1-15 (1978).

/3/ Carreau, P.J., Ph.D. Thesis (University of Wisconsin, Madison 1968).

MEASUREMENT OF THE SECOND NORMAL STRESS DIFFERENCE BY DIFFERENT METHODS WITH A ROTATIONAL VISCOMETER

M. Seeger and W. Heinz

Brabender OHG
Postfack 350162, D-4100 Duisburg 1, Federal Republic of Germany

With the NORMAL-F-SENSOR of the rheometer RHEOTRON it is possible to measure in connection with a cone-plate-system normal forces which occur in visco-elastic material. With the normal force F the first normal stress difference v_1 is calculated in an easy way:

$$v_1 = \frac{2F}{\pi r^2} \tag{1}$$

r = radius of cone

We want to report here that by special techniques of operation not only the first but the second normal stress difference v_2 too can be measured by the RHEOTRON.

Measurements of the second normal stress difference have been performed till now mostly with a pressure method. Holes are drilled in the cone or the plate. Very small pressure transducers are located in the holes. With the measured pressures the second normal stress difference can be calculated. But this method is expensive and the holes produces the so-called "pressure-hole-error".

We used the formulas of Jackson and Kaye[1] at one hand and the method and formula of Ginn and Metzner[2] at another.

We performed our measurements with a material consisting of 50 % Polyisobutylen and 50 % mineral oil. The flow curve of this material (tangential stress versus shear rate) shows, that the material is non-Newtonien, namely structural-viscous.

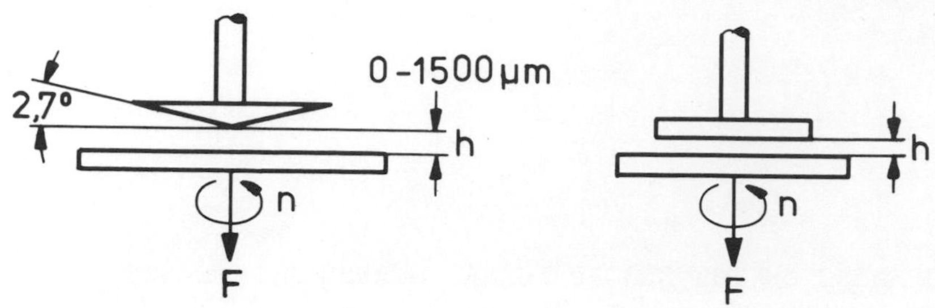

Fig. 1. Scheme of the measuring systems

First we proceeded according to Jackson and Kaye. This is a one-geometry-method by using a cone and plate system with variable distance between cone-tip and plate and variable rotational speed (figure 1). The formula of Jackson and Kaye:

$$(2)\quad v_2 = -\frac{\alpha}{\pi r}\left(\frac{\partial F}{\partial h}\right)_{h\to 0} - D\,\frac{dv_1}{dD}$$

α= cone-angle; h = distance; D = rate of shear

To see the demands of formula 2, we divide v_2 by v_1 and get finally formula 3:

$$(3)\quad \frac{v_2}{v_1} = -\frac{\alpha r}{2F(h=0)}\left(\frac{\partial F}{\partial h}\right)_{h\to 0} - \frac{n}{F(h=0)}\left(\frac{\partial F}{\partial n}\right)_{h\to 0}$$

Which are the demands of this formula concerning the measurements?
We must determine the normal force F as a function as well of the rotational speed n as of the distance h. The both functions in the form of curves are to be differentiated graphically at the points n respectively and h = 0.

According to this programm we have performed the measurements of the normal forces of the material. Figure (2) shows the type of the measurements. Here normal forces as function of the distance were determined with a fast recorder. The x-axis is the time-base, in y-direction you see the normal forces. The peaks of the curves were taken into account in the calculations. The arrow marks the point of the switching-off of the drive. Behind those points the curves show the asymtotic decrease of the normal force during rest.

The measurements with variing speeds gave similar results.

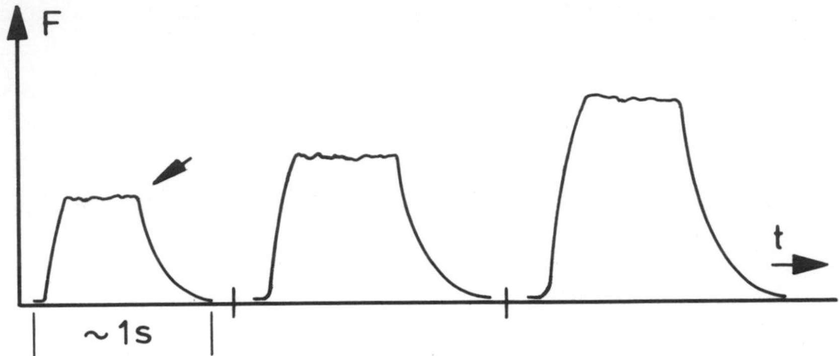

Fig. 2. Type of the recorder diagramm

We saw an instability of the measurements by time at shear rates higher than 100 [s^{-1}]. For the normal forces here were rather constant only during a short period after each beginning. Subsequent there was a decay. Consequently the measurements were transient and only the first hundred milliseconds after each beginning were taken into account.

Figure (3) shows curves calculated with these data, namely the normal forces as function of the distance between cone-tip and plate, with the rotational speeds as parameter.

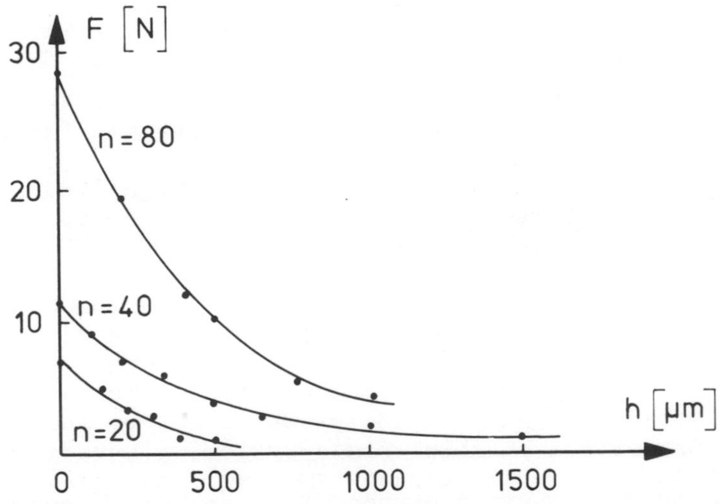

Fig. 3. Normal forces versus distance cone-plate
n = (min^{-1})

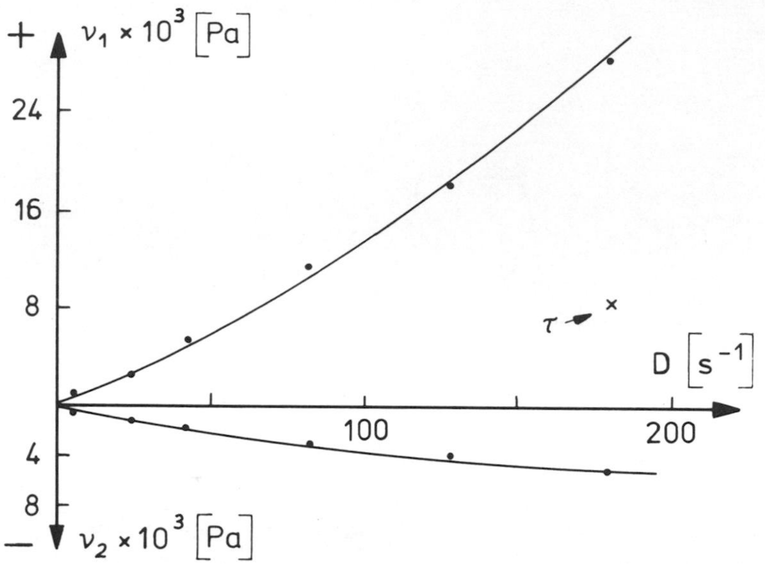

Fig. 4. 1. and 2. normal-stress-difference versus rate of shear; the shear-stress τ is marked at one point.

Now we have the basis to calculate the first normal-stress difference v_1 with formula (1) and the second normal-stress difference v_2 with formula (3).
Figure 4) represents the results:
the first and second normal-stress difference of our test material determined by the method of Jackson and Kaye. v_1 is positive and grows more than proportional with growing shear rate. This is a well known behaviour.
The second normal-stress difference v_2 is negative, it is much smaller than the first normal-stress-difference and it grows less than proportional.

The second part of our report deals with the determination of the second normal-stress-difference v_2 by the method of Ginn and Metzner. This is a two-geometry-method: the material is measured with a cone- and plate-system with distance 0 between cone tip and plate to get v_1 and it is measured too with a parallel plate-system with variable distances to get the difference $(v_1 - v_2)$. The parallel plates we used with the NORMAL-F-SENSOR had a diameter of 50 mm, the distances were between 200 and 1.000 µm (figure 1). (4) is the formula given by Ginn and Metzner:

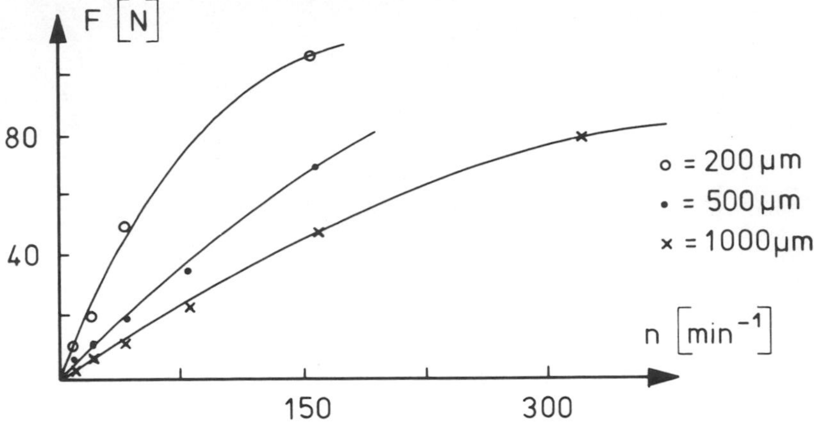

Fig. 5. Measurements with parallel plates, with distances as parameter

$$(4)\quad (\nu_1 - \nu_2)_{D(r)} = \frac{2F}{\pi r^2}\left[1 + \frac{1}{2}\,\frac{d\ln F}{d\ln D(r)}\right]$$

were the normal force F is determined with parallel plates.

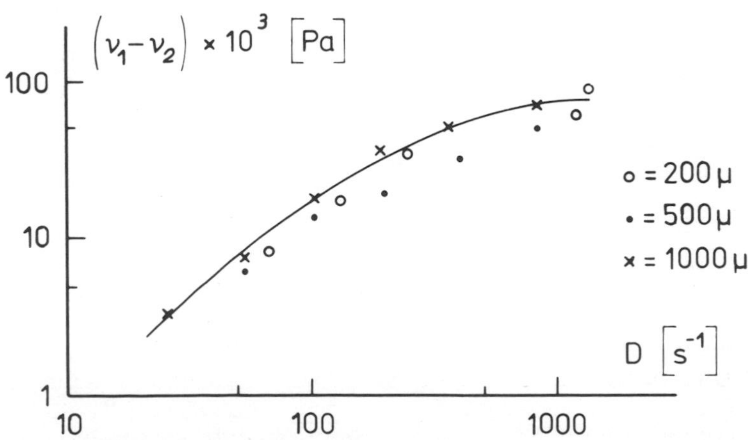

Fig. 6. Determination of $v_1 - v_2$ with parallel plates, with various distances

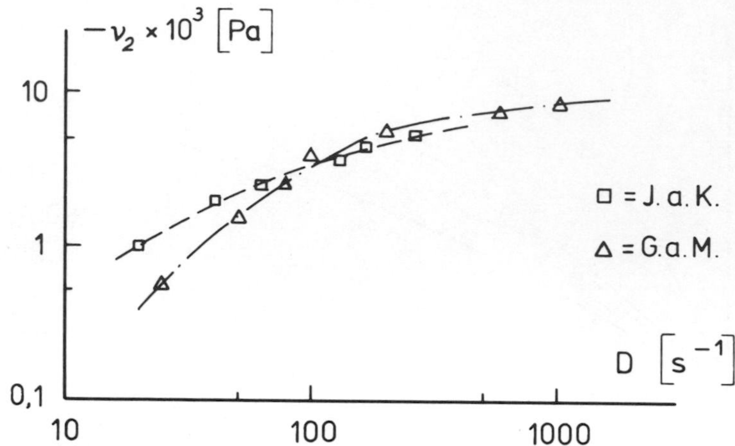

Fig. 7. 2. normal-stress difference of a mixture of polyiso-butylene-mineral-oil, determined with two methods: according to Jackson a. Kaye (J.a.K.) and Ginn a. Metzner (G.a.M.)

$$D_{(r)} = \frac{\Omega r}{h} \tag{5}$$

is the rate of shear at the rim with radius r of the plate. Ω is the angular speed. The results with the parallel plates are given in figure (5). The distance h is the parameter. From these curves we have calculated with formula (4) the values of figure (6). The results should be invariant in respect to the distance between the plates. This is only partial valid. The points gotten at a distance of 1000 µm are connected. Figure (7) shows our final result. The conformity of the two methods is not very good on one side but on the other it is well known, that $-v_2$ is difficult to be measured in any case. It is our impression that the method of Jackson and Kaye is superior. Furthermore it has the advantage to use only one geometry, that means one measuring system.

References:

1. R.A. Jackson and A. Kaye, Chapter 4.5: Basic equatious for other rheogoniometer flow in: "Rheometry", K. Walters, John Wiley and sons, New York (1975)
2. R.F. Ginn and A.B. Metzner, Chapter 4.3: Basic theory for torsional flow in: "Rheometry"

NORMAL STRESS MEASUREMENTS ON VISCOELASTIC FLUIDS USING HOLOGRAPHIC INTERFEROMETRY TO MEASURE THE FREE SURFACE IN THE WEISSENBERG EFFECT

M.F. Hibberd and H.G. Hornung

Department of Physics, Australian National University
Canberra, A.C.T. 2600, Australia

INTRODUCTION

Results obtained to date for viscometric flows of simple fluids have not provided a very detailed picture of the normal stresses for these fluids. The normal stress ratio N_2/N_1 is generally agreed to lie in the range -0.1 to -0.4 for moderately concentrated polymer solutions, but some experiments show a larger range. Most difficulty is experienced measuring N_2, due mainly to its nature (being the difference in stress between the two directions normal to the flow). The use of the shape of the free surface as a measure of the state of stress in a fluid has been explored on a limited scale over the past 30 years[1]. Theoretical work by Bohme[2,3] has shown that, in restricted circumstances, it is possible to obtain expressions for the shape of the free surface of a second order fluid being sheared between two rotating solid surfaces of revolution. The experiments described here are an attempt to obtain the separate values of $\Psi_{1,0}$ and $\Psi_{2,0}$ from measurements of the Weissenberg effect in spherical symmetry (see eqn. 4), and compare them with the value of $(\Psi_{1,0} + 4\cdot\Psi_{2,0})$ from a cylindrical geometry.

THEORY

We consider steady isothermal flow of an incompressible fluid with velocity vector given by $\underset{\sim}{v} = (\dot{\gamma}x_2,0,0)$ in a Cartesian coordinate system with position vector (x_1,x_2,x_3) and shear rate $\dot{\gamma}$. For a simple second order fluid at small shear rates we can define the viscometric functions σ, N_1 and N_2 in terms of the non-zero components of stress:

$$\sigma_1(\dot{\gamma}) = \sigma_{12} = \eta_0\dot{\gamma} + O(\dot{\gamma}^3)$$

$$N_1(\dot{\gamma}) = \sigma_{11} - \sigma_{22} = \Psi_{1,0} \cdot \dot{\gamma}^2 + O(\dot{\gamma}^4) \qquad (1)$$

$$N_2(\dot{\gamma}) = \sigma_{22} - \sigma_{33} = \Psi_{2,0} \cdot \dot{\gamma}^2 + O(\dot{\gamma}^4)$$

where $\Psi_{1,0}$ and $\Psi_{2,0}$ are termed the first and second normal stress difference coefficients. Bohme[3] examined the flow of such a fluid between two rigid surfaces of revolution rotating with small different angular velocities about their common axis. Attention was restricted to "slow" flows, defined by the conditions

$$\text{(Reynolds no.)} \qquad Re = \rho\, VL/\eta_0 << 1 \qquad (2)$$

$$\text{(Weissenberg no.)} \qquad We = (|\Psi_{1,0}|+|\Psi_{2,0}|)\, \dot{\gamma}/\eta_0 << 1$$

where ρ is the fluid density and V and L are characteristic velocity and length scales. For these conditions he considered the flow to be made up of a primary creeping flow, characterised by circular streamlines perpendicular to the axis of rotation, which induces a field of inertial forces and normal stresses causing a smaller secondary flow (not in circular cylindrical symmetry). The general streamfunctions thus obtained were combined with particular boundary conditions to give explicit expressions for the shape of the free surface. Geometries considered in this paper are i) coaxial circular cylinders with the inner one rotating and ii) half-filled concentric spheres with the inner sphere rotating about an axis perpendicular to the fluid surface. At each value of polar radius R the normalised surface deflection Z is related to the measured change in surface height h, and the normal stress coefficients by the equations;

$$Z_c(R) = g \cdot h_c(R)/\Omega_1{}^2 = A_c(R) \cdot (\Psi_{1,0} + 4 \cdot \Psi_{2,0}) + C_c(R) \qquad (3)$$

$$Z_s(R) = g \cdot h_s(R)/\Omega_1{}^2 = A_s(R) \cdot \Psi_{1,0} + B_s(R) \cdot \Psi_{2,0} + C_s(R) \qquad (4)$$

where the subscripts c and s refer to cylindrical and spherical symmetry, Ω_1 and R_1 are the angular velocity and radius of the stirrer, and the A_c, C_c, A_s, B_s and C_s are complicated functions of R with coefficients dependent on ρ, R_1 and the radius of the bowl R_2. The restrictions of small Re and We placed on the flow by the theory mean that low shear rates and weakly elastic fluids must be used. Under these conditions the changes in the surface shape are small and the rise of the liquid next to the rod due to surface tension can be neglected provided that the h used in equations (3) and (4) is the change in the surface height due to stirring. Equation (3) has been derived assuming semi-infinite cylinders and thus end effects must be accouned for experimentally.

REAL TIME HOLOGRAPHIC INTERFEROMETRY

A hologram[4] is essentially a recording of the interference between two light beams which are derived from a single monochromatic coherent light source. In our experiments a variable ratio beam-splitter is used at the output of a 100mW He-Ne laser to obtain the two beams. Each beam is expanded through a spatial filter to improve its spatial coherence. When recording a hologram, one beam illuminates the object and is scattered onto a photographic plate; the other (reference) beam illuminates the plate directly. If the plate is developed and reilluminated by the reference beam, the light scattered from the plate produces a 3-D virtual image of the original object.

In the actual experiments a hologram is taken of the bottom of the bowl through the unstirred viscoelastic liquid being investigated. After development the hologram is replaced in the same position in which it was exposed using a simple six-point contact plate holder. When the hologram and the bottom of the bowl are reilluminated, the virtual image of the object and the object itself coincide exactly. If the liquid is then stirred, there will be changes in the optical path length, to the bottom of the bowl (due to the different refractive indices of air and liquid). The two images of the bottom of the bowl (one virtual, one real) will then be at slightly different positions and so interference fringes will be observed. If the illuminating and viewing directions are near normal to the fluid surface, then a change in fringe number of m for a liquid of refractive index n will be related to the change in surface height h by the equation: $h = \lambda m/(n-1)$, where λ is the wavelength of the illuminating light. Changes in the surface height in the range 0.3 μm to 60 μm can be resolved.

EXPERIMENTAL APPARATUS AND PROCEDURE

The sensitivity of the fluid surface and holographic setup to building noise presented a major problem. For this reason all the experiments are performed on a 1.6m by 2.7m reinforced concrete table of webbed construction weighing 1½ tons. It is supported on five inflated car inner tubes, connected via capillary restrictions to larger air reservoirs to provide optimum damping and a resonant frequency of about 1 Hz. The fluid stirrer consists of a base on which the spherical (or cylindrical) bowl is accurately located. The stirring rod, which has been ground to a fine finish, runs directly on lightly oiled nylon bushes pressed into two bearing plates which are supported above the bowl by three posts. The lower bearing plate also supports a front silvered mirror which turns the beam through 90°. The radius of both bowls is 74.5 mm, the inner sphere has a radius of 9.0 mm and the cylindrical rod has a radius of 9.3 mm. The fluid in the cylindrical bowl has a maximum depth of 85 mm and the end of the rod is 9.3 mm from the bottom of the bowl. The drive system consists of a small sewing machine motor controlled with a stabilised D.C. power supply. It drives a pulley on the stirring rod via a 1.6 mm diameter neoprene belt. The stirring speed is measured to within ±0.02 rad/sec using an electronic circuit to count holes around the periphery of the main pulley. Typical stirring speeds are from 0.5 to 10 rad/sec. Holograms are made with exposures of about 2 seconds on Agfa-Gevaert 8E75 plates. Photographs of the fringe patterns are then taken over as large a range of stirring speeds as possible. For various experimental reasons the illuminating beam is slightly divergent and the illuminating and viewing directions are not normal to the fluid surface. This complicates the interpretation of the fringe pattern. A simple computer program is used to obtain the true surface profile at each stirring speed, and then the values of $\Psi_{1,0}$ and $\Psi_{2,0}$ by fitting equations (3) and (4) to the data using a least squares technique.

RESULTS AND DISCUSSION

An initial check of the system was made with a light oil (fluid A, see table 1) which was believed to be Newtonian. The experimental surface shapes obtained are shown in figs. 1 and 2, together with the theoretical profiles for a Newtonian fluid. Agreement is fairly close and the results in table 1, from the least squares fitting, show the possibility of only very small normal stresses. These may be due to some additives, and a pure low molecular weight white oil, Primol 355, is to be tested soon which should provide more conclusive results. Although Re is quite high for the faster speeds (up to 6) there appears to be no significant variation from the theory.

Figures 3 and 4 show the results for a 30,000cst. silicone oil (fluid B) which has a constant viscosity for $\dot{\gamma}$ up to 100 sec^{-1}. The theoretical fit to the spherical data provides the ratio $-N_2/N_1$ as 0.31 with an error of only ±3%. The error is small because of the method of fitting over the whole curve. This means that a shift of the experimental points, by an amount equal to their estimated error,

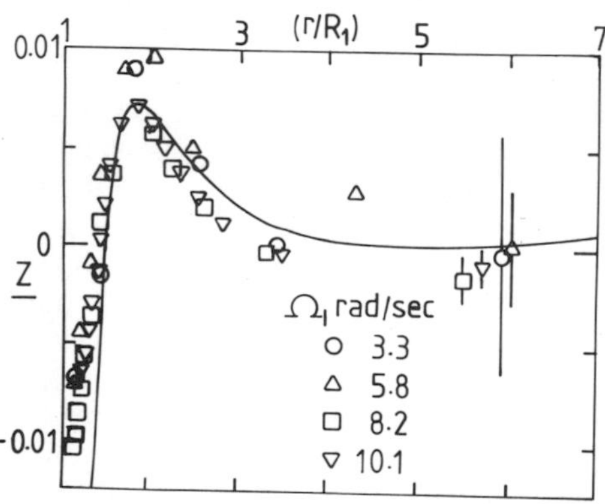

Fig 1. Fluid A spherical data, solid line —— Newtonian fit.

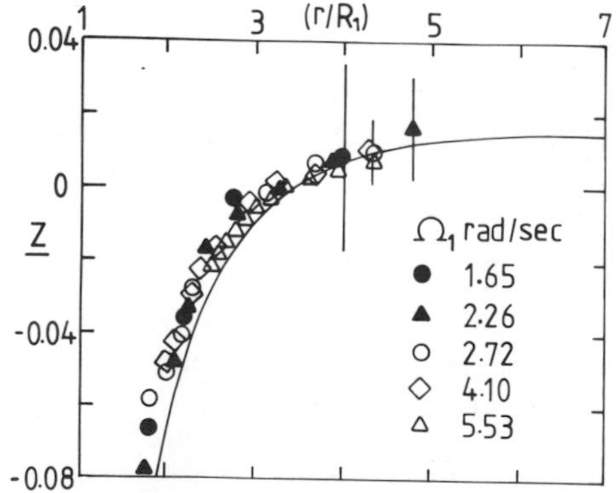

Fig 2. Fluid A cylindrical data, solid line —— Newtonian fit.

Table 1. Fluid properties (25°C) and results for $\Psi_{1,0}$ and $\Psi_{2,0}$ from fitting surface profiles to equations (3) and (4).

Properties and Results	Fluid A Perfecto Medium T68 Turbine Oil Castrol Australia	Fluid B Dow Corning 200 Silicone fluid 30,000 cst.
Density, ρ ($kg.m^{-3}$)	876	975
Viscosity, η_0 (Pa.s)	0.12	32
Refractive index, n	1.4832	1.4035
Spherical data $\Psi_{1,0}$ ($Pa.s^2$)	0.024 ± 0.006	1.40 ± 0.18
$\Psi_{2,0}$ ($Pa.s^2$)	-0.006 ± 0.002	-0.44 ± 0.06
$-\Psi_{2,0}/\Psi_{1,0}$ (ratio)	0.25 ± 0.02	0.31 ± 0.01
Cylindrical data		
$\Psi_{1,0} + 4\cdot\Psi_{2,0}$ ($Pa.s^2$)	0.023 ± 0.010	0.105 ± 0.015
Range of Reynolds no.	1.0 to 6.2	0.004 to 0.02
Range of Weissenberg no.	0.002 to 0.020	0.07 to 0.32

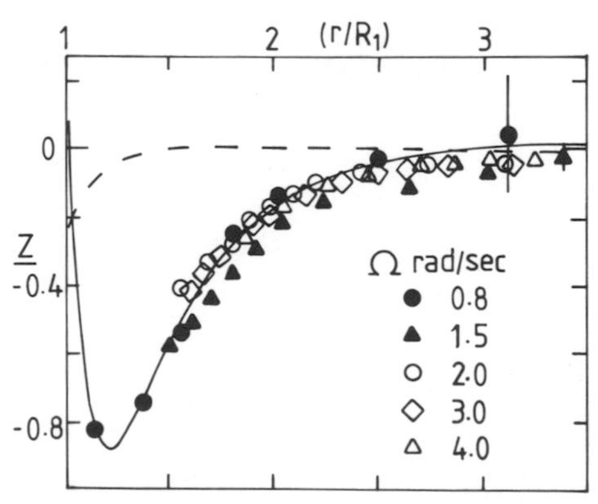

Fig 3. Fluid B spherical data,
——— best fit to equation 4,
– – – $\Psi_{1,0} = \Psi_{2,0} = 0$

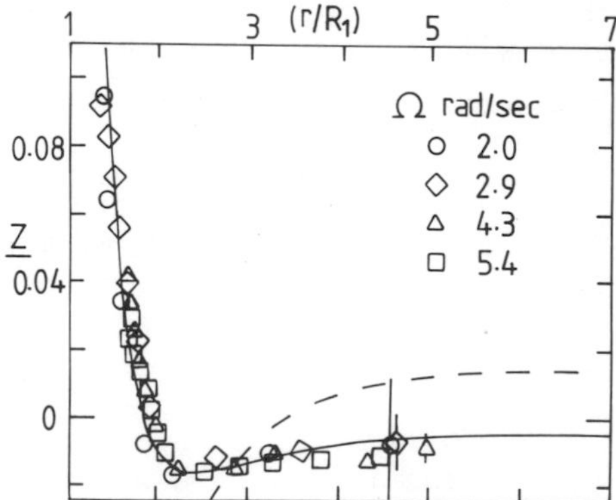

Fig 4. Fluid B cylindrical data,
——— best fit to equation 3,
– – – $(\Psi_{1,0} + 4\cdot\Psi_{2,0}) = 0$.

will cause a change in $\Psi_{1,0}$ of say 10%, but to maintain the shape of the curve, $\Psi_{2,0}$ changes by a similar amount in the same direction. The value of 1.4 kg/m obtained here for $\Psi_{1,0}$ compares well with a value of 0.9 kg/m reported by Chhabra[5] for measurements on a different batch of the same fluid using a Weissenberg rheogoniometer ($\dot{\gamma} < 5sec^{-1}$). The normal stress ratio of 0.31 is higher than many reported values for polymer solutions, but agrees with a result of 0.3 ± 0.1 reported[6] for the NBS non-linear fluid No. 1 (polystyrene) in its second order region, suggesting perhaps that there is a decrease in $-N_2/N_1$ with increasing shear rates.

The cylindrical results for the silicone oil however are inconsistent with the spherical data. Least squares fitting of the cylindrical data shows $(\Psi_{1,0} + 4 \cdot \Psi_{2,0})$ equal to 0.11 whereas the spherical data gives -0.36 for this linear combination. The reasons for this discrepancy are not fully understood. The surface profiles obtained here are believed to be accurate records of the surface as photographed; the results are reproducible and similar discrepancies have observed for other silicone oils. The effect on the results of possible lack of symmetry in the experimental apparatus have been investigated[7] but none is very large. The values of We and Re are sufficiently small that the second order theory should apply. Note that although Z appears to be independent of Ω_1 this is only a necessary, but not sufficient condition for second order behaviour. There remains a possibility that end effects in the cylindrical case are significantly disturbing the flow compared to that predicted by the theory. This is to be investigated further, and measurements of N_1 are to be made on a conventional rheometer to check the spherical results for $\Psi_{1,0}$.

The technique described in this paper is fairly complicated, and limited to weakly elastic fluids at low shear rates (i.e. small Re and We), however we believe that further experiments using this method will provide some important normal stress data.

REFERENCES

1. G.S. Beavers and D.D. Joseph, Experiments on free surface phenomena, J. non-Newtonian Fl. Mech., 5, 323-353 (1979).
2. G. Böhme, "A theory for secondary flow phenomena in non-Newtonian fluids", Europ. Space Res. Organ. Report No. TT 157 (1974).
3. G. Böhme, Rotation-symmetrische Sekundarstromungen in nicht-Newtonschen Fluiden, Rheol. Acta, 14 669-698 (1975).
4. R.J. Collier, C.B. Burckhardt and H.L. Lin, "Optical Holography", Academic Press, New York (1971).
5. R.P. Chhabra, private communication.
6. M. Keentok, A.G. Georgescu, A.A. Sherwood and R.I. Tanner, J. non-Newtonian Fl. Mech., to appear.
7. M.F. Hibberd, Ph.D. thesis, A.N.U., Canberra, in preparation.

CONTINUOUS RHEOMETRY : A NEW PROCESS FOR ON-LINE MEASUREMENTS AND CONTROL FOR LABORATORY AND INDUSTRY

Dr. André Kepes

Polymer Consultant

22, av. Prise d'Eau F 78110 LE VESINET FRANCE

INTRODUCTION

Tremendous progresses were accomplished in experimental rheometry in the past 10 years, involving electronics, microprocessors, digitalization, solid-state memory and computerizing. It is therefore extremely unfortunate, that no progress at all was made in the field of mechanics and instrument geometry : end effects, gap setting and calibration errors, sample manipulation and related hazards. And yet, those factors are responsible of poor reliability, errors of 10% and more in rotational rheometry. The new technique described below claims to have brought mechanics to the same performance as solid-state electronics, and thus having made desappear all major uncertainties in experimental rheometry.

RHEOPROCESSOR : A NEW TECHNOLOGY

The process here-under described is based on a unique design, leading to a continuous on-line measuring device, where the sample is changed automatically, without disturbing system equilibrium nor introducing human errors.

Fig. 1 shows schematically the core of the instrument, named Rheoprocessor, with its centered double-bicones forming the gap-space. The central fixed bicone conducts the fluid through its axial hole towards the gap, while the external bicone, shaped like a reversed cup, is rotated by applying a programmed torque, either constant or variable.

Geometry

According to the bicones' profiles, the gap can be parallel, producing a torsional shear similar to a parallel plate device, or, when adjacent cones have common apex, there is a constant shear-rate as in cone and plate viscometers.

The double-bicone geometry features major advantages which we have to focus first.
- There is no need of a guard-ring, which means often : stagnation, cross-linking, decomposition, friction with the so-called fixed part.
- Furthermore, there is no edge-effect in the Rheoprocessor, responsible of huge errors and aleatory results. In commonly, the relationships between torque and speed, the external diameter appears at the third power in cone and plate, and at the forth power in parallel-plate geometry. It is therefore easy to evaluate the disastrous edge-effects : at the beginning of the measurement, there is too much fluid in the gap, its free surface taking the shape of a barrel ; later on, with the action of gravity, centrifugal force, metal wetting and thermal contraction (of the fluid and the rotating shafts), its diameter decreases to an unknown final value.
- Small radial and axial excentricities of the moving part have no effect with concentric bicones : the resulting error is of the third order. In cone and plate geometry, the gap-thickness is of the first order : after initial adjusting, loose mechanics, thermal dilatations, Weissenberg normal forces make it difficult to secure gap stability all along the measurement.
- Air is nerver in contact with the fluid sample in the Rheoprocessor : no oxydation, decomposition or cross-linking can occur during the filling and the measurement, even at high temperatures as required in melt viscometry.

Filling the Gap : Introduction of Fluid Sample

In this respect, there are few comments to make, so obvious are the advantages of the Rheoprocessor. Every experimentator knows about the major disturbance caused by sample replacement opening the device, cleaning the chamber, recharging, setting the gap, waiting for thermal equilibrium compromise the reliability of the best instrument, not to speak of oxygen pollution. These hazards are completely elimiated by automatic filling.

But, even more important is the ability of the Rheoprocessor to make closed-loop industrial on-stream analysis and control for process monitoring and optimization.

By means of a multiple way admission valve, quick fluid replacement without disturbance is easily performed. Besides operating versatility, this has another important consequence : the perfect control of hazardous drift effects. By alternating the sample under study with a standard, or a calibrated fluid and comparing sucessive measurements, the Rheoprocessor definitely eliminates this frequent u ncertaintly of classical rheometers.

Imposed shear-stress operation

In rotational rheometry, speed both constant or oscillatory is generally imposed to the main rotating shaft, while the torque is measured on the "fixed" part. The last is therefore equiped with elastic suspension and force tranducers. The opposite choice was made for the Rheoprocessor : a torque of determined value is apaplied to the moving bicone, and the resulting speed is measured on the same axis, the other part being really fixed. This means greater robustness and some extra advantages :

- No mechanical speed-box is necessary
- No deterioration can occur when accidental blocking happens
- Adherence of the fluid is maintained without slipping
- Thermal dissipation is automatically controlled : if viscosity increases, thermal effect drops with decreasing speed
- Less lag effect is to be feared in oscillatory measurements.

The applied torque may be either constant, aperiodic or periodic ; in the last case, its signal may be sinusoidal or even square : electrically produced torque is easy to control and can be changed instantaneously. Incremental encoder gives extreme accuracy in speed measurement ; it calculates the desired parameters, displays and prints data and results. A feed-back signal is delivered for monitoring industrial processes.

METHODS AND EXPERIMENTAL TECHNIQUES WITH THE RHEOPROCESSOR

Five main rheological processes are to be mentioned ; two of them are completely new.

1° - Steady state, constant shear stress rheometry. This is a classical technique in cone and plate rheometry. This is a classical technique in cone and plate rheometry. The fluid has to be maintained in the gap during applying the torque ; speed is read after its stabilization. The next sample is then injected into the Rheoprocessor.

2° - Transient regime. Same equipement, but used, with aperiodic shear stress and simultanious recording of the speed. Torque can be suddenly applied or stopped, or its value changed. If the sample is elastic for small stresses, as some gels are the displacement instead of the speed can be measured in order to determine elastic modulus and for yield stress.

3° - Viscoelasticity. Oscillatory dynamic measurements are to be performed in this case. Sinusoidal stress (or some other periodic signal) is applied and the resulting amplitude and phase shift are recorded. Both components of the complex modulus are then calculated by the usual methods.

4° - The "steady transient" cross flow regime. This unique process is characterized by the continuous flowing of the sample through the Rheoprocessor while performing measurement. Each flow unit (molecules, segments, particles) is under stress only for a limited time, depending on the flow rate and the gap volume. In presence of finite relaxation times, no steady orientation or configuration can be attained during the passage. Experiments at different flow rates (and possibly different shear stresses) will then give extensive information about transient propoerties, especially for large relaxation times. Very good stability and accuracy has to be expected, as heat dissipation will not cause temperature shifts of the fluid, which is progressively replaced all along the process.

5° - Automatic Melt Indexer . The corresponding version of the Rheoprocessor claims to solve the ever lasting problem of continuous control and monitoring in polymer production. The core-geometry is of the parallel-cones type in this case.

The torsional flow produces in the material a velocity gradient which linearly increasis with its distance from the axis. A similar phenomenon happens in capillary flow, where the shear stress is a linear function of the distance.

Now, if we apply to the bicone a torque of such constant value, that it produces, at its maximum diameter, exactly the same stress as prescribed by ASTM (or other standard) for the capillary wall of Melt Indexers, the sample will be submitted in absolute value, to a substantially same spatial distribution of stress as in capillary extrusion. The speed of the bicone is then simply proportional to the Melt Index.

But this melt-indexer goes much farther. By easy programming of a torque half of the preceeding value, and then the double of it. (or by any other factor) we immediatly obtain the slope and the radius of the curve at the very same point of the stress shear rate function.

Those are quite useful parameters for producers and fabricatiors. They were not easy to obtain before even in laboratories : they now may be automatically delivered in production-control.

FEEDING THE RHEOPROCESSOR

The above described instrument may be fed by all flowable substances, under slight pressure : liquids, polymer melts (pure or filled), emulsions, pastes, gels, mud, slurries, foams, etc. Even abrasive suspensions can go through without damaging it.

The input pipe can be branched directly to the instrument entry commanded only by a simple valve, in order to stop the stream for steady flow measurements. Fluid can be delivered by all types of source : reactor, processor or simply a container under inert gas pressure.

But experimentators will appreciate the appropriate Feeding Unit specially designed for the Rheoprocessor.

Feeding Unit

The F.U. is a compact programmable fluid circulator composed of 3 main parts, each one with independent command.

1° - A five way rotating valve S selects between 5 permanently connected sources, the desired one. They are, for instance, the fluid to be studied, the standard fluid, the cleaning was or resin, a heavy solvent and an inert gas (generally nitrogen).

2° - A 4 position valve V is reacked by the fluid coming from the selector S. Its purpose is visible on Fig. 2.

- The first position, D, is the direct passage towards the Rheoprocessor.
- The second, I1, connects the source to the a compartment of the pump (see below), the b compartment being connected to the Rheoprocessor.
- The third position, I2, inverts the 2 preceeding connections.
- In the fourth, M, the pump is transformed in an internal mixer by short circuiting the two compartments of the pump body through the V valve.

3° - The double acting pump's cross section is schematically shown on Fig.2 ; its toroidal chamber is divided into 2 compartments a and b by a rotating piston.

When the switch valve V is in the I1 position, the volume of the a compartment a, connected to the source, increases from 0 to its maximum value, while b, going from its maximum to 0, injects its fluid content into the Rheoprocessor. A moment later, after switching from I1 to I2 and simultaneous reversion of the pump drive, b is filled with the fluid. The same as before or another one, depending on the selector's position - and a expulses its content towards the rheometer.

The volume change is proportional to the angular displacement of the pump's driving shaft : this allows precise control of the injected volume as well as of the flow rate. For segmential steady state measurements, the piston may be stopped in every position, after having pumped just the required volume into the Rheoprocessor.

For transient or continuous non-sequential measurements, the pump maintains a permanent flow rate across the Rheoprocessor's body. The inversion at each run's end takes less than 1 sec and the run may last about 10 minutes before each inversion.

Internal Mixing

This function is highly desirable in research as well as in investigations about blends and alloys. The pump compartment, starting from 0 volume, sucks a certain amount of the first material, controlled by the angular displacement of the pump's shaft. After selecting the second source, one aspires the second fluid and so on,till the chamber is completely full. Then the V valve is switched in the M position, which makes direct connection between the a and b compartments, and the mixing begins.

The melt is pushed back and forth 10 or 20 times from one side to the other across the valve, whose orifice is properly adjusted. When mixing is over, the homogenized mass is pushed either into the Rheoprocessor for immediate measuring, or backwards in a source container to be used later on.

The different components of the blend and additives are exactly vibrated and the entire operation takes place without any pollution. An extra sample can be extraded from the mixier for solid state testing.

It is known that thermoplastic alloys often develop improved properties, as proved by their fast growing market in the U.S.

In some cases, through cleaning of all the system may be necessary. It is easely made by the pump : one source is fed by a specially filled low-molecular weight plastic or wax : cleaning can be completed by rinsing with a suitable solvent

CONCLUSION

The Rheoprocessor equiped with its Feeding Unit, forms the first automatic Rheological station for scientists as well as for industrialists. Its design guarantees unique performances, the highest degree in precision, outgent, reliaility and robustness.

It may help research, development and on-stream control in many Industries concerned with Rheology, such as Plastics, Food and Drug, Comestics, Paints and Inks, Cements and all Chemicals.

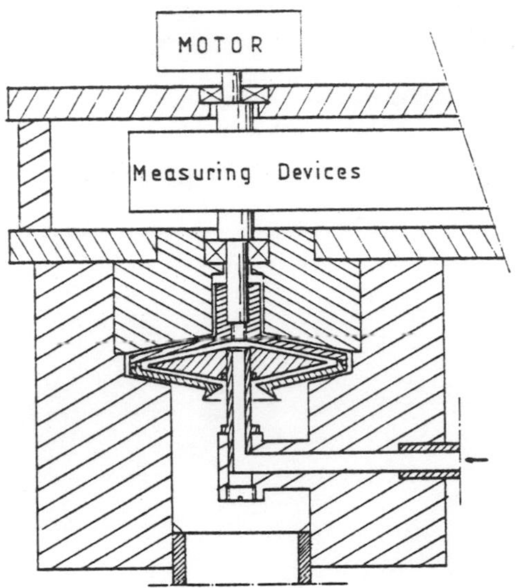

Fig. 1 Rheoprocessor's Principle

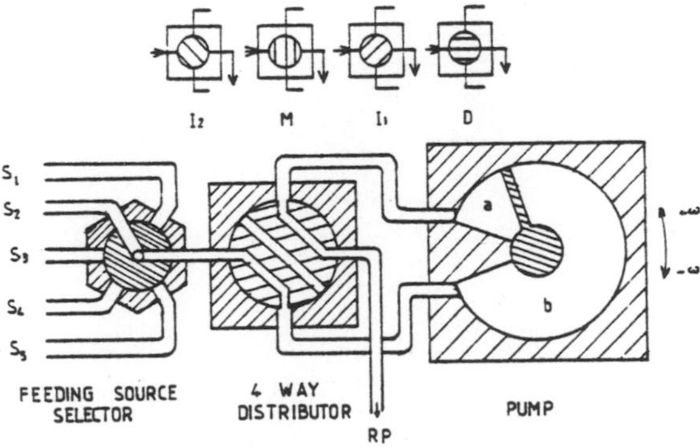

Fig. 2 Feeding System

SLIT RHEOMETRY OF BPA-POLYCARBONATE

M.G. Hansen and J.B. Jansma

General Electric, Corporate Research and Development
Chemical Laboratory, P.O. Box 8
Schenectady, New York 12301 U.S.A.

INTRODUCTION

The first work on slit rheometry was conducted by the group at Central Laboratory TNO, Delft, The Netherlands.[1-3] In their work, they reported on velocity profiles and pressure measurements as well as the influence of slit aspect ratio. They experimentally determined that an aspect ratio of 10:1 or greater was sufficient to provide good results. Flush mounted pressure transducers were inserted into the die to measure the axial pressure drop. These mechanical measurements were complemented by optical measurements of melt flow birefringence in the 1-3 plane of the usual coordinate system of slit rheometry. Han and co-workers[4-9] also constructed a slit type apparatus using flush mounted pressure transducers. Measurements were presented of wall normal stress distribution in the axial direction with the presence of an "exit pressure." This pressure is the extrapolated value of the wall normal stress to the die exit, but is not equal to the primary normal stress difference as shown by Walter's[10] analysis of slit die rheometry. Slit rheometry has two conflicting practical limits to producing good results. Walter's analysis assumes that the flow in the slit, where pressure measurements are being made, is two-dimensional and is fully developed. Thus, a long die is favored so that all pressure transducers are positioned to produce a constant axial pressure gradient. But as was presented by Robens and Winters,[11] the temperature of the melt increases towards the end of the die due to viscous dissipation. Thus, a short die is favored for isothermal tests.

Leblanc[12,13] constructed a pressure driven slit rheometer with three flush mounted pressure transducers. He presented data for a

4-arm di-block polymer of styrene-butadiene. Fully developed flow was not attained in the measurement section since the axial pressure gradient was not constant. Thus, the analysis presented to determine viscosity from these data is not valid. This is one example of not meeting one of the practical limits of slit rheometry to produce good results.

From both an experimental and theoretical point of view, the weakest assumption in Walter's analysis is that of assuming fully developed flow exists up to the exit of the die. Boger and Denn[14] have written a critical appraisal of the slit method of normal stress measurements. They discuss the problem of velocity rearrangement near the exit and develop insight into the magnitude of its effect on normal stress measurements. Further effort is still needed to fully solve this problem.

The purpose of our work is to gain rheological knowledge in an efficient manner of "engineering thermoplastics," i.e., polymers with high aromatic content in the main chain. This is to be done at processing shear rates and a maximum operating temperature of 425°C and maximum operating pressure of 15,000 psi. To meet our objectives, we have developed a slit-die rheometer.

APPARATUS

The measurements presented in this paper were obtained with a slit-die rheometer of our own design and construction. It was designed to replace an Instron load frame model capillary rheometer (type MCR).[15] The general configuration of our rheometer is illustrated in Figure 1. This configuration is similar in general assembly to the rheometer of Robens and Winter.[11] A 3/4 inch diameter single-screw extruder is used as a melt source to fill the barrel with molten polymer and to purge the rheometer of entrained air.

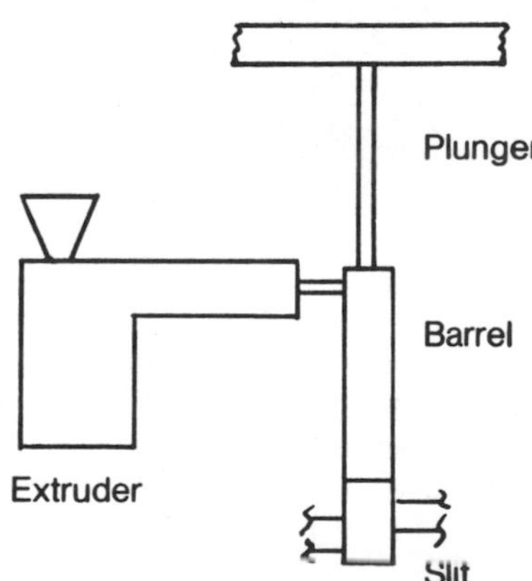

Figure 1 Schematic of Slit Rheometer

The bore of the barrel is 3/8 inch in diameter and the slit is of the same width. Four shims with thicknesses of .020, .030, .040 and .059 inches are used to establish the slit height. The barrel, slit and slit shims were all made from air hardened machine tool steel and all polymer contacting surfaces and sealing surfaces were ground to 16 rms, or better, surface finish. These metal-metal seals have performed well to date at a maximum operating pressure of 11,000 psi.

Four flush mounted pressure transducers are mounted axially along the die with two transducers on either side of the die. The pressure transducer faces are 9/32 inch in diameter and are mounted flush to the die face within ±.0005 inch, thus avoiding any question of "hole error" effects.[16,17] The pressure transducers are Dynisco model PT422-A with digital readout amplifiers. These gauges were calibrated using a thermally controlled deadweight tester designed and constructed by Golba.[18] All gauges were then calibrated to a maximum temperature of 300°C and a maximum pressure of 10,000 psi, or full scale. The digital readout was found not to be accurate. Hence, the analog output of 0-1 vdc full scale was measured with a digital voltmeter. It was determined that hysteresis and thermal dependence were negligible and that all gauges were linear in the upper 90% of full scale. Non-linearity of pressure versus voltage output did occur in the lower 10% of full scale. This calibration method enabled us to greatly expand downward the useful measurement range of a given pressure transducer.

The rheometer is divided into four temperature controlled zones. The four heaters are controlled by proportional-integral-differential controllers from Leeds and Northrup, model 6430. The temperature of the rheometer was monitored at five axial locations and it was determined that the axial temperature variation of the rheometer over the four zones was 1°C after 36 hours of adjusting to a set point.

RESULTS

Our first effort in quantifying the operation of the slit rheometer was to evaluate a silicone fluid (General Electric Silicone Products, SE-30). Two different Rheometrics^R Dynamic Spectrometers and one Mechanical Spectrometer were used with cone-and-plate geometry to determine complex shear viscosity at 22°C. These measurements, which agreed well among all three machines, were than compared with steady shear viscosity measurements from slit rheometry. Figure 2 shows these data and it should be noted that the data overlap between 1 and 10 sec^{-1} for the different measurement methods.

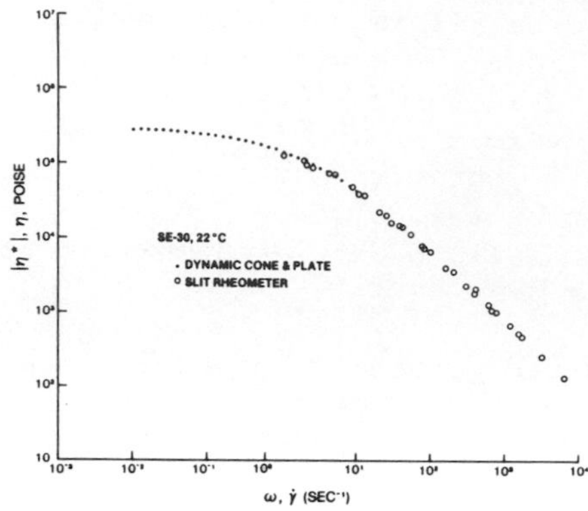

Figure 2 Viscosity of Silicone Fluid

In the introduction we stated that our purpose in constructing a slit rheometer was to gain rheological knowledge of "engineering thermoplastics." The first such resin to be investigated was linear BPA-polycarbonate of four different molecular weights. The $\overline{Mw}$ values are as follows: sample A, 24,100; sample B, 24,800; sample C, 29,700 and sample D, 37,100. Just as with silicone fluid, our first task in studying BPA-polycarbonate was to determine if steady shear viscosity from cone-and-plate measurements, absolute value of complex viscosity and steady shear viscosity determined from slit rheometry would agree, when measured at the same temperature. Figure 3 presents typical results for sample A at 250°C. Contrary to our good agreement with silicone fluid, it is noted that for all four polycarbonate polymers that the slit rheometer measurements are less than those of cone-and-plate measurements. It is further noted that this disagreement is greatest at lower slit rheometer shear rates. Initally, we were not able to account for this disagreement, but concluded that what was occurring was that the polycarbonate was degrading in the slit rheometer during measurement. For isothermal tests, the polymer resides in the rheometer for greater and greater lengths of time as the shear rate becomes progressively lower. In all cases the agreement between slit measurements and cone-and-plate measurements is best at the highest shear rates. We have now solved this problem by reducing the average residence time of the melt in the rheometer (compare square symbols to diamond symbols in Figure 3). The agreement of slit measurements with cone-and-plate measurements are very good. Thus, to obtain results which are absolute, it is of paramount importance to minimize polymer degradation during slit rheometry measurements. For those measurements where degradation was at a minimum, the exit pressure was experimentally negligible.

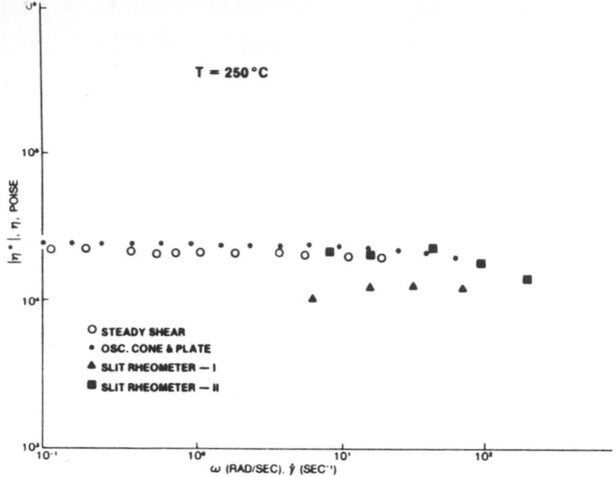

Figure 3 Viscosity of BPA-Polycarbonate, Sample A

CONCLUSIONS

We have demonstrated that the slit rheometer, which we have designed and constructed, can produce absolute rheological results which agree within experimental error with measurements from cone-and-plate rheometers. Furthermore, it is possible to produce absolute rheological results with a slit rheometer for materials which thermally degrade.

REFERENCES

1. R. Eswaran, H. Janeschitz-Kriegl and J. Schijf, Rheol. Acta. 3:83 (1963).

2. J.L.S. Wales, J.L. den Otter and H. Janeschitz-Kriegl, Rheol. Acta. 4:146 (1965).

3. J.L. den Otter, J.L.S. Wales and J. Schijf, Rheol. Acta. 6:205 (1967).

4. C.D. Han, M. Charles, and W. Philippoff, Trans. Soc. Rheol. 13:453 (1969); ibid., 14:363 (1970).

5. C.D. Han, J. Appl. Polym. Sci. 15:2567 (1971).

6. C.D. Han, T.C. Yu and K.U. Kim, J. Appl. Polym. Sci. 15:1149 (1971).

7. C.D. Han, Trans. Soc. Rheol. 18:163 (1974).

8. C.D. Han, "Rheology in Polymer Processing," Academic, New York (1976).

9. C.D. Han and D.A. Rao, J. Appl. Polym. Sci. 24:225 (1979).

10. K. Walters, "Rheometry," Halstead Press, England (1975).

11. G. Robens and H.H. Winter, Kunststofftechnik 13:61 (1974).

12. J.L. Leblanc, Polymer 17:236 (1976).

13. J.L. Leblanc, Rheol. Acta. 15:654 (1976).

14. D.V. Boger and M.M. Denn, "Capillary and Slit Methods of Normal Stress Measurements," presented at the Golden Jubilee Society of Rheology Meeting, Boston, MA (1979).

15. Instron Corporation, Canton, MA 02021.

16. J.M. Broadbent, A.S. Lodge and D.G. Vale, Nature (London) 217:55 (1968).

17. J.M. Lipson and A.S. Lodge, Rheol. Acta. 7:364 (1968).

18. Joseph C. Golba, Jr., General Electric Company, Polymer Physics and Engineering Branch, Corporate Research and Development Center, Schenectady, New York 12301.

NON-NEWTONIAN VISCOSITY AT HIGH STRESSES

A. Eastwood and G. Harrison

Glasgow University
Department of Electronics and Electrical Engineering
Glasgow, G12 8QQ, Scotland

INTRODUCTION

It has long been established that in polymer melts and solutions the viscosity is a non-linear function of shear rate, the value falling from the 'Newtonian' or low-shear-rate limiting value as the shear rate is increased. Attempts to observe directly the non-Newtonian behaviour of simple liquids and lubricants in continuous shear at room temperature have met with little success, however, due principly to the considerable heat generated even in a very thin fluid film at the high shear rates required. Non-Newtonian effects have been observed in oil films under elastohydrodynamic conditions, where the fluid is subject to high pressures, and it has been deduced that these effects occur above a critical shear-stress level of between 1 and 100 MPa, depending on the pressure and the molecular weight of the liquid (Hirst and Moore, 1978).

The problems of heat generation at the high shear rates necessary with low viscosity liquids can be overcome if measurements can be carried out near the glass transition temperature, where the viscosity is approaching a value of 10^{12}Pa s. The required shear stresses of 1 MPa and above can then be obtained at shear rates of the order of $10^{-6}s^{-1}$. The heat generated, which is proportional to $\eta\dot{\gamma}^2$ is thus reduced by many orders of magnitude and becomes negligible.

The measurements reported here, of non-Newtonian flow in low-molecular-mass materials, were carried out using a specially constructed viscometer in which shear stresses in the range 0.5 to 50 MPa can be applied.

EXPERIMENTAL SYSTEM

The viscometer is of the coaxial cylinder "Pochettino" type, in which the motion of the inner cylinder is along the axis of cylindrical symmetry. The sample chamber, which acts as the outer cylinder, is attached to a base plate, which is mounted above a thermostat bath, with the sample chamber immersed in the bath liquid (Fig.1). The inner cylinder, which takes the form of a thin rod, is attached to the end of a lever which is mounted above the base plate. The viscometer is operated by applying a load to the end of the lever arm, and monitoring the movement of the inner cylinder as it is pulled vertically through the sample. The lever arm has a 5:1 mechanical advantage, and is pivoted using a pair of ball races. The movement of the inner cylinder was monitored using a linear variable differential transducer (L.V.D.T.), the output voltage was measured using a digital voltmeter. Two different L.V.D.T.'s were used and were calibrated against a precision micrometer. Both were found to be linear to better than 2.5°/₀ over a travel of 4 mm with a resolution of ±0.3 nm.

The sample chamber itself consists of a glass U-tube. The bottom of the U is filled with mercury and the sample floats above this in one of the arms. This method of containing the sample avoids end effects due to non-viscometric flows at the end of the inner cylinder. This has been confirmed by making measurements with various sample lengths in the range 0.8 cm. to 4 cm. Summing all the errors due to movement inaccuracies, length variations, and temperature fluctuations gives an overall error of about 15°/₀ at worst and 9°/₀ at best.

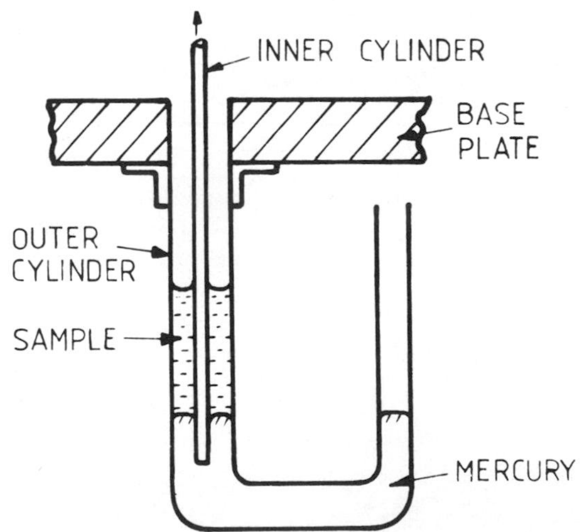

Fig. 1: Schematic Diagram of Viscometer

THEORY

In this viscometer the only movement is along the axis of cylindrical symmetry, and so the equations of motion reduce to the equation of continuity of stress, τr = constant; and $\dot{\gamma} = dv/dr$. τ is the shear stress and $\dot{\gamma}$ is the shear rate at radius r, v is the velocity in the z direction. The velocity of the inner cylinder relative to the outer cylinder (radius r_o) is obtained by integrating the shear rate between the inner and outer radii. If the viscosity is independent of the shear stress, i.e. the liquid is Newtonian, then the shear rate at the inner cylinder is

$$\dot{\gamma}_I = \tau_I/\eta = v\Big/\left\{r_I \ \ell n \ (r_o/r_I)\right\} \qquad (1)$$

In the viscometer a load W is applied to the inner cylinder so that there is a reference stress τ_I at the radius r_I of the inner cylinder, given by

$$\tau_I = Wg/2\pi r_I L \qquad (2)$$

where L is the sample length and g is the acceleration due to gravity. Equations (1) and (2) can be applied to the measured values of velocity and load to determine the viscosity $\eta = \tau_I/\dot{\gamma}_I$. In the case of a liquid whose viscosity varies with shear stress, this procedure results in an "apparent Newtonian viscosity". These values can be corrected if the form of the variation of viscosity with stress is known.

The Eyring (1936) expression for the high stress viscosity is

$$\eta = \eta_o a\tau/\mathrm{Sinh}(a\tau) \qquad (3)$$

where a is related to the pressure dependence of the viscosity. Using a series expansion for Sinh $a\tau$, the velocity of the inner cylinder is given by

$$v = \left(\tau_I r_I/\eta_o\right)\left\{ \ell n(r_o/r_I) + \frac{(a\tau)^2}{2.3!}\left[1 - (r_I/r_o)^2\right] + \ldots\right\}$$

With a thin rod for the inner cylinder, $r_I/r_o \ll 1$ and so the second order terms in (r_I/r_o) and above can be neglected. For the stress values used, neglecting terms of $(a\tau)^6$ and above introduces errors of less than 2.5%. This expression for the velocity can be used in equation (1) to obtain an expression for the "apparent Newtonian viscosity"

$$\eta_a/\eta_o = 1\Big/\left\{1 + \frac{(a\tau_c)^2}{12}\left[1 + \frac{(a\tau_c)^2}{40}\right]\Big/\ell n \ (r_o/r_I)\right\} \qquad (4)$$

RESULTS

A study has been made of four glass-forming materials, which have glass transition temperatures, T_g, which were accessible using conventional refrigerated thermostat baths. Three samples were low-molecular-mass polystyrenes, obtained from Pressure Chemical Co. Inc., U.S.A. Details of these materials have been published previously (Gray, Harrison and Lamb, 1977), the molecular mass and T_g values are reproduced here, Table 1. The remaining material was a six-ring poly-phenyl ether (PPE), trade name OS 138, supplied by Monsanto Corporation, U.S.A. This material has a T_g of -20.5°C.

The viscometer was initially used to measure the low stress (Newtonian) viscosity of OS 138 and PS 580 at several temperatures. These results were compared with results obtained using a parallel-plate torsional creep viscometer, similar to that described by Plazek (1968). The inner cylinder used for these experiments had a radius of 0.7875 mm. The radius of the outer cylinder was 8.85 mm. The results obtained from the two techniques agreed in all cases to within the measurement accuracy of ± 15°/o over the region where the two techniques overlapped, and reproducible viscosity values could be obtained using the new viscometer up to viscosities of about 10^{13}Pa s.

High-stress measurements were made over a range of temperatures close to the glass transition temperature for each sample. For all samples the values of apparent viscosity calculated using equations (1) and (2) decreased from the low-stress (Newtonian) value with increasing shear stress, when a critical stress has been exceeded. The results are shown in Fig.2; measurements at different temperatures have been normalised by dividing the viscosity values by the low-stress value at each temperature of measurement.

Obtaining data at high stresses proved very difficult, as the stresses applied were approaching the tensile strength of the materials. If the stress was applied rapidly, cracks would appear, which had the effect of reducing the sample length, and so the velocity of the inner cylinder increased markedly. In nearly all the measurements where the viscosity was lower than the Newtonian value cracking would at some time occur. As soon as this happened the measurement was terminated and the viscosity calculated only for the period when the sample appeared homogeneous.

Table 1. Properties of polystyrene samples.

Name	Molecular Mass	$T_g/^{\circ}C$
PS 580	580	-20
PS 2000	2000	43
PS 4000	3500	63

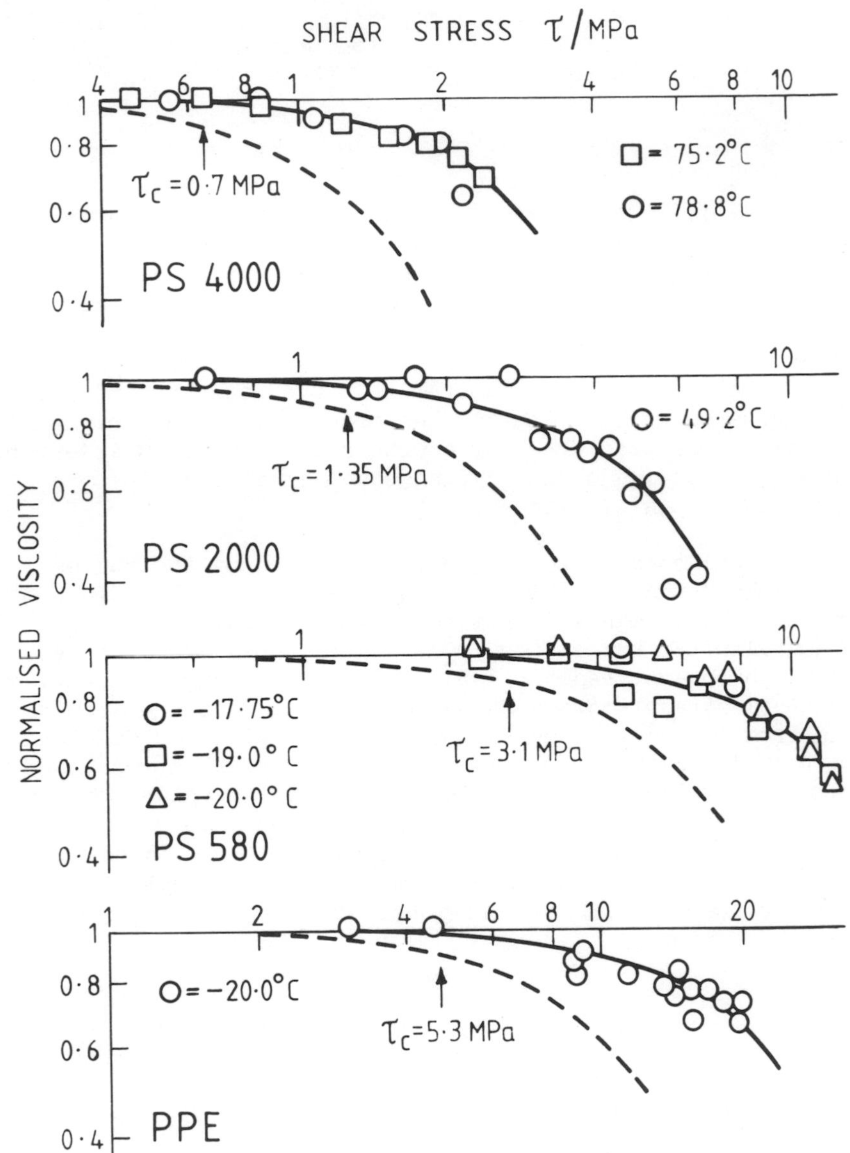

Fig.2. Normalised "apparent viscosity" of polystyrenes (PS) and poly-phenyl ether (PPE) plotted to a base of shear stress —— prediction of Eq.(4), ---- Eq.(3).

Between each viscosity measurement the sample was heated to about $(T_g + 80)^oC$ so that any residual stresses were fully relaxed. The maximum applied stress was limited to about 20 MPa by the onset of cracking, and by the lack of adhesion between the liquid and the smooth surface of the inner cylinder of the viscometer.

DISCUSSION

The experimental values of the apparent Newtonian viscosity are compared in Figure 2 with the values predicted using equation (4), shown by the solid curves. The dashed curves show the corresponding true viscosity variation given by the Eyring equation (3). The best fit to the data is obtained by an empirical adjustment of the theoretical curve along the shear stress axis, equivalent to varying the parameter a. Over the stress range that could be covered without the samples cracking, the experimental points follow the predicted curve within the experimental error, supporting the choice of the Eyring equation to describe the variation of viscosity with shear stress.

A critical shear stress τ_c may be defined by the expression $a\tau_c = 1$, when $\text{Sinh}\,a\tau_c = 1.175$, and $\eta/\eta_o = 0.85$. The values of τ_c determined in this manner, from the best fit of the curves to the experimental data, range from 5.3MPa in the poly phenyl ether, to 0.7MPa in the polystyrene PS 4000. In the polystyrene samples, the critical stress shows a strong dependence on the molecular mass, falling from 3.1MPa for the PS 580, to 1.35MPa for PS 2000 and 0.7MPa for the PS 4000. These values of critical stress, and the dependence on molecular weight, are close to the values of limiting shear stress observed in similar liquids by workers using techniques operating at high pressures (Blair and Winer, 1979; Hirst and Moore, 1978). The technique described here provides a simple way of determining the critical, or limiting, shear stress in liquids at atmospheric pressure.

REFERENCES

Blair, S. and Winer, W.O., 1979, Trans.A.S.M.E. J.Lub.Tech., 101: 251-257.

Eyring, H., 1936, J.Chem.Phys.4: 283-291.

Gray, R.W., Harrison, G. and Lamb, J., 1977, Proc.Roy.Soc.Lond. A356: 77-102.

Hirst, W. and Moore, A.J., 1978, Proc.Roy.Soc.Lond. A360: 403-425.

Plazek, D.J., 1968, J.Polymer Sci.A-2, 6: 621-638.

FORCE BALANCE CAPILLARY RHEOMETER

Jean-Pierre Chalifoux
E. A. Meinecke

Institute of Polymer Science
The University of Akron
Akron, Ohio 44325

INTRODUCTION

Capillary rheometry is extensively used to determine the flow properties of polymers. The most commonly used expression to determine the shear stress at the wall is:

$$\tau_{wa} = \Delta P/4(L/D) \tag{1}$$

with ΔP, the pressure drop across the die, L its length and D its diameter. Eqn. 1 is based upon the assumption that the pressure gradient is constant over the length of the die as shown by the dotted line in Fig. 1.

It has been shown, however, that the pressure profile along the length of the die consists of three regions: (solid line in Fig. 1): a) one arising from entrance effects ($\Delta P_{ent.}$), b) one resulting from the flow of the polymer in the "fully developed flow" region and c) one possibly arising from the rearrangement of the velocity profile at the exit (exit effect)[1].

This type of pressure profile has been determined with the help of pressure transducers implanted in the wall of the capillary[2-4]. The total normal stress thus measured is given by:

$$S_{rr}(R,z) = p(R,z) + \tau_{rr}(R,z) \tag{2}$$

with p(R,z) the hydrostatic pressure at the wall and τ_{rr} (R,z) the normal stress arising from the visco-elasticity of the polymer. In order to determine τ_w, it had to be assumed that the deviatoric component of the normal stress τ_{rr} is constant over

Table 1. Dimensions of Two-part Dies

Die	L_b (cm)	L_b/D	L_t (cm)	L_t/D
A	1.55	4.88	4.47	14.05
B	3.12	9.81	2.90	9.12
C	4.71	14.81	1.31	4.12
D	5.90	18.55	0.12	0.38

All dies: Brass, D = 0.318 cm, L/D = 18.93

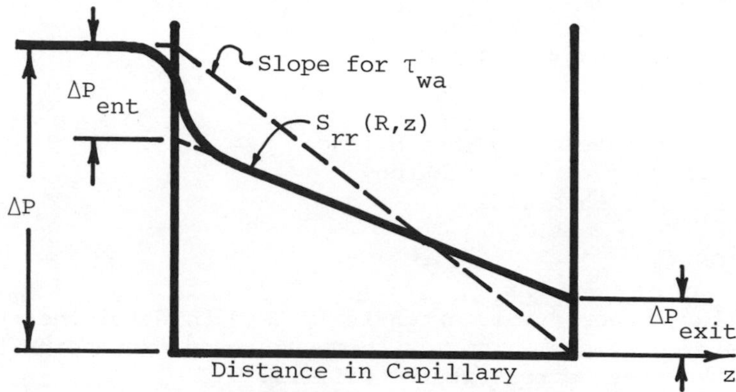

Fig. 1. Pressure Inside a Capillary as a Function of Position

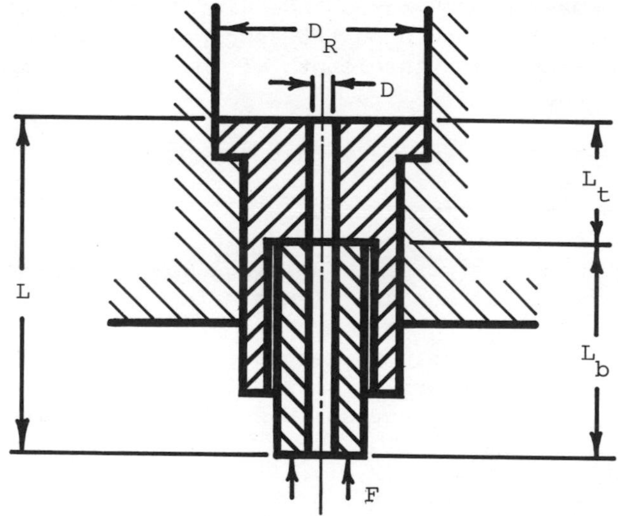

Fig. 2. Schematic of Two-Part Die

the fully developed flow region[5]. With this assumption (i.e., $\partial\tau_{rr}/\partial z = 0$) one obtains the true shear stress at the wall as:

$$\tau_w = (-\partial S_{rr}/\partial z)(D/4) = (\partial p/\partial z) \quad (D/4) \tag{3}$$

There is some evidence, however, that the true shear stress at the wall may be constant over the length of the die[6].

A further problem is the definition of the "fully developed flow region" over which τ_{rr} is assumed to be constant[7]. Criteria used were either that ΔP_{exit} (or the amount of die swell) on the one hand, or S_{rr} on the other, became constant with increasing L/D. The first criterion is often met at L/D = 20, the second one at L/D = 1.

The discrepancy in interpretation could possibly be indirectly resolved using Bagley's approach[8] to obtain the true slope of the pressure gradient inside the die by using dies of different lengths. However, a newly developed method, force balance rheometry[9], can give direct shear stress at the wall data.

EXPERIMENTAL

Force balance rheometry is performed with a two part die as shown in Fig. 2. The bottom die is free to move axially within the top one. During steady state extrusion a force F maintains contact between the two elements. While the polymer is extruded, the force is reduced gradually until an equilibrium is reached for which the force F_e is equal to the summation of all shear stresses within the bottom part of the die:

$$F_e = \pi D \int_o^L \tau_w(z)dz \tag{4}$$

Whenever τ_w is constant over some distance along the length of the die, F_e is expected to be proportional to L_b. This line will intersect the origin if τ_w is constant up to the exit. It is thus possible to obtain information about the changes of τ_w close to the exit without the necessity to perform experiments in that region.

A set of dies with the dimensions shown in Table 1 were used. The tests were carried out with a constant speed rheometer using a low density polyethylene (Tenite 800A, Eastman Chemical Products) having a density of 0.917 g/cm^3 and a melt index of 1.7.

RESULTS AND DISCUSSION

In order to test the capabilities and accuracy of this new device, experiments were performed under conditions where the entrance and exit effects are expected to be negligible, i.e., at

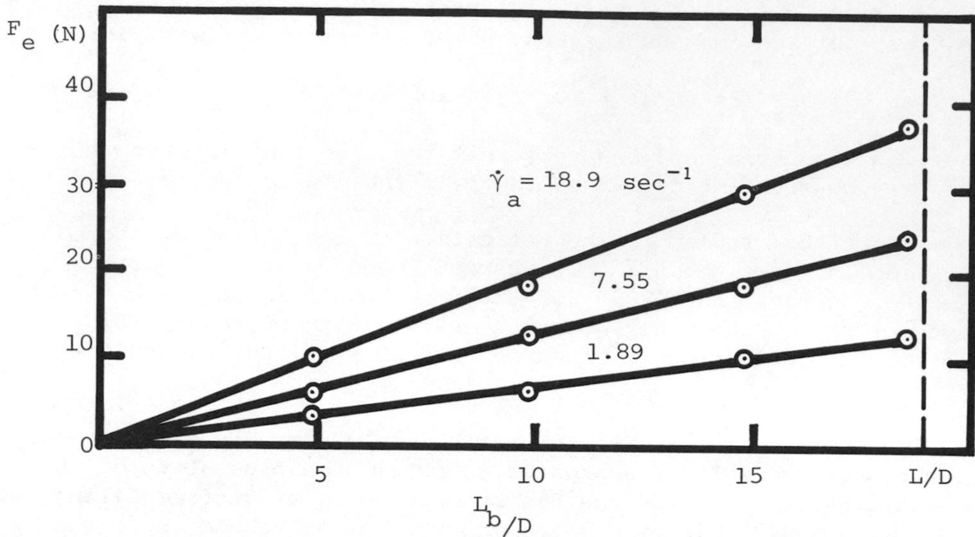

Fig. 3. Equilibrium Force F_e vs. Length of Lower Die for Three Shear Rates.

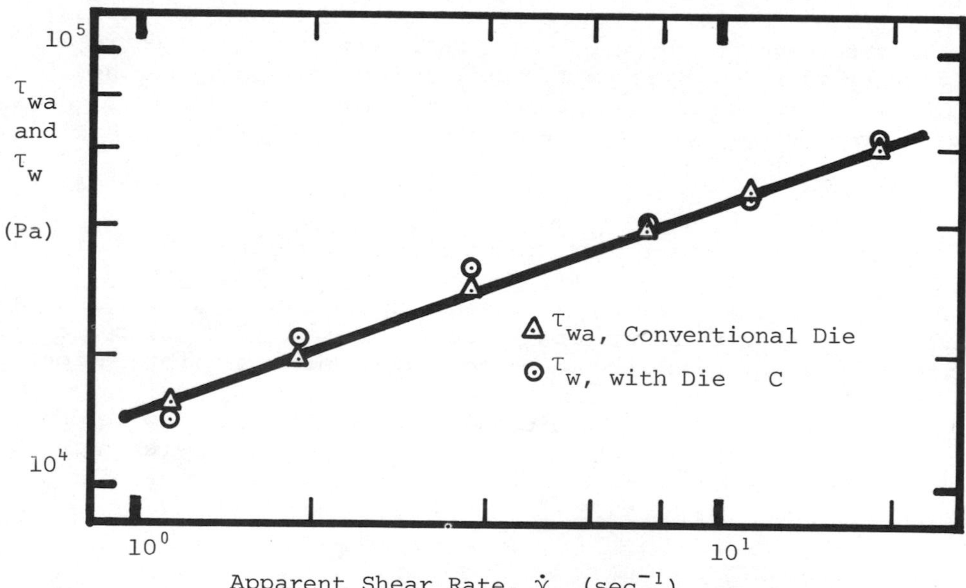

Fig. 4. Apparent Shear Stress and True Wall Shear Stress vs. Apparent Shear Rate for LDPE at 160°C.

low shear rates and a small diameter ratio for the die and the reservoir ($D:D_R = 1.3$).

Values of the equilibrium force F_e (which is proportional to the average shear stress at the wall according to Eqn. 5) have been plotted vs. L_b/D in Fig. 3 for three different shear rates. It can be seen that the data points fall on straight lines going through the origin for all shear rates. This finding indicates that the true shear stress at the wall is independent of the position inside the die. If the shear stress at the wall were not constant up to the exit, the straight lines would not intersect the origin. Whenever data are of this form, measurements with only one die have to be performed to obtain τ_w.

The shear stress at the wall from the F_e values have been plotted vs. the apparent shear rate in Fig. 4. Plotted also is the shear stress obtained with the help of Eqn. 1. Both sets of data are identical over the shear rate region investigated, in other words, entrance and exit effects are not present under the present extrusion conditions. Furthermore, the excellent agreement between the two sets of data show the accuracy this device is capable of.

It is interesting to note that no pressure drop due to entrance or exit effects is present in spite of the fact that over the range of shear rates employed, the material was not Newtonian. The slope of the τ_w vs. $\dot{\gamma}_a$ line in Fig. 4 is 0.48. Also, extrudate swell ranged from 1.7 to 1.9, well in excess of the "Newtonian value" of 1.12.

REFERENCES

1. B. A. Whipple and C. T. Hill, Velocity Distribution in Die Swell, AIChE J., 24:665 (1978).
2. B. C. Sakiadis, Equilibrium Flow of a General Fluid Through a Cylindrical Tube, AIChE J., 8:317 (1962).
3. C. D. Han, M. Charles, and W. Philippoff, Measurement of the Axial Pressure Distribution of Molten Polymers in Flow Through a Circular Tube, Trans. Soc. Rheol., 13:455 (1969).
4. K. Funatsu and Y. Mori, On the Viscoelastic Flow of Polymer Melts in the Nozzle and Reservoir, in: "Proc. Int. Congr. Rheol.", 5th (S. Onegi, ed.), Vol. 4, p. 537. University Park Press, Baltimore, Maryland (1970).
5. C. D. Han, Flow of Molten Polymers Through Circular and Slit Dies, in: "Rheology of Polymer Processing" Academic Press, New York (1976).

6. S. Okubo and Y. Hori, Shear Stress at Wall and Mean Normal Stress Difference in Capillary Flow of Polymer Melts, J. Rheol. 23:625 (1979).
7. C. D. Han and M. Charles, A Criterion for Fully Developed Flow of Polymer Melts in a Circular Tube, AIChE J., 16:499 (1970).
8. E. B. Bagley, End Corrections in the Capillary Flow of Polyethylene, J. Appl. Phys., 28:624 (1957).
9. J. P. Chalifoux and E. A. Meinecke, A New Method to Determine the Shear Stress in Capillary Flow by Force Balance. J. Polym. Sci., Polym. Letters, in print (1980).

SLIP EFFECT IN VISCOSITY MEASUREMENT OF GASES AT LOW PRESSURE WITH AN OSCILLATING-DISK VISCOMETER

K.Yoshida, Y.Kurano and M.Kawata

National Research Laboratory of Metrology,
1-4, 1-Chome, Umezono, Sakura-Mura, Niihari-Gun
Ibaraki 305, Japan

INTRODUCTION

In viscosity measurement of gases at very low pressure with oscillating-disk viscometer, it is necessary to consider the effect of slip at the surface of the disk.

The slip effect is examined by changing the distance between the disk and the fixed plates in the suspension system of the oscillating-disk viscometer[1] for four gases (helium, neon, nitrogen and carbon dioxide) in range of 1 to 1×10^5 Pa ($=7.5 \times 10^{-3}$ mmHg to 1 atm) in pressure and at 25°C in temperature.

EXPERIMENTAL METHOD AND APPARATUS

In measurements of viscosity with the oscillating-disk viscometer, the viscosity is given by the following equation for practical application[2]

$$\eta = K\,(\lambda/T - \lambda_0/T_0) = K'b\,(\lambda/T - \lambda_0/T_0) \quad \text{----------(1)}$$

where η is the viscosity of the gas, λ, λ_0 and T, T_0 the logarithmic decrements and periods of the oscillation in the gas and vacuum, respectively, K a constant of the apparatus, b the distance between the oscillating disk and the fixed plate, $K' = K/b$.

At low pressure in which the mean free path of the gas molecule l is comparable with the distance b, a correction term on the effect of slip should be added to Eq.(1).

$$\eta_c = \eta\ (1 + \Delta b/b) \qquad (2)$$

where η_c is the viscosity corrected for slip effect, and Δb called here the slip correction which is expressed as the increment of the distance b.

From Eqs.(1) and (2), we obtain

$$b/\eta = b/\eta_c - \Delta b/\eta_c \qquad (3)$$

and if the relation between b and b/η is linear for same pressure, the values of $1/\eta_c$ and $\Delta b/\eta_c$ can be obtained from the slope and the intercept in the straight line, respectively.

By the application of this relation, the value of Δb, which gives the information on the slip effect, and also the value of η_c are examined.

The viscometer used in this experiment is described in the previous paper[1].

The suspension system is located in a vacuum vessel with a fused quartz window, and the oscillating disk is suspended by a strand of fine fused quartz and placed between two fixed plates. The essential part of the suspension system is shown in Fig.1. The distance between the disk E and the plate C can be changed by replacing the spacers F. The replacing of spacers is done by the use of the guide bars B and the adjusting screws G. The adjustment

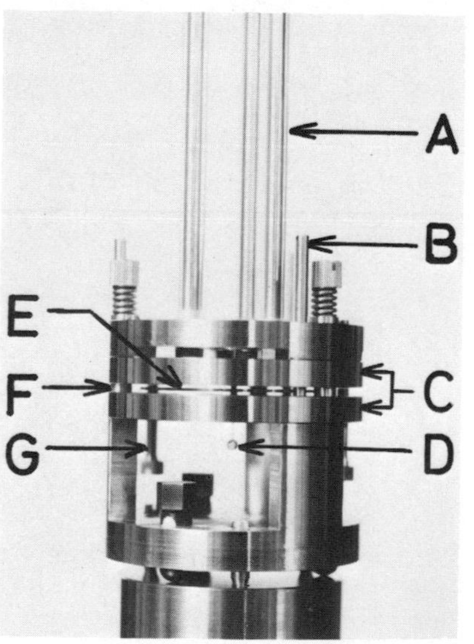

A: SUPPORTING COLUMN
B: GUIDE BAR
C: FIXED PLATES
D: MIRROR
E: OSCILLATING DISK
F: SPACER
G: ADJUSTING SCREW

Fig. 1 Suspension system.

Table 1. Characteristics of the suspension system

Oscillating disk (Material: Super invar)	
Radius 30.001 mm,	Mass 20.907 g
Thickness 0.916 mm	
Suspension strand (Material: Fused quartz)	
Radius 0.036 mm,	Length ca.190 mm
Distance between	
the two fixed plates	2.162mm, 2.472mm, 3.050mm
the disk and the plate, b=	0.623mm, 0.778mm, 1.067mm

for positioning the disk mid-way between the plates is done by the fine adjusting screw on the top of the suspension system. The surface of the disk and the plates have been carefully lapped to a roughness of 3 μm. Two fixed plates and the spacers are made of stainless steel.

The characteristics of the suspension system are shown in Table 1.

The vacuum vessel, made of stainless steel, has the capacity of 4.2×10^3 cm^3. To reduce the pressure in the vessel about 10^4 to 10^{-3} Pa, a mechanical vacuum pump and an oil diffusion pump are used. Low pressures below atmospheric are measured by an alphatron type and an ionization type vacuum gages. The alphatron vacuum gage has been calibrated for each gas used with an accuracy of ± 5% in the pressure range of 10^5 to 1 Pa by using a high precision MacLeod gage especially designed by the authors. The automatic pressure controller combined with the mechanical vacuum and the oil diffusion pumps is used for keeping a constant pressure within ± 5% at pressures below 5×10^2 Pa in the vessel.

The logarithmic decrement of the disk oscillation is obtained by automatically measuring the angular velocities near zero rest position of the oscillation with the aid of electronic techniques[1].

The period of the disk oscillation is also measured electrically every half period by using the reflected beam of the mirror at the rest position.

EXPERIMENTS AND RESULTS

Helium, neon, nitrogen and carbon dioxide have been prepared to examine on the effect of slip at low pressures, and also the effect in terms of the mean free path of the gas molecules.

The experiments were done for the distances b of 0.623, 0.778 and 1.067 mm, and the measurements of these distances were done with a cathetometer having an accuracy of ± 4 μm in the standard deviation.

At first, the relation between the pressure and the viscosity η according to Eq.(1) was experimentally examined for each gas at 25°C. The constant of the apparatus K has been determined for each distance b by using nitrogen gas whose viscosity is $177.73\text{x}10^{7}$ Pa.s at 25°C and $1.013\text{x}10^{5}$ Pa (1 atm). The logarithmic decrement λ_0 and period T_0 of the oscillation in vacuum were measured at $2.7\text{x}10^{-3}$ Pa, and the values observed, $\lambda_0=1.097\text{x}10^{-4}$ and $T_0=25.684$s, were used to determine the constant K and the viscosity η.

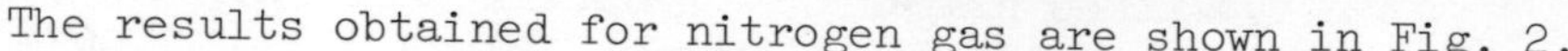
The results obtained for nitrogen gas are shown in Fig. 2.

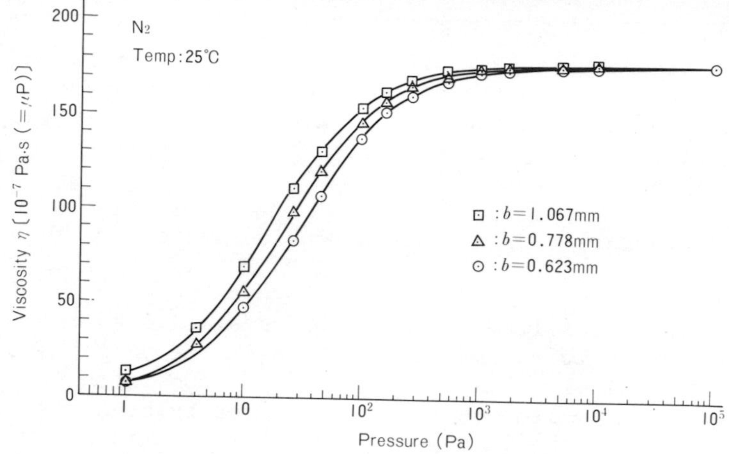

Figure 2 Plot of Viscosity η vs. Pressure for Nitrogen

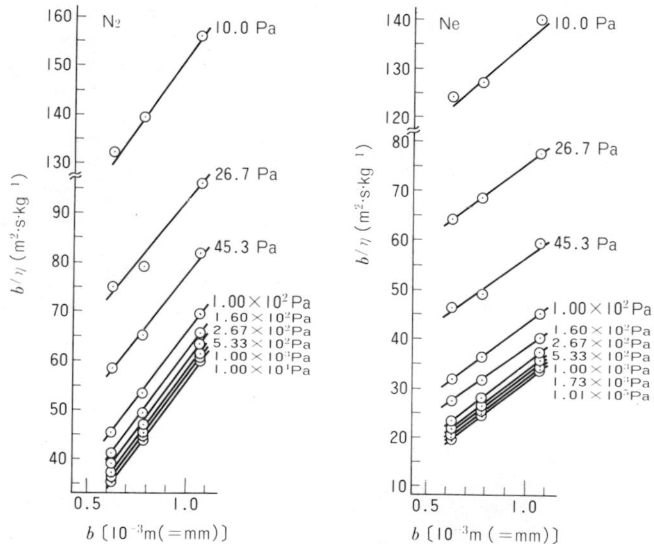

Figure 3 Plot of the Observed Value of b/η vs, the Distance b for Nitrogen and Neon

Fig. 2 shows the value of η increases with the gas pressure, and with the distance between the disk and the plate b. The same results are also obtained for the other gases.

From the data obtained, the relation between the observed value of b/η and the distance b are examined, and shown in Fig. 3 for nitrogen and neon gases. It is found that the relation is almost linear for each gas. Therefore, using Eq.(3), the value of slip correction Δb and the viscosity η_c corrected for slip effect can be calculated.

The values of $\Delta b/b$ obtained for nitrogen and neon gases are shown in Table 2.

Table 2. The values of $\Delta b/b$ for nitrogen and neon (%)

Pressure (Pa)	N_2				Ne			
	Δb (mm)	b= 0.623 mm	b= 0.778 mm	b= 1.067 mm	Δb (mm)	b= 0.623 mm	b= 0.778 mm	b= 1.067 mm
1.00×10^3	0.023	3.7	2.9	2.2	0.042	6.8	5.4	4.0
5.33×10^2	0.051	8.1	6.5	4.8	0.066	10.6	8.5	6.2
2.67×10^2	0.077	12.4	9.9	7.2	0.102	16.4	13.1	9.6
1.00×10^2	0.202	32.4	26.0	18.9	0.416	66.8	53.5	39.0

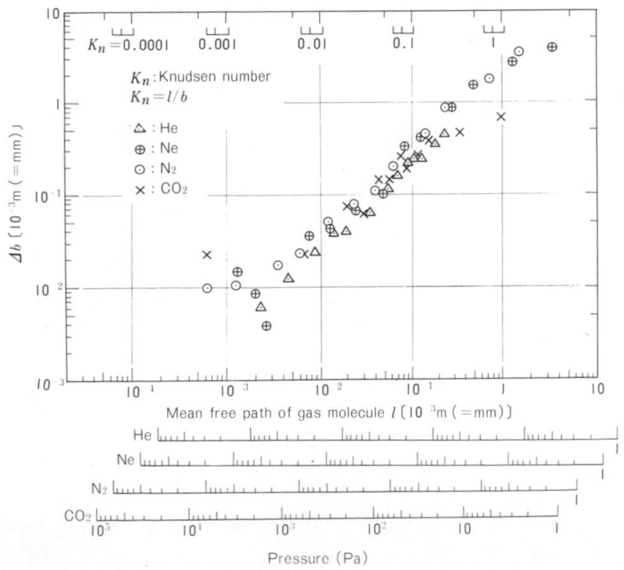

Figure 4 Plot of Slip Correction Δb vs. Mean Free Path of Gas Molecule l for HElium, Neon Nitrogen and Carbon Dioxide

The plots of the slip correction Δb vs. the mean free path of the gas molecule[4] l, and the viscosity η_c vs. the pressure for each gas are shown in Fig. 4 and 5, respectively.

From these experimental results, it is found that the slip effect increases with the mean free path of the molecules, and that the viscosity, η_c, corrected for slip effect keeps a constant in the pressure which corresponds Kundsen number of $K_n \leq 0.1$.

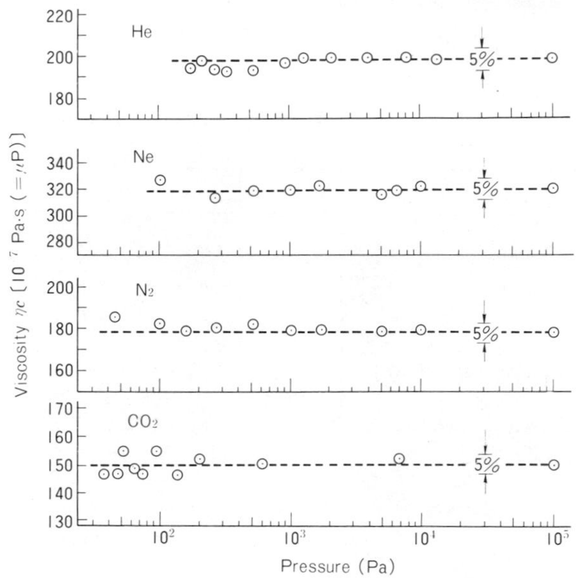

Figure 5 Plot of Viscosity η_c vs. Pressure for Helium, Neon Nitrogen and Carbon Dioxide (at 25°C)

References

1. K.Yoshida, Y.Kurano, and M.Kawata, An oscillating-disk viscometer with automatic measuring system for the decrement of oscillation, Proc. VIIth Int. Congr. Rheol. 470 (1976).
2. D.G.Clifton, Measurements on the viscosity of krypton, J. Chem. Phys. 38: 1123 (1963).
3. K.Kurase, H.Kobayashi, and M.Kawata, An absolute measurement of nitrogen viscosity by capillary method, Proc. VIIth Int. Congr. Rheol. 480 (1976).
4. G.W.C.Kaye, and T.H.Laby, " Tables of Physical and Chemical Constants," Thirteenth Ed., Longmans, London (1966).

A MODIFIED PULSATILE FLOW APPARATUS FOR MEASURING FLOW ENHANCEMENT IN COMBINED STEADY AND OSCILLATORY SHEAR FLOW

J. M. Davies and A. Chakrabarti

Plymouth Polytechnic
Drake Circus
Plymouth
Devon, England

INTRODUCTION

An experimental investigation is carried out to determine the pulsatile flow properties of polymeric liquids in a straight tube of circular cross-section. Barnes et al[1] have essentially used an apparatus in which the mean pressure gradient is controlled and the pulsed pressure amplitude and mean flow rate is measured. In the specially constructed apparatus described in this paper, the mean flow rate is controlled and the pulsed flow rate and mean pressure gradient is measured.

Barnes et al[1] have shown experimentally that when a sinusoidal pressure gradient is superimposed onto a steady flow of viscoelastic liquid in a straight tube,the mean flow rate is increased for the same mean pressure gradient. In particular their results showed that the mean flow rate increased with pulsatile frequency when the ratio of the pulsatile pressure amplitude to the steady pressure gradient was kept constant. These results were in contrast to their theoretical predictions[1] based on the Oldroyd 4 constant model. A Goddard Miller model has also been used[2] in trying to resolve this problem but the mean flow rate prediction showed a similar trend to those of the Oldroyd model. An interesting theoretical analysis has been carried out by Phan-Thien[3] using a simple Generalised Maxwell model. This model predicts an increase in mean flow rate with frequency and therefore shows the same trend as the experimental results. It would appear therefore, that the more complicated models are unable to predict combined steady and oscillatory flow behaviour in a straight tube. To resolve the matter we have decided to obtain more experimental data over a wider range of pulsatile frequencies. A special apparatus had to be constructed for this purpose.

In the conventional apparatus the pulsed pressure amplitude was measured. However, at large frequencies this measurement can be subject to error. The pressure tappings on the test pipe are connected to the pressure transducer by narrow tubes which contain the test liquid. Due to the viscosity of the liquid the pulsed movement in the connecting tubes is restricted and can, therefore, lead to an underestimation of the pulsed pressure value. The mean pressure measurement is not subject to this error. In the modified apparatus we measure the pulsed flow rate amplitude and therefore the pulsed pressure measurement is no longer required.

THEORY

Since the measured quantities are different for each apparatus considered, the theory will also be different. We demonstrate this point by considering the theory for the Generalised Newtonian model.

1. Theory for Conventional Apparatus

In this type of apparatus the pulsed pressure gradient is a measured quantity. Therefore we begin the analysis by writing

$$\overline{P} = \overline{P}_s \, [1 + \varepsilon \, \mathrm{Re}\{e^{i\omega t}\}] \tag{1}$$

where $\overline{P}_s$ is the steady pressure gradient, ω the pulsed frequency, and ε the ratio of pulsed pressure amplitude to the steady pressure gradient. Re{ } denotes the real part of the quantity contained within the brackets.

The Generalised Newtonian model is given by

$$\tau_{ik} = \eta(\dot{\gamma}) e^{(1)}{}_{ik} \tag{2}$$

where τ_{ik} is the shear stress, $e^{(1)}{}_{ik}$ the rate of strain tensor and η the viscosity which is a function of shear rate $\dot{\gamma}$. Using a perturbation analysis for small ε in the equation of motion the percentage increase in mean flow rate is given by

$$I_Q = -25\varepsilon^2 \, \frac{\int_0^{\dot{\gamma}_w} \left[\tau^4(\dot{\gamma}_o)\, \tau''(\dot{\gamma}_o) \Big/ \tau'^2(\dot{\gamma}_o)\right] d\dot{\gamma}_o}{\int_0^{\dot{\gamma}_w} \tau^2(\dot{\gamma}_o)\tau'(\dot{\gamma}_o)\dot{\gamma}_o \, d\dot{\gamma}_o} \tag{3}$$

where $\dot{\gamma}_o$ is the steady state shear rate, $\dot{\gamma}_w$ is the steady shear rate at the tube wall and $(')$ refers to differentiation with respect to $\dot{\gamma}_o$. Fluid inertia effects have been neglected in the derivation of equation (3).

In the case of a power law inelastic model, equation (3) reduces to

$$I_Q = 25\varepsilon^2 \frac{(1 - n)}{n} \tag{4}$$

where n is the power law viscosity index. Both equations (3) and (4) show the increase in mean flow rate to be independent of the pulsatile frequency. Therefore any dependence of the experimental mean flow rate results on the pulsatile frequency must be due to the elasticity of the fluid.

2. Theory for Modified Apparatus

Since the pulsed flow rate is measured in this type of apparatus we begin the analysis by writing the flow rate as

$$Q = Q_s[1 + \delta \, \mathrm{Re}\{e^{i\omega t}\}] \tag{5}$$

where Q_s is the steady flow rate and δ is the ratio of pulsed flow rate amplitude to the steady flow rate. Using the Generalised Newtonian model given in equation (2) and assuming δ is small, the increase in mean pressure gradient is given by

$$I_p = 25\varepsilon^2 \frac{\left[\int_0^{\dot\gamma_w} \dot\gamma_o \, \tau_o^2 \, \tau_o' \, d\dot\gamma_o\right]^2 \int_0^{\dot\gamma_w} \left[\tau_o^4 \, \tau_o'' / \tau_o'^2\right] d\dot\gamma_o}{\left[\int_0^{\dot\gamma_w} \tau_o^3 \, d\dot\gamma_o\right]^3} \tag{6}$$

Using the power law inelastic model equation (6) reduces to

$$I_p = -25 \, \delta^2 \, n(1 - n) \tag{7}$$

It can be see that equations (6) and (7) have completely different forms to those of equations (3) and (4). A viscoelastic analysis would also be expected to give a completely different expression for I_p for the modified apparatus.

Apparatus

The modified apparatus is shown in Fig 1. The steady flow rate is produced by the movement of a piston in a cylindrical container and is forced down by an Instron machine. The steady state flow rate can then be calculated from the piston speed. The pulsed flow rate is generated by a vertical rod which is made to oscillate by

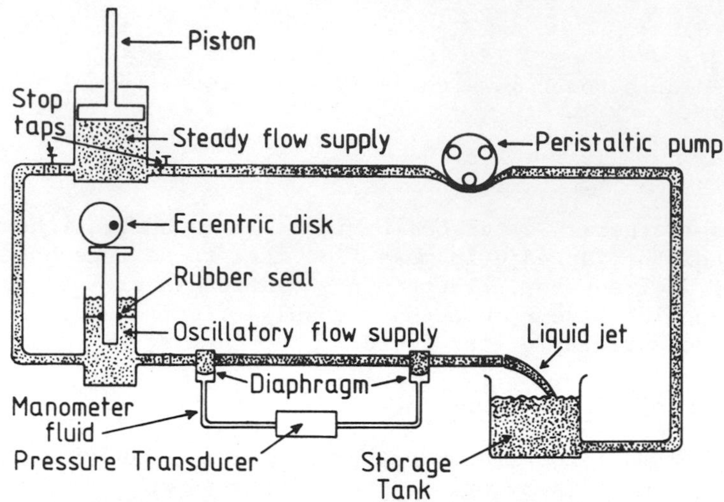

Fig 1 Schematic diagram of modified pulsatile flow apparatus

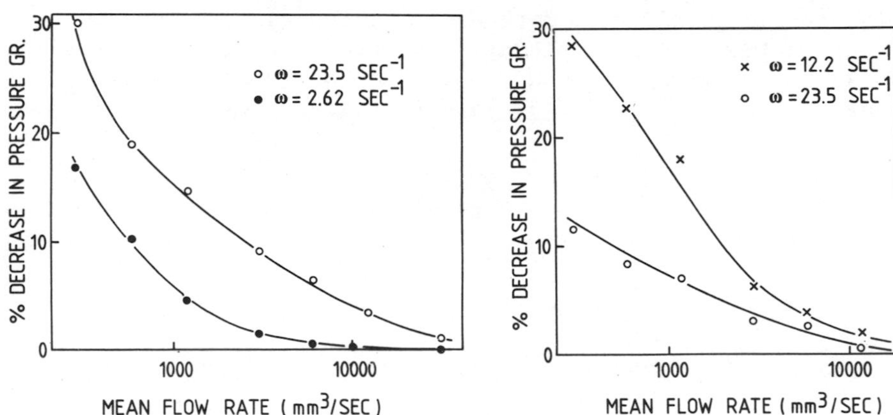

Fig 2 Percentage decrease in mean pressure gradient vs mean flow rate on log base for a 2% polyacrylamide solution where pulsed flow rate amplitude = 309ω mm^3/sec.

Fig 3 Percentage decrease in mean pressure gradient vs mean flow rate on log base for a 2% polyacrylamide solution where frequency x pulsed flow rate amplitude = 3814 mm^3/sec.

means of an eccentric disk connected to a variable speed motor. The combined steady and oscillatory flow then passes through the stainless steel test pipe which contains two pressure tappings. These pressure tappings are connected to the pressure transducer by means of thin plastic tubes containing manometer fluid. The test liquid was separated from the manometer fluid by flexible diaphragms placed at the test pipe pressure connections. At the end of a test run the liquid from the storage tank was returned to the steady flow container by means of a peristaltic pump. This method of refilling the system avoided the need for dismantling the steady flow container after each test run.

In order to obtain an accurate pulsed flow rate measurement it was important that no air bubbles were trapped in the system. For this reason various parts of the apparatus were constructed from a perspex material so that the presence of air bubbles could be detected.

Experimental Results

Preliminary experimental results have been carried out on the modified apparatus for a 2% polyacrylamide solution. The results for percentage decrease in mean pressure are presented in Fig 2 for a constant pulsed flow rate amplitude. We note that there is a larger decrease in mean pressure gradient with increasing pulsatile frequency. For comparison with theoretical results the data must be plotted for a fixed ratio of pulsed flow rate amplitude to the steady flow rate value. This requires the product of the pulsed flow rate amplitude and the pulsed frequency to remain constant. For this case the percentage decrease in mean flow rate results are presented in Fig 3. There is now a smaller decrease in mean flow rate with increasing frequency.

References

1. H. A. Barnes, P. Townsend and K. Walters. On Pulsatile Flow of Non-Newtonian Liquids, Rheol Acta 10: 517 (1971).

2. J. M. Davies, S. Bhumiratana and R. B. Bird. Elastic and Inertial Effects in Pulsatile Flow of Polymeric Liquids in Circular Tubes, Jn. Non-Newt. Fluid Mech.3:237 (1977/1978)

3. N. Phan-Thien On Pulsating Flow of Polymeric Fluids, Jn. Non-Newt. Fluid Mech. (1978).

NON-ISOTHERMAL ROTATIONAL VISCOMETRY

(AN ANALYTICAL EQUATION FOR TREATMENT OF EXPERIMENTAL DATA)

A.Ya.Malkin*
A.M.Stolin**

*Institute of Plastics
Moscow, USSR
**Institute of Chemical Physics
Chernogolovka Branch
Academy of Sciences,USSR

In the viscometry of highly viscous liquids (particularly concentrated polymer solutions and polymer melts) it is necessary to take into account heat dissipation, leading to non-isothermal flow. Thus the general treatment of a viscometric experiment must rely on a joint solution of the heat and force balance equations. Such an approach to the model of the rotational viscometer has been repeatedly discussed (see for example Refs. 1-4). For this model the possibility of hydrodynamic thermal explosion has been established 4,5 . But the non-isothermal method for viscosity measurements has not been used because of the absence of simple and ready-for-use expressions for treatment of experimental data as the known formulae are too complicated.

The goal of this communication consists in deducing an analytical expression based on the solution of the problem of non-isothermal Couette flow. The solution must make it possible to use non-isothermal rotational viscometry for the direct determination of the viscosity of highly viscous liquids.

Let us consider the plane Couette flow (with a very narrow gap between the cylinders) and let the viscosity-vs.-temperature dependence be presented by the equation:

$$\eta(T) = \eta_0 \exp[-\kappa(T-T_0)] \qquad (1)$$

where η_0 is the viscosity at the temperature T_0, K is the temperature-viscosity coefficient and T_0 is the initial temperature.

The equation of stationary heat conductivity with an internal heat generation source can be written as:

$$\lambda \frac{d^2T}{dy^2} + \eta(T)\left(\frac{d\vartheta}{dy}\right)^2 = 0 \qquad (2)$$

where ϑ is the velocity which depends on the transverse coordinate y; $d\vartheta/dy$ is the velocity gradient and λ is the heat conductivity coefficient.

The gap between the cylinders is equal to 2h. The axis y is directed across the gap. One of the cylinders (let it be the inner one at $y=-h$) does note move and the other (at $y=h$) rotates under the action of the constant stress τ_0. In the further we shall consider the case of isothermal walls: $T(-h) = T(h) = T_0$

From the force balance equation one can get the formula for the velocity distribution:

$$\vartheta(y) = \tau_0 \int_0^y \frac{dy}{\eta(T)} \qquad (3)$$

Substituting the expression into Eq.(2) and using the dimensionless parameters $\theta = \kappa(T-T_0)$ and $\xi = y/h$ one can get the following equation for $\theta(\xi)$:

$$\frac{d^2\theta}{d\xi^2} + \delta \exp\theta = 0 \qquad (4)$$

where $\delta = \kappa\tau_0^2 h^2/\lambda\eta_0$ is the dimensionless parameter representing the intensity of the heat output.

The solution of Eq.(4) under the above-mentioned boundary conditions is available 6 . It has the form:

$$\theta = \ln a - \ln\left[\cosh^2\left(\sqrt{\frac{a\delta}{2}}\,\xi\right)\right] \qquad (5)$$

where a is the constant determined from the transcedent equation:

$$a = \cosh^2(\sqrt{a\delta/2}) \tag{6}$$

On intergrating Eq.(3) with account taken of formula (5) one can obtain an expression for the velocity:

$$\frac{\vartheta}{h\tau_0/\eta_0} = \sqrt{\frac{2a}{\delta}}\left[\tanh\left(\sqrt{\frac{a\delta}{2}}\,\xi\right) + \tanh\left(\sqrt{\frac{a\delta}{2}}\right)\right] \tag{7}$$

Formulae (5) through (7) constitute the solution of the direct task. It means that for a liquid with known properties (i.e.,values of κ , λ and η_0) under definite boundary conditions (i.e.,the parameters τ_0, T_0 and h) ane can calculate the value of δ . Then from Eq.(6) it is possible to find a and to determine the temperature and velocity profiles.

But we are mainly interested in the inverse viscometric problem: how to find the viscosity η_0 corresponding to the temperature T_0 and the viscosity-temperature coefficient κ if we measure the values of τ_0, ϑ_0 and T_0 at the boundary surface (values of h and λ are also supposed to be known). The true viscosity value η_0 does not coincide with the "average" (or apparent) viscosity η_m , which can be estimated (quite correct if thermal effects are ignored) as:

$$\eta_m = 2h\tau_0/\vartheta_0 \tag{8}$$

In non-isothermal flow the stress τ_0 is always less than in the isothermal case at the same pre-set velocity ϑ_0 and therefore $\eta_0 > \eta_m$.

As the initial viscosity η_0 is included in the dimensionless factor δ the latter can be rearranged in order to contain the experimental data only. Then

$$\delta = \frac{\kappa\tau_0^2h^2}{\lambda\eta_0} = \frac{\kappa\tau_0^2h^2}{\lambda\eta_m}\left(\frac{\eta_m}{\eta_0}\right) = \frac{\delta_0}{4\eta'} \tag{9}$$

where $\eta' = \eta_0/\eta_m$ and $\delta_0 = 2\kappa h\tau_0\vartheta_0/\lambda$.

From Eq.(7) at $\xi = 1$ the relationship between and the stress (expressed via the dimensionless factor

δ) is expressed thus:

$$\frac{\vartheta_0 \eta_0}{h \tau_0} = 2\sqrt{2\frac{a}{\delta}}\, \tanh\left(\sqrt{\frac{a\delta}{2}}\right) \tag{10}$$

Using the transcedent equation (6), we can find the dimensionless viscosity:

$$\eta' = 2\sqrt{\frac{2a\eta'}{\delta_0}}\, \tanh\left(\frac{1}{2}\sqrt{\frac{a\delta_0}{2\eta'}}\right) \tag{11}$$

Expanding the right-hand side of Eq.(11) into the sum of powers of its argument, one can get for small values of the factor δ :

$$\eta' \approx 2\sqrt{\frac{2a\eta'}{\delta_0}}\left(\frac{1}{2}\sqrt{\frac{a\delta_0}{2\eta'}} - \frac{1}{48}\frac{a\delta_0}{\eta'}\sqrt{\frac{a\delta_0}{2\eta'}}\right) = a\left(1 - \frac{a\delta_0}{24\eta'}\right) \tag{12}$$

If we take into account that Eq.(6) yields the following approximate relationship:

$$a \approx 1 + \frac{\delta_0}{8\eta'} \tag{13}$$

and substitute this into Eq.(12), we shall have:

$$\eta' = \left(1 + \frac{\delta_0}{8\eta'}\right)\left[1 - \left(1 + \frac{\delta_0}{8\eta'}\right)\frac{\delta_0}{24\eta'}\right] \approx 1 + \frac{\delta_0}{12\eta'} \tag{14}$$

and finally:

$$\eta' \approx 1 + \frac{\delta_0}{12} \tag{15}$$

In dimensional terms it has the form:

$$\eta_0 = \frac{2h\tau_0}{\vartheta_0}\left(1 + 0{,}083\,\frac{K}{\lambda}\,\tau_0 2h\vartheta_0\right) \tag{16}$$

This is an approximate relationship. The computer solution of Eq.(11) gives the following, more exact (within 1%) expression for the viscosity:

$$\eta_0 = \frac{2h\tau_0}{\vartheta_0}\left(1 + 0{,}1\,\frac{K}{\lambda}\,\tau_0 2h\vartheta_0\right) \tag{17}$$

The last formula shows the influence of the main factors in a non-isothermal experiment and makes possible a direct estimation of the viscosity by means of experimental values of τ_0 and ϑ_0 when one uses a viscometer with an isothermal wall at $T = T_0$.

Measurements of τ_0 at some pre-set values of ϑ_0 allow one to find the ratio (K/λ). If the thermal conductivity coefficient λ is known, the method proposed here for data treatment in the case of non-isothermal Couette viscometry allows us to find not only η_0 but the temperature coefficient K as well

The results of the investigation and the proposed method of treatment of experimental data can be generalized for the viscometry of non-Newtonian liquids. As shown in 7 it leads to some changes in the form of the dimensionless factor δ , all analytical results for the dimensionless viscosity η' remaining unchanged.

REFERENCES

1. S.A.Bostangijan, A.G.Merzhanov and S.I.Khudyaev,J. Appl.Mech.and Techn.Phys.(in Russian),No.5,45(1965).
2. J.Gavis and R.L.Laurence,Ind.and Rng.Chem.Fund., 7,232(1968).
3. A.M.Stolin, S.A.Bostangijan and N.V.Plotnikova, in "Heat-Transfer",vol.7,261,Minsk,1976.
4. A.G.Merzhanov and A.M.Stolin, Reports of the USSR Academy of Sciences(in Russian),198,1291(1971).
5. A.G.Merzhanov, A.P.Pocetselskii, A.M.Stolin and A.S.Schteinberg, Reports of the USSR Academy of Sciences(in Russian), 210, 52(1973).
6. D.A.Frank-Kamenetskii, Diffusion and Heat Transfer in Chemical Kynetics (in Russian),"Nauka" Publ., Moscow,1967.
7. J.Gavis and R.L.Laurence, Ind.and Eng.Chem.Fund., 7,525(1968).

A NEW RESEARCH RHEOMETER

J. M. Starita

Rheometrics, Inc.
2438 U.S. 22
Union, NJ 07083

Just a decade ago Rheometrics introduced the Mechanical Spectrometer.[1] Sophisticated rheological measurements are now common in most large academic, industrial and government laboratories. These measurements are being made over an ever widening range of materials and test conditions. Rheologists are now beginning to push present instruments to their design limitations. One example is measurements at ultra-low shear rate. It is extremely valuable for molecular characterization to obtain limiting viscosity and normal stress coefficients for polymer melts. This often requires shear rates below 0.01 s^{-1}. At the opposite extreme, ver fast transient measurements are needed for determination of short relaxation times and for identifying the rubbery plateau. Furthermore, not only shear but tensile deformations are required for many applications. To deal with the range of materials encountered in many laboratories much wider ranges in stress detection are required: from water-like fluids to rigid solids.

To meet these increased demands we have recently designed a new research rheometer, the RMS System Four. Early in our design work we realized that it would not be possible to cover such a wide range of applications with a single drive or a single stress transducer. For example, for the very low axial run out needed for ultralow shear rates on polymer melts a high precision air bearing should be used. But the inertia of such a system will be too high for high rate transient testing. And the stiff torque-normal force load cell required for melts will not be sensitive enough for low viscosity systems. Such considerations led us to a modular concept for the System Four.

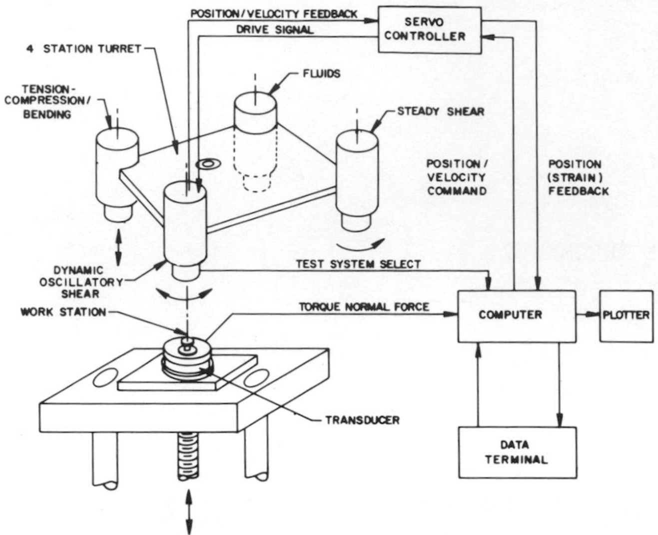

Fig. 1. Schematic diagram of the RMS System Four.

Fig. 1 shows a schematic diagram of the new instrument. The upper part consists of a turret with up to four drive or transducer units. By rotating the turret these units can be brought over the test area. They are then lowered onto a precision ground locating ring of 250mm diameter. This ring provides repeatable location to ±50 µm with quick (5 min) interchange of units. The ring is part of the rigid main frame of the System Four. The axial stiffness of this frame at the point of normal force application is 10 nm/N.

Four stress detection transducers covering different ranges are mounted on a stiff platform. The platform can translate up and down by means of ball screws and a stepper motor. With this control system gap or sample length can be set and maintained to ±1 µm throughout a test.

Two environmental control systems are used. Forced gas convection provides -150 to 400°C control of melts and solids. With nitrogen or other inert gas, oxygen concentration can be maintained below 100 ppm. For more volatile materials a chamber with a saturated solvent environment can be used. Its temperature is controlled from 0 to +80°C with liquid recirculation.

Below we describe three of the four test systems and the central digital control system in more detail. The tensile test system will be included in an extended version of this paper.

Steady Shear

As indicated above, there is strong interest in carrying out steady shear measurements at low rates on very viscous materials. This requires a drive system with good average and instantaneous speed control and low axial run out. Axial noise can have a large effect on normal force measurements of viscous melts[2]. Because of the long time required in these measurements, temperature control and transducer stability are also critical.

Very low rotation speeds present a particular design challenge. We have developed a unique control system which combines the advantages of digital and analog devices. Insteady of generating a series of step pulses as with an encoder, we generate a sequence of sine waves. This combined with our digital control system, provides absolute as well as instantaneous speed accuracy of 0.1% over the range from 10^{-6} rad/s to 100 rad/s.

The drive system is mounted on a large air bearing, resulting in axial run out on the test fixture of <50nm. Stability of the high stiffness torque and normal force transducer is maintained by circulation of temperature controlled liquid through the transducer. Torque drift is less than 10^{-4} Nm/h (<1 gm_fcm/h). Temperature using the forced convection system described above has a stability of $\pm 0.2^{o}$C.

Typical torque data with the steady shear mode of the RMS System Four is shown in Fig. 2. This high molecular weight low density polyethylene shows stress overshoot even at 0.06 s^{-1} shear rate. The steady shear and normal stress data are plotted in Fig. 3. We see that both parallel plate and cone and plate data agree with each other as well as with the data reported by Meissner on this same sample [3,4]. Note that shear rates down to 10^{-4} s^{-1} were achieved and the zero shear rate limit of viscosity was accurately determined.

Transient Shear

Linear viscoelastic measurements are particularly valuable for characterization of material structure. Since these methods use small amplitude deformation they can be employed to study both solids and liquids and for materials undergoing changes with time such as curing, degradation or crystallization. Viscoelastic methods, like sinusoidal oscillations and stress relaxation, are especially helpful in covering a wide range in time scale quickly,[5] characterizing the full relaxation spectra for a complex material. However, in order to make these methods practical, automatic data analysis is desirable. For example, to determine the dynamic moduli from the sinusoidal torque and position signals, digital cross correlation methods should be used.[6] These are built into the System Four microprocessor.

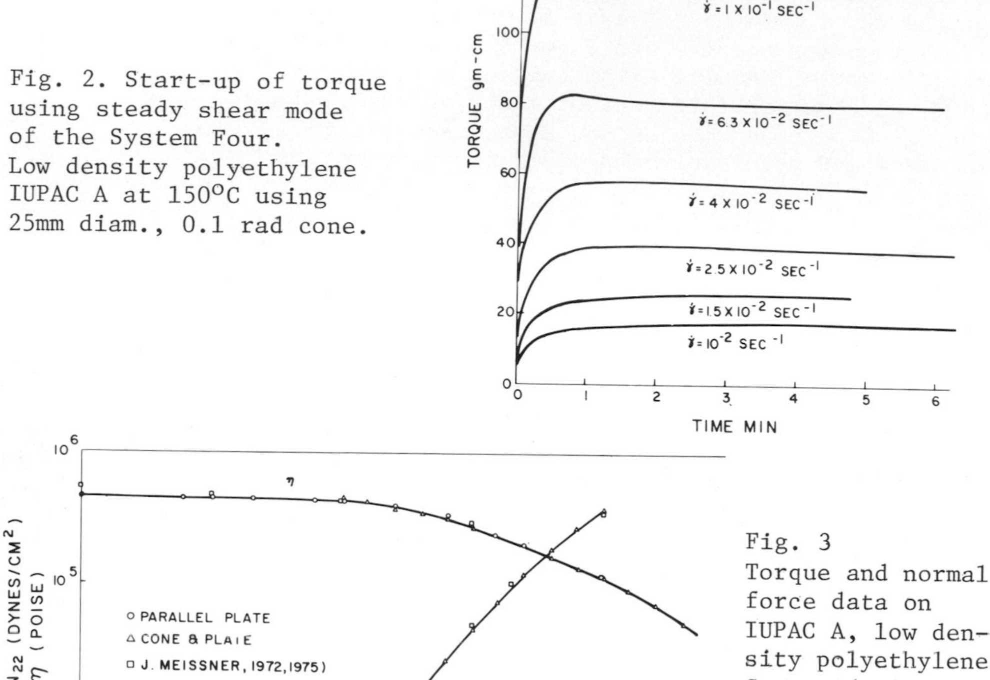

Fig. 2. Start-up of torque using steady shear mode of the System Four. Low density polyethylene IUPAC A at 150°C using 25mm diam., 0.1 rad cone.

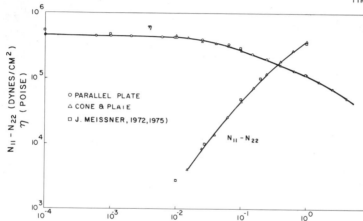

Fig. 3 Torque and normal force data on IUPAC A, low density polyethylene. Cone and plate and parallel plates compared to that reported by Meissner.[3,4]

To measure at high frequency, the short time end of the relaxation spectra, a low inertia motor must be used with a high gain position control system. This is available in the transient shear package, the second mode of the RMS System Four. A special limited rotation motor with a magnesium shaft coupled to light weight test fixtures gives 5ms rise time. This is shown in Fig. 4 where strain in a cone and plate test is plotted vs. time. Such fast response permits determination of the relaxation modulus, G(t), over a rather wide time scale in a single test. This is illustrated for a relatively fast relaxing polydimethylsiloxane melt in Fig. 5. We see that relaxation from 0.01 to 10 s is recorded covering nearly a 1000 fold drop in modulus. Measurement at such short relaxation times has generally not been possible in the past. The transient shear mode can also cover a range of 10^{-3} to 500 rad/s in sinusoidal oscillations, all under automatic programming.

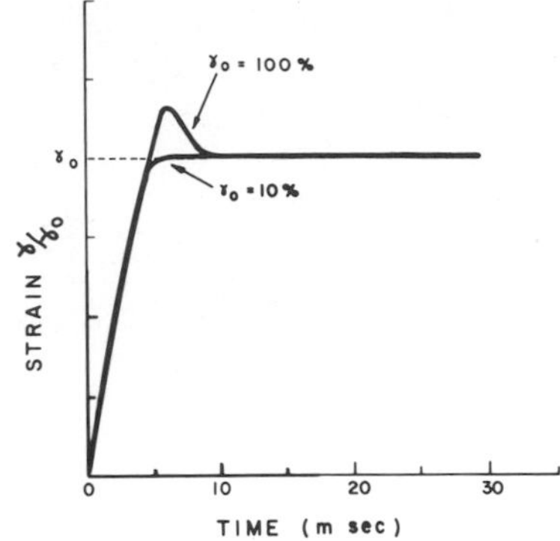

Fig. 4. Normalized strain response with the transient shear system 100% strain represents 0.1 rad. rotation.

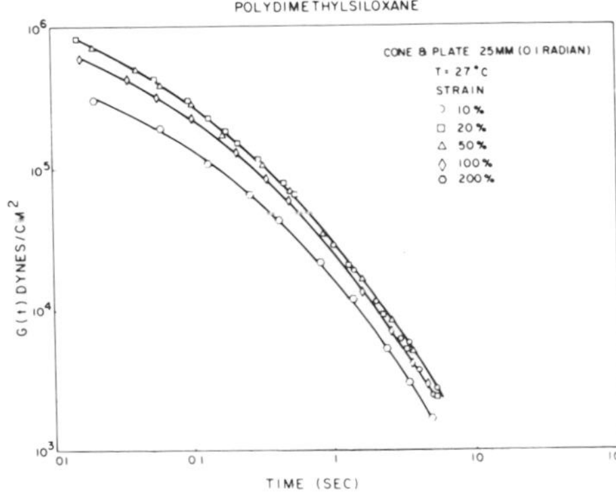

Fig. 5. Shear relaxation modulus plotted vs. time at several strain levels for a polydimethylsiloxane melt.

Fig. 5. also shows the influence of large deformation on the relaxation modulus. At strains greater than 50% G(t) decreases. Wagner and Laun show how such large strain relaxation experiments can be used in a relatively simple way to predict other nonlinear viscoelastic behavior.[7,8]

Fluids

There is a growing interest in making detailed rheological measurements on low viscosity non-Newtonian materials such as coatings, inks, emulsions, paints and biological fluids. There is often a need to characterize these materials at very high shear rates to simulate process experience. Low shear rate viscosity and low frequency dynamic data are important for identifying yield stress, structure recovery and molecular size. Environmental control is a particular problem because of the high volatility of solvents used. Thus, some of the special requirements for an ideal fluids test system are a sensitive stress transducer, a wide range of shear rate and control over solvent evaporation.

This is achieved in the fluids mode of the System Four. For better stability of the transducer, it is placed in the turret and the drive motor on the lower platform. Driving from the bottom also means that the outer cylinder rotates in couette flow. This stabilizes the flow against the formation of Taylor vortices at high rotation speed.

Driving from the bottom also permits temperature control by circulating a liquid in direct contact with the rotating fixture. A rotating seal around the bottom of the motor shaft allows the bath to remain stationary. Inside the test chamber absorbent pads filled with the appropriate solvent are used to maintain a solvent saturated environment.

Fig. 6 shows an application of this system to study the coagulation of whole blood. Sinusoidal oscillations were used to follow the aggregation of blood cells during clotting. McIntyre[9] has related the shape of the G' rise curve to clotting mechanisms. The data were analyzed and plotted automatically using the "cure" subroutine of the System Four microprocessor.

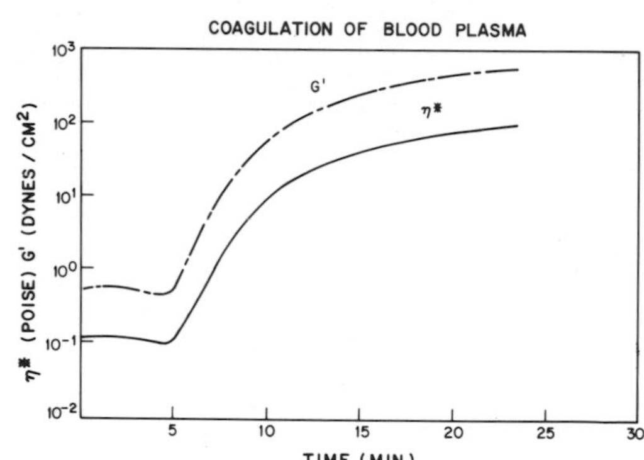

Fig. 6.
Dynamic modulii vs. time for blood undergoing coagulation. Data was collected using the fluids system.

References

1. C. W. Macosko and J. M. Starita, Soc. Plast. Eng. Journ. 27, no. 11:38 (1971).
2. N. Adams and A. S. Lodge, Phil. Trans. A265:149 (1964).
3. J. Meissner, J. Appl. Polymer Sci. 16:2877 (1972).
4. J. Meissner, Pure Appl. Chem. 26:553 (1975).
5. J. D. Ferry, "Viscoelastic Properties of Polymers." 2nd ed., John Wiley, New York (1970).
6. W. M. Davis and C. W. Macosko, Polym. Eng. Sci. 17:32 (1977); C. W. Macosko, R. G. Garritano and J. M. Starita, Soc. Plast. Eng. Tech. Papers 24:2 (1978).
7. M. H. Wagner, Rheol. Acta 15:136 (1976).
8. H. M. Laun, Rheol. Acta 17:1, 138 (1978).
9. L. V. McIntyre, Ann. Rev. Fluid Mech. in press (1980).

USE OF ANNULAR FLAT PLATES IN THE MODIFIED BALANCE RHEOMETER TO MEASURE NORMAL STRESSES

T.E.R. Jones and J.M. Davies

Plymouth Polytechnic
Drake Circus, Plymouth
Devon, England. PL4 8AA

INTRODUCTION

Various workers (Walters[1], Ginn and Metzner[2]) have utilised torsional flow in the determination of the viscometric functions of elastic liquids.

By modifying a conventional rheogoniometer Walters[3] has shown that the instrument can be used to study elastico viscous squeeze film flow. It is observed by Walters that by adding a rotational component to conventional squeezing flow[4] the gap between the plates reaches an equilibrium value at which the rotational normal forces balance the applied load. This work led to the development of the Torsional Balance Rheometer[1] which utilises conventional torsional flow. The novel feature of the rheometer is that the total normal force F and rotational speed Ω, is fixed and the equilibrium gap between the plates is measured.

In this paper we shall describe a modification to the Balance Rheometer in which the total normal force and equilibrium gap is fixed and the speed of rotation is measured by means of a variable speed motor. We shall show that this modified rheometer has certain advantages over the Balance Rheometer. In using this modified instrument we shall utilise three different flow geometries; the parallel plate and cone and plate geometries shown in Fig 1 and the annulus parallel plate geometry shown in Fig 2.

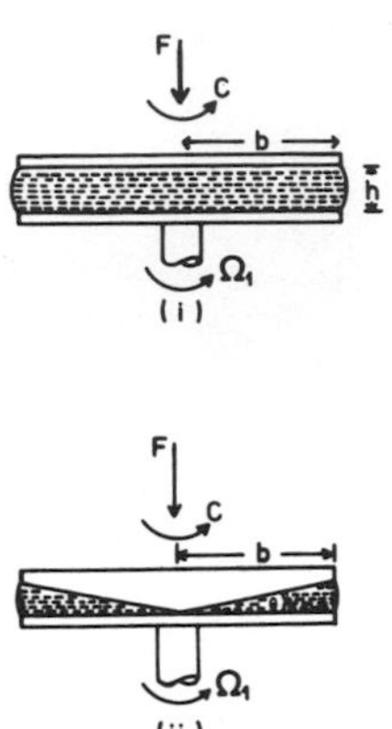

Fig 1 Schematic diagram of
(i) torsional flow
(ii) cone and plate flow

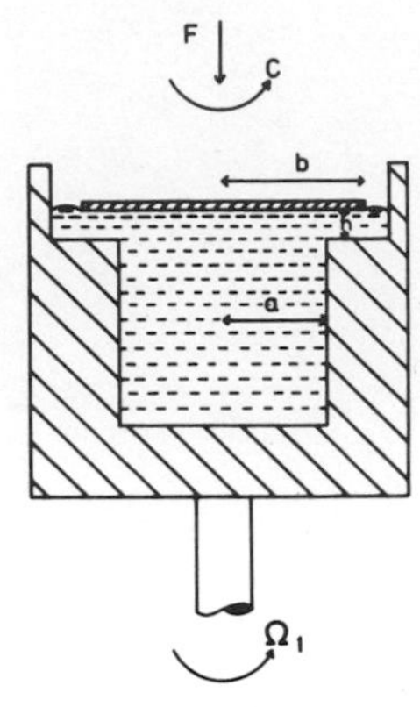

Fig 2 Schematic diagram of annular plate flow.

INSTRUMENTATION

Various modifications were made to the conventional rheogoniometer. These involved removing the torsion head of the rheogoniometer and replacing it by the system shown in Fig 3. The measuring device consists of two parts. The first part is fixed to the main body of the instrument and contains two transducers which monitor the vertical displacement of the upper platen. One of the transducers is used to measure the equilibrium gap set by means of adjustment screws. The other transducer is used to measure very small displacements about the equilibrium gap by setting the transducer to its most sensitive scale. This is necessary because our method of measurement requires the exact plate speed at which the normal forces of the liquid balance the applied load for the set gap. This arrangement differs from the single transducer employed by Walters[1] to measure the gap variation.

The second part consists of a moveable frame which is attached to the top platen and is free to move in a vertical direction through the air bearing. To this frame were attached armatures associated with the two transducers mentioned previously. The weight of the moveable frame was approximately 400g. In order to measure total normal forces below this value it was necessary to attach a pulley system to the moveable frame. By placing weights in the pan or on the frame the effective load could be varied from zero to 10Kg.

The changes referred to above were concerned with normal stress

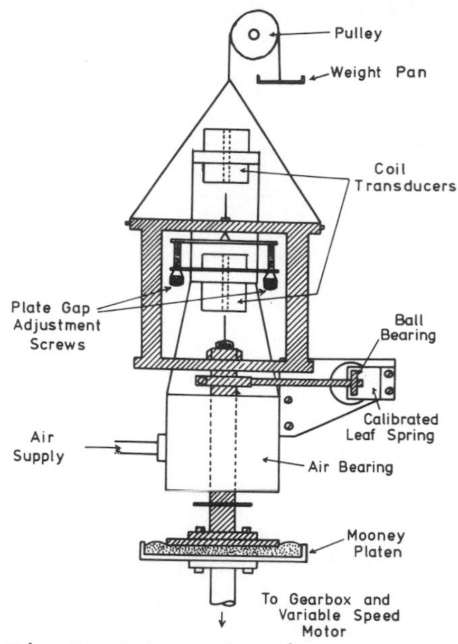

Fig 3 Schematic diagram of modified rheogoniometer

Fig 4 Variation of lower platen speed with gap between the plates

measurement. In order to measure shear stress the Walters[1] modification was carried out.

Theory

The normal stress equation for torsional flow is given by

$$\nu_1 - \nu_2 \Big|_{\dot{\gamma}_b} = \frac{2F}{\pi b^2}\left[1 + \frac{1}{2}\frac{d \ln F}{d \ln \dot{\gamma}_b}\right] \tag{1}$$

$$\text{where } \dot{\gamma}_b = \frac{\Omega_1 b}{h} \tag{2}$$

is the shear rate at the rim. F is the total normal force, Ω_1 is the angular velocity of the bottom plate, b is the radius of the plates and h is the gap (see Fig 1(i)), ν_1 and ν_2 are the first and second normal stress differences respectively. Direct interpretation of experimental results is not possible. However normal stress data can be produced by evaluating the slope at various points of the experimental curve and substituting in the appropriate expression.

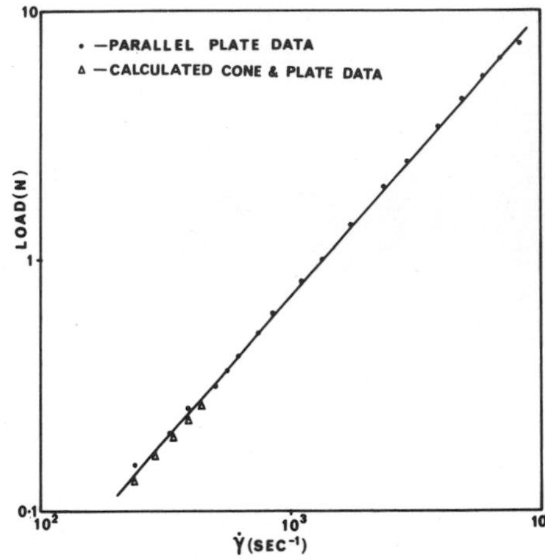

Fig 5 Variation of applied load with shear rate for a 1% solution of polyisobutylene (B200) in dekalin

Fig 6 Variation of applied load with shear rate for a 2% solution of B200 in dekalin

In the case of a cone and plate system the relevant equation is given by Walters[1]

$$\nu_1 \ (\dot{\gamma}) \ = \ \frac{2F}{\pi b^2} \tag{3}$$

$$\text{where} \quad \dot{\gamma} \ = \ \Omega_1/\Theta_o \tag{4}$$

is the shear rate. Θ_o is the angle between the cone and the plate. For this system, shown in Fig 1, a direct interpretation of the experimental results is now possible.

We shall now consider the theory for an annular flow situation shown in Fig 2. We consider cylindrical polar co-ordinates (r,Θ,z) the z axis being along the normal axis to the plates such that the lower annulus and upper plates are given by $z = 0$ and $z = h$ respectively. The top plate is assumed to be of radius b. The inner radius of the annulus is a.

Since the shear rate in the reservoir is small compared with the annular shear rate we can assume that the isotropic pressure is constant and the extra normal stresses are zero throughout the reservoir. For the force calculation to be consistent we must therefore assume that the annular second normal stress difference is negligible. We shall also assume that edge effects on the inner boundary of the annulus can be ignored.

Therefore the total normal force on the plate is given by

$$F = - \int_a^b P_{(zz)} \Big|_{z = h} 2\pi r dr - \pi a^2 P_{(zz)} \Big|_{r = a} \tag{5}$$

It can be shown by assuming ν_2 is small that

$$\pi b^2 \dot{\gamma}_b \nu_1 \Big|_{\dot{\gamma}_b} - \pi ab \dot{\gamma}_a \nu_1 \Big|_{\dot{\gamma}_b} \tag{6}$$

$$= \frac{d}{d\gamma_b} (\dot{\gamma}_b{}^2 F)$$

In principle by the method of induction it is possible to calculate ν_1

$$\text{If } F = c\dot{\gamma}_b{}^s \text{ then } \nu_1 = K\dot{\gamma}_b{}^s$$

$$\text{where } K = \frac{(s + 2)b^{s+1}c}{\dot{\gamma}_b[b^{s+1} - a^{s+2}]} \tag{7}$$

and c, s & K are power law constants. Therefore if applied load F is power law with respect to $\dot{\gamma}_b$ then it is a comparatively simple procedure to calculate ν_1 the first normal stress difference.

EXPERIMENTAL RESULTS

For the torsional flow geometry (shown in Fig 1(i)) equation (1) shows that ν_1 is a function of the rim shear rate and hence a function of Ω_1/h. Therefore if we double the set gap we would expect the measured speed to increase by a factor of two for the same applied load. This linear relationship provides us with a means of checking our experimental method. Experimental results for a 1% solution of polyisobutylene in Dekalin are shown in Fig 4 and confirm this linear relationship. These curves serve to provide confidence in the experimental technique.

We shall now present normal stress data for various solutions of polyisobutylene in Dekalin for the three geometries considered. Fig 5 shows a good argument between the results obtained from the parallel plate system and expected torsional load values calculated from cone and plate data.

Fig 6 contains applied load results for parallel plate and annular plate systems for a 2% solution of polyisobutylene in Dekalin. The parallel plate results are higher than the annulus results, as expected. It can be shown that by using equations (1) and (6) normal stress data can be calculated and, to within experimental error, agreement is obtained.

Disadvantages of the Method

1. Immediate interpretation not possible for parallel plate and annular plate systems.

2. Torsional flow yields the normal stress combination $\nu_1 - \nu_2$ and annular flow yields a complicated combination of ν_1 and ν_2.

3. The instrument can only be used successfully for mobile liquids.

Advantages of the Method

1. From the practical point of view we have found our system easier to operate than conventional rotational instruments.

2. High shear rate data can be obtained by using very narrow gaps in the parallel plate and annular plate systems.

3. The zero gap can be set accurately.

4. The cone and plate system can be used by utilising our modification of the Balance Rheometer.

5. The Mooney platen is no longer essential.

Acknowledgements

We would like to thank Dr. H. t'Sas of Hecules BV, Holland and Mr. W. O'Callaghan of Yewpalm, Watford for all the help and encouragement we have received during the preparation of this work.

References

1. K. Walters and D.M. Binding, Torsional-Balance Rheometer, Jn of Non-Newt. Fluid Mech. 1 : 277 (1976).

2. R.F. Ginn and A.B. Metzner, Proc. 4th Int. Congress on Rheology Part 2, Interscience 1965, Trans. Soc. Rheol. 13 : 429 (1969).

3. K. Walters, G. Brindley and J.M. Davies, Elastico-viscous squeeze films, Jn. of Non-Newt. Fluid Mech. 1 : 19 (1976).

4. K. Walters, D.M. Binding and J.M. Davies, Superimposed Rotation, Jn of Non-Newt Fluid Mech. 1 : 269 (1976).

ON THE TORQUE AND ENERGY BALANCES FOR THE FLOW BETWEEN E(CCENTRIC) R(OTATING) D(ISCS)

H.A. Waterman

Department of Applied Physics
Twente University of Technology
Enschede - The Netherlands

INTRODUCTION

Rheometrical flow systems in which a small perturbation is superimposed on a rigid rotation have been the subject of considerable interest in recent years, both from the experimental and theoretical points of view. Yet there are some problems with respect to the flow field in these rheometers which are not satisfactorily solved. This paper will mainly deal with the orthogonal rheometer, but similar problems arise with other rheometers based on the same principle. The velocity field in ERD flow as realized in the orthogonal rheometer was first proposed by Blyler and Kurtz[1]. It was shown, however, that this velocity field leads to some fundamental problems. First of all there is the so-called "stress power paradox"[2,3], expressing the paradoxal observation that, although energy is dissipated in the sample, no power seems to enter into it from the apparatus. In the second place it was pointed out[4,5] that the total torque acting on the sample differs from zero. This led these authors to the assumption that a torsional flow was superimposed on the primarily assumed flow. Finally, the stress field resulting from the proposed flow field[1], also when an additional torsional flow field is superimposed on it, does not obey the boundary conditions at the free edge. In order to overcome this latter difficulty it was assumed[4] that there is an edge zone in which the flow field is not of the form postulated. Jongschaap and Mijnlieff[6] showed that if the non-isotropic stress field in the bulk continues till the free edge, both the torque balance and the energy balance are satisfied. In particular it appears that in this case the power is supplied to the sample by the free boundary. To satisfy the boundary conditions at the free surface these authors also assume the existence of an edge zone, but in view of the foregoing they postulate that this edge zone plays a dominant role on the power supply to the sample. Since no power can enter the sample through the free edge, their hypothesis is that the energy dissipated in the sample is supplied partly or totally by the upper plate through the boundary between the edge zone and that plate.

FLOW FIELD IN ERD FLOW

In fig. 1a a schematical view of the orthogonal rheometer is given. The sample, with thickness h, is contained between two discs of radius R. The upper disc is driven, f.i. counter-clockwise, by a motor and the other one is free to rotate in an almost frictionless bearing. The axes of rotation of the two discs are not coincident, but are separated by a distance a. Until recently it was commonly assumed that under these conditions the lower plate rotates with the same angular velocity as the upper plate. The velocity field in the sample can then by written in the form[1]:

$$\underline{v} = (-\omega y + \omega\psi z,\ \omega x,\ 0), \tag{1}$$

where $\psi \equiv a/h$. From (1) it is seen that particles turn circles in horizontal planes around the line MM'. In deriving this equation inertia effects are ignored. Inertia effects will not be taken into account in this paper since it will merely complicate matters without adding anything essential to the understanding of the problems. The rate of strain tensor $\underline{D}$ and the stress tensor $\underline{T}$ for this flow field in the region of linear viscoelastic response have been given by several authors. For the stresses $\underline{t}_u$ and $\underline{t}_\ell$ in the fluid at the upper and lower plate it is found:

$$\underline{t}_u = \underline{T}\cdot\underline{n}_u = (\psi G''(\omega),\ \psi G'(\omega),\ 0), \tag{2}$$

$$\underline{t}_\ell = \underline{T}\cdot\underline{n}_\ell = (-\psi G''(\omega),\ -\psi G'(\omega),\ 0) \tag{3}$$

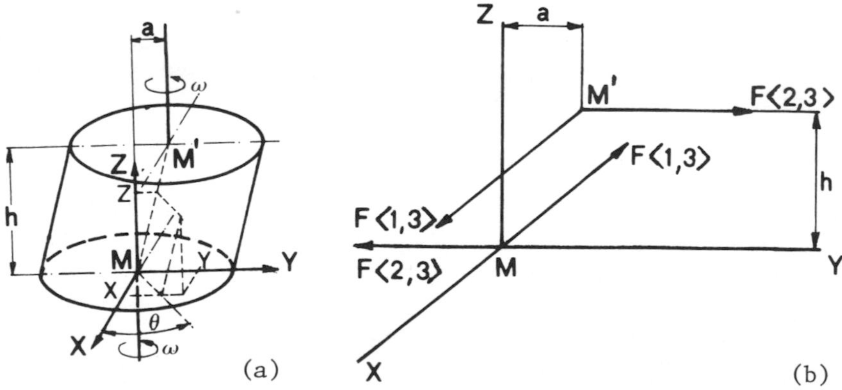

Fig. 1. a) geometry of ERD flow; b) plate forces

$$\underline{M} = \pi R^2 \psi h (G'(\omega), 0, -\psi G''(\omega)). \tag{4}$$

So we see that the x and z components of the torque are not compensated. In order to compensate $\underline{M}\langle z\rangle$ it was assumed[4,5] that there is a velocity lag $\Delta\omega$ between the angular velocities ω_2 and ω_1 in a torsional flow. It is clear, however, that this torsional flow does not compensate the $\underline{M}\langle x\rangle$ component of the torque. Only for a Newtonian liquid $\underline{M}\langle x\rangle = 0$ and hence the torque is compensated. In this case the velocity lag $\Delta\omega$ between the discs required to compensate $\underline{M}\langle z\rangle$ is found[4,5]:

$$\Delta\omega = 2\omega(a/R)^2. \tag{5}$$

The rate of energy dissipation W is found[6]:

$$W = \pi R^2 h \omega \psi^2 G''(\omega). \tag{6}$$

On the other hand it is found from the flow field that no power is supplied to the fluid by either the upper and lower plate[6]. If, however, a torsional flow according to eq. (5) is superimposed on the original flow field (1), this flow will introduce an extra shear stress $\tau_{\theta z}$, which for a Newtonian fluid amounts to:

$$\tau_{\theta z} = \eta_0 r \Delta\omega/h \tag{7}$$

and the power supplied by the upper plate to the sample is then found from

$$P = \int_0^{2\pi}\int_0^R \eta_0 \omega \Delta\omega \; r^3/h \; d\theta \, dr = \pi R^4 \omega \eta_0 \Delta\omega/2h. \tag{8}$$

Substitution of (5) into (8) leads to:

$$P = \pi R^2 h \omega \psi^2 G''(\omega). \tag{9}$$

So it is seen that by the introduction of a torsional flow according to (5) both the torque and energy balances for a Newtonian fluid are satisfied. For a general linear viscoelastic liquid, however, the torque balance is violated in so far that $\underline{M}\langle x\rangle \neq 0$.

From (1) is is seen that for $\psi = 0$ the sample performs a rigid rotation. So for $\psi << 1$ the flow field will be a superposition of a small perturbation on a rigid rotation. It is inviting therefore to consider the flow field in a coordinate system rotating with the same angular velocity ω as the plates. In the skew cylindrical coordinates r, θ, and z, zie fig. 1a, the coordinates of a material point can be written, with $\theta = \omega t + \phi$:

$$\left.\begin{aligned} x &= r\cos(\omega t + \phi) \\ y &= r\sin(\omega t + \phi) + \psi z \\ z &= z \end{aligned}\right\}. \tag{10}$$

Now consider a frame rotating counter-clockwise with angular velocity ω and which for $t = 0$ coincides with the spatially fixed frame. In this frame the velocity field reads:

$$\underline{v}^* = \psi\,\omega\, z(\cos \omega t,\; -\sin \omega t,\; 0). \tag{11}$$

The rate of strain tensor $\underline{D}^*$ is found from $\underline{D}^* = \underline{\Omega} \cdot \underline{D} \cdot \underline{\Omega}^T$, where $\underline{\Omega}$ is the transformation tensor between both frames.
In the region of linear viscoelastic response it is found for the non-zero components of the stress tensor[2]:

$$\left.\begin{aligned} T^*_{13} = T^*_{31} &= + G'(\omega)\sin \omega t + G''(\omega)\cos \omega t \\ T^*_{23} = T^*_{32} &= + G'(\omega)\cos \omega t - G''(\omega)\sin \omega t \end{aligned}\right\}. \tag{12}$$

From (12) the stresses $\underline{t}^*_u$ and $\underline{t}^*_\ell$ in the fluid at the upper and lower plate become:

$$\left.\begin{aligned} \underline{t}^*_u &= \underline{T}^* \cdot \underline{n}^*_u = \psi(T^*_{13},\, T^*_{23},\, 0) \\ \underline{t}^*_\ell &= \underline{T}^* \cdot \underline{n}^*_\ell = \psi(-T^*_{13},\, -T^*_{23},\, 0) \end{aligned}\right\}. \tag{13}$$

Since the rate of energy dissipation is frame-indifferent, it is given by (7). For the power supplied to the sample by the upper and lower plate it is found from (11) and (13):

$$P_u = \int_{S_u} \underline{v} \cdot \underline{t}\; dS = \pi R^2 h \omega \psi^2 G''(\omega) \tag{14}$$

$$P_\ell = \int_{S} \underline{v} \cdot \underline{t}\; dS = 0. \tag{15}$$

So it is found that in the rotating frame there is no "stress power paradox" at all and no torsional flow is required to remove it. The torque balance, however, is still violated.

THE CORRESPONDENCE PRINCIPLE

It is well-known[2, 7] that in ERD flow the deformation can be described as due to a superposition of two harmonic shear deformations polarized perpendicular to each other, both with amplitude a, but showing a phase shift of $\pi/2$ radians. So the deformations in the sample are harmonic and the correspondence principle can be applied, i.e. the solution can be found by solving the analogous problem with the same boundary conditions for a linear elastic material and substituting the complex moduli into the final relations. This leads to the theorem[7].

If inertia effects are ignored, the magnitude of the force on a linear viscoelastic fluid performing a nearly rotational flow is found from the elasticostatic response of a linear elastic solid of the same geometry, showing a relevant modulus $|M^*(\omega)|$ for the deformation which turns the flow from rotational into nearly rotational. The direction of the force lags the direction of the deformation by the angle $\delta = \tan^{-1} M''(\omega)/M'(\omega)$ in the direction of rotation of the apparatus.

In order to solve the elasticostatic problem for the orthogonal rheometer one has to find the stress field in a cylindrical bar with radius R and length h, where $h \ll R$ due to a displacement of one end face over the distance a with respect to the other one in such a way that the end faces remain parallel and a constant distance h apart.
Since no tractions are applied at the free surface, this is not a state of simple shear. This was already noted by Kearsley[8], who stressed that the flow between ERD is not controlled. Unfortunately no exact analytical solution for this problem is present. Is is more illustrative therefore to consider the case of a long slender cylinder. This applies to

THE REVOLVING ROD APPARATUS OF KIMBALL AND LOVELL

The basic principle underlying the orthogonal rheometer was published as early as 1923 by Kimball[9]. In the apparatus based on this principle[10] the sample, a rod with a diameter of 0.5 inch and a length of about 1 meter, was supported in two ball bearings. The rod was rotated by an electric motor at one end. The outer bearing was located so that more than half of the rod overhung. On the other end of the rod there was another ball bearing, on which a frame loaded with a dead weight Q was hung. From a theoretical investigation it was shown that as a result of internal friction in the bar material the loaded end should be deflected out of the x-z plane over the angle ϕ, where $\tan \phi = \Lambda/\pi$, see fig. 2. Since $\Lambda = \pi \tan \delta$, the angle of deflection equals the loss angle. This behaviour was confirmed by the experimental results on various solids.

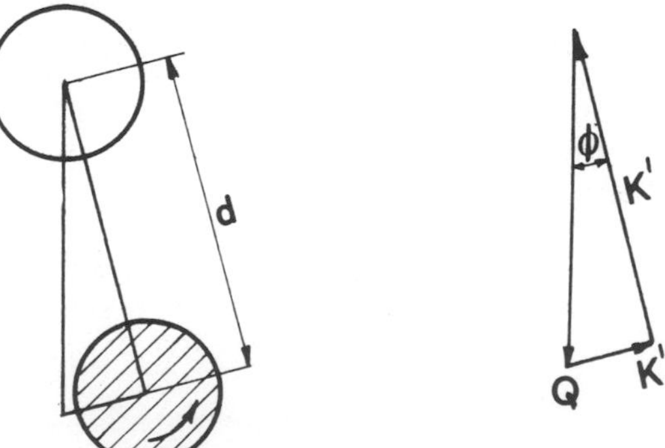

Fig. 2. View of weighted end of rod showing side-ways deflection due to viscoelastic behaviour

From fig. 2 it appears that the theorem stated before is confined. The force lags the deformation over the angle δ in the direction of rotation of the bar, while the magnitude of the deformation is: $d = Q \cos \delta / C\, M(\omega) = Q/C|M^*(\omega)|$. Note that the in-phase component of the force has the direction of the displacement. The elastico-static problem to be solved here is the deflection of a bar clamped at one end and loaded at the other end by a dead weight Q. For simplifying the calculations body forces will be ignored. The relevant equations are:

$$\mathrm{div}\, \underline{T} = 0 \tag{16}$$

$$(1+\nu)\nabla^2 T\langle ij\rangle + T_{pp,ij} = 0, \tag{17}$$

where

$$T_{pp,ij} \equiv \frac{\partial^2 T\langle pp\rangle}{\partial x_i\, \partial x_j}. \tag{18}$$

(18) are the Beltrami compatibility equations, in which body forces are not taken into account.
Further we have the boundary condition

$$\underline{t} = \underline{T} \cdot \underline{n} = 0 \tag{19}$$

on the cylindrical surface.
The problem was solved by Saint Venant[11] by using a semi-inverse method. The solution reads:

$$T\langle 11\rangle = T\langle 22\rangle = T\langle 12\rangle = 0 \tag{20}$$

$$T\langle 33\rangle = -Q(\ell - z)x/T \tag{21}$$

$$T\langle 13\rangle = \frac{(3+2\nu)Q}{8(1+\nu)I}\left(R^2 - x^2 - \frac{1-2\nu}{3+2\nu}y^2\right) \tag{22}$$

$$T\langle 23\rangle = -\frac{(1+2\nu)Q\, xy}{4(1+\nu)I}, \tag{23}$$

where $I \equiv \pi R^4/4$.
(20) - (23) present an exact solution of the problem, satisfying the boundary conditions.

From the stress field found some important conclusions can be drawn. In the first place it is seen that the stress tensor shows two non-zero oof-diagonal components, $\underline{T}\langle 13\rangle$ and $\underline{T}\langle 23\rangle$. Both components are functions of the x and y coordinates in such a way as to satisfy the boundary conditions at the free surface. In the second place the stress tensor has a non-zero diagonal component which takes care of the torque balance of the $M\langle z\rangle$ component of the torque.

In order to convert the revolving rod apparatus into an instrument which essentially is an orthogonal rheometer the rod is twisted over the angle $-\phi$, the ball bearing at the far end is fixed

and the dead weight is removed. In the final situation the forces on the ball bearing are a force along the y axis due to the elastic force in the sample and a force in the x direction due to the viscous response.
In fact this step, which is a main contribution to the development of the present orthogonal rheometer from the revolving rod apparatus, was taken by Gent[12].

DISCUSSION

The problems encountered with ERD flow are a consequence of the fact that this flow is not controlled. In principle ERD flow is a controllable flow, since the necessary surface tractions can be supplied by geometrical constraints on the boundary without reference to the properties of the material[8]. Controllable flows offer the possibility of producing a known flow in a material with unknown properties. As shown in this paper, in a rotating frame the power is supplied to the sample by the upper plate. If the ERD flow were controlled by the addition of an elastic cylindrical confining wall[8], the power input in an Eulerian frame would take place via the surface tractions on the cylindrical boundary[6]. These tractions would also take care of the torque balance. Since in reality these constraints are absent, the resulting flow depends on the properties of the material. This flow must ensure the torque balance through the appearance of normal forces and compensate any energy dissipation in the sample through the appearance of a torsional flow.

REFERENCES

1. L.L. Blyler, jr., S.J. Kurtz, J. Appl. Phys. 11: 127 (1967)
2. M. Yamamoto, Japan. J. Appl. Phys. 8: 1252 (1969)
3. J.F. Hutton, in: Theoretical Rheology, p. 365 (London 1975)
4. P. Payvar, R.I. Tanner, Trans. Soc. Rheol. 17: 449 (1973)
5. W.M. Davis, C.W. Macosko, AIChe J. 20: 600 (1974)
6. R.J.J. Jongschaap, P.F. Mijnlieff, Rheol. Acta 15: 623 (1976)
7. H.A. Waterman, Rheol. Acta 15: 444 (1976)
8. E.A. Kearsly, J. Res. NBS C 74: 19 (1970)
9. A.L. Kimball, Phys. Rev. 21: 703 (1923)
10. A.L. Kimball, D.E. Lovell, Phys. Rev. 30: 948 (1927)
11. J. Saint Venant, J. Mathémat. (Lionville), series 2, vol. 1 (1856)
12. A.N. Gent, Brit. J. Appl. Phys. 11, 165 (1960)

A COMPUTERIZED TORSIONAL PENDULUM FOR MEASURING THE DYNAMIC MECHANICAL PROPERTIES OF POLYMERS

Michael Seeger

Brabender OHG
Postfach 350162
D-4100 Duisburg 1

Introduction

The mechanical properties of polymers are largely determined by the motion of long and short segments of the main chain and of side chains.
The average time required for moving a chain segment is known as the relaxation time, which is correlated to a frequency of the applied field where the loss of energy retains its maximal value. Rising the temperature diminishes the relaxation time. Consequently the relaxation process can be observed also by variation of temperature at a constant frequency. In polymers there exists a whole spectrum of relaxation times, resulting an increased linewidth and often more than one dispersion region.
Because of its universal applicability dynamic mechanical testing is widely used. This test is usually performed in a wide range of temperature and in a narrow range of frequency. It is difficult to vary the frequency in a corresponding wide range. For many glass transitions, a decade of the frequency corresponds to a temperature shift of about 5-10°C. For solid polymers, the torsional pendulum is most frequently applied and described in many standards, as DIN 53445, ISO R537 and 4665, ASTM D2236 [1-6].

Design of the Instrument

Improvements over conventional techniques were made by designing the BRABENDER Torsionautomat System Lonza, which has the following features:
1) Free Torsional Oscillations.
Such devices are relatively simple to design and to operate, suitable for a wide variety of

specimens (also for very hard samples and small mechanical loss factors). The impact to the sample is extremely small (maximum measuring angle 3° corresponds to a maximum shear amplitude $\gamma = 10^{-3}$), thus nonlinear effects of the amplitude of deformation can be neglected as well a s viscous heating effects.
2.) Suspension of the Oscillating Disk.
In experimental setups according to the Schmieder-Wolf pendulum [1] the sample is supporting the inertia member and is consequently under permanent stress. The suspension on a wire avoids an excessive load on the specimen. Thus, the frequency can be adapted between 0.1 and 10 Hz by altering the interia mass without an influence on the stress at the sample [7]. The elongation of the sample is measured continuously at the balance, which enables also to evaluate static glass transitions from the change of the heat expansion coefficient. This corresponds to a dilatometric measurement [8].
3.) Complete Automation of the Control, Measuring and Evaluation.
The test is relatively simple, but very time consuming, since runs are required at small increments of temperature (1°C). Also a data reduction is necessary for getting the final result. For this reason, a complete automation of the whole test was achieved by means of a desk computer (8 K Byte RAM) and a corresponding high resolution interface. The automation does include also the data storage and graphic plotter representation. This has been described in a previous paper [8].

Brief Review on the Theory

The dynamic mechanical behaviour of a viscoelastic material can be described by the complex shear (torsion) modulus $G^* = G' + i\,G''$.
This is based on the assumption that there is only a narrow frequency band, which is true in the approach of a sinoidal low damped oscillation.
For a combined system of specimen, inertia mass and a wire the following equations are derived [4, 6, 8]:

$$(1) \qquad G' = 4\pi^2 a I \cdot \left\{ f^2 \left(1 - \frac{\Lambda^2}{4\pi^2}\right) - f_0^2 \right\}$$

$$(2) \qquad G'' = 4\pi a I f^2 \cdot \Lambda$$

The measured quantities are
f = resonant frequency of the oscillation with sample
Λ = logarithmic decrement of the combined system.

f_o (resonant frequency of the wire), I (in ertial mass) and the size factor a are entered into the computer. For specimen with rectangular cross section and $H < B$ (L = length, B = width, H = thickness) it is
$a = 3\,L / B.H^3\,(1 - 0.63\,H/B + 0.052\,(H/B)^5)$

Results and Discussion

Polymethylmethacrylate (PMMA) of molecular weight $\simeq 10^6$ was chosen to demonstrate the potential of this test and to study the influence of experimental conditions (e.g. the frequency) on the results. The samples with different thickness were cut from one piece and were thermally conditioned around the glass temperature. Before testing the thickness H and width B were measured exactly, while the length L between the two clamps was always 50 mm. A moment of inertia of 82 kg mm^2 (f_o = 0.17 Hz) was selected and the oscillations were initiated at 1°C (1 min) intervals.

Fig. 1 shows the graphic representation of a test with a sample of H = 0.555 mm and B = 9.84 mm carried out in the temperature range between -170 and +180°C. G', G" and the measured quantities f and Λ are plotted logarithmically versus the temperature (°C).

The extension is plotted at the lower linear part (scale from -1 to +1 %; 0 % elongation at 20°C).

The diagram exhibits the α-transition (glass-transition) at 132°C. Since measurements of $\Lambda > 2.5$ were difficult, a curve was calculated by using a least square polynomial approximation, and the maximum of the drawn Λ was evaluated. G" has one maximum at 125°C (at 1 Hz) and a second maximum at 20°C (at 2.35 Hz), the latter corresponding to the β-transition due to the relaxation of the carbomethoxy side chain.

The shoulder at -100°C may be correlated to the H_2O-transition 5 . G' decreases from $2.7 \cdot 10^9$ to 10^6 Pa, while the frequency varies from 3 to 0.185 Hz.

Fig. 2 illustrates the results from the same material with H = 1.59 mm and B = 9.85 mm between -160 and +200°C. Due to the greater thickness the frequency was about 13 Hz at the beginning and 0.3 Hz at the end.

The α-transition peak has shifted to 138°C.

For G" the α-maximum is again at 125°C (4 Hz) while the β-peak appears now at 40°C (10 Hz).

For calculating the frequency-temperature variation of the β-transition an Arrhenius equation can be applied. This has been used before with a forced torsion oscillator at frequencies of 0.1, 1 and 10 Hz [9] .

With $a_T = f/f_1$ we calculate for the temperature transformation:

$$T_1 = T \,/\, \left(1 + \frac{2.303\,RT}{\Delta H} \cdot \log \frac{f}{f_1}\right) \qquad 3)$$

Through this equation the values of G', G", Λ etc. measured at the temperature T with the frequency f are shifted to the temperature T_1 with the frequency f_1.

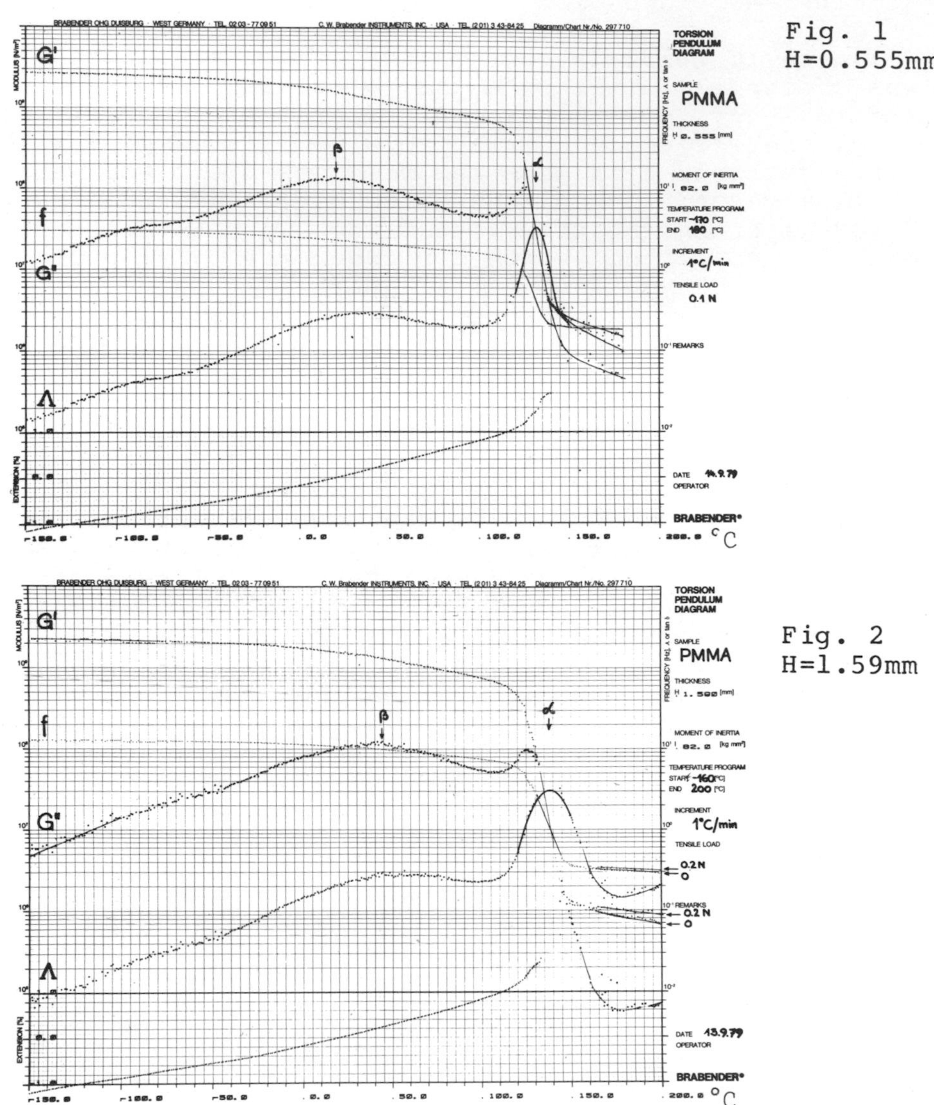

Fig. 1
H=0.555mm

Fig. 2
H=1.59mm

From the above data we estimate an activation energy $\Delta H \simeq 14$ kcal/mole, which is close to reported values (17-21) [5,9]. In Fig. 3 the functions G', G" and Λ from the data in Figs. 1 and 2 were transformed according to Eq. (3) by setting for both tests a uniform frequency f_1 = 1 Hz. Now the β-transition has moved to 5°C. The transformed Λ-values of both tests are coincident in the region of the β-peak. There is less agreement in the region of the H_2O-transition since the activation energy might be different.

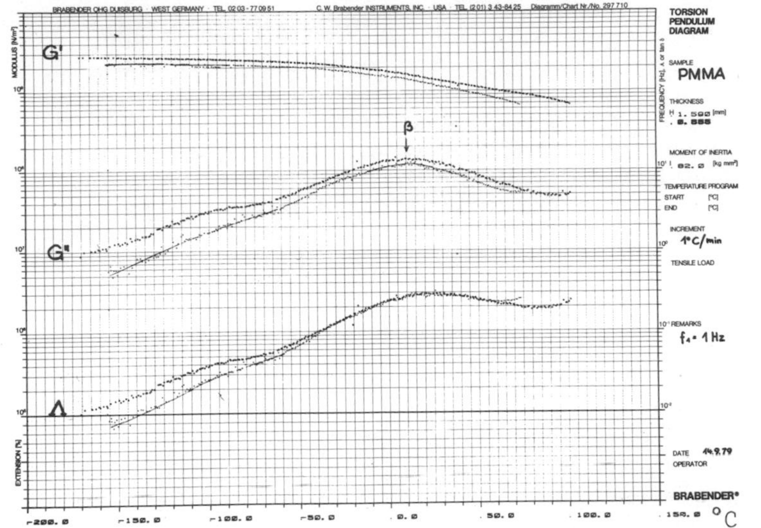

Fig. 3 Transformed data to f_1=1Hz

For G' and G" a gap of about 15 % is between the two measurements H = 0.555 mm (upper curves) and H = 1.59 mm (lower curves). Tests with other sample sizes have evidenced that end corrections by the clamps might be responsible. For thin samples the clamps can cause a stiffening effect which redues the effective length of the specimen and leads to a higher apparent modulus [10].

The frequency-temperature variation of the α-(glass) transition will be better described on the basis of the WLF equation [11].

$$(4)\quad \log a_T = -\, c_1 (T - Tr)/(c_2 + (T - Tr))$$

where c_1 and c_2 are constants and Tr is a reference temperature.
The following formula for small shifts, which replaces Eq. (3), was derived by applying Eq. (4) twice for T and T_1 and using for both temperatures a mutual reference Tr:

$$(5)\quad T_1 \simeq T - \frac{c_2}{c_1} \cdot \log\left(\frac{f}{f_1}\right) \cdot \left(1 + (T - T_r)/c_2\right)^2$$

There is a nonlinear increase of the temperature shift as T becomes larger than T_r. This relation was checked with a NBR rubber (H = 1.235 mm) by using different I (43, 163, 278 kg mm^2). Master curves of G and tan $\delta = G''/G'$ could be drawn by applying Eq.(5) to the different data and setting f_1=1 Hz. For the constants c_1 and c_2, the reported values 17.44 and 51.6 were used and for T_r the static (dilatometric) glass transition temperature Tg was chosen [8].

This type of equation seems to be applicable also to the PMMA α-transition. Up to now the best fit was obtained with $c_2 / c_1 \simeq 3$, $c_2 \simeq 10\text{-}15$ and Tr = 120°C. This study is subject of a comprehensive report [12].

The small changes of the modulus in the soft region can be correlated to the change of the frequency by twisting a sample under prestress. This effect has been tabulated in the 1965 issue of DIN 53445 and could be verified quantitatively.

Conclusions

These measurements were made with the same moment of inertia I. Normally one would adapt I to H in order to get about the same variation of frequency. Although the shifts within the range of free torsional oscillations (0.1-10 Hz) are relatively small, they are detectable due to the high resolution of our instrument.
It is obvious that only the complete automation with a computer, which enables also further treatment of data, can comply with the actual demands on this type of test.

References

1 K.Schmieder and K.Wolf, Kolloid Z. 127, 65(1952) and 134, 149 (1953)
2 W.Kuhn and O.Kunzle, Helv.Chim.Acta 30, 839(1947)
3 L.E.Nielsen, Rev.Sci.Instr. 22, 690 (1951)
4 K.H.Illers and H.Breuer, Kolloid Z. 176, 110 (1961)
5 N.G.McGrum, B.E.Read, G.Williams: Anelastic and dielectric effects in polymeric solids, Wiley New York 1967
6 Ch.Fritzsche, Kunststoffe-Plastics 24, 17 (1974)
7 Ch.Fritzsche, B.Höchli, K.Moser, Kunststoffe 64, 675 (1974) (English translation available)
8 M.Seeger, Proceedings of the International Rubber Conference, Venice/Italy 1979, p. 442 - 450
9 W.M.Davis, C.W.Macosko, Polym. Eng. Sci. 17, 32 (1977)
10 C.J.Nederveen, J.F. Tilstra, J. Phys.D.Appl.Phys., 4, 1661 (1971)
11 M.L.Williams, R.F.Landel, J.D.Ferry, J.Am.Chem.Soc. 77, 3701 (1955)
12 M.Seeger, paper prepared for Kautschuk + Gummi, Kunststoffe

THE MEASUREMENT OF DYNAMIC SHEAR PROPERTIES OF POLYMER MELTS WITH A RHEOVIBRON VISCOELASTOMETER

T. Murayama

Monsanto, Research Triangle Park, U.S.A.

(Abstract)

The Rheovibron viscoelastometer is useful for obtaining dynamic tensile mechanical properties of films and fiber over a wide temperature range of -160°C to 250°C in an atmosphere at 0% relative humidity. In recent years a modification which makes possible measurement on materials in the shear and bending mode has been reported.

In this paper a new method for investigation of dynamic shear mechanical properties of polymer melts is presented. The oscillating unit and the sample holders of the Rheovibron are modified. The shear grips consist of sample holding unit and a shearing plate with 15 cm. long rod. A analysis of the instrument to obtain the dynamic shear moduli of polymer melts is given, as well as a procedure for determining the internal friction of melts as a function of temperature.

THE MEASUREMENT OF THE ELONGATIONAL VISCOSITY OF POLYMER SOLUTIONS

J. Ferguson and M. K. H. El-Tawashi

University of Strathclyde
Department of Fibre Science and Textile Technology
204 George Street
Glasgow, G1 1XW, Scotland

A number of years ago the Chemical Engineering and Technology committee of the British Science Research Council decided that there was a lack of commercial instrumentation for the measurement of elongational viscosity. Following an assessment of the field three instruments were selected for development. The long range intention was that these instruments should become commercially available. This paper describes one of these instruments and discusses the results which can be obtained from it. The instrument itself was constructed at the SRC's Rutherford Laboratory as a result of joint discussions on design and the modification of the original.

THE ELONGATIONAL VISCOMETER

Figure 1 shows a line drawing of the instrument. It is identical in concept to the original version[1], however a number of advances have been made which make it both easier to operate and more versatile rheologically.

The fluid under investigation is placed in a stainless steel reservoir A, and pressurised at 0-30kNm^{-2} by air, nitrogen or any suitable gas. A solenoid valve system maintains the pressure accurately at the present value. The fluid is forced through the thin walled stainless steel tube B. A guide rod is intended to prevent accidental damage to the tube. The fluid leaves the tube through a nozzle C. This can be removed for ease of cleaning and a range of nozzle sizes can be used. The fluid falling vertically from the nozzle is caught on a rotating stainless steel drum D and elongated. Drum speed is accurately controlled through a direct drive motor. Both the motor and drum

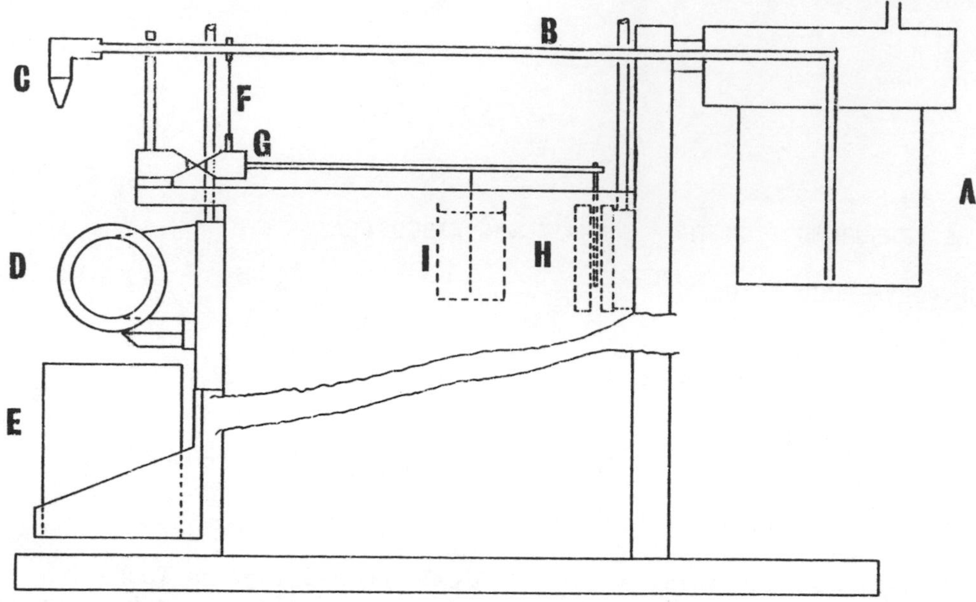

Fig. 1. Spin-line elongational viscometer

are mounted on a slide attachment which enables the fluid filament length to be varied at will up to 25 cm.

The load on the fluid imposed by the elongation causes the tube to deform. This motion is transmitted by a fine metal strip F to a pivot device G. At the end of the hollow aluminium tube which is held at one end by the pivot, is a core rod for the capacitance transducer H. This provides a mechanical amplification of approximately 12x and a further electronic amplification when the signal is passed to the recorder. The transducer H is mounted on a vertical slide which considerably improves ease of operation as the system when pressurised causes the tube B to exhibit a "Bourdon-Tube" effect causing a large shift in the zero position of the core. Using the slide this can be quickly reset. A scraper removes the fluid from the drum and a receptacle E collects it for reuse.

The viscometer is mounted on a base plate with levelling feet and the whole instrument is contained within an environmental chamber. Temperature of operation is ambient to 100°C. Thermostatting io achieved hy a proportional controller which feeds a series of heating elements mounted on the reservoir top and in the side walls of the viscometer. In order to cut down vibrations, a damper pot is attached to the aluminium tube. The instrument and environmental chamber are mounted on a concrete block supported by anti-vibration feet.

Two methods are in use for measuring filament shape. A vertically travelling microscope can be employed to take direct diameter readings at fixed distances along the filament or a photographic method can be used. For long filament lengths it is necessary to construct a composite photograph. From these readings filament diameter can be calculated and hence stress and rate of elongation.

Theoretical. The equations used and the assumptions made have been detailed in previous publications[1,2]. We assume the equation of motion $\rho\frac{DV}{Dt} - \nabla T = f$ (1)

where $V=(U,0,V)$ is the velocity vector, T the stress vector and f the body force density vector. If flow is steady in the Eulerian sense,

$$\frac{\sigma(t)}{\rho V} = g(t_L - t) + V - V_L + \frac{(F - F_s)}{Q} \quad (2)$$

where σ is the axial tensile stress; t is time of elongation; F_s is the surface force due to surface tension and Q = mass flow rate. Subscript L indicates values at distance L.

$$\text{Rate of Strain, } \dot{\epsilon}(t), = \frac{\partial V(t)}{\partial Z} \quad (3)$$

This is calculated in practise by computer curve fitting of the filament diameter followed by differentiation.

$$\text{Total Strain, } \epsilon(t), = \ln(V/V_0) \quad (4)$$

A computer programme has been set up to carry out the calculations so that the effects of gravity and inertia are eliminated. Surface tension forces and aerodynamic drag are assumed to be negligible. A correction can also be made for the effect of relaxation from shear flow in the nozzle using the equation

$$\sigma(t) = \hat{\sigma}(t) - \hat{\sigma}_0 \exp(-t/t_0) \quad (5)$$

where $\sigma(t)$ is true axial stress, $\hat{\sigma}(t)$ is the apparent value of stress; t_0 is the relaxation time and $\hat{\sigma}_0$ the average normal stress as measuredusing a Weissenberg Rheogoniometer at the shear rate existing in the nozzle.

The use of the last equation is now a matter of some debate. It has been claimed that relaxation effects from shear can be ignored. The value of "instantaneous" elongational viscosity (η_{el}) can now be calculated from the equation

$$\eta_{el} = \sigma(t)/\dot{\epsilon}(t) \quad (6)$$

RESULTS AND DISCUSSION

Hypothetical Fluids. Previous experimental work[2,3] has shown that η_{el} plotted against strain rate or total strain can vary in several different ways. In general it appears to increase slowly at low strain rates then fall to a minimum and finally increase rapidly. This last transition appears to be associated with a change from a viscosity dominated to an elasticity dominated deformation. In order to investigate the effect of filament profile on the η_{el} vs total strain curves, a series of hypothetical fluid profiles were assumed. Figure 2 gives three examples. Profile A is typical of many experimentally observed fluids. Emerging from the nozzle it shows post extrusion swelling. The fluid then thins down under the elongational strain imposed by the rotating drum. In profile B the post extrusion swelling is larger, the initial shape being circular in profile, followed by a fairly rapid fall in diameter. In contrast profile C shows moderate post extrusion swelling followed by an initially rapid reduction in diameter. However as total strain increases the rate of change of diameter falls until the filament is almost, but not quite, of a constant cross section. In order to process the data from these hypothetical fluids a constant rate of flow and fluid density were assigned.

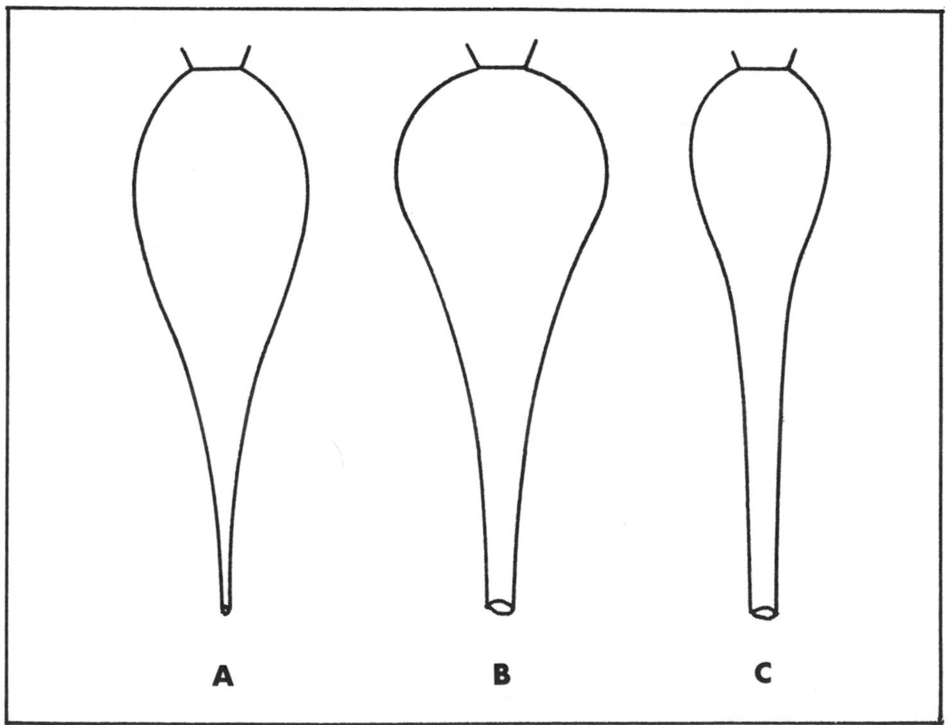

Fig. 2. Fluid profiles

Using equations 1 to 6 elongational viscosity was then calculated. Values of diameter were taken every millimetre starting at the point of maximum post extrusion swelling. In addition the fluids were each given relaxation times of zero, 7.6×10^{-3}s, 7.6×10^{-2}s and 7.6×10^{-1}s.

Figure 3 gives the variation of η_{el} with total strain for the three fluids. The curves shown were obtained using a relaxation time of 7.6×10^{-3}s. Fluid A was Newtonian at low strain rates then η_{el} fell rapidly, apparently approaching a minimum at high strains. The behaviour of fluid B was in sharp contrast. Newtonian at low strain rates, η_{el} then rose rapidly to a sharp peak before falling at large values of total strain. Fluid C was different again. An initial fall in η_{el} was followed by a sharp rise which suggests the presence of a maximum total strain in this system.

The effect of relaxation time was of considerable interest. The difference between ignoring the relaxation from shear and of using the equation 5 correction was very small. Only when the larger values of relaxation time were used was there a significant fall in η_{el} produced. In general however the shape of the η_{el} vs total strain curve was unchanged.

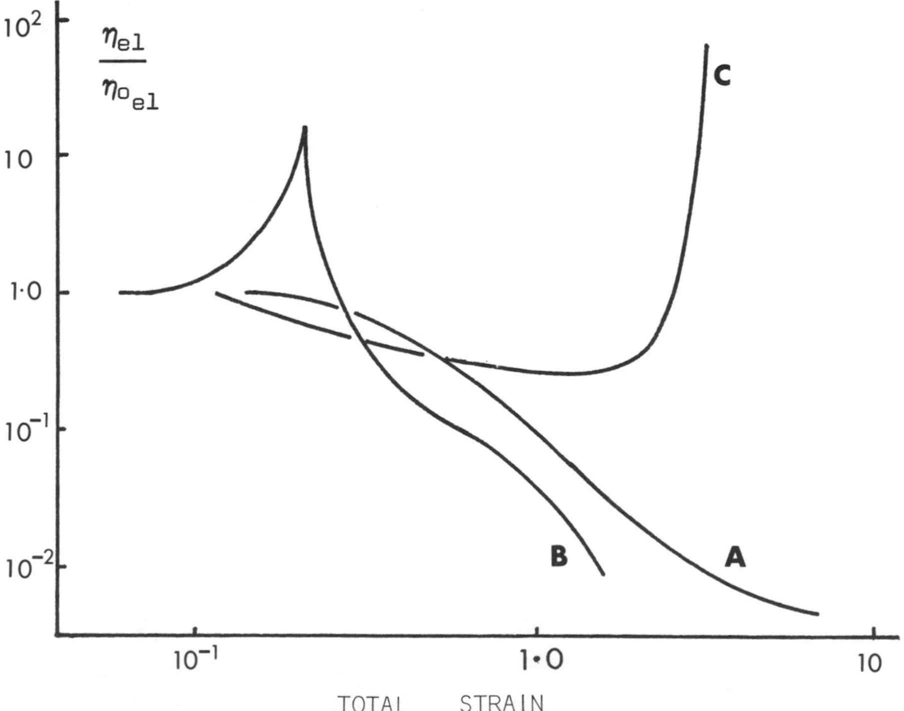

Fig. 3. Variation of relative elongational viscosity with total strain for hypothetical fluid profiles

The implications of curve C, Figure 3, and of the shape of the fluid filament which produced it, curve C, Figure 2 are of considerable importance and merit some attention. Where a filament on extension is tending towards a constant diameter it will be highly spinnable. The rheological conditions which produce such an effect have been discussed by various authors[4]. Figure 3 shows clearly that such a fluid will be "strain hardening". That is its elongational viscosity or modulus rises faster than its cross section decreases. This is however not necessarily true. We have observed at very high strain rates that it is possible to produce a filament that is parallel sided in profile. Since it is being elongated it must be assumed that the rate of elongation is constant along its length. That is, it is displaying a solid rubber-like deformation. If this is the case it is possible that η_{el} does not rise rapidly at high strains and strain rates, but undergoes a transformation from viscous to elastic deformation. Previous work[2] has suggested this may be the case. It remains however to carry out the experimental work to determine which of the two theories is correct. This can be done by photographing particles moving in the threadline.

Real Fluids. A range of fluids have so far been examined at 21° using the instrument. Generally these exhibit the characteristics shown in Figure 3. For example, a 0.5% by weight solution of polyacrylic acid in ethylene glycol showed η_{el} Newtonian at low total strains then falling (cf curve A, Figure 3). Increasing the polymer concentration to 1% produced a sharp peak in η_{el} as total strain increased, followed by a fall (cf curve B). Finally a 2% by weight solution of polyethylene oxide in water gave η_{el} that fell to a minimum then rose sharply as total strain increased. (cf curve C).

REFERENCES

1. J. Ferguson and N. E. Hudson, A new viscometer for the measurement of apparent elongational viscosity, J.Phys.,E, 8:526 (1975).
2. N. E. Hudson and J. Ferguson, Correlation and molecular interpretation of data obtained in elongational flow, Trans.Soc.Rheol., 20:265 (1976).
3. G. R. McKay, J. Ferguson and N. E. Hudson, Elongational flow and the wet spinning process, J. Non-Newtonian Fluid Mech., 4:89 (1978).
4. C. J. S. Petrie, "Elongational Flows", Pitman, London (1979).

THE USE OF EXTENSIONAL RHEOMETRY TO ESTABLISH OPERATING PARAMETERS FOR STRETCHING PROCESSES

G.H. Pearson, R.C. Connelly

Eastman Kodak Company, Research Laboratories

(Abstract)

The extensional rheology of several high-viscosity polymer melts has been studied using a rod-pulling technique. The limits of uniform extensibility can be correlated with the relaxation characteristics for these polymers and predicted to a first approximation by the BKZ constitutive equation. The implication and application of this work to the definition of proper process parameters will be discussed.

POLYMER SOLUTIONS

DILUTE SOLUTION PROPERTIES OF POLYSTYRENE, POLYMETHYL-METHACRYLATE, AND THEIR COPOLYMERS

S. K. Ahuja

Joseph C. Wilson Center for Technology
Xerox Corporation
Rochester, New York

INTRODUCTION

The conformational and thermodynamic properties of flexible homopolymers and copolymers can be described by an unperturbed average dimension of a polymer chain, the short range interaction parameter, and an excluded volume effect in a given environment, the long range interaction parameter.[1,2] The two independent parameters can be determined from experimental data of the molecular weight effects on the second virial coefficient, the mean square statistical radius, and the intrinsic viscosity of the chain.[2,3] An unperturbed average dimension of a copolymer chain (A) is a simple addition of proportional, unperturbed dimensions of constituent homopolymers. An excluded volume parameter (B) of a copolymer has a quadratic dependence on composition with repulsive interaction between the two constituent units.[4,5]

The dilute solution viscosity of a polymer obeys Arrhenius-type laws, and the activation energy depends on both molecular weight and concentration. The activation energy parameter, independent of solute concentration and solvent, is 1-2 orders of magnitude lesser for flexible chains than for stiff chain macromolecules.[6]

EXPERIMENTAL

The polystyrene samples were obtained from Pressure Chemical Company, Pittsburgh, Pa. The polymethyl methacrylate samples were obtained from Scientific Polymer Products, Ontario, N.Y. Styrene-methylmethacrylate copolymers (S/MMA) were synthesized by T. Smith of Xerox Corporation, using suspension polymerization. The composition was determined by pyrolysis gas chromatography and proton magnetic resonance spectrometry. The number and weight averages were determined by gel permeation chromatography. The dilute solution viscosities in the concentration range of 0.25 to 2.0 g/dl were measured by A. Mukherji of Xerox Corporation at 25°, 35°, and 45°C in toluene using Cannon-Ubbelohde capillary viscometer.

RESULTS AND DISCUSSION

The viscosity of dilute polymer solutions varies with temperature according to an Arrhenius-type law,

$$\eta = A \exp(Q/RT).$$

The apparent activation energy Q is dependent upon the polymer solvent system, the solution concentration, and the molecular weight of the polymer.[6]

The apparent activation energy (Q) can be related to the flexibility of polymer chains in dilute solutions,

$$Q = Q_o + K_e Mc,$$

where Q_o is a constant dependent on the solvent; K_e depends both on the nature of the solvent and of the polymer; and M and c are the molecular weights of the polymer and concentration of solution, respectively.[6]

Figure 1 shows that the coefficient K_e for polystyrene decreases with increasing molecular weight, M_n. There is a change in slope around $M_n = 25{,}000$, close to the molecular weight where polystyrene chains entangle. In Figure 2, K_e for polymethyl methacrylate increases with increasing M_n The K_e values for polystyrene are positive compared to the negative values for polymethyl methacrylate. The

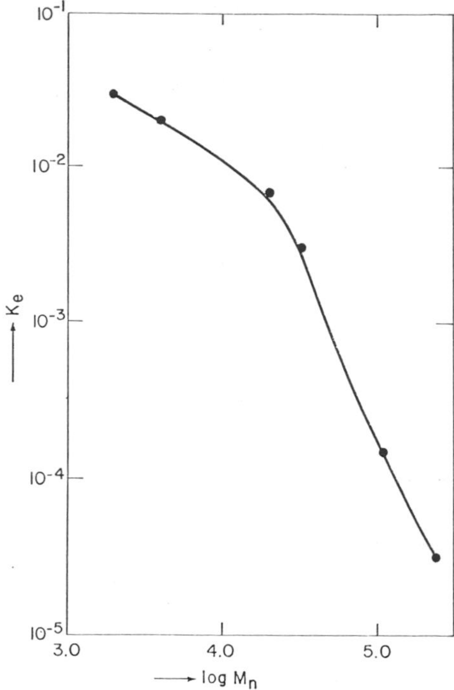

FIGURE 1. Activation energy constant K_e and molecular weight of polystyrene

decrease in K_e with the molecular weight of polystyrene is more significant than the increase of K_e with the increase in the molecular weight of polymethyl methacrylate. These differences are due to the chain flexibility of polystyrene versus polymethyl methacrylate. Due to the presence of polar groups, polymethyl methacrylate has highly flexible chains. Figure 3 shows that 15% methacrylate monomer concentration in the copolymer significantly reduces K_e and, therefore, flexibility of the copolymer chain.

The long range parameter B can be determined from Kurata Stockmeyer viscosity plots:[2]

$$(\eta)^{2/3}/M^{1/3} = K^{2/3} + 0.363\ \Phi_o B[g(\alpha\eta)\ M^{2/3}/(\eta)^{1/3}].$$

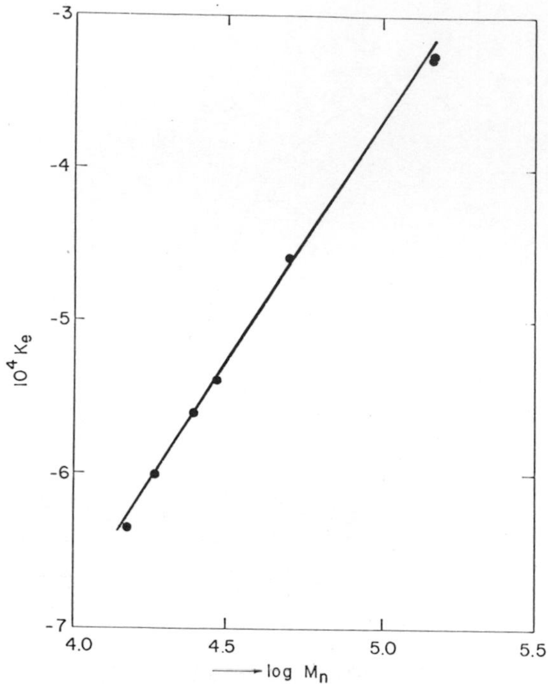

FIGURE 2. Activation energy constant K_e and molecular weight of poly (methyl methacrylate)

where $g(\alpha\eta)$ is a coefficient dependent on $\alpha\eta$, the expansion factor is:

$$g(\alpha\eta) = 8\alpha\eta^3 (3\alpha\eta^2 + 1)^{2/3}$$

$$\alpha\eta^3 = (\eta)/KM^{1/2}.$$

The short-range parameter A can be obtained from K, the constant determined from Kurata Stockmeyer viscosity plots, and Φ_o, the viscosity constant:[2]

$$K^{2/3} = \Phi_o^{2/3} A^2, \ [\Phi_o - 2.81 \times 10^{21} \text{cgs unit}].$$

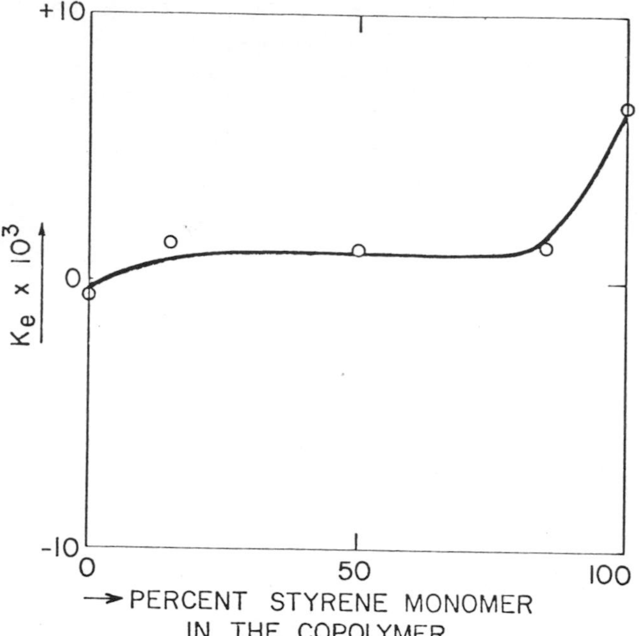

FIGURE 3. Activation energy constant K_e and concentration of styrene in styrene methyl methacrylate copolymer

From the short range (B) and the long range (A) parameters, the excluded volume factor Z can be obtained as:

$$Z = \left(\frac{B}{A^3}\right) M^{1/2}.$$

Table 1 contains Mark-Hauwink Constants (K_o) and short range (A) and long range (B) parameters for polystyrene and polymethyl methacrylate. Both Mark-Hauwink constants and short range parameters for polystyrene decrease in magnitude from 25°C to 45°C. The short range parameter reflecting unperturbed chain dimensions becomes less flexible. Further increase in temperature does not bring about any change in the short range parameter. The long range parameter B and excluded volume factor Z for polystyrene show an increase in magnitude from 25°C to 45°C.

TABLE 1

MARK-HOUWINK CONSTANT (K_o), LONG RANGE (B), SHORT RANGE (A) PARAMETERS

Material	Temperature	Mark-Houwink Constant (K_o)	Long Range (B)	Short Range (A)
		(8.0×10^{-4})*	(107×10^{-30})*	(7.0×10^{-9})*
Polystyrene	25°C	6.7×10^{-4}	121×10^{-30}	6.1×10^{-9}
	45°C	2.8×10^{-4}	278×10^{-30}	4.6×10^{-9}
	65°C	3.1×10^{-4}	263×10^{-30}	4.8×10^{-9}
Polymethyl	25°C	2.7×10^{-4}	39.0×10^{-30}	4.5×10^{-9}
Methacrylate		(5.0×10^{-4})*	(60×10^{-30})*	(2.3×10^{-9})*
	45°C	2.4×10^{-4}	38.2×10^{-30}	4.4×10^{-9}
	65°C	2.3×10^{-4}	35.8×10^{-30}	4.3×10^{-9}

* These are the published values from Kotaka et al.[5] Note that the differences in PMMA between the published values and our values in K_o, B, and A are due to poly-dispersity in our samples, known in literature.

Berry proposed a relationship for B, $B = B_o(1-\theta/T)$, where B_o is a parameter proportional to Flory's entropy of mixing,[1] θ is the temperature at which osmotic second virial coefficient A_2 vanishes,[1] and T is any temperature. Thus B would increase with increasing temperature above θ temperature. As temperature is increased from 45°C to 65°C, the long range parameter B shows no further increase. On the contrary, polymethyl methacrylate shows no change in either short range or long range parameters. Polymethyl methacrylate chains are highly flexible because of polar graphs and, therefore, increasing temperature does not change their flexibility.

REFERENCES

1. P.J. Flory, "Principles of Polymer Chemistry," Cornell University Press, Ithaca, NY, 1953.

2. M. Kurata and W.H. Stockmayer, Fortschr. Hochpolym. Forsch. 3:196 (1963).

3. G.C. Berry, J. Chem. Phys. 44:4550 (1966).

4. W.H. Stockmayer, L.D. Moore, Jr., M. Fixman and B.N. Epstein, J. Polymer Sci. 16:517 (1955).

5. T. Kotaka, Y. Murakami, and H. Inagaki, J. Phy. Chem. 72:829 (1968).

6. N. Aclenei and I.A. Schneider, Europ. Polym. J. 12:849 (1976).

CORRELATIONS FOR RELAXATION-TIMES FOR MONODISPERSE POLYSTYRENE SOLUTIONS

P.Attane, P.Le Roy, J.M.Pierrard, G.Turrel

Université Scientifique et Médicale de Grenoble
France

(Abstract)

Relaxation tests after cessation of steady flow at constant shear rate $\dot{\gamma}$, were performed in a broad range of $\dot{\gamma}$ (10^{-3} to 10^{2} sec^{-1}) with monodisperse polystyrene solutions in dibutyl-phthalate (Molecular weight, M, ranging from 140000 to 1300000 and concentration, c, from 5% to 40%). The apparent relaxation time of shear-stress curves in the range 2×10^{-2} to 10^{2} sec. Modification of our Weissenberg Rheogoniometer allow us to determine these shear stress without mechanical coupling. For all $\dot{\gamma}$ the bracking time is less than 16 m sec.

From these experimental curves one can deduce the transient viscosity $\eta(t; \dot{\gamma})$ for $t > \tau_m$. τ_m depends on the bracking time but also on the time dependence of the velocity field near the rigid boundary.

Then, using a corotational Maxwell model, two relaxation times are inferred from $\eta(t; \dot{\gamma})$. The largest one, $\tau_2(\dot{\gamma}, M, c)$ is characteristic of the behaviour near the rest state. The shortest one $\tau_1(\dot{\gamma}, M, c)$ is characteristic of the behavior of entanglements subjected to large shear rates.

The well-known correlations as $\eta_0 \propto M_w^{3.4}$; $J_c^{\circ} \propto c^{-2}$ are verified. As usual, the elastic compliance $J_c = N_1(\dot{\gamma})/2\,\dot{\gamma}^2 \eta^2(\dot{\gamma})$ is found to be dependent of $\dot{\gamma}$. We show that the another measure of elasticity: $\left[d\eta(t; \dot{\gamma})/dt\right]_{t=0}$ is independent of $\dot{\gamma}$.

(Note that in the linear theory for $\gamma \to 0$, $\left[d\eta(t;\gamma)/dt\right]_{t=0} = G_{(o)}$).

We present then the variation of the two relaxation times with shear rate, molecular weight and concentration. For instance, τ_2 is found to be independent of γ as expected for a parameter characteristic of the behavior near the rest state.

HYDROPHOBIC EFFECT ON THE INTRINSIC VISCOSITY OF GLOBULAR PROTEINS

Ruth Tanner and ChoKyun Rha

Food Materials Science and Fabrication Laboratory
Department of Nutrition and Food Science
Massachusetts Institute of Technology
Cambridge, Massachusetts

INTRODUCTION

The viscosity of a protein depends on the conformation which is a result of the hydrophobic effect. Most of the hydrophobic side chains are buried in the interior of globular proteins in solution and thus determine the conformation of proteins (Tanford, 1962; Nemethy and Scheraga, 1962; Kauzmann, 1959). Therefore, the viscosity of a globular protein changes when a conformational change of a protein is induced.

This change in geometry of a globular protein as it is effected by a ligand, and as measured by the intrinsic viscosity, can give a measure of the degree of hydrophobicity of the protein. In addition, estimated dimensions of the globular protein-ligand complex make it possible to propose a structure of the complex that is hydrodynamically most likely in solution.

EXPERIMENTAL

Bovine serum albumin (BSA) and ovalbumin (OA) were reacted with the ligand, sodium dodecyl sulfate (SDS). Protein concentration ranged from 0.5 to 2% wt in phosphate buffer (pH 7.2, ionic strength 0.026) and the sodium dodecyl sulfate-protein solution ratios varied by weight from 0.01 to 2.6 in 0.01M β-mercaptoethanol. Viscosity measurements were made in Cannon-Fenske viscometers with flow times of 200-500 sec. The binding data were obtained by the equilibrium dialysis method of Ray et al. (1966).

RESULTS AND CONCLUSIONS

The reduced viscosity of the SDS-BSA and SDS-OA complexes generally increases more rapidly as the SDS-protein ratio or protein concentrations are increased (Figure 1). Using the intrinsic viscosities obtained from Figure 1 and from the water insoluble dye data (Steinhardt, Scott, and Birdi, 1977) the increase in intrinsic viscosity of the protein-detergent complex was examined in terms of the ratio of SDS/protein, and the shape factor. The increase of the axial ratio of the complex is mainly represented in the change in the principle axis (Table 1).

The change in the intrinsic viscosity on conformation occurs periodically as the gram ratio of SDS to protein is increased continuously, (Figure 2) as observed through optical rotatory dispersion (Reynolds, et al. 1967). Using the data on sequential change in intrinsic viscosity or association sequence, the amount of SDS per gram of protein required to make a given change in the intrinsic viscosity can be related to hydrophobicity (Table 2). Each association represents the amount of SDS necessary to either partially unfold the protein which is accompanied by a corresponding change in the intrinsic viscosity or the amount of SDS required to saturate the binding sites at a particular unfolded state which is not accompanied by a corresponding change in the intrinsic viscosity. The change in the axial ratio within a sequence gives a measure of

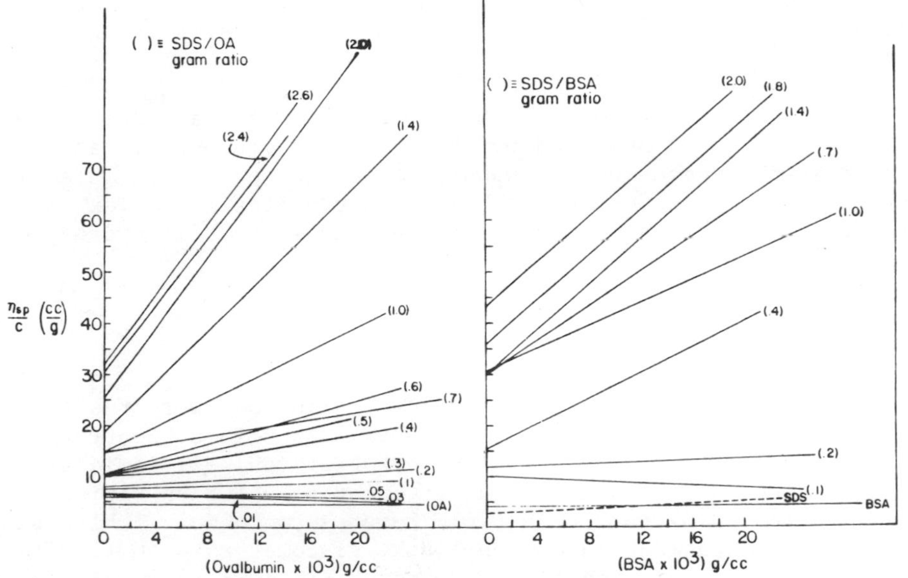

Fig. 1. Reduced Viscosity of Bovine Serum Albumin-Sodium Sulfate and Ovalbumin-Sodium Dodecyl Sulfate Complexes.

the change in conformation. The change in the residue ratio within a sequence normalizes the molecular weight variation for different proteins. The rate of these changes gives a measure of the relative hydrophobic character of the various stages of the unfolded protein.

Table 1. Intrinsic Viscosity and Hydrodynamic Shape of Ovalbumin Related to the Hydrophobic Effect.

$[\eta]$	δ_2^T	δ_2^B	V	a/b	$\bar{\nu}$	$\frac{AAR}{SDS}$	a(A°)	b(A°)
4.31	0	0	4.0	3.4			57	17
6.29	0.03	0.025	5.7	4.9	4	99	126	25
6.00	0.05	0.04	5.4	4.7	6	66	121	26
7.54	0.10	0.08	6.6	5.6	13	30	137	25
8.10	0.20	0.12	6.9	5.8	19	21	141	24
8.50	0.25	0.16	7.0	5.9	25	16	142	24
10.1	0.30	0.20	8.0	6.6	31	13	154	23
10.3	0.40	0.22	8.1	6.7	35	11	155	23
10.3	0.50	0.29	7.7	6.5	45	9	151	23
10.2	0.60	0.32	7.5	6.3	50	8	149	24
14.0	0.72	0.39	7.8	6.5	63	6	153	24
14.0	1.0	0.58	8.8	7.2	91	4	164	23
17.9	1.4	0.80	10.0	7.9	126	3	177	22
25.3	2.0	1.2	12.0	9.1	190	2	194	22
33.5_R	>2.6	1.4_R	15.0	11.0	220	<2	215	21

$[\eta]$ Intrinsic Viscosity, cc/g
δ_2^T g SDS(total)/g protein
δ_2^B g SDS(bound)/g protein
AAR Amino Acid Residue
R Data from Reynolds, Tanford (1970a)
a/b Axial ratio, assuming prolate ellipsoid
$\bar{\nu}$ Molar ratio (moles SDS bound/mole protein)
Vol Volume, assuming Prolate Ellipsoid
V Shape Factor [Mehl, et al. (1940) and Sadron (1953)], assuming (a) partial specific volume of the protein = 0.748 [Kuntz & Kauzmann, (1974)], (b) hydration = 0.33 gH_2O/g Protein from NMR [Kuntz (1971)] (c) specific volume of SDS = 0.866 [Anacker, Rush and Johnson (1964)].

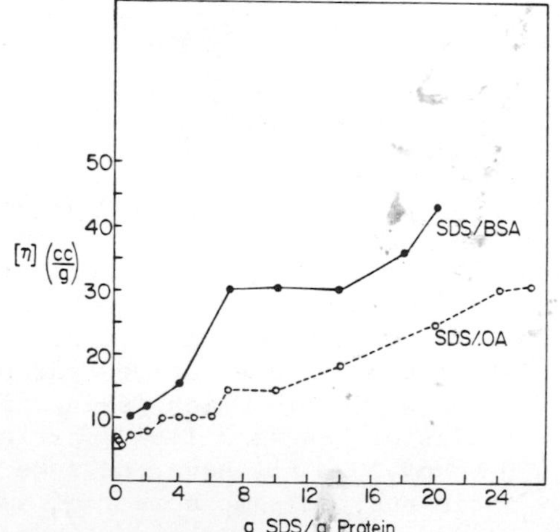

Fig 2. Sequential Increase in Intrinsic Viscosity of Sodium Dodecyl Sulfate Complexes of Bovine Serum Albumin and Ovalbumin.

Table 2. Relative Hydrophobicity*.

Protein	Association Sequence	$\bar{\nu}$	ΔSDS^B (g)	$\Delta(a/b)$	$\Delta(\frac{AAR}{SDS})$	$\frac{\Delta(a/b)}{\Delta(AAR/SDS)}$
Ovalbumin	1→2→3	63	0.39	6.3	2.2	0.29
Bovine Serum Albumin	1→[2]	100	0.44	1.91	4.0	0.48

* $\frac{\Delta(\text{Axial ratio})}{\Delta(\text{Amino acid residues/SDS molecule})}$

Table 3. General Mechanism for the Protein-SDS Complexes

Protein	MW	a(A°)[1]	AAR/SDS[2]	AAR/A°
Ovalbumin	45,000	157	1.7	2.5
Bovine Serum Albumin	69,000	231	1.7	2.5
β-Lactoglobulin	18,400	66	1.8	2.5
Chymotrypsinogen	25,700	87.4	1.9	2.8

[1] Axial length at 1.4g SDS/ g protein (Reynolds, Tanford)
[2] At 1.4g SDS/ g protein

The largest relative hydrophobic effect for ovalbumin is shown at the binding level of 0.4 g SDS/g protein coinciding with the first of the two major equilibrium binding levels generally accepted for proteins. The relative hydrophobic ratios (RHR) of BSA and ovalbumin at an association sequence which represents the same binding levels shows that BSA has a relative hydrophobicity approximately 1.7 times that of ovalbumin which is in agreement with the well known characteristics of BSA as a lipid carrier. Globular proteins (M.W. 18,000 to 69,000) all have the same number of amino acid residue to SDS ratio and the same number of amino acid residues per unit length in the SDS-protein complex at the same maximum binding level . Therefore, a simple relationship is not obvious or is difficult to establish between bound SDS molecules per number of hydrophobic amino acids which varies from 38-46% in the proteins (Table 3).

Table 4. Proposed Molecular Conformation of Ovalbumin-SDS Complex.

Complex: 1.4g SDS/g ovalbumin

Conformation: Prolate ellipsoid
a = 215 A°, b = 21 A°

Residue Ratio: 2 amino acid residues/SDS molecule

	Length (A°)	Volume (A°^{3})
Sodium dodecyl sulfate	20	23.6
Extended peptide/unit	3.6	1684
α-helix peptide/unit	1.5	829
2 extended peptides plus 1 SDS		2.39×10^3
2 α-helix peptides plus 1 SDS		1.68×10^3
Ovalbumin-SDS complex		4.01×10^5
Ovalbumin-SDS complex:		
55% extended peptide/unit		2.6×10^5
45% α-helix peptide/unit		1.5×10^5

With the change in the intrinsic viscosity interpreted by the conformational change related to the hydrophobic effect, a structure of the SDS-protein complex can be proposed with the dimensions from the intrinsic viscosity measurement and the residue ratio. The ovalbumin-SDS complex, at its maximum weight ratio binding level (1.4 g SDA/gram ovalbumin) corresponding to the ratio of two amino acid residues per SDS molecule (Table 4) was selected for the structural analysis. The volume of the SDS-OA complex calculated assuming a completely extended chain as well as assuming 100% alpha helix indicates that a portion needs to be helical or folded in order to satisfy the dimensions and volumes calculated from the intrinsic viscosity data. This is in agreement with the suggestion made by Reynolds and Tanford (1970b) for other protein-SDS complexes, and also as supported by the suggestion of Tanford, et al. (1964).

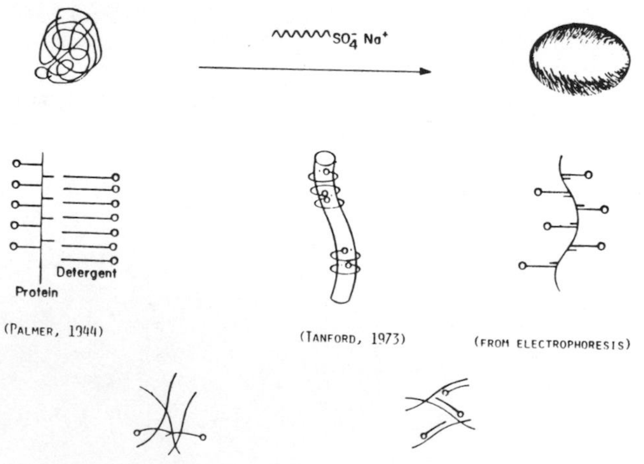

Fig 3. Proposed Protein-SDS Complex

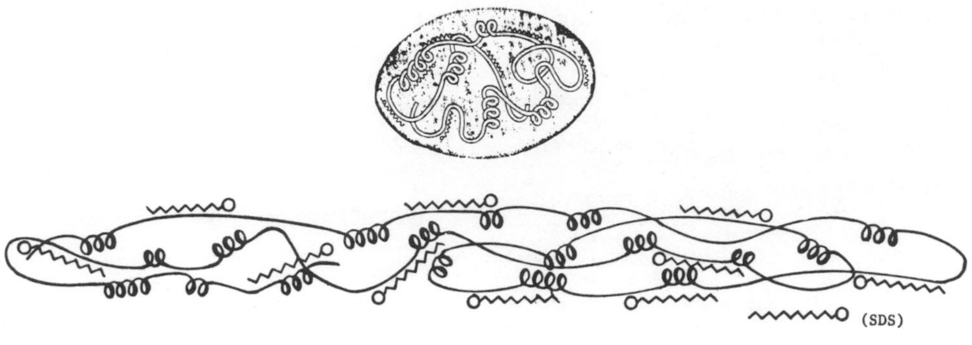

Fig.4. Proposed Structure for Albumin-SDS Complexes.

Based on the estimated dimensions, the structures proposed for the SDS-protein complex (Palmer, 1944; Tanford; 1973) can be evaluated. Structure 1 and 3 (Figure 3) postulate the SDS molecule being associated perpendicularly to the peptide backbone. Since the length of SDS molecule is approximately the same as the minor axis of complex itself, to satisfy this condition, the width of the complex needs to be about two times greater than that as calculated from the intrinsic viscosity. Strucure 2 (Figure 3) proposes a sheath-like arrangement which would allow for surface interaction but this complexing scheme does not allow for sufficient surface area necessary to accomodate all of the negatively charged head groups of SDS. The surface area of the prolate elipsoid proposed is not sufficient to accomodate this proposed structure and the required SDS to amino acid ratio. Therefore, a SDS-protein complex structure which satisfies the required geometry and supports the intrinsic viscosity data is proposed and given in the schematic representations 4 and 5 in Figure 3. The proposed structure would allow the SDS some mobility in the complex and provide a more random, fluid association while meeting the axial, helical, and charge distribution requirements. Figure 4 illustrates a more detailed diagram of the proposed structure for the globular protein/SDS complex. A structural analysis of the SDS-protein complexes in solutions employing laser light scattering experiments is currently in progress to confirm the proposed complex structure and to determine the complexing mechanisms for globular proteins and ligands in solution (Tanner et al.).

REFERENCES

Anacker, E. W., Rush, R. M., and Johnson, J.S., 1964, J. Phys, Chem., 68:81.

Kauzmann, W., 1959, Advan. Protein Chem. 14:1.

Kuntz, I. D., 1971, J. Am. Chem. Soc., 93:2514.

Kuntz, I. D. and Kauzmann, W., 1974, "Advances in Protein Chemistry", C. B. Anfinsen, J. T. Edsall and F. M. Richards, eds., Vol. 28, p. 239, Academic Press.

Mehl, J. W., Oneley, J. L., and Simha, R., 1940, Science, 92:132.

Nemethy, G. and Scheraga, H. A., 1962, J. Chem Phys., 36:1773.

Palmer, K. L., 1944, J. Phys. Chem., 48:12-21.

Ray, A., Reynolds, J., Polet, H., and Steinhardt, J., 1966, Biochemistry, 5:2606.

Reynolds, J.A., Herbert, S., Palet, H., and Steinhardt, S., 1967, Biochemistry, 3, 943.

Reynolds, J.A. and Tanford, C., 1970a, J. Biol. Chem., 245:5163.

Reynolds, J. A. and Tanford, C., 1970b, J. Biol. Chem., 245:5164.

Sadron, C., 1953, Progr. in Biophys, and Biophys. Chem., 3:254-260.

Steinhardt, J., Scott, J., and Birdi, K., 1977, Biochemistry, 16:723.

Tanford, C., 1962, J. Am. Chem. Soc., 84:4260.

Tanford, C., 1973, "The Hydrophobic Effect: Formation of Micelles and Biological Membranes," John Wiley & Sons, New York, p. 139.

Tanford, C., Nozaki, Y., Reynolds, J., and Makino, S., 1974, Biochemistry, 13:2375.

Tanner, R. E., Herpigny, B., Rha, C. K., and Chen, S. H., August, 1980, 180th National Meeting, ACS, Symposium on Colloidal Chemistry, San Francisco, California.

ACKNOWLEDGEMENT

This study was partially supported by NIH Biomedical Research Support Grant No. 87804 from the Office of the Provost, M.I.T.

LINEAR VISCOELASTIC PROPERTIES OF POLYBUTADIENE SOLUTIONS

G. Marin , W. W. Graessley*, J.P. Montfort and Ph. Monge

Université de Pau - 64000 PAU (France)

*Northwestern University - EVANSTON Illinois 60201 (U.S.A.)

INTRODUCTION

The study of the viscoelastic properties of polymer solutions has a double interest : from a theoretical point of view, the variations of the elastic (limiting compliance, plateau modulus) and viscous (zero-shear viscosity) parameters as a function of polymer concentration, are a definite test for the various theories on the relaxation and diffusion of long and flexible chains. From a practical point of view, such studies will allow to better understand the effects of the addition of plastifiers on the melt rheology of commercial polymers (viscosity, die swell, glass transition temperatures, etc ...).

1) SAMPLES - EXPERIMENTAL

The polymers used in this work are anionic 1-4 polybutadienes having a very narrow molecular weight distribution (polydispersity index $P < 1.1$). The four linear samples (M = 11 000 , 200 000 , 230 000 , 340 000 and 350 000) have been synthetized in the Chemical Engineering Dpt at Northwestern University (EVANSTON , Illinois, U. S. A.). The two star branched samples (4 star , M = 200 000 ; 3 star , M = 100 000) have been synthetized by L.J.Fetters at Akron University (Ohio, U.S.A.).

We have studied the linear viscoelastic properties of 46 solutions of these polymers in various solvents : the glass transition temperatures Tg of theses solvents were above (aromatic oil) , below (tetradecane) and equal to (low molecular weight P.B.) the Tg of polybutadiene.

The linear viscoelastic properties were determined with a Rheometrics Mechanical Spectrometer. The storage and loss moduli G' and G'' were measured as functions of frequency ω (10^{-3} to $100\ s^{-1}$) using the eccentric rotating disks (E.R.D.) method.

Whenever sample viscosity was too low (η_o < 100 poises) for this "complete" characterization, the viscosity only was measured in cone-plate geometry (steady flow). In the case of the (PB 11 000 - oil) series, viscosities were measured (± 0.2 poises) using a Cannon-Ubbeholde Capillary viscosimeter.

All measurements have been made at room temperature (25.0 ± 0.5°C), except for the temperature factor measurements of PB and oil.

2) RESULTS

The three characteristic parameters for linear viscoelasticity were determined for every solution

- The zero shear viscosity

$$\eta_o = \lim_{\omega \to 0} \frac{G''(\omega)}{\omega}$$

- The limiting compliance

$$J_e^o = \lim_{\omega \to 0} \frac{G'(\omega)}{[G''(\omega)]^2} = \frac{1}{\eta_o^2} \lim_{\omega \to 0} \left[\frac{G'(\omega)}{\omega^2}\right]$$

- The plateau modulus

$$G_N^o = 2\pi \int_o^a \frac{G''(\omega)}{\omega}\, d\omega$$

integral of the low frequency peak G'' (log ω).

1°) The zero shear viscosity η_o :

In order to get the intrinsic behavior of the polymer, it is necessary to correct the rough data from solvent effect, i.e to get the variations of viscosity vs polymer volume fraction at same free volume (or same monomeric friction coefficient ξ). This is in fact a correction for the increase (or decrease) of Tg due to solvent addition.

It has been possible to make these corrections only in the case of - PB in oil - solutions.

The theories based on a free volume concept allow to express in a consistent way the temperature dependence for the viscosity of glass-forming polymers.

$$\eta_o = C\ e^{1/f}$$

f being the free volume fraction, and C a constant characterizing the material.

The temperature dependence of the free volume fraction can be expressed as :

$$f = \alpha_f\ (T - T_\infty)$$

$$\log \eta_o = \log C + \frac{B}{2\,303\ (T - T_\infty)}$$

In the case of polymers, $C(M) \propto M^{3.4}$, B and T_∞ being independent of chain length for long chain ($M > 5\,000$ for PB). The free volume dependence of solutions on temperature could be represented by the equation :

$$f_{solution} = \varphi f_1 + (1-\varphi) f_2 + k\ \varphi(1-\varphi)$$

φ being the polymer volume fraction, f_1 and f_2 being the free volumes of polymer and solvent at the same temperature, and k a constant.

This equation is able to fit the temperature dependence for the viscosity of solutions, and the variations of Tg with the composition of solutions.

Figure 1 represents the corrected variations of viscosity at same free volume, for linear and star branched PBs. The viscosity of linear polymers was found to vary as $\varphi^{4.0}$ in the entangled regime, and $\varphi^{1.0}$ in the Rouselike regime (low polymer concentration or short chains).

2°) <u>The limiting compliance J_e^o</u> :

We have reported on Fig.2 the variations of the compliance vs polymer volume fractions for the - linear PB in oil - solutions. In the entangled regime (high polymer concentration), J_e^o was found to vary as $\varphi^{-2.25}\ M^{0.0}$, independently of solvent nature (same variations have

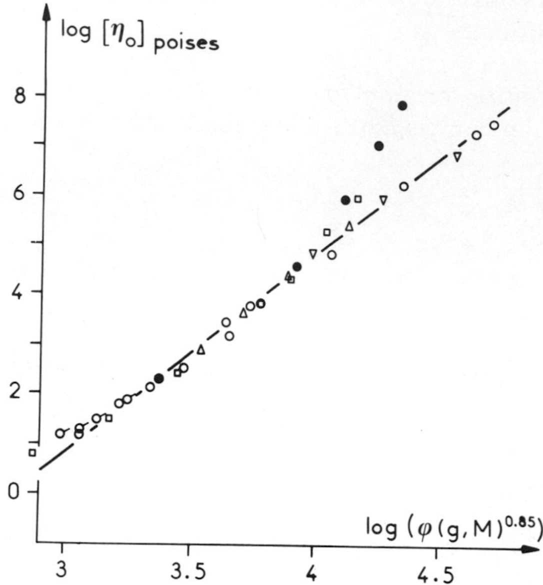

FIG. 1 Variations of corrected viscosities vs polymer volume fraction

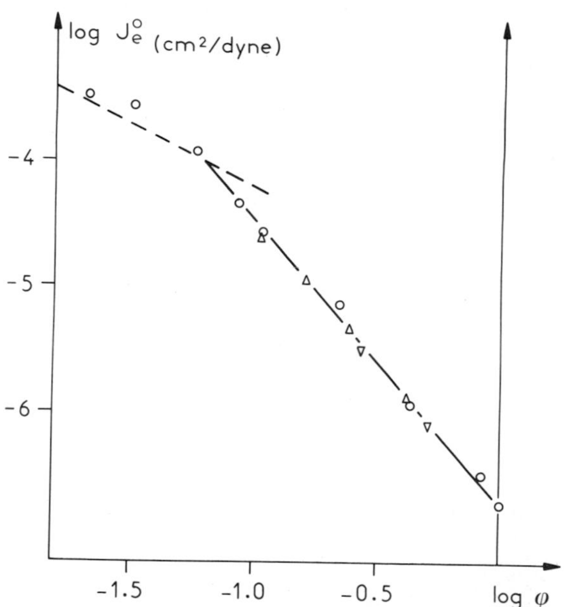

FIG. 2 Variations of compliance vs polymer volume fraction

been found in tetradecane and low molecular weight P.B.). At low concentrations, the variations follow the Rouse theory ($J_e^o \propto \varphi^{-1.0} M$).

On the contrary, the limiting compliance of branched P.B. solutions is very close to the theoretical Rouse value in the whole concentration range.

3°) The plateau modulus G_N^o :

The variations of G_N^o vs polymer volume fraction are exactly the inverse of the J_e^o dependence :

$$G_N^o \propto C^{2.25} M^{0.0}$$

C O N C L U S I O N

The variations of the compliance and plateau modulus of polymer solutions with concentration are found to depart from the previously assumed mean-field behavior ($G_N^o \propto \varphi^{2.0}$ and $J_e^o \propto \varphi^{-2.0}$).

It has been conjectured that the osmotic pressure and plateau modulus could be a measure of the density of segment-segment contacts in the good solvent case ; recent - theoretical and experimental - works on the thermodynamics of polymer solutions seem to agree partly with the picture found here. However, this parallel can be drawn only in the semi dilute - good solvent case, and we are observing no differences between semi dilute and concentrated regime, and between good and θ or near θ solvents, as far as G_N^o and J_e^o are concerned. All these conclusions offer a new challenge for rheologists-both theoretical and experimental.

A COMPARISON OF THE BEHAVIOR OF CONCENTRATED POLYISOBUTYLENE SOLUTIONS AND THE PREDICTIONS OF A NEW CONSTITUTIVE EQUATION

L. J. Zapas

National Bureau of Standards, Washington, U.S.A.

(Abstract)

A new constitutive equation is being proposed which contains as a special case the Bernstein, Kearsley and Zapas theory of an elastic fluid. In this equation, the non-linear moduli are corrected by a stiffening factor which depends on the previous strain history. The equation was applied in the description of the simple shearing behavior of a 19.3% solution of polyisobutylene (L-100) in cetane at 23°C. Having obtained the surface $H(\gamma, t)$ of the single step stress relaxation response over a wide range of time t and strain γ, we proceeded to calculate the contribution of the siffening factor for a suddenly applied shear experiment where the rate of shear was 11.1 s^{-1}. From all these data, we calculated the behavior as a function of time for rates of shear from 22.2 to 111 s^{-1}. The comparison with experimental data was excellent. Very good agreement was obtained for stress relaxation after cessation of flow. The first normal stress difference behavior was also compared favorably for the same type of strain histories.

The two steps stress relaxation experiments where the first step was twice as big as the second step showed also a good agreement between theory and experiment for both shear and first normal stress behavior.

Consequences due to the presence of the stiffening factor will be discussed for experiments where data are not as yet available.

CHARACTERIZATION OF CONCENTRATED SYSTEMS:

CONSTRAINT AND COMPRESSIBILITY EFFECTS

Juan A. Menjivar and ChoKyun Rha

Food Materials Science and Fabrication Laboratory
Department of Nutrition and Food Science
Massachusetts Institute of Technology
Cambridge, Massachusetts

INTRODUCTION

While the rheology of isolated macromolecules in dilute solutions has been extensively studied, concentrated macromolecular systems have received little theoretical consideration.[1] Theories developed for isolated macromolecules are usually not applicable to concentrated solutions, mainly because factors such as; intermolecular interactions, spacial constraints, molecular entanglements and aggregation are not considered.[1] These factors are interdependent and their effects need to be analyzed and considered.

This study is aimed at investigating the effect of intermolecular interactions in the rheology of macromolecules. Globular proteins have been selected as model macromolecules because they have well defined molecular domains and do not entangle, contain a variety of non-polar, polar and ionizable groups, and show unique self-association properties. Therefore, globular proteins are suitable models to examine the effects of intermolecular interactions.

A model that describes the rheological behavior of globular proteins has been proposed.[2] Three hydrodynamic parameters are derived to characterize the solution behavior of macromolecules, The parameters are evaluated and the model is then tested to analyze the relative importance of shape and electrostatic forces in the rheology of proteins.

DESCRIPTION AND RATIONALE OF THE MODEL

Generally, shear viscosity of colloidal dispersions increases with concentration of particles and exponentially above a certain concentration[3]. In particular, globular proteins have consistently indicated that above a "characteristic concentration" (Cch) shear viscosity no longer increases linearly but exponentially [4,5]. As concentration increases further, shear viscosity increases rapidly to infinity representing a crowding limit, which has been defined earlier as the "critical concentration" (C_{cr})[5].

This phenomenon is schematically represented by the model shown in Fig. 1, along with a generalized viscosity-concentration curve for macromolecules. The model visualizes each macromolecule composed of a viscoelastic "molecular core" and an interactive volume". The "molecular core" is the volume of a swollen quaternary structure whose viscoelasticity is represented by radially distributed voigt models. The viscoelasticity of the molecular core is due to its assymetry[6], potential interactions with other molecules[7], and its eventual compressibility[8]. The core is surrounded by an "interactive volume", represented by dashed lines in Fig. 1. This includes the effects of hydrodynamic and/or potential intermolecular interactions.

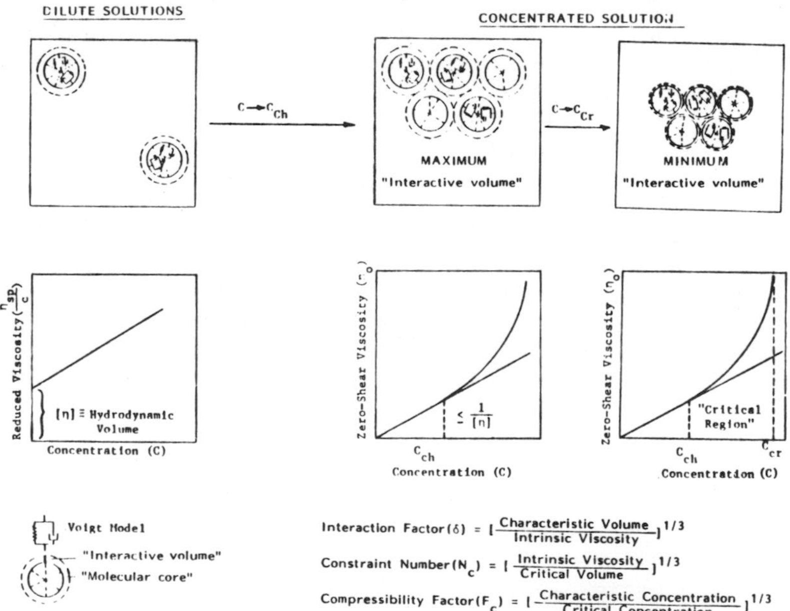

Fig. 1. Descriptive Model of Rheological Behavior of Macromolecules.

Fig. 1 shows that these molecular processes can be characterized by; the intrinsic viscosity ([η]), at the limit of infinite dilution where molecules act independently of each other. As concentration increases the "interactive volume" of molecules start to interact and, depending on its size and nature, C_{ch} will be either similar or lower than $1/[\eta]$. Table 1 shows that C_{ch} is lower for more assymetric and aggregated structures. Electrostatic forces also lower C_{ch} from the expected ($1/[\eta]$) values. Thus, C_{ch} measures the maximum "interactive volume" and is dependent on molecular shape, aggregate geometry and electrostatic charges. Above C_{ch}, molecules start to pack and crowding effect becomes important. Maximum packing concentrations (C_{cr}) will differ depending again upon molecular shape, aggregate shape and nature, and electrostatic charges. Rods pack densely while aggregated structures pack loosely (Table 1). Hence C_{cr} is a measure of the minimum "interactive volume".

In order to isolate the effects of shape and aggregation, three dimensionless hydrodynamic parameters are defined in Fig. 1. They are:
1. *the interaction factor* (δ) which provides an upper limit of the interaction radius of a macromolecule. Table 1 shows that the interaction factor (δ) of non-interacting spheres and rods is 1.0, if hydrodynamic interactions and particle collisions are neglected. For an aggregated spherical structure δ is 1.16 and it is expected that for assymetric aggregates δ will be larger than 1.16. In addition, electrostatically charged molecules will show a $\delta>1$, the magnitude of the deviation depending on the double layer thickness;

Table 1. Hydrodynamic Parameters of Reference Shapes and Globular Proteins.

PARAMETER	SPHERES			RODS (L/D=4)			GLOBULAR PROTEIN		
	Hexagonal Close	Random Close	Spherical Aggregates Random Close	Uniaxial Hexagonal Close	Uniaxial Simple Cubic	Three-Dimensional Random	BSA	Lysoz.	β-Lactog.
Intrinsic Viscosity (ml/g)	2.5	2.5	2.5	4.6	4.6	4.6	3.7	2.7	3.4
Char. Concentration (g/dl)	40.0	40.0	25.5	21.6	21.6	21.6	17	12	10
Char. Volume (ml/g)	2.5	2.5	3.9	4.6	4.6	4.6	5.9	8.3	10.0
Crit. Concentration (g/dl)	74.0	63.7	40.6	90.7	78.5	62.5	68.7	56.5	58.8
Crit. Volume (ml/g)	1.35	1.57	2.46	1.10	1.27	1.60	1.5	1.8	1.7
Interaction Factor	1.00	1.00	1.16	1.00	1.00	1.00	1.17	1.45	1.43
Constraint Number	1.23	1.17	1.01	1.61	1.53	1.42	1.35	1.14	1.26
Compressibility Factor	1.23	1.17	1.17	1.61	1.53	1.42	1.58	1.66	1.81

2. *the constraint number (N_c)* measures the lower limit of the interaction distance between molecules at C_{cr}. Table 1 shows that rods have the capability of forming more ordered structures than spheres and then generate larger N_c values. Spherical aggregates in constrast give low N_c values. Rods with large length to diameter (L/D) ratios are likely to form loose random structures and have the lowest N_c values;
3. *the compressibility factor (F_c)* represents the capacity of the "interactive volume" to be compressed between C_{ch} and C_{cr}. This will depend on the assymetry of the hydrodynamic unit, thus, randomly packed spherical aggregates are less compressible compared with rods packed in uniaxial hexagonal close arrangements (Table 1).

The hypothesis is then that interpretation of these hydrodynamic parameters provides a first approximation of the nature of macromolecular interactions; in this study protein-protein interactions in concentrated solutions.

MATERIALS AND METHODS

Shear viscosity was determined for bovine serum albumin (BSA) (Sigma Chemical Co., St. Louis MO) in phosphate buffer at pH 5.6, 7.2, 10.0 and ionic strength 0.15, 0.15, 0.01 repectively, and lysozyme (Sigma Chemical Co., St. Louis, MO) at pH 7.2 and ionic strength 0.01. Cannon-Fenske capillary viscometers were used to measure the shear viscosity of the solutions. Apparent shear rates ranged from 1400-14 sec for 1% to 45% protein solutions.

The characteristic concentration (C_{ch}) was determined using Einstein-Simha's equation[3] for dilute suspensions of assymetric particles and the shape factor (ν) was determined. Mooney's equation[9] for concentrated suspensions was used to determine the critical concentration (C_{cr}) and the shape parameter (k). Finally, the general equation (1) for dilute colloidal dispersions,

$$\eta_r = \exp \{[\eta]\ \phi\} \qquad (1)$$

where η_r is the relative viscosity of the colloidal dispersion, $[\eta]$ the intrinsic viscosity or shape factor of the colloidal particles or solute, and ϕ the volume fraction of colloidal particles or solute, has fitted and the corresponding intrinsic viscosity $[\eta]$ determined.

EXPERIMENTAL RESULTS

Figure 2 shows the viscosity-concentration curves for BSA, lysozyme and data on β-lactoglobulin at pH 7.0, ionic strength 0.04 from the literature[4]. Neither pH or ionic strength affected the

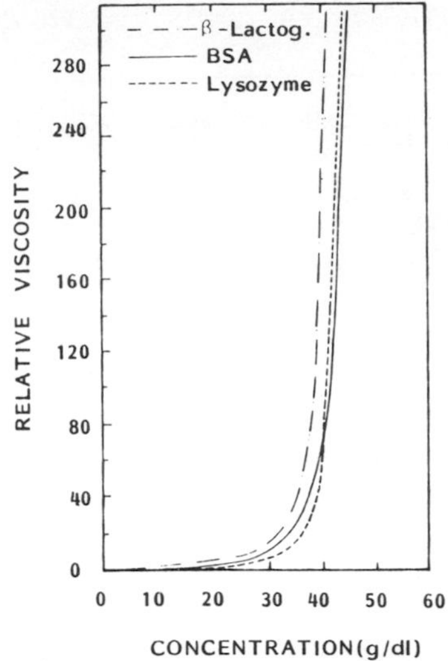

Fig. 2. Viscosity-Concentration Curves for Globular Proteins

rheological behavior of BSA; apparently the initial salt content of approximately 1.3 mg/g albumin[10] is sufficient to shield the effect of surface charge on viscosity. Tanford and Buzzell showed that pH and ionic strength in the ranges 4.3 to 10.5 and 0.01 to 0.5 respectively, had very small effect on the intrinsic viscosity of BSA; [η] changing from 3.6 to 4.3 ml/g. Squire et al.[12] have determined that about 12% of BSA molecules in a 2% solution of this protein are in the form of association products, 10% being oligomers. The viscosity-concentration curve for BSA is then characteristic of protein molecules that form aggregates in solution.

Lysozyme molecules, in contrast, double their intrinsic viscosity at low ionic strength (0.01M), compared with that at high ionic strengths (~0.3M)[13]. In fact, intrinsic viscosity changes from approximately 3.0 ml/g to 6.0 ml/g, considerably more than BSA [13]. This phenomenon is observed for lysozyme solutions at any pH value between 3 and 7. Therefore, the viscosity curve of lysozyme is then characteristic of protein molecules that show significant electrostatic repulsion.

The calculated model parameters are presented in Table 1 for each of the three proteins. Table 2 lists the shape factors, ν, k, and [η], obtained by fitting the three equations mentioned above.

Table 2. Shape Factors for Selected Proteins.

PARAMETER	Protein		
	BSA	Lysozyme	β-Lactoglobulin
Experimental	3.2	4.5	2.8
Einstein-Simha's	7.5	4.4	9.0
General Equation	5.0	4.3	6.7
Mooney's Equation	3.7	2.5	3.5

DISCUSSION OF RESULTS

The viscosity-concentration curves in Fig. 2 show only small differences between the behavior of each of the proteins. A more sensitive analysis is required to distinguish between the factors governing the rheology of proteins.

Comparison of the three parameters; δ, N_c and F_c (Table 1), for lysozyme and BSA, indicated that lysozyme molecules start to interact at significantly lower concentrations ($C_{ch} \sim 12$ g/dl) than expected from $C_{ch} \sim 1/[\eta] \sim 19$ g/dl. Therefore the interaction factor ($\delta \sim 1.45$) is larger than 1 due to the electrostatic forces. As concentration is increased above C_{ch}, lysozyme molecules do not approach as closely as the other two proteins; the constraint number ($N_c = 1.14$), is 15% lower than BSA and close to the $N_c = 1.17$ for randomly packed spheres (Table 1). The compressibility factors (F_c) show no significant difference between these two molecules, however the compression of the interactive volumes have a different nature in each case; electrostatic repulstions for lysozyme and assymetry of the aggregated structure for BSA, this may be confirmed by elasticity measurements[7]. The larger constraint number of BSA indicates that it tends to pack closer than lysozyme above C_{ch} due to the aggregation effect. This causes C_{ch} to be lower than expected. However the aggregates may not be highly assymetric since the interaction factor (δ) is 1.17.

Interesting is the interpretation of Pradipasena and Rha's[4] data on β-lactoglobulin using this model. The interaction factor for β-lactoglobulin ($\delta = 1.43$) is similar to that of lysozyme ($\delta = 1.45$), however the constraint number ($N_c = 1.66$) is higher supporting the existence of rod-shaped aggregates. Therefore, β-lactoglobulin shows an early deviation from linearity (i.e. lower C_{ch} than expected), a rather compressible interactive volume ($F_c = 1.81$ is the most compressible of the 3 molecules) and closely packed. In fact β-lactoglobulin is well known by its capacity for self-association[14].

The interpretation of these hydrodynamic parameters is further supported by the shape factors (Table 2). The shape factor of lysozyme (~4.4) remains basically constant up to the characteristic concentration (C_{ch} ~ 24%), and at higher concentrations its shape factor approaches that of spheres (~2.5) due to the increasing proximity of molecules. On the other hand, BSA and β-lactoglobulin show larger shape factors than those of their isolated molecules.

CONCLUSIONS

1. A model that describes the rheology of macromolecules in solution for a wide range of concentrations has been proposed. Three hydrodynamic parameters that characterize the state of macromolecules have been derived.
2. The hydrodynamic parameters have been determined for model globular proteins. These parameters were significantly different for aggregated and non-associated electrostatically charged proteins.
3. The hydrodynamic parameters, δ, N_c and F_c, are sensitive to and are capable to differentiate between aggregation and electrostatic effects in the rheology of proteins.
4. The interpretation of the hydrodynamic parameters can be substantiated with Einstein-Simha's equation[3] and the generalized equation (1) for concentrations below $C_{ch} = 1/[\eta]$ and Mooney's approach[9] for concentrations above C_{ch}.

REFERENCES

1. R. B. Bird, O. Hassager, R. C. Armstrong, and C. F. Curtiss, "Dynamics of Polymeric Liquids," vol. II New York (1977).
2. J. A. Menjivar, M. O. Mitchell, A. M. Hollenbach, and C. K. Rha, 50th Meeting of The Society of Rheology, Boston, MA (1979).
3. H. L. Frisch. and R. Simha, in: "Rheology," vol. I, F. R. Eirich, ed., New York (1956).
4. P. Pradipasena, and C. K. Rha, Polym. Eng. Sci. 17:861(1978).
5. J. A. Menjivar, and C. K. Rha, Rheol. Acta, in press.
6. H. A. Scheraga, J. Chem. Phys. 23:1526(1965).
7. T. G. M. Van De Ven, and R. J. Hunter, J. Colloid Interface Sci. 68:135(1979).
8. B. Gavish, Biophys. Struct. Mech. 4:37(1978).
9. M. Mooney, J. Colloid Sci. 6:162(1951).
10. J. E. Cohn, L. E. Strong, W. L. Hughes, D. J. Mulford, J. N. Ashworth, M. Merlin, and H. L. Taylor, J. Am. Chem. Soc. 68:459(1946).
11. C. Tanford, and J. G. Buzzell, J. Am. Chem. Soc. 60:225(1956).
12. P. G. Squire, P. Moser, and C. T. O'Konski, Biochem. 7:4261(1968).
13. M. Komatsubara, K. Suzuki, H. Nakajima, and Y. Wada, Biopolymers 12:1741(1973).
14. R. Townend, and S. N. Timasheff, Arch. Biochem. Biophys. 63:482 (1956); J. Am. Chem. Soc. 79:3613(1957), 82:3161, 3168(1960).

BROAD SHEAR RANGE VISCOMETRY OF HIGH POLYMER SOLUTIONS:

POLYSTYRENE AND POLYISOBUTENE IN DECALIN

T. Shimida, P.L. Horng, R.S. Porter

Department of Polymer Science and Engineering,
Materials Research Laboratory, The University of
Massachusetts, USA

(Abstract)

The flow behavior at 25°C for decalin solutions of polystyrenes (PS) of molecular weights up to 7 x 10^6 and polyisobutenes (PIB) of molecular weights up to 3 x 10^6 was examined over a wide range of shear rate up to about 10^5 sec^{-1} for several polymer concentr ations. Measurements involved conventional glass capillary viscometers, a Rheometric Mechanical Spectrometer, and custom-built high-shear concentric cylinder viscometer. Low shear Newtonian viscosities for the higher concentration entanglement region follow the forms $\eta_o = CM^{0.57}$ for PS and $\eta_o = CM^{0.68}$ for PIB. Gel permeation chromatography was used to determine polymer molecular weight, its distribution, and possible change on shearing. Shear degradation was observed for polymers at high values of concentration, molecular weight, and shear stress.

SHEAR RATE DEPENDENCE OF THE ASSOCIATION OF HIGH MOLECULAR WEIGHT MACROMOLECULES IN DILUTE SOLUTION

Marie-Noëlle Layec-Raphalen(*), Claude Wolff(**)

(*) Laboratoire d'Hydrodynamique Moléculaire - Faculté des Sciences, 6, av Le Gorgeu F 29283 Brest Cédex
(**) Laboratoire de Mécanique et Rhéologie - E.N.S.I.T., 11, rue A. Werner F 68093 Mulhouse Cédex

INTRODUCTION

Dilute solutions of flexible, high molecular weight macromolecules in nearly θ solvents have been demonstrated to have an increase of their relative viscosity above a critical shear rate, depending on the concentration[1,2]. In the same time, a non linear dependence of the reduced viscosity η_{sp}/c is observed. Such a variation may be explained by an association process[3,4,5,6]. In this paper, we intend to apply the procedure described in our recent work[5], at different shear rates to the following materials which have shown shear thickening[2] : Polystyrenes in decalin and polyethyleneoxide WSR301 in water. Calculations have been carried out in dimerization and multimerization cases. But only the dimerization appears to be compatible with the experimental results. The molecular weights of the polystyrene samples are listed in table II. We have made the assumption that the Huggins constant k_H is nearly independent on the degree of association.

According to[5], we obtain the following expression of the reduced viscosity η_{sp}/c.

$$\eta_{sp}/c = \langle\eta_{sp}/c\rangle + k_H c \langle\eta_{sp}/c\rangle^2 \quad (1)$$

where $$\langle\eta_{sp}/c\rangle = [\eta]_1 (\sqrt{1+8\varepsilon} - 1) \left[1 + (\sqrt{1+8\varepsilon} - 1)\frac{r}{2}\right] /4\varepsilon \quad (2)$$

with $r = [\eta]_2/[\eta]_1$, $\varepsilon = Kc/M_1$, $K = [M]_2/[M]_1$

The subscripts 1 and 2 are relative respectively to the monopolymer of molecular weight M_1 and dipolymer of molecular weight

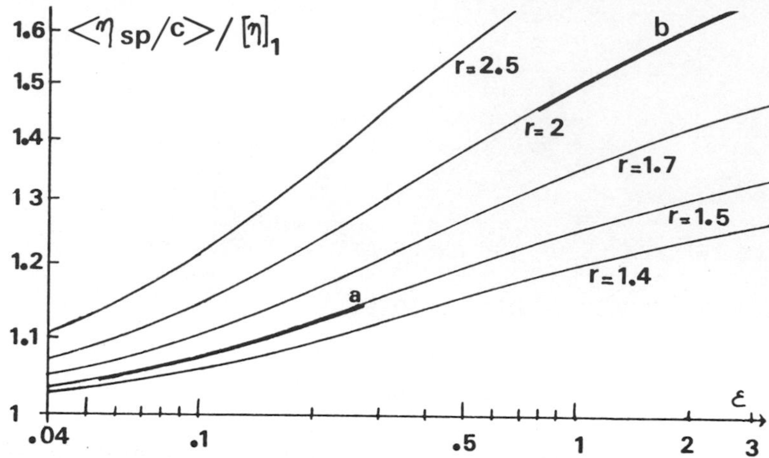

Fig. 1 Fit of the curves $<\eta_{sp}/c>_{exp} = f(c)$ for polystyrene ($M_W = 3{,}7\ 10^6$) to the theoretical curves $<\eta_{sp}/c>/[\eta]_1 = f(\varepsilon)$
a) $G = 13260\ s^{-1}$: $r = 1{,}5$; $[\eta]_1 = 164\ cm^3/g$; $K/M = 110$
b) $G = 26820\ s^{-1}$: $r = 2$; $[\eta]_1 = 120\ cm^3/g$; $K/M = 1550$

$M_2 = 2\ M_1$. $[\eta]$ is the limiting viscosity number, K the association constant and $[M]$ the molar concentration.

The best fit of the experimental curve $<\eta sp/c> = f(c)$ to the theoretical nest of curves $<\eta_{sp}/c>\ /\ [\eta]_1 = f(\varepsilon)$ (with r as parameter gives quickly the values of $[\eta]_1$, K/M_1 and r (fig. 1). The experimental values of $<\eta_{sp}/c>$ are calculated from the values of η_{sp}/c as solution of equation (1), assuming a definite value of the Huggins constant k_H (.4 for polystyrene and .35 for polyethylene-oxide).

RESULTS

The results have been obtained at 25°C in capillary viscometers of the Ostwald type connected to a pressure regulation allowing the wall shear rate G to vary between 300 and 50 000 s^{-1}.

By use of the Bagley procedure [7], we have established that these dilute solutions don't show measurable end effects.

<u>Polyethylene oxide Polyox WSR 301 in distilled water</u>

The concentrations ranged from 40 to 400 ppm and the corresponding relative viscosities from 1,05 to 1,40.

Because of the important curvature of the η_{sp}/c versus c curves at low concentrations and high shear rates and the resulting imprecision in the fitting, we have used only the results obtained for

$G < 8\ 000\ s^{-1}$. The data obtained are listed in table I.

The intrinsic viscosity of the monopolymer at G = 0 is $[\eta]_{1,0} = 1040\ cm^3/g$. $[\eta]_{1,r}$ is the relative intrinsic viscosity of the monopolymer $[\eta]_{1,r} = [\eta]_{1,G}/[\eta]_{1,0}$.

These results suggest the following considerations :

1) The parameter r is independent on the shear rate G. Because of its high value it must be assumed that the associated macromolecule is elongated. The most probable conformation is therefore an end-to-end association which is also compatible with the flow birefringence results[8] The deformation exists even at low shear rate, perhaps owing to the low internal viscosity of this polymer[9].

2) The ratio K/M_1, i.e the association constant K varies sharply with the shear rate according approximatively to the law (fig. 2).

$$K/M_1 = (K/M_1)_s + AG^x$$

where $(K/M_1)_s = 400$ is the value in the absence of flow, $A = 8{,}5\ 10^{-7}$, $x = 2{,}5$.

3) The shear rate dependence of the quantity $G/(1-[\eta]_{1,r})$ is linear [10]. The slope of the straight line is related to the limit of the intrinsic viscosity at $G \to \infty$, $[\eta]_{1,\infty}$. Therefrom $[\eta]_{1,\infty}/[\eta]_{1,0} = 0{,}30$.

This value is lower than the usual values of the relative intrinsic viscosity limits of flexible coils submitted to laminar flow[11]. It confirms the anisotropy of the polyethylene oxide particule in water. If we assume a prolate ellipsoïdal configuration, this value of $|\eta|_{1,\infty}/|\eta|_{1,0}$ leads to an aspect ratio p of about 8[12]. The value of the aspect ratio deduced from the flow birefringence measurements[8] is lower but it can be noticed that this technique is much more sensitive to polydispersity.

Table 1 Values of $[\eta]_1$, K/M_1 and r for PEO WSR 301

G	$[\eta]_1$	K/M_1	r	$G/(1-[\eta]_{1,r})$
1 000	920	380	3,5	8 700
2 000	830	550	3,5	9 900
3 000	750	800	3,5	10 800
4 000	690	1 200	3,5	11 900
5 000	655	1 670	3,5	13 500
6 000	630	2 410	3,5	15 200
8 000	580	5 000	3,5	18 100

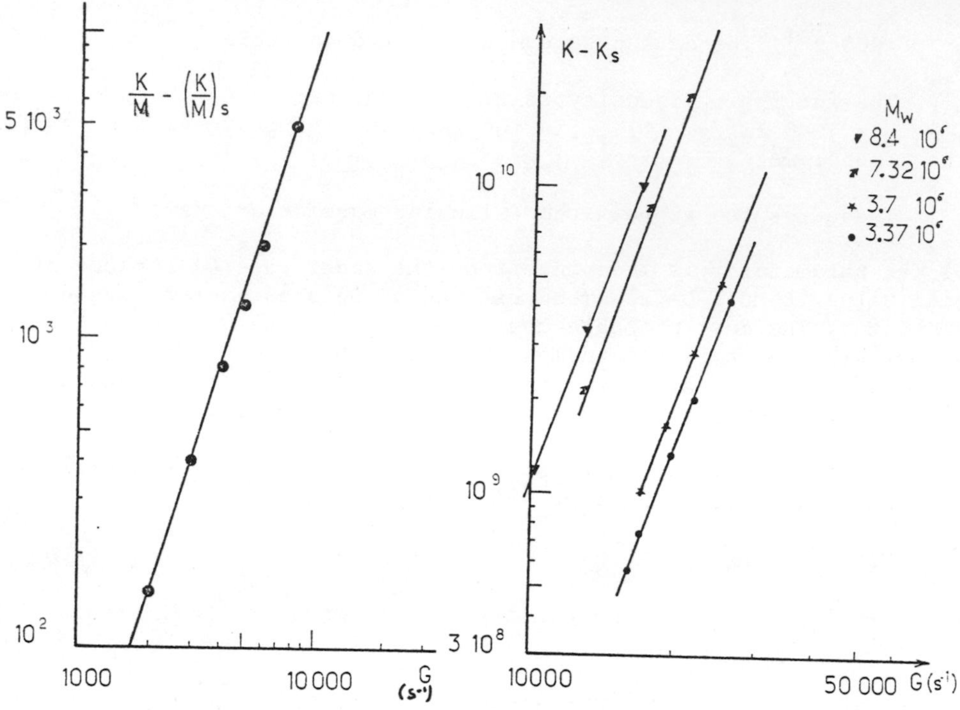

Fig. 2 Variation of the association constant with the shear rate G for Polyox WSR301

Fig. 3 Variation of the association constant with the shear rate G for the 4 polystyrenes.

Polystyrenes in cis-trans decalin

The shear thickening and η_{sp}/c versus c data for the four polystyrenes used have been published earlier[2]. The concentrations have been choosen so that the relative viscosities ranged from 1,06 to 1,5. The results are listed in table II.

The association constant K depends on the shear rate G as :

$$K - K_s = AG^x \qquad \text{(fig. 3)}$$

The exponent x = 4,3 is independent on the molecular weight.

On the contrary of Polyox, the parameter r varies with the shear rate from 1,4 at low shear rate to 2,1 at high shear rate. These values suggest that the association of polystyrene molecules is only slightly anisotropic and that the molecular deformation increases with the shear rate.

Table 2 Characteristics of the 4 polystyrenes

M_w	$[\eta]_1$	$K_s x 10^{-7}$	$A x 10^{10}$	x	a	$[\eta]_\infty/[\eta]_o$
$3,37\ 10^6$	168	0	4,4	4,3	1,60	0,38
$3,7\ 10^6$	215	7	6	4,3	1,60	0,38
$7,32\ 10^6$	265	146	45	4,3	1,79	0,44
$8,4\ 10^6$	260	101	70	4,3	1,70	0,41

The value of the aspect ratio p deduced from the intrinsic viscosity limit $[\eta]_{1,\infty}/[\eta]_{1,0}$, according to a rigid ellipsoid model is about p = 4. This value is comparable to the one deduced from the critical concentration appearing in shear thickening assuming that this concentration corresponds to joined ellipsoids[2].

The molecular weight dependence of the association constant K is approximatively

$$K - K_s = 1,18\ 10^{-29}\ M_1^3\ G^{4,3}$$

in the range of the experimental values.

CONCLUSION

It has been shown that the non linear concentration dependence of the reduced viscosity η_{sp}/c can be explained in term of association. For the two systems we studied, the association is limited to dimerization. The associated molecule of Polyox WSR 301 in water appears strongly anisotropic even at low shear rates whereas the polystyrene molecule is only slightly anisotropic. The association constant K varies sharply with the shear rate. The exponent x of the power law $K = K_s + AG^x$ is 2,5 for Polyox and 4,3 for polystyrenes. The molecular weight dependence of the association constant obeys approximatively to a power law with 3 as exponent for polystyrene.

1. O. Quadrat, M. Bohdanecky, and P. Munk, Influence of Thermodynamic quality of a solvent upon Non-Newtonian viscosity of Poly(methylmethacrylate) solutions, J. Polym. Sci., Part C, 16:95 (1967)
2. M. N. Layec-Raphalen, and C. Wolff, On the shear thickening behaviour of dilute solutions of chain macromolecules, J. N. N. Fl. Mech., 1:159 (1976)
3. L. H. Frisch, and R. Simha, in Rheology, F. Eirich Editor, Acad. Press, N.Y, vol. I, p. 525 (1956)
4. H. G. Elias, Association of synthetic polymers, Intern. J. Polym. Mater., 4, 3-4:209 (1976)
5. C. Wolff, A. Silberberg, Z. Priel, and M. N. Layec-Raphalen Influence of the association of macromolecules in dilute solution on their reduced viscosity, Polymer, 20:281 (1979)
6. M. Bohdanecky, B. Lanska, J. Sebenda, and Z. Tuzar, Dilute solutions of Nylon 12-III Intermolecular and Intramolecular association of end-groups, to be published in Europ. Polym. J.
7. E. B. Bagley, End corrections in the capillary flow of Polyethylene, J. Appl. Phys., 28,5:624 (1957)
8. A. Dandridge, G. H. Meeten, M. N. Layec-Raphalen, and C. Wolff, Flow birefringence of dilute solutions of Polyethylene oxide of high molecular weight at high shear rate, Rheol. Acta, 18:275 (1979)
9. M. C. Kohn, Energy storage in drag reducing Polymer solutions, J. of Polym. Sci., 11:2339 (1973)
10. C. Wolff, Sur une représentation asymptotique linéaire des mesures de viscosité intrinsèque non-newtonienne, J. chim. Phys-, 1174 (1962)
11. C. Wolff, Comparaison des courbes de viscosité intrinsèque non-newtonienne obtenues avec différents types de polymères, J. Phys. (Suppl. n° 10), 32:C5a-263 (1971)
12. Y. Layec and C. Wolff, Sur la viscosité non-newtonienne des solutions d'ellipsoïdes rigides, Rheol. Acta, 13: 696 (1974)

ULTRASONIC VELOCITIES AND RAO FORMALISM IN POLYMER SOLUTIONS

R.P. Singh, G.V. Reddy, S. Majumdar, and
Y.P. Singh

Polymeric Materials Laboratory
Materials Science Centre
Indian Institute of Technology
Kharagpur-721302, India

INTRODUCTION

Ultrasonic absorption and dispersion studies in polymer solutions (Dunn et al., 1969; Matheson, 1971; Pethrick, 1973) and in solid polymers (Phillips and Pethrick, 1977-78) have been the subjects of extensive research activity in the last decade. Through these investigations considerable progress has been made in the understanding of inter- and intramolecular interactions, modes of motion, and conformational changes in polymers.

In ultrasonic investigations, the experimentally determinable parameters such as ultrasonic velocity and ultrasonic absorption coefficient are correlated with the molecular parameters of the system under study. Ultrasonic velocity measurements are much simpler, faster, and more accurate than absorption measurements; hence they offer a promising method of characterizing polymer solutions in polymer processing.

Recently Von Krevelen (1976) has pointed out that since the Rao constant (Rao, 1940, 1941) is an additive quantity, it can be evaluated for a particular polymer from the constituent group contributions. Hence the Rao constant can be used for structural analysis of polymers. However for the evaluation of Rao constant, the ultrasonic velocity and density in solid samples are necessary. Ultrasonic velocity and density measurements are much simpler in polymer solutions. If some correlation exists between polymer solution and solid state measurements, solution studies can also be utilized for polymer analysis. A correlation has been established by making comprehensive ultrasonic velocity and density measurements in a number of polymer solutions (Reddy and Singh, 1980; Reddy et al., 1980). The present paper indicates a simple method of polymer analysis based on Rao formalism.

EXPERIMENTAL

Ultrasonic velocity measurements have been carried out by the interferometric technique (Hubbard, 1931) employing an ultrasonic interferometer (Mittal Enterprises, New Delhi). The experimental cell has a double-walled jacket, and thermostated water is circulated through it from an ultrathermostat (Model U10, VEBMLW Prüfgeräte-Werk, GDR) with thermal stability of ±0.05°C. The experimental frequency is 2 MHz and the accuracy of the velocity measurements is better than ±0.5%. Densities of all solutions are measured by using a specific gravity bottle and are accurate to the fourth decimal place. The polymers and solvents which have been studied are given in Table 1.

Table 1. List of Polymer and Solvent Systems

Polymer	Solvent
Polyvinylpyrrolidone (high-molecular-weight, BDH, U.K.)	Double-distilled water
Polymethylmethacrylate (high-molecular-weight, BDH, U.K.)	Toluene
Polyacrylamide (BDH, U.K.)	Double-distilled water
Polyacrylic acid (BDH, U.K.)	Double-distilled water
Polyethyleneoxide (Polyox, BDH, U.K.)	Double-distilled water
Polyvinylalcohol (BDH, U.K.)	Double-distilled water
Polyvinylacetate (BDH, U.K.)	Chloroform
Polyvinylchloride (RAPRA, U.K.)	Tetrahydrofurone

RESULTS AND DISCUSSION

The empirically established Rao relationship has been found to be applicable to a large number of liquids (Beyer and Letcher, 1969), liquid mixtures (Nomoto, 1958), and solid polymers (Schuyer, 1958; Chattopadhyay and Singh, 1980). The Rao constant is given by

$$R = \frac{M}{\rho} v^{1/3} \tag{1}$$

where M, ρ, and v are, respectively, the molecular weight, density, and ultrasonic velocity of the system under study. For binary liquid mixtures, the Rao constant can be calculated by the Nomoto relation (Nomoto, 1940)

$$R_m = R_1X_1 + R_2X_2 \tag{2}$$

where R_m, R_1, and R_2 are the Rao constants for the liquid mixture and the constituent liquids, respectively. X_1 and X_2 are the mole fractions of the constituent liquids of the mixture.

Table 2. Comparison of Theoretical and Experimental Values of the Rao Constant

Polymer solution	Conc., wt.%	Ultrasonic velocity, m/sec	Density, kg/m^3	Rao constant Expt. $\frac{kg}{mole}\frac{m^3}{kg}(\frac{m}{sec})^{1/3}$	Rao constant Theor. $\frac{kg}{mole}\frac{m^3}{kg}(\frac{m}{sec})^{1/3}$
Polyvinylpyrrolidone in water	20	1591.2	1027.3	0.245	0.245
Polymethylmethacrylate in toluene	20	1359.8	890.0	1.143	1.151
Polyox coagulant in water	5	1524.0	1009.5	0.211	0.212
Polyox WSR-750 in water	5	1529.6	1008.0	0.211	0.212
Polyacrylamide in water	4	1522.8	1010.6	0.211	0.212
Polyacrylic acid in water	5	1519.6	1020.0	0.210	0.212
Polyvinylchloride in THF	2	1288	889.2	0.878	0.881
Polyvinylalcohol in water	2	1487.6	1004.7	0.207	0.208
Polyvinylacetate in chloroform	2	1028.4	1408.7	0.849	0.841

In evaluating the Rao constant of the solution, equation (1) may be used by replacing M by M_{sol}

$$M_{sol} = M_s X_s + M_p X_p \tag{3}$$

where M_{sol}, M_s, and M_p are the molecular weights of the solution, solvent, and polymer. X_s and X_p are the mole fractions of the solvent and solute or polymer in the solution. The value of the Rao constant of the solution may also be calculated from equation (2) theoretically. It has been found that the two values of the Rao constants are almost equal within experimental error, as shown in Table 2 for a number of polymers.

The consequences of this analysis are far-reaching. The Rao constant is thus given by

$$R_1 X_1 = R_m - R_2 X_2 \tag{4}$$

for solid polymers, from equation (3), in terms of the ultrasonic velocity and density data in solution (R_m) and solvent (R_2 and X_2). Now, the Rao constant being an additive quantity, its value for a polymer can be determined from the contributions of the constituent groups. Thus from the simple measurements of ultrasonic velocity and density of polymer solution and solvent, one may get information which may be helpful in the structural analysis of polymers.

This approach has been extended to solutions of polybenzimidazoles in formic acid (Maiti et al., 1980) and to solutions of the polyester-imide copolymer condensates (Das et al., 1980). It has been found that this approach is justified even in such complicated systems.

ACKNOWLEDGMENT

The authors gratefully acknowledge the financial support of the Department of Science and Technology, Government of India.

REFERENCES

Beyer, R.T., and Letcher, S.V. 1969, "Physical Ultrasonics," Academic Press, New York.

Chattopadhyay, S., and Singh, R.P., 1980, Rao Constants of Polymers of Differing Structures (under publication).

Das, S., Singh, R.P., and Maiti, S., 1980, Ultrasonic Studies and Rao Formalism in Polycondensates (communicated to Polymer Bulletin).

Dunn, F., Edmonds, P.D., and Fry, W.J., 1969, Absorption and Dispersion of Ultrasound in Biological Media, in: "Biological Engineering," H.P. Schwan, ed., McGraw-Hill Book Co., New York.

Matheson, A.J., 1971, "Molecular Acoustics," Wiley Interscience, London.

Maiti, A.K., Singh, R.P., and Banerjee, P.C., 1980, Ultrasonic Effects in Formic Acid Solutions of Polybenzimidazoles (under publication).

Nomoto, O., 1958, Empirical Formula for Sound Velocity in Liquid Mixtures, Phys. Soc. Japan, 13:1528.

Pethrick, R.A., 1973, Acoustic Studies of Polymer Solutions, J. Macromol. Sci., C-9:91.

Phillips, W.J., and Pethrick, R.A., 1977-78, Ultrasonic Studies of Solid Polymers, J. Macromol. Sci., C-16:1.

Rao, M.R., 1940, A Relation between Velocity of Sound in Liquids and Molecular Volume; Ind. J. Phys., 14:109.

Rao, M.R., 1941, Velocity of Sound in Liquids and Chemical Constitution, J. Chem. Phys., 9:682.

Reddy, G.V., and Singh, R.P., 1980, Ultrasonic Velocities and Rao Constants of Aqueous Solutions of Polyvinylpyrrolidones; Ultrasonic Velocities and Rao Constants of Polymethylmethacrylate in Toluene. Acustica, 5 (in press).

Reddy, G.V., Majumdar, S., and Singh, R.P., 1980, Ultrasonic Velocities and Rao Formalism in Solutions of Polymers of Differing Structures. Acustica (communicated).

Schuyer, J., 1958, Molar Sound Velocity of Solid, Nature, 181:1394.

Von Krevelen, W.D., and Hoftyzer, P.J., 1976, "Properties of Polymers," Elsevier Scientific Publishing Co., Amsterdam, p. 269.

THE IMPORTANCE OF ENERGETIC INTERACTIONS ON THE VISCOELASTIC PROPERTIES OF POLYMER SOLUTIONS

Werner-Michael Kulicke

Institute of Chemical Technology

Technical University of Braunschweig, F.R.G.

INTRODUCTION

The rheological behaviour of polymer fluids can be described on a molecular basis using the entanglement concept[1]. In this case it is explicitely assumed that there are mechanical entanglements (above a minimum concentration and/or molecular weight) caused by the polymer chains, which are responsible for the behaviour in the linear and non-linear viscoelastic region. It has been shown in the past that a quite great number of rheological properties can be theoretically explained on this basis.

However, in the present work it will be shown that energetic interactions can influence the material functions of polymer fluids so strong, that they are in contrast to experimental facts for most other polymer fluids. This will be shown by using an empirical rule (Cox-Merz). Also the entanglement concept is not able to explain the rheological behaviour in those cases.

Results in the linear viscoelastic region

The zero-shear viscosity of mixtures of polyacrylamides (PAAm) with the sodium salt of poly-(acrylic acid) (PAAcNA) has been compared with the zero-shear viscosity of poly(acrylamide-co-sodium acrylates) (PAAm/AAcNa). The PAAm and PAAcNa were prepared in our laboratory by solution polymerization. The copolymers are prepared by saponification of the PAAm using NaOH. The zero-shear viscosity from these samples are measured in dilute solutions. The used solvent was 0.5 n NaBr-solution. By plotting the zero-shear-viscosities devided by $\eta_{0,PAAm}$ of different mixtures of PAAm ($M_\eta = 1,3\cdot 10^6$ g/mol) and PAAcNa ($M_\eta = 1,6\cdot 10^6$ g/mol) in a concentration range of

0,1 and 0,3 % one will get a linear plot over the whole range of mixtures (22-100 mol% acrylamide). Only in concentrated systems (c > 5%) a maximum can be observed. But by plotting the zero-shear viscosity devided by $\eta_{O,PAAm}$ of PAAm/AAcNa of different copolymer degrees one can observe a maximum at a copolymer degree of about 35 mol% acrylamide. At 0 and 100 mol% acrylamide the viscosity ratio $\eta_O/\eta_{O,PAAm}$ is one at least in a concentration range of 0-4 wt%. In case of the intrinsic viscosity the maximum at about 35 mol% acrylamide of $[\eta]/[\eta]_{PAAm}$ = 1,9 and increases by increasing concentration[2]. Therefore, even in the presence of high amounts of electrolytes (NaBr) the copolymers have higher viscosities than the corresponding mixtures of homopolymers.

Infrared-spectroscopy results show that H-bonds exist: in the mixtures and in the copolymers. The reason for the different behaviour in the viscosities of dilute solutions as a function of mol% acrylamide can only be understood when one assumes that in the copolymers are intramolecular H-bonds and in the mixtures only intermolecular H-bonds exist. This is supported by the fact that by increasing the concentration (> 5%) also in the mixture a maximum can be observed. The chain expansion therefore is caused by H-bonds between R - COO^- --- H_2N - R. This H-bond therefore is stronger than the R - C(-NH_2) = O---H_2N - C(=O) - R in the PAAm.

Results in the non-linear viscoelastic region[3]

Determination of prestationary stress functions (overshoot-measurements) for characterizing viscoelastic fluids have found an increasing interest in the last years. By measuring the shear stress (σ_{12}) as well as the normal stress ($\sigma_{11} - \sigma_{22}$) as a function of time it is shown that in case of particel solutions (dilute solutions) the transition to a steady value monotonically increases. In case of concentrated solutions above a critical shear rate $\dot{\gamma}$ the stress goes through a maximum (overshoot) smoothly declining to steady state value. Speaking in terms of the entanglement concept this is caused by the entangled chains which are not longer able to slip off. To study this, a cone-and-plate geometry was used with a cone angle of 2,3^o and a radius of r=5 cm. In case of polystyrene (PS7-decalin solutions (c=11,5 wt%) the normally found behaviour can be observed. This is that the overshoot increases and that the time t_m reaching the maximum decreases by increasing the shear rate. But at relatively low shear rates the solution flows out of the gap. In case of the higher molecular weight ($M = 2 \cdot 10^6$ g/mol) the solution leaves suddenly the gap at $\dot{\gamma}$ = 25 sec^{-1} and using the lower molecular weight (M = 670 000 g/mol) at $\dot{\gamma}$ = 625 sec^{-1}. At the temperature of 23 ^oC which measurements were made, decalin is at theta solvent for PS. The exponent, a, of the Mark-Houwink-equation is very close to a = 0,5. This means that the solution behaves pseudoideally. The solution behaves like a rubber in the critical shear rate range $\dot{\gamma}$ = 10 - 25 sec^{-1}. May be energetic polymer-polymer interactions which arise from the π-electron-system of the phenylic side groups are the reason for this unusual behaviour.

On the other hand we studied in the same cone-and plate geometry polyacrylamides (PAAm) in H_2O, formamide and ethylenglycole. In aqueous solutions the normally expected behaviour can be observed up to the highest achievable shear rate $\dot{\gamma} = 1000\ sec^{-1}$. The quantitative values of the characteristic values (t_m, t_{st} etc.) with respect to shear and normal stress always differ.The characteristic times t_m and t_{st} (time to reach the steady valus) shift to smaller values and the overshoot of the shear and normal stress increases with increasing shear rate. However in formamide and ethylenglycole solutions an important difference occurs. The stress functions at the onset of steady state flow of a 1% PAAm/ethylenglycole solution ($M_w = 4{,}7 \cdot 10^6$ g/mol) are up to $\dot{\gamma} = 100\ sec^{-1}$ qualitatively similar to those observed for aqueous solutions. Increasing the shear rate to $\dot{\gamma} = 158\ sec^{-1}$ a transitional response with two overshoots can be seen. The first is similar to those observed before and is caused by mechanical entanglements and a second, which is much broader, showing gel fracture, which is caused by energetic interactions (assoziation). Only the energetically controlled overshoot does not recover in less than several hours. At higher shear rates the polymer solution flows out of the gap. Similar results are observed in PAAm/formamide solutions. Resting periods of several hours are insufficient to obtain the overshoot of a breaking gel again. At higher shear rates the solution flew out of the gap suddenly. This is the first time that overshoots caused by mechanical entanglements and overshoots caused by energetic interactions can be distinguished[3].

Relation between steady state and dynamic rheology[4]

It is well known that an empirical rule exists to get shear viscosity informations when only linear viscoelastic data are available and the reverse, the so called Cox-Merz-rule. This empiricism predicts that the magnitude of the complex dynamic viscosity $|\eta^*|$ should be comparable with the shear viscosity η also at high values of frequencies ω respectively shear rates $\dot{\gamma}$.

Experimental results show that this is a good approximation for different polymers, molecular weights, structures and molecular weight distributions. But only in case of homogeneous PAAm/H_2O-solutions we found deviations between $|\eta^*|$ and η at relatively high $\dot{\gamma}$ and ω. We suggest that hydrogen bonds may be the reason for this behaviour. Hydrogen bonds influence the internal structure of the solved PAAm-molecules in water so strongly that they differ in their dimensions and rheological behaviour in contrast to experimental facts for most other polymer fluids. In case of steady shear flow first the hydrogen bonds are destroyed so that the slope in the η vs $\dot{\gamma}$ function is steeper, because the hydrogen bond junctions are reduced. In dynamic measurements we suggest that we measure mainly the chain flexibility between two junctions and there is no possibility to distinguish between junctions caused by hydrogen bonds or by mechanical entanglements.

CONCLUSION

Therefore energetic interactions can influence the solution structure of polymer molecules so strong, that the material functions will deviate from the usual expected and theoretically predicted behaviour. To use furthermore the entanglement concept for polar systems we suggest to introduce a new term which is responsible for junctions caused by energetic interactions and therefore distinguishable from mechanical entanglements.

References

1) W.W. Graessley, Adv. Polym. Sci. 16 (1974) 1

2) R. Heitzmann, Dissertation, Inst. f. Chemical Technology, T.U. Braunschweig 1977

3) W.-M. Kulicke and J. Klein, and R.S. Porter, Angew. Macromol. Chemie 76/77 (1979) 151

4) W.-M. Kulicke, R.S. Porter, Rheol. Acta (in preparation)

EFFECTS OF POLYMER CONCENTRATION AND SHEAR RATE ON DIFFUSION IN POLYSACCHARIDE SOLUTIONS

D.W. Hubbard, F.D. Williams, and G.P. Heinrich

Michigan Technological University

Houghton, Michigan USA

INTRODUCTION

The structure of a polymer solution depends on the type of polymer, on the polymer concentration, and on the shear rate existing in the fluid. Shearing polymer solutions can cause dramatic effects such as polymer degradation when a certain threshold shear stress is exceeded (Leopairat and Merrill, 1978). Under less severe conditions, experiments show that other critical or yield phenomena can occur in following polymer solutions where major perturbations of molecular shape and coil volume or pervaded volume can occur. Peterlin (1970) showed that molecular extension occurs in flow systems causing enormous increases in the radius of gyration of the dissolved polymer.

Polymers in solution can be swollen or solvated. Data suggest that critical phenomena related to mass transport in polyelectrolyte solutions occur at polymer concentrations at which all "free" solvent just disappears. This critical polymer concentration, which can be calculated from the hydrodynamic radius of the polymer molecule, depends on shear rate and on solution composition. Diffusion experiments performed in flowing polymer solutions should give insight into the structure of the solution.

Numerous data for the diffusion of solutes in polymer solutions have been reported. Diffusion of dissolved gases has been studied by Astarita (1965), by Metzner (1965), by Quinn and Blair (1967), by Zandi and Turner (1970), and by Grief et al (1972). Diffusion of organic solutes has been studied by Burns (1960), by Clough et al (1962), and by Astarita (1966).

The diffusion of ionic species has been studied by Arvia, et al (1968), by Fortuna and Hanratty (1972), by McConaghy and Hanratty (1977), and by Rickard et al (1978). Several different experimental methods--failing films, laminar jets, dissolving walls, and electrochemical methods using stationary and moving electrodes--have been employed. In many cases, the effect of polymer concentration on the diffusion coefficient was discussed, but the effect of shear rate was not determined or could not be calculated from the data reported.

Two other papers report data which show the effect of shear rate on the diffusion coefficient. Wasan et al (1972) measured the diffusion coefficient for dissolved oxygen in aqueous solutions of several different types of polymers by observing mass transfer in a thin film flowing down a cylindrical wall. The shear rate was varied by varying the flow rate. Their data show that the diffusion coefficient increases as the shear rate at the wall increases. Kumar and Upadhyay (1980) measured the mass transfer rate of benzoic acid dissolving from a tube wall into one percent carboxymethyl-cellulose aqueous solution flowing through the tube in laminar flow. The shear rate was varied by changing the flow rate. Over a shear rate range from 1000 to 8000 1/s, the data showed that the diffusion coefficient does not depend on shear rate.

EXPERIMENTAL METHODS

The diffusion of ferricyanide ions in aqueous xanthan gum solutions was studied using the volammetric electrochemical method described by Fortuna and Hanratty (1972). The electrolyte contained 0.005 molar ferricyanide, 0.005 molar ferrocyanide, 0.1 molar potassium chloride, and 200 to 5000 ppm polysaccharide. The polymer used had the trade name Kelzan XC® (Kelco Company, Houston, Texas, USA). This material has a molecular weight of approximately 2×10^6.

The electrolyte was pumped through an annular spece between two concentric cylindrical electrodes. The inner electrode was a nickel plated brass rod 9.5 mm in diameter with an active length of 252.4 mm. The outer electrode was a piece of nickel pipe 15.8 mm in diameter. Upstream and downstream from the active electrode surface, were calming sections 217 mm long to ensure that fully developed laminar flow was obtained. The Reynolds numbers for the experiments were in the range from 1 to 1000.

The inner electrode was made the cathode, and ferricyanide ions were reduced there. The reverse reaction took place at the outer electrode, so there was no net change in solution composition during the course of an experiment.

The cathode polarization was measured as a function of the cell current, and the limiting current was determined. The diffusion coefficient was calculated using the Graetz-Leveque equation.

The rheological behavior of the solutions was described by the power law model and the rheological parameters were determined using a capillary tube rheometer. The flow behavior index, n, ranged from 0.77 for solutions containing 200 ppm polymer to 0.3 for solutions containing 5000 ppm polymer. The consistency index ranged from 0.18 $g/cm\ s^{n-2}$ for solutions containing 200 ppm polymer to 39 $g/cm\ s^{n-2}$ for solutions containing 5000 ppm polymers.

RESULTS

The diffusion coefficients measured for ferricyanide ions in the polymer solutions were approximately five to twenty times smaller than those measured with no polymer present. Figure 1 shows how the diffusion coefficient depends on shear

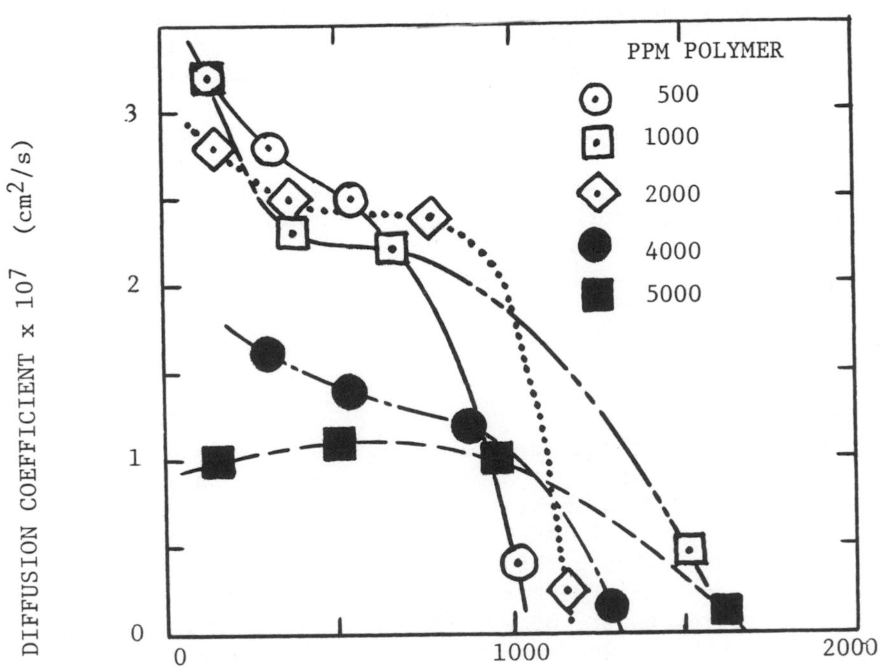

Fig. 1 Effect of Shear Rate

rate at different polymer concentrations. There seems to be a critical shear rate or yield value at 600 1/s. At lower shear rates, the diffusion coefficient decreases slowly as shear rate increases. At higher shear rates, the diffusion coefficient decreases sharply. At these shear rates, the polymer molecules could be extended rather than coiled. This would expose more charged binding sites, and some of the ferricyanide ions might be bound decreasing the concentration of electroactive ferricyanide ions. The diffusion coefficient measured would be smaller than that which would be measured at lower shear rates when the polymer molecules are more tightly coiled.

The data can be cross-plotted as shown in Figure 2 to indicate the effect of polymer concentration. The critical shear rate is again evident in this figure and apparently has a value between 600 and 700 1/s. At lower shear rates, there is a minimum at a polymer concentration of approximately 1000 ppm--a value having the same order of magnituude as the critical polymer concentration calculated to be 600 ppm for the particular polysaccharide studied. At low shear rates and at polymer concentrations below this critical value, the curves lie close together and have a uniform slope. As the shear rate increases above the critical value the curves diverge from each other and the minimum appears at lower polymer concentrations. This is taken to indicate that the critical polymer concentration depends on shear rate which in turn suggests that the critical polymer concentration depends on the extension of polymer molecules in the solution. At polymer concentrations above the critical concentration, all the curves are quite close together. The diffusion coefficient increases with polymer concentration up to a concentrtion of 2900 ppm where a sharp decrease occurs.

CONCLUSION

Electrochemical methods can be used to measure diffusion coefficient for ions in polymer solutions, and the results of such measurements can indicate something about the polymer solution structure. In a flowing polyelectrolyte solution, the shear causes the polymer molecule to extend exposing charged sites. Ion-polymer binding then causes changes in the electrochemical behavior. The electrochemical diffusion data are difficult to interpret, because the effects of solvation and ion-polymer binding are included along with ordinary diffusion. The effects of shear rate and polymer concentration may indicate something about the solvent-ion-polymer interactions in the solution.

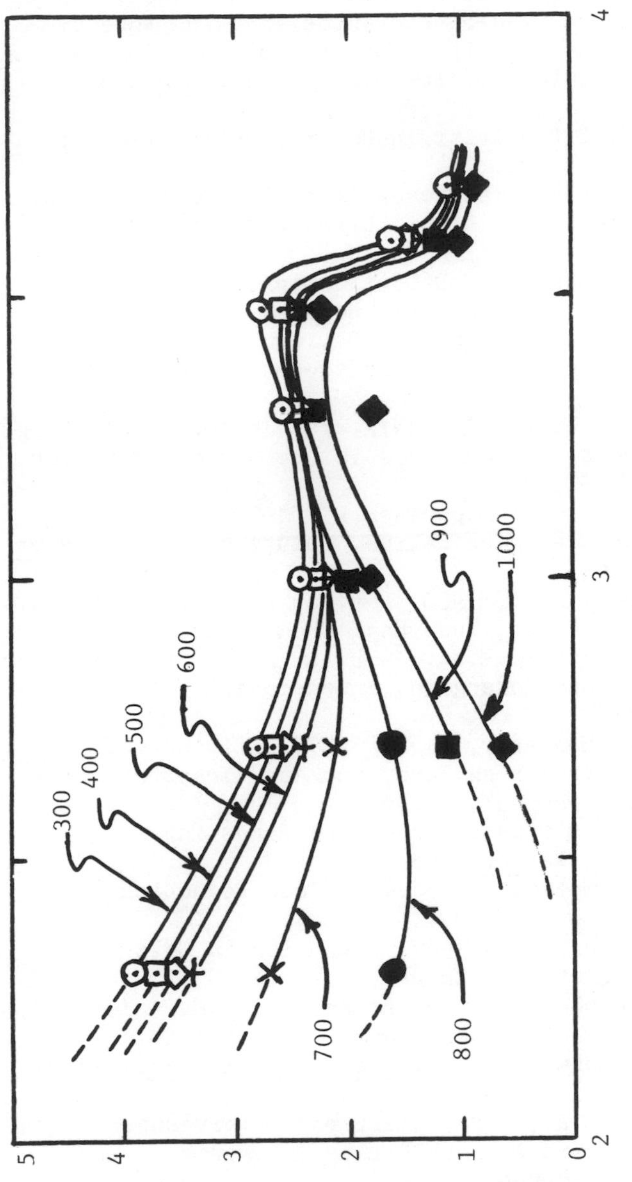

Fig. 2. Effect of polymer concentration of several values of shear rate from 300 1/s to 1000 1/s

REFERENCES

Arvia, A.J., Bazan, J.C., and Carrozza, J.S.W., 1968, The Diffusion of Ferro- and Ferricyanide Ions in Aqueous Potassium Chloride Solutions and in Solutions Containing Carboxymethylcellulose Sodium Salt, Electrochimica Acta, 13:81.

Astarita, G., 1965, Diffusivity in Non-Newtonian Liquids, Ind. Eng. Chem. Fund, 4:236.

Astarita, G., 1966, Mass Transfer from a Flat Solid Surface to a Falling Non-Newtonian Liquid Film, Ind. Eng. Chem. Fund, 5:14.

Burns, K.S., 1960, A Study of Diffusion in Viscous Liquids, M.S. thesis, Univ. of Delaware, Newark.

Clough,S.B., Read, H.E., Metzner, A.B., and Behn, V.C., 1962, Diffusion in Slurries and Non-Newtonian Fluids, AIChE Journal, 8:346.

Fortuna, G. and Hanratty, T.J., 1972, The Influence of Drag Reducing Polymers on Turbulence in the Viscous Sublayer, J. Fluid Mech., 53:575.

Grief, R., Cornet, I., and Kappesser,R., 1972, Diffusivity of Oxygen in a Non-Newtonian Saline Solution, Int. J. Heat. Mass. Tr., 15:593.

Kumar,S. and Upadhyay,S.N., 1980, Laminar Flow andd Diffusion in Non-Newtonian Fluid, Ind. Eng. Chem. Fund, 19:75.

Leopairat, P. and Merrill, E.W., 1978, Degradation of Polymer in Dilute Solutions Undergoing Extensional Flow, Polymer Preprints, 19:418.

McConaghy, G.A. and Hanratty, T.J., 1977, Influence of Drag Reducing Polymers on Turbulent Mass Transfer to a Pipe Wall, AIChE Journal, 23:493.

Metzner, A.B., 1965, Diffusive Transport Rates in Structural Media, Nature, 208:267.

Peterlin, A., 1970, Molecular Model of Drag Reduction by Polymer Solutes, Nature, 227:598.

Quinn, J.A. and Blair, L.M., 1967, Diffusion in Gels and Polymeric Solutions, Nature, 214:907.

Rickard, K.A., Hubbard, D.W., and El Khadem, S.H., 1978, Shear Rate Effects on Mass Transfer and Diffusion in Polysaccharide Solutions, AIChE Symp. Series No. 182, 74:114.

Wasan, D.T., Lynch, M.A., Chad, K.J., and Srinivasan, N., 1972, Mass Transfer into Dilute Polymeric Solutions, AIChE Journal, 18:928.

Zandi, I. and Turner, C.D., 1970, The Absorption of Oxygen by Dilute Polymeric Solutions: Molecular Diffusivity Measurements, Chem. Eng. Sci., 25:517.

NON-NEWTONIAN EFFECTS IN DILUTE SOLUTIONS DUE TO HETEROGENEITY

C.Elata, H.Ram

Technion, Haifa, Israel

(Abstract)

Dilute polymer solutions often show a remarkable and unexpected non-Newtonian flow behavior, which can only be explained by assuming hydrodynamic interaction between the macro-molecules.

Such interaction may occur in dilute solutions, if the molecules are sufficiently stretched by the flow but also in solutions which are heterogeneous and consists of long strings of concentrated solution dispersed in the solvent.

In the following it is shown that such strings may be present in "fresh" polymer solutions, which are dilute only on the average. The strings may be made visible by various means. Experimental observations are explained by assuming a fluid model based on heterogeneity due to the presence of strings of concentrated solut ions. The non-Newtonian flow characteristics will depend on the dimensions and concentration of the strings which, in their turn, depend on the manner of preparation of the solution. From theoret ical considerations the similarity between the flow of dilute and concentrated solutions can also be explained.

Experimental results of the flow of sodium oleate (soap) solutions through a porous medium are presented, showing an apparent onset, maximal and subsequent decreasing non-Newtonian effect with increasing strain rate. These results for low Reynolds number flows of sodium oleate solutions are similar to those of macro-molecular solutions, just as their drag reduction characteristics in turbulent shear flow are similar. The non-Newtonian effects of sodium oleate solutions are thought to be due to the stringlike structure formed by their miscelles, another indirect support of the proposed model.

AN ANALYSIS OF RHEOLOGICAL BEHAVIORS OF POLYMER SOLUTION WITH BEADS-SPRING MODEL

Yuan-Tse Hsu
Inst. of Chemistry
Academia Sinica
Beijing (China)

P. Schümmer
Inst. f. Verfahrenstechnik
RWTH Aachen
Turmstraße 46
D-5100 Aachen

INTRODUCTION

The beads-spring model of Rouse and Zimm has successfully predicted certain viscoelastic properties for a number of polymer solutions both dilute and concentrated. One of the basic refinements to Rouse model was added by Zimm for hydrodynamic interaction (HI) origninating from the perturbation of solvent hydrodynamics within the coil. The constitutive equation of Rouse-Zimm model was constructed, also the necessity of exactly solving the Zimm's HI problem was recognized and the exact eigenvalues were produced through number of submolecules N=300 by Lodge and Wu. /1/ Nevertheless, because of the complexity of the expression of HI, the analytical expressions of rheological properties of polymer solution in general cases have not yet appeared. In present paper we establish these expressions from the exact numerical solution of Zimm's eigenvalue problem partially by comparison with experimental results. It should be mentioned that this investigation is limited to the range of low shear rates.

ROUSE-ZIMM MODEL

The memory function of Rouse-Zimm model is written as:

$$\mu(t-t')=nkT\sum_{p=1}^{N}(1/\tau_p)\exp(-(t-t')/\tau_p) \qquad (1)$$

where n is polymer concentration (cm^{-3}), k is Boltzmann constant, T is absolute temperature, τ_p is the relaxation time of p-th mode of submolecular motion,

$$\tau_p = (fb^2/6kT)/\lambda_p \tag{2}$$

where f=friction coefficient of a submolecule, b^2=mean square end-to-end distance of submolecule, λ_p=p-th eigenvalue. For shear flow by standard procedure we may write down the constitutive equation:

$$S_{12} = \dot{\gamma} \int_{-\infty}^{t} \mu(t-t')(t-t')\,dt' = \dot{\gamma} nkT \sum_{p=1}^{N} \tau_p \tag{3}$$

(S_{ik} are components of stress tensor contributed by polymer, $\dot{\gamma}$ means the shear rate)

$$S_{11} = \dot{\gamma}^2 \int_{-\infty}^{t} \mu(t-t)(t-t)^2\,dt' = 2\dot{\gamma}^2 nkT \sum_{p=1}^{N} \tau_p^2 \tag{4}$$

$$S_{22} = S_{33} = 0$$

In order to solve these rheological expressions one must know the discrete relaxation time spectrum. This reduces to Zimm's eigenvalue problem.

ANALYTICAL EXPRESSIONS OF RHEOLOGICAL PROPERTIES

It is known, that the Zimm's HI problem can be reduced to solving the eigenvalues of a NXN symmetric matrix.

$$h^* = (12\pi)^{-\frac{1}{2}} (f/\eta_s b) \tag{5}$$

is a single parameter of the matrix.
By using IMSL routine programm of solving NXN symmetric real matrix (EIGRS) we obtain the eigenvalues with h^*=0.00-0.30, through N=400.

Intrinsic viscosity According to (3),

$$[\eta] \equiv (\eta_P/\eta_s c)_{c\to 0} = K_1 \sum_{p=1}^{N} \lambda_p^{-1}/N$$

$$\text{with } K_1 = \tilde{N} fb^2/6\eta_s M_s \tag{6}$$

where $\tilde{N}$ is Avogadro constant, M_s is molecular weight of a submolecule. Fig. 1 shows curves of $(\Sigma\, 1/\lambda_p)/N$ as a function of N. It is notable, that from h^*=0.01 to 0.30 and N > 10 the curves in Fig. 1 approach straight lines, that corresponds almost all possible cases of coiled polymer - solvent systems. By means of regression analysis one can obtain the exact linear equations with its correlation coefficient better than 0.9999. These lines cross in the same point. The relative deviation of its coordinate is 6,2%. Thus, we have:

$$\log\left(\sum_1^N \frac{1}{\lambda_p}/N\right) - \log A = a(\log N - \log B) \tag{7}$$

with dimensionless universal constants A=1.60, B=7.82.
From (6), (7), we obtain:

$$[\eta] = K M^a \quad \text{with} \quad K = \frac{AK_1}{(BM_s)^a} \tag{8}$$

For dispersed system, the concentration $c=(n/\tilde{N})\int_0^\infty M\phi(M)\,dM=\ \bar{M}_n/N$ where $\phi(M)$ is number function of MWD. By assuming the additivity of molecular contribution to η_p, we write:

$$[\eta]=(Nfb^2/6\eta_s\bar{M}_n)\int_0^\infty\sum_1^N\lambda_p^{-1}\ \phi(M)dM,\ \text{with}\ \sum_1^N\lambda_p^{-1}=AB^{-a}N^{a+1} \quad (9)$$

Thus, $[\eta] = K\ \bar{M}_\eta^a$ with $\bar{M}_\eta\equiv(\int_0^\infty M^{a+1}\phi(M)dM/\int_0^\infty M\phi(M)dM)^{1/a}$ (10)

Eq. (10) is the well known Mark-Houwink emperical equation for real polymer solutions. In the limit case of Rouse ($a=1, N\to\infty$), Eq. (10) becomes: $[\eta]=K\bar{M}_w$. ($\bar{M}_w$ is weight average molecular weight).

Normal stress coefficient Similarly, diagrammatizing $\sum_1^N(1/\lambda_p)^2$ to N, we also obtain a group of lines ($N\geq 10$), the linearity of which seems very well with correlation coefficient $R\geq 0.9999$. The equation of these lines is:

$$\log\ (\sum_1^N 1/\lambda_p^2) = \beta\log N + \log E \quad (11)$$

where E is a constant dependent on h^*.

From (11) for $\dot{\gamma}\to 0$ we obtain:

$$\Psi_1\equiv(S_{11}-S_{22})/\dot{\gamma}^2=K'M^\beta\ \text{with}\ K'=nEf^2b^4/18kTM_s^\beta \quad (12)$$

From the slope data in Fig. 1 and (11) we can write with correlation coefficient R=0.9994:

$$\beta = 2.06\ a + 1.90 \quad (13)$$

Approximately, $\beta=2(a+1)$.

It is noteworthy to write down the following equations for polydispersed system:

$$\Psi_1 = K'\int_0^\infty M^\beta\ \phi(M)\,dM \quad (14)$$

In the case of free draining: (a = 1)

$$\Psi_1=(9.3\times10^{-4}cNf^2b^4/kTM_s^4)\bar{M}_w\bar{M}_z\bar{M}_{z+1} \quad (14a)$$

In the case of nondraining: (a=0.5)

$$=(5.6\times10\ c\tilde{N}f^2b^4/kTM_s^3)\bar{M}_w\bar{M}_z \quad (14b)$$

Steady state compliance J_e^o J_e^o depends on the viscosity of solution η_o. In concentrated or bulk systems $\eta_o=\eta_p$, we have

For free draining:

$$J_e^o=(1/2\eta_o^2)\Psi_1=(0.4/cRT)\bar{M}_z\bar{M}_{z+1}/\bar{M}_w \quad (15)$$

For nondraining:

$$J_e^o=(0.3/cRT)\bar{M}_w\bar{M}_z/\bar{M}_\eta \quad (15a)$$

In very dilute solution $\eta_o\doteq\eta_s$, one can also write formulas for J_e^o.

From (14)-(15a) we know, that the normal stress and the recoverable compliance are very sensitive to MWD, especially to the tail at high molecular weights. Furthermore, the effect becomes more significant with decreasing of HI (i.e. with increasing of solvent power).

In concentrated solution of good solvent eq. (15) is the same as the formula suggested by Ferry based on Rouse's theory. The formulas with HI are remain to be examined with experiments.

MAXIMUM RELAXATION TIME play main role to viscose quantities (proportional to λ_{min}^{-1}) and the elastic (to λ_{min}^{-2}).

From the data of eigenvalues one can find the independence of contributions of τ_{max} on N (as N≥10). It means that the contribution of whole relaxation spectrum can be estimated by τ_{max} or λ_{min} with a constant factor slightly changing upon HI.

ESTIMATION OF THE PARAMETERS IN EXPRESSIONS

The Flory constant $\tilde{\emptyset}$ can be obtained from the model by assuming mean square end-to-end distance of chain molecule $r^2 = N\,b^2$ in (6).

$$\tilde{\emptyset} \equiv [\eta]\,M/(r^2)^{3/2} = (\pi^3/3)^{1/2}\tilde{N}h^* \sum_{p=1}^{N} (\lambda_p^{-1})/N^{3/2} \tag{16}$$

Generally $\tilde{\emptyset}$ depends on N (i.e.M), but in Θ condition r_Θ^2 is unperturbed dimension of the chain, $\tilde{\emptyset}_\Theta$ should be constant.

After careful calculation we know that in Θ condition or equivalently nondraining case both of which are characterized by $\tilde{\emptyset}_\Theta$ = const. and a=0.5, h_Θ^* should equal to 0.255 with $\tilde{\emptyset}_\Theta = 2.82\text{x}10^{23}$ in the range of N=10-400, we can write the submolecular parameters as:

$$f_\Theta/\eta_s = 4.91\ b_\Theta \tag{17}$$

and $b_\Theta/M_s^{\frac{1}{2}} = (r^2/M)^{\frac{1}{2}} = 1.53\text{x}10^{-8}\ K_\Theta^{1/3}\ (\text{in}(dl/g)^{1/3})$ (18)

According to (20), we have:

$$b_\Theta = 5.46\text{x}10^{-7}\ K_\Theta 1/3 \tag{18a}$$

In non-Θ solvents the parameters of submolecule change depending on the hydrodynamic interaction. From (11) we have:

$$\log(K/K_\Theta) = (0.5-a)\log(7.82M_s) + \log(h^*\alpha_s^3/h_\Theta^*) \tag{19}$$

where expansion factor $\alpha_s = b/b_\Theta$, K, K_Θ and a are measurable quantities. Writing down the measured data of log (K_Θ/K) to (a-0.5) for 10 sorts flexible chain polymer solutions /2/, we find they are fall on the same line through zero point with slop equal to 4.0. (see Fig. 2).

Consequently we obtain:

$$\text{Log}\ (K_\Theta/K) = 4(a-0.5),\ M_s = 1.28\text{x}10^3,\ h^*\alpha_s^3 = h_\Theta^* = 0.255 \tag{20}$$

This significant results strengthens the basic

assumption-all random coils of linear chains behave dynamically much the same. Besides that through regression we have:

$$a=(11.3h^{*}+1.10)^{-1/2} \quad (21)$$

Thus,

$$b=b_{\theta}\alpha_{s}=5.46\times10^{-7}(2.88\ a^{2}/1-1.1a^{2})\ K_{\theta}^{1/3} \quad (22)$$

$$f/\eta_{s}=2.68\times10^{-6}\ (1-1.1a^{2}/2.88a^{2})\ K_{\theta}^{1/3} \quad (23)$$

THE PREDICTION OF RHEOLOGICAL QUANTITIES

Because all the submolecular parameters are expressed with measured quantities, we can estimate the rheological properties with K_{θ}, a and M or $\emptyset(M)$, or vice versa. Some of the important rheological quantities are listed as follow:

Steady shear

$$[\eta]=(RT/\eta_{s}\bar{M}_{n})\sum_{p=1}^{N}\tau_{p}\doteq 100K_{\theta}(\bar{M}_{\eta}\times10^{-4})^{a} \quad (24)$$

$$=2.13\times10^{-14}c\,\eta_{s}^{2}K_{\theta}^{2}E/kT\bar{M}_{n}(1.28\times10^{3})^{2a}\int_{0}^{\infty}M^{2a}\emptyset(M)dM \quad (25)$$

Small amplitude oscillation

$$\eta'(\omega)=\eta_{s}+nkT\sum_{p=1}^{N}\frac{\tau_{p}}{1+\omega^{2}\tau_{p}^{2}}=\eta_{s}+nkT\sum_{p=1}^{N}\frac{m\lambda_{p}}{\lambda_{p}^{2}+m^{2}\omega^{2}} \quad (26)$$

$$\eta''(\omega)=nkT\sum_{p=1}^{N}\frac{\omega m^{2}}{\lambda_{p}+m^{2}\omega^{2}} \quad (27)$$

where $m=1.33\times10^{19}\eta_{s}K_{\theta}/kT$

EXPENSION FACTOR

It was suggested that HI and $\alpha_{\eta}\ (=([\eta]/[\eta]_{\theta})^{1/3})$ are not independent. From (29), we have:

$$\log\alpha_{\eta}=1/3(a-0.5)(\log M-4)=1/3((11.3h^{*}+1.10)^{-0.5}-0.5)(\log M-4) \quad (28)$$

The exactness of (33) is fully proved in Fig. 2. For dispersed polymer system M should become $\bar{M}_{\eta}$. It should be mentioned, the originally defined expansion factor is $\alpha=(r^{2}/r_{\theta}^{2})^{1/2}$, which is obtained from (29) with the assumption of uniform expansion of segment: $\alpha=\alpha_{s}$. Thus, $\alpha=(2.88a^{2}/1-1.1a^{2})^{1/3}=(0.255/h^{*})^{1/3}$ (29)
(29) and (28) show the relation between flow and equilibrium properties and present a estimation method from one to another. The application and comparison with experiments will be reported separately.

ACKNOWLEDGMENT

The authors would like to thank Deutsche Forschungsgemeinschaft for the support and Alexander von Humboldt-Stiftung for a fellowship and Mr J. Röpke for the help in computing.

REFERENCES

/1/ A.S. Lodge and Y.J. Wu, Constitutive equations for polymer solutions derived from the bead-spring model of Rouse and Zimm, Rheol. Acta, 10 (1971) 539-553

/2/ J. Brandrup and E.H. Immergut (Eds.), Polymer Handbook, Intersci. Pub., New York, (1965) iv-1-iv-72

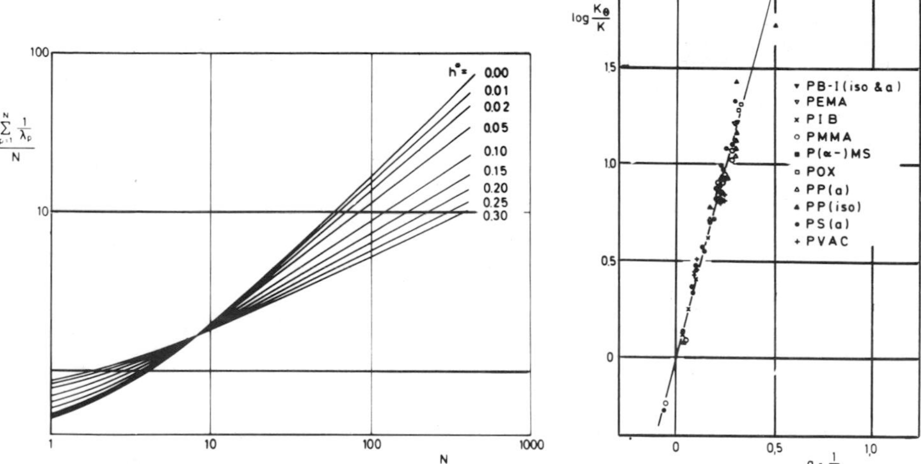

Fig.1 $\sum_{p=1}^{N} \lambda_p^{-1}/N$ as a function of N, data from computing

Fig. 2 Relationship of K and a for 60 polymer-solvent systems. (From /2/)

A RHEOLOGICAL INVESTIGATION OF A SEMI-STIFF CHAIN AROMATIC POLYAMIDEHYDRAZIDE IN DILUTE SOLUTION: ULTRA-HIGH MODULUS FIBERS AND LIQUID CRYSTALLINE POLYMER SOLUTIONS

L.L. Chapoy and N.F. la Cour

Instituttet for Kemiindustri, The Technical University of Denmark
DK-2800 Lyngby, Denmark

INTRODUCTION. The production of ultra-high modulus polymeric fibers (1), having a tensile strength comparable to that of steel wire, is based on the ability to achieve a very high degree of long range molecular order (2) during the spinning process. The realization of such high degrees of orientation can be facilitated by solution spinning polymers which are capable of forming a lyotropic, nematic mesophase, i.e. a liquid crystalline solution. The spinning and stretching processes orient the liquid crystal such that the chains are uniaxially oriented with respect to the fiber axis. The solvent is then rapidly removed so as to freeze in the high degree of orientation.

The polyterephthalamide of p-amino-benzhydrazide, X-500 is a semi-stiff chain polymer. While it does not form a mesophase in the quiescent state at concentrations as high as 20 g/dl i.e. the solution is not birefringent in the unperturbed state, it none the less gives rise to high modulus fibers.

As a possible explanation of this anomaly, Ciferri and co-workers (3) have proposed a theory which incorporates the mechanical energy contribution from flow processes into the free energy criteria for the isotropic-liquid crystal concentration driven phase transition. Briefly the theory indicates that a mesomorphic phase will be formed under conditions of flow at a concentration which is less than the critical concentration for such a transition in the quiescent state. This effect becomes more perceptable as the free energy contribution due to flow becomes more pronounced. It is, thus, implied that X-500 is on the verge of forming a mesophase in the quiescent state, but it is only through the action of the flow field that such a transition can occur.

The occurrence of an isotropic-anisotropic phase transition will be accompanied by changes in the physical properties of the sample. For polymers forming a lyotropic mesophase in the quiescent state, the zero shear viscosity has been shown to pass through a maximum at the critical concentration characterizing the phase transition (4, 5). The critical concentration has been found to decrease with increasing shear rate. Similar effects have been noted for the concentration-shear rate dependence of the viscosity for the lecithin lyotropic mesophase transition (6). For X-500 such a maximum in the viscosity is only observed at moderate rates of shear (the so-called B. effect (4)), thus supporting the concept that the transition is flow induced.

Furthermore, changes in slopes are found in double logarithmic plots of viscosity-concentration (4) and birefringence-concentration (7) at constant shear rate at the same concentration at which the viscosity goes through the maximum. Such slope changes are to be expected (8), however, from fundamental considerations of the onset of entanglement phenomena in polymer solutions of finite concentration. For polyacrylonitrile, PAN, a flexible chain-polymer, which exhibits a diminished B. effect, a change in slope was observed (4) which was more pronounced than that for X-500, but which occurred at a concentration considerably below the onset of its B. effect. This slope change was obtained at a shear rate which was low enough so as not to produce a B. effect and was therefore attributed to entanglement phenomena.

It is thus difficult to unequivocally equate the reported slope change in the case of X-500 to a flow induced phase transition. With this in mind, the dilute solution (.06 - 3.0 g/dl) rheological properties of X-500 have been investigated as a function of concentration and shear rate using the Rheometrics Mechanical Spectrometer. The semi-stiff nature of the X-500 chain gives rise to the expectation that chain overlap and entanglement phenomena will occur at concentrations considerably less than those customarily found for flexible chain polymers and that a second slope change at a lower concentration will be found. The objective of this work is thus to distinguish unambiguously between changes in viscosity to be expected from entanglements and those resulting from a possible phase transition.

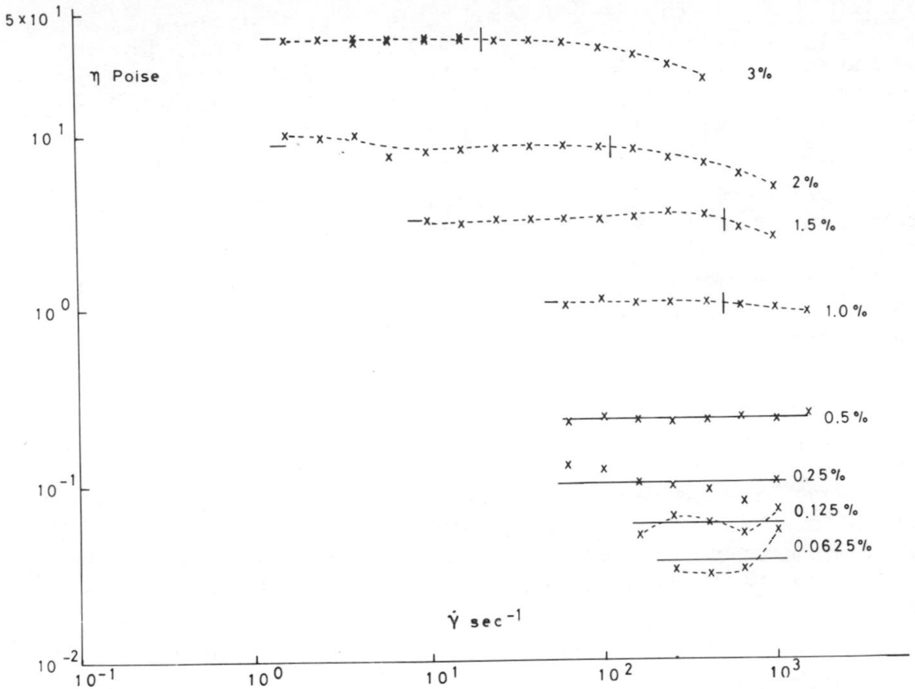

Fig. 1: A double logarithmic plot of viscosity, η, versus shear rate, $\dot{\gamma}$, for the concentrations as indicated (% = g/dl).

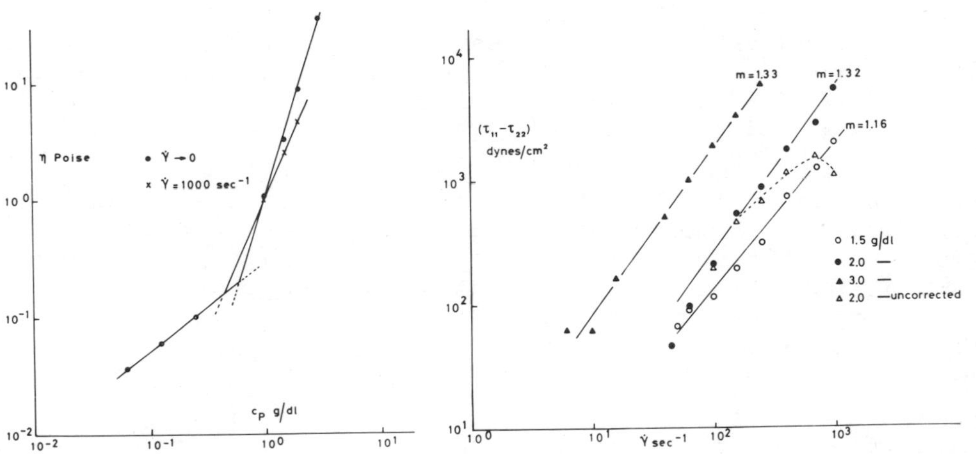

Fig. 2: A double logarithmic plot of η versus the polymer concentration, C_p.

Fig. 3: A double logarithmic plot of the first normal stress difference $(\tau_{11} - \tau_{22})$ versus shear rate for the concentrations as indicated.

EXPERIMENTAL. The X-500 employed was courteously supplied to us by A. Ciferri, Duke University, Durham, North Carolina, USA. It had an intrinsic viscosity, $[\eta]$ = 5.94 dl/g in dimethylsulfoxide at 25°C corresponding to a viscosity average molecular weight of 50500, close to that used previously (4, 7, 9).

Steady state shear viscosity and normal stress measurements as a function of shear rate and concentration were obtained using the Rheometrics Mechanical Spectrometer in the cone and plate geometry. The cone and plate aggregate had a radius of 36 mm and a cone angle of 2.3°. Measurements were obtained in dimethylsufoxide at 22°C.

RESULTS AND DISCUSSION. The shear viscosity for X-500 as a function of shear rate is given in Fig. 1. For the low concentrations, the solutions appeared to be Newtonian over the whole shear rate range investigated. At higher concentrations, the viscosity is Newtonian at shear rates less than that indicated by the vertical line. The curves do not cross indicating that these data are too dilute to exhibit B. effect behaviour as previously observed (4).

The shear viscosity for X-500 as a function of polymer concentration is given in Fig. 2 for the Newtonian regime, i.e. $\dot{\gamma} \to 0$, and for $\dot{\gamma}$ = 1000 sec^{-1}, the highest shear rate measured. The abrupt change in slope occurs at some critical concentration C_p^+. For the Newtonian data the concentration dependence of the viscosity changes from $C_p^{.76}$ for $C_p < C_p^+$ to $C_p^{3.2}$ for $C_p > C_p^+$. This is in excellent agreement with entanglement theory where a change from $C_p^{1.0}$ to $C_p^{3.4}$ at some critical concentration is predicted (8). Note the following experimental facts which also indicate that the value of C_p^+ as observed here is related to the entanglement phenomenon:

1. The semi-stiff nature of the X-500 chains as indicated by the large value of the intrinsic viscosity implies that this molecule has a large hydrodynamic volume and that the chain entanglement phenomenon will be operative at a very low concentration. C_p^+ is found here to be ~ 0.6% while that for PAN, a flexible chain polymer is ~ 8% (4).

2. C_p^+ ~ 0.6% as measured here for X-500 is significantly less than the critical concentration previously reported (4), C_p^* ~ 3%, which was attributed to the phase transition.

3. C_p^+ for X-500 occurs significantly below the onset of the B. effect analogously to that for PAN.

4. Poly-p-phenyleneterephthalamide, PPT, a polymer known to form a lyotropic mesophase in concentrated solution, shows (10) an unusually high concentration dependence of $C_p^{6.8}$ immediately below the critical concentration at which the lyotropic mesophase is formed. The high slope in the case of PPT could in the context of the discussion here be compared with the increase in the slope in excess of 3.4 for X-500 at the concentration where the B. effect becomes operative (4). It has been well documented that pretransitional order exists in mesophase formation, but that it is not easily detected by the usual means (11, 12).

It appears, thus, that $C_p^+ \sim 0.6\%$ marks the onset of entanglements as predicted by theory, while $C^* \sim 3\%$ marks the concentration at which a flow induced phase transition to the lyotropic mesophase can occur.

The agreement between the flow birefringence data and the viscosity data (4, 7) is not unexpected in light of the theory of birefringence (14) which predicts appromixate proportionality between the measured birefringence and the solution viscosity. If the method were sensitive enough to measure such small values of the birefringence that would result from the concentrations and shear rates employed here we would expect a corresponding effect at $C_p^+ \sim 0.6\%$.

The results for the normal stress measurements are shown in Fig. 3. At low concentrations, the measured normal stresses are negative due to the domination of inertial effects (14). The data shown were corrected for inertial effects where necessary. The effect of the correction is shown for the $C_p = 2$ g/dl data. One should thus bear this correction in mind when examining normal stress data for changes due to orientational effects as have been predicted (15). The data showed power law behaviour, i.e. $(\tau_{11} - \tau_{22}) \propto \dot{\gamma}^m$. The values of m are shown in Fig. 3 and are found to be close to 1 rather than 2 which is to be expected in the Newtonian limit. It has been noted that normal stresses for anisotropic fluids generally depend on $\dot{\gamma}$ rather than $\dot{\gamma}^2$ (15). This fact could also be interpreted in terms of the previously noted B. effect: that shearing X-500 in the concentration range of 2-3 g/dl produces a lyotropic mesophase.

It is well documented (16) that order exists in the isotropic phase for example for alkanes just over their melting point and for thermotropic liquid crystals just over their clearing temperature. This order can be detected by various methods such as depolarized light scattering or magnetic birefringence, i.e. the measurement of the Cotton-Mouton Coefficient. It is proposed that these methods be applied to investigate X-500 in the quiescent state as a function of concentration to seek evidence of pre-transitional order in these solutions.

CONCLUSION. The rheological investigations performed here are consistent with the notion that solutions of X-500 are capable of forming flow induced mesophases under shear flow. An absolute proof of this, however, will be difficult to obtain from rheological measurements alone since the action of the flow will always produce an anisotropic fluid, and changing rheological properties which need not be associated with the liquid crystalline state. Some alternative experimental procedures to investigate this problem are proposed.

REFERENCES

(1) A. Ciferri, Int.J.Polym Mater., 6:137 (1978).
(2) L.L. Chapoy, D. Spaseska, K. Rasmussen and D.B. DuPré, Macromolecules, 12:680 (1979).
(3) G. Marrucci and A. Ciferri, J.Polym.Sci., Polymer Lett. Ed., 15:643 (1977).
(4) B. Valenti and A. Ciferri, J.Polym.Sci., Polymer Lett.Ed., 16:657 (1978).
(5) G. Kiss and R.S. Porter, J. of Polym.Sci., Polymer Symp. 65, 193 (1978).
(6) R.W. Duke and L.L. Chapoy, Rheologica Acta, 15:548 (1976).
(7) D.G. Baird, A. Ciferri, W.R. Krigbaum and F. Solaris, J.Polym.Sci., Polymer Phys.Ed., 17:1649 (1979).
(8) T.G. Fox and V.R. Allen, J.Chem.Phys. 41:344 (1964).
(9) J.J. Birke, J. Macromolecular Sci. - Chemistry, A7:187 (1973).
(10) D.G. Baird and R.L. Ballman, J. of Rheology, 23:505 (1979).
(11) P.G. de Gennes, The Physics of Liquid Crystals, Oxford University Press, London, 1974.
(12) E.B. Priestly, P.J. Wojtowicz and P. Sheng, Introduction to Liquid Crystals, Plenum Press, New York, 1974.
(13) V.N. Tsvetkov, Chapter XIV in Polymer Reviews, Volume 6: Newer Methods of Polymer Characterization, Interscience Publishers, New York, 1964.
(14) J.D. Huppler, E. Ashare, and L.A. Holmes, Trans. Soc. of Rheol., 11:159 (1967).
(15) D.G. Baird, Chapter 7 in Liquid Crystalline Order in Polymers, A. Blumstein, Ed., Academic Press 1978.
(16) E.W. Fischer, G.R. Strobl, M. Dettenmaier, M. Stamm, and N. Steidl, Discussions of the Faraday Society, NO. 68, 1979, Contribution 68/1.

SHEAR DEGRADATION OF POLY(VINYL ACETATE) IN TOLUENE SOLUTIONS BY HIGH-SPEED STIRRING

S.H. Agarwal, R.S. Porter

Materials Research Laboratory, Polymer Science and Engineering Department, University of Massachsetts USA

(Abstract)

A poly(vinyl acetate) (PVAc) of Mw 750,000 and Mw/Mn 5.10 in toluene solution was sheared in a Virtis-60 Homogenizer. The polymer concentration was 3.0 - 12.0 gm/100 ml and test temperature was 10 $\pm$ 0.5°C. The extent of degradation was measured by gel permeation chromatography (GPC). It was concluded that on shearing (i) the molecular weight decreases rapidly at the beginning of shearing, and thereafter decreases ever more slowly towards a limiting value, (ii) the molecular weight distribution is narrowed, (iii) no degradation occurs up to 5000 rpm and thereafter increases with stirring speed, (iv) degradation is more at lower concentrations but concentration is not a sensitive variable, and (v) the chain scission occurs randomly. The Mark-Houwink relationship for PVAc in THF at 25°C was derived as $[\eta]=2.47\times10^{-4}\times Mv^{0.644}$.

APPARENT VISCOSITY CHARACTERISTICS OF GUAR GUM SOLS

Shiba C. Naik

Department of Chemical Engineering
Regional Engineering College
Rourkela 769008, India

INTRODUCTION

Guar plants (Cyamopsis tetragonoloba) are found on the Indian subcontinent. Guar gum, a nonionic natural hydrocolloid, is obtained from the refined endosperms of guar seeds. Technically a galactomannan, it is more commonly referred to as a mannogalactan. When dispersed in water, it hydrates to provide sol (colloidal dispersion). The sol's viscosity remains unaffected in the 1.0 - 10.5 pH range. It is used as a potable water clarifying agent, a gangue depressant in the beneficiation of several ores, and a filter aid for filtering ore concentrates. For all such purposes, sol is generally used in the concentration range from 2.5 to 5.0 kg/m^3. Unfortunately, the literature does not provide any information on apparent viscosity properties of guar gum sols. This paper reports some useful information on apparent viscosity properties.

EXPERIMENTAL

Hercules-made guar gum THI having molecular weight 220,000 was used in this work. The concentrations of the sols were 2, 3, 5, and 6 kg/m^3 of distilled water. Sols in such a concentration range are used by industries. The calculated amount of powder was added slowly, over a period of 15 minutes, into the vortext of an agitator agitating distilled water at 300 rpm. Then the sols were heated at 80°C for 15 minutes. Smooth, uniform sols were obtained and stored in screw-top plastic bottles. Shear stress τ vs shear rate $\dot{\gamma}$ measurements were made 12 hours after preparation of sols.

This ensured complete hydration. The measurements with one sample were completed within a week. Weekly check runs were made. There was no disagreement between the initial and check runs over a period of six weeks.

A standard Ferranti-Shirley cone-and-plate viscometer was used for the measurements of shear stress over a shear rate from 412 to 20,600 s^{-1}, and a temperature range from 25 to 70°C. A vapor shield was used to prevent evaporation of water from the sols. Circulation of temperature-regulated water through the viscometer plate enabled control of temperature with an accuracy of ±0.1°C. The results were quite reproducible.

RESULTS AND DISCUSSION

Guar gum sols are highly pseudoplastic, exhibit no measurable thixotrophy, and have no yield stress.

Out of several models[1] proposed to correlate τ with $\dot{\gamma}$, the power law model of Ostwald-de Waele,

$$\tau = k\dot{\gamma}^n \tag{1}$$

is the simplest one. In Equation (1), k and n are consistency coefficient and power law exponent respectively. Equation (1) fits data quite well and enables evaluation of k and n. Figure 1 illustrates some plots of τ vs $\dot{\gamma}$ in logarithmic coordinates.

The effect of temperature T on k can be described by an Arrhenius type equation

$$k = k_0 \exp (E/RT) \tag{2}$$

where E, k_0, and R are activation energy, frequency terms, and gas constant respectively. Figure 2 is based on Equation (2). Naik et al.[2,3] found such a relationship rigidly valid for hydroxyethyl and carboxymethyl cellulose solutions. Using Equation (2), E and k_0 are estimated.

E seems to depend on concentration C of the sols. A model of the type

$$E = \alpha_1 C^{\beta_1} \tag{3}$$

correlates them satisfactorily (Figure 3). Also, k_0 and C can be correlated by an empirical equation

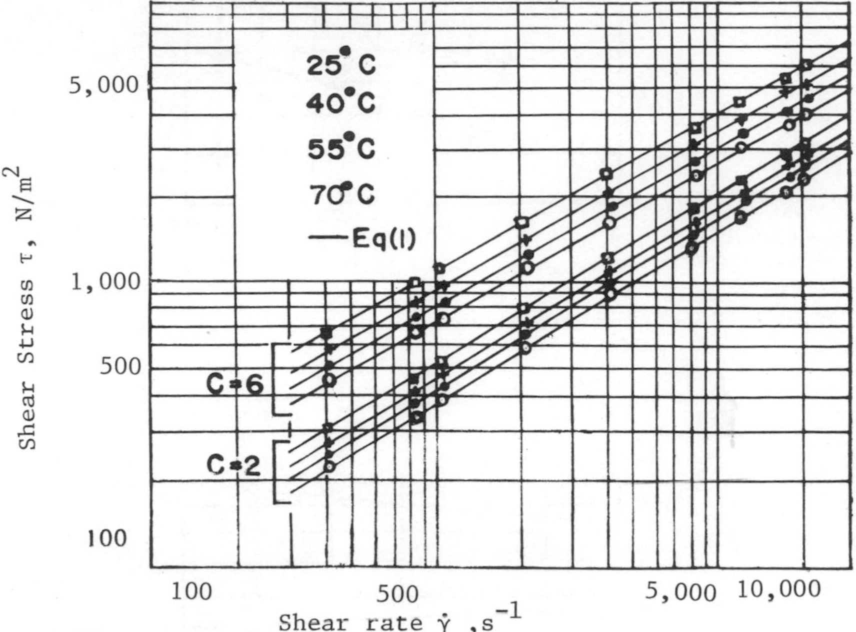

Fig. 1. Shear stress vs shear rate curves for two samples.

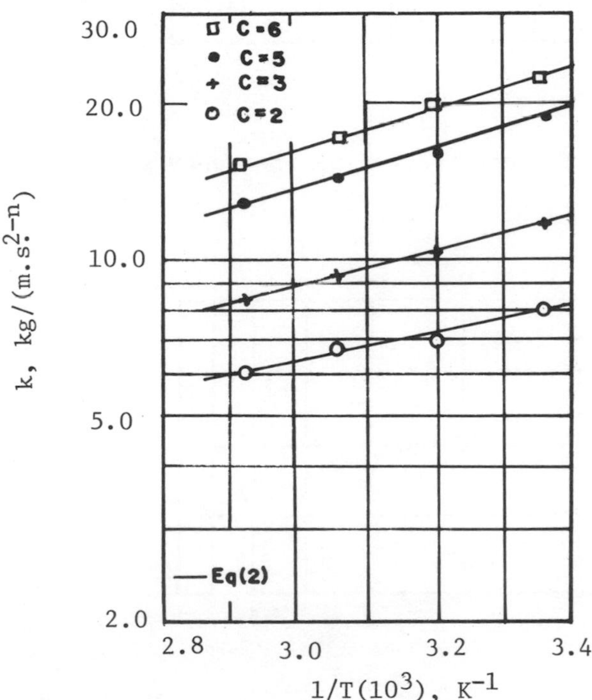

Fig. 2. Effect of temperature on consistency coefficient.

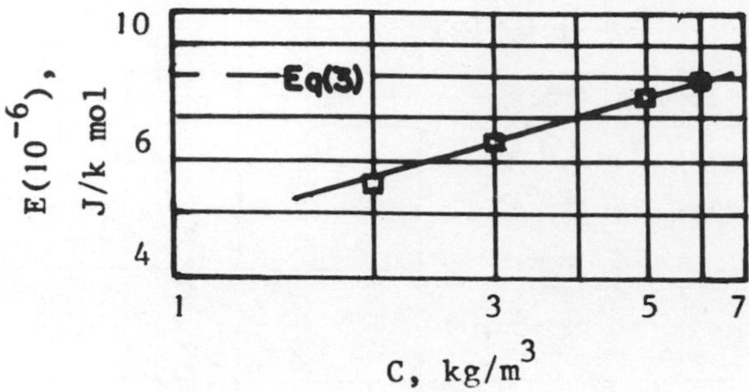

Fig. 3. Effect of concentration on activation energy

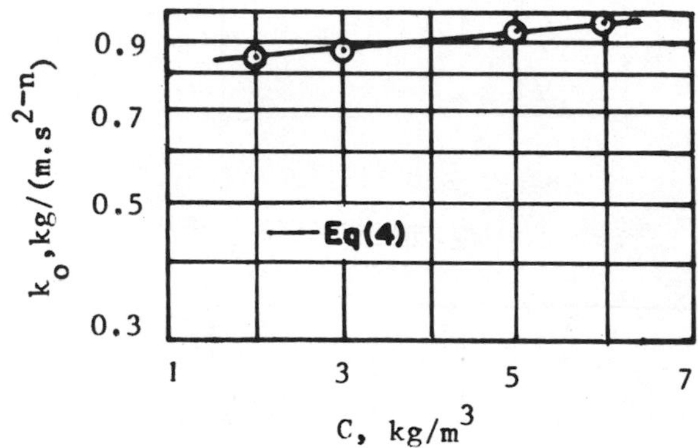

Fig. 4. Effect of concentration on frequency term

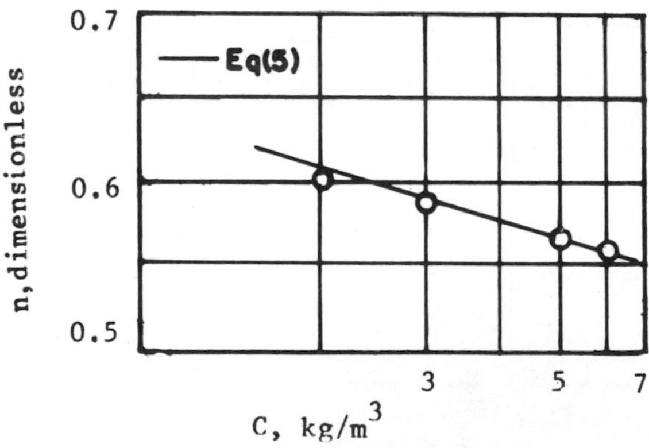

Fig. 5. Effect of concentration on power law exponent

$$k_0 = \alpha_2 \exp(\beta_2 C) \tag{4}$$

Figure 4 is based on Equation (4)

The power law exponent n depends on C and is very little influenced by T. Its effect on C can be described by a simple model:

$$\exp(n) = \alpha_3 C^{\beta_3} \tag{5}$$

Figure 5 illustrates the results. Average values of n are used and the maximum deviation of actual values is 3.0%.

All the expressions are valid over the experimental conditions studied in this work. When these expressions are used, the maximum deviation between experimental results and calculated values becomes 2.0%. α and β are characteristic constants of the sols. For any given sol concentration, E, k_0, and n can be evaluated from Equations (3), (4), and (5) respectively when the values of α and β are known (Table 1). Then, values of E and k_0 can be substituted into Equation (2) to obtain k at any desired T. Finally, values of k and n can be used in Equation (1) to obtain τ vs $\dot{\gamma}$ data. Thus, apparent viscosity $\mu_a(\tau/\dot{\gamma})$ can be estimated for any given C, T, and $\dot{\gamma}$ values.

Table 1. Value of Characteristic Constants

Equation	Figure	Value of constants
(3)	3	$\alpha_1 = 4.5557\ (10^6)$, $\beta_1 = 0.3054$
(4)	4	$\alpha_2 = 0.7895$, $\beta_2 = 0.0332$
(5)	5	$\alpha_3 = 1.8688$, $\beta_3 = -0.0366$

REFERENCES

1. W.L. Wilkinson, Chapter 1 in "Non-Newtonian Fluids," Pergamon Press, London (1960).
2. S.C. Naik, J.F.T. Pitman, and J.F. Richardson, The rheology of hydroxyethyl cellulose solutions, Trans. Soc. Rheol, 20:639 (1976).
3. S.C. Naik, J.C. Lee, and J.F. Richardson, The rheology and aging characteristics of sodium carboxymethylcellulose solutions, Can. J. Chem. Eng. 55:90 (1977).

POLYMERS IN ENHANCED OIL RECOVERY - SOLUTION STRUCTURE AND VISCOSITY OF WATER-SOLUBLE VINYLPOLYMERS

Joachim Klein

Institute of Chemical Technology

Technical University of Braunschweig, F.R.G.

INTRODUCTION

The development of chemical systems for enhanced oil recovery is essentially an economical problem, where the total costs of chemicals, their formulation and field application has to be balanced against the price of oil to be recovered. As in many other cases the polymers applied in the very early attempts are performing quite satisfactorily and are difficult to replace by later developments.[1,2] Many industrial and academic research activities however are an indication that there is still a need to develop improved polymers where the following parameters have to be considered: solubility in water and salt brine, viscosity,elasticity, adsorptivity, filterability, shear stability, temperature stability, compatibility to other chemicals as surfactants. The search for new polymers has to follow rational lines of chemical and physical nature to develop an understanding of the relation between composition, solution structure and efficiency of a given polymer. In this contribution viscosity as one of the basic physical parameters has been selected to allow for a first evaluation of performance in EOR application.

The above mentioned "established" polymers are a) saponified polyacrylamides and b) polysaccharids from bacterial synthesis. Both types of polymers have been improved considerably throughout the last years on the basis of industrial development work.In this contribution the group of vinyl-polymers will be discussed specifically, which include the modification of polyacrylamides, the synthesis of substituted maleicacid copolymers and the synthesis of sugar substituted poly-methacrylesters. The selection of this polymers was based on the concept, to obtain polymers with extended chain configuration and possibly enhanced intermolecular interaction.

The solution structure will be discussed according to two factors, namely (1) conformation of individual chain and (2) interaction from one chain to another. While (1) applies to the dilute solution case both (1) and (2) have to be considered in the interpretation of more concentrated solutions.

CHAIN CONFORMATION

Since absolute informations about the configuration of a polymer chain in solution are difficult to obtain, polyacrylamid (PAAm) will be used as reference substance to evaluate the structure of the other polymers. This procedure finds its justification with regard to two reasons: a) The overall structure of PAAm in dilute solution can be described as flexible statistical coil[3] (a physical argument). b) PAAm is one of the standard materials in EOR (an application oriented argument).

Chain stiffening from comonomer interaction

If the viscosity of copoly(acrylamide/sodiumacrylate)-obtained from polymeranalogeous reaction- is compared to PAAm of the same degree of polymerization, always a higher value is observed also in salt solution[4]. This observation holds also for extrapolation to zero polymer concentration and has therefore to be considered as an intrinsic property of the copolymer chain. Based on an in-

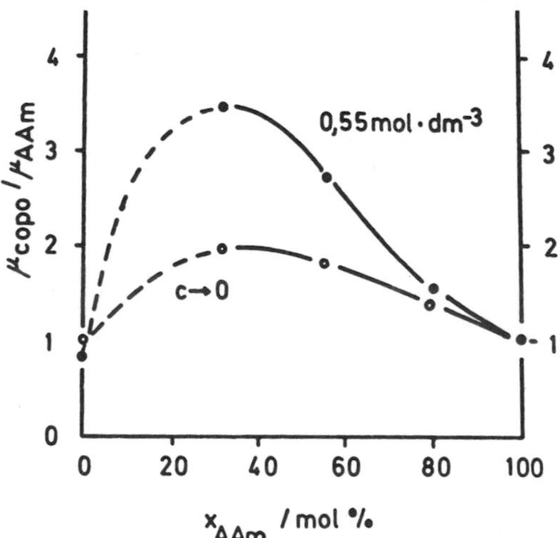

Fig.1 Viscosity of copoly(acrylamide/sodium acrylate) for different compositions of a given chainlength as compared to PAAm in dilute (c→0, i.e. [η]) and concentrated (c=0.55 mol·dm^{-3}) solution. (T=25°C, $\bar{M}_w$=470.000, solvent: 0.5 m NaBr in H_2O)

tensive characterization study[5,6] - including the composition dependence of the Mark-Houwink exponent, the sedimentation coefficient and calorimetric data - the concept of coil extension due to chain stiffening was established: The chemical argument is the cyclization of adjacent amide and carboxylic side groups due to hydrogen bonding.

The fact that other acrylamide copolymers - as e.g.(AAm/Vinylsulfonic acid) - or physical mixtures of PAAm and sodium polyacrylate did not reveal this increased viscosity levels[2] can be seen as additional evidence for the specific interpretation given above.

Chain extension from short chain branching

Another possibility to achieve some degree of chain extension might be the incorporation of short side chains (combelike structure) due to a steric hindrance and decrease in segmential rotation.

Such polymers were prepared starting from an alternating ionic (methylvinylether/maleic acid) copolymer ($Na^+:H^-=1:1$). Hydrophobis (PEG) and hydrophobic (n-alcanols) esters of different chainlength showed an increased viscosity level in aqueous as well as in saline solution. despite of the limited degree of ester-substitution. The chain length of all polymers in table 1 is $\bar{P}_n$ 2000 and a corresponding PAAm sample would have a viscosity of about 1.2 m·Pas; Otherwise spoken the samples No.3 and No.5 show the same viscosity effect in salt solution as a PAAm of a fivefold chain length ($\bar{M}_w=1.7\cdot10^6$). As can be seen from sample No.1 a complete interpretation of the viscosity of those polymers in comparison to PAAm would require the consideration of ionic contributions as well[2,7].

Table 1: Dilute solution (C_p = 1 g/dm^3) viscosity of various side chain substituted (methylvinylether/maleic acid) copolymers (X_E = mol % ester formation; u_{H_2O} = viscosity in water, u_S = viscosity in 10 g/dm^3 NaCl solution; T = 20 °C)

No	Polymer	X_E mol %	u_{H_2O}	u_S
1	GAN-H	0	16.1	1.56
2	GAN-PEG 300	3	22.7	1.67
3	GAN-PEG 600	9	47.9	1.91
4	GAN-Octanol	17	19.8	1.54
5	GAN-Dodecanol	16	22.5	1.95
6	GAN-Hexadecanol	9	7.6	1.30

Chain stiffening from bulky side groups

Another possible mechanism for chain expansion of vinyl-polymers may be the incorporation of bulky side groups, again causing hindered rotation of backbone chain segments. Following to this concept methacrylicacidester-monomers with mono- or oligosaccharide residues were synthesized and used for free radical polymerization to form watersoluble polymers[9]. Poly-(3-O-methacryloyl-D-glucopyranose) and poly-(6-O-methycryloyl-D-galactopyranose) are typical examples of such monosaccharide polymers. Oligosaccharide polymers have been prepared using an oligomer mixture with an average of 4 glucose units. So far viscosity levels of 2 m Pas (at c_p=1%) have been obtained and more characterization work has been done to evaluate the performance of this type of polymers.

In concluding this section it should be mentioned that many other polymers have been synthesized during this study: Various copolymers of AAm and nonionic comonomers, methacryl-ester-monomers plus ionic comonomers, homo- and copolymers of cationic nature. None of those polymers showed exceptional viscosity levels[2].Studies in another laboratory centimed our findings of the chain stiffening effects in copoly(acrylamide/sodiumacrylate) Furthermore comblike polymers were prepared by grafting of AAm onto a polysaccharide (Dextran)[8].

CHAIN INTERACTION

With increasing polymer concentration interaction from one polymer chain to another becomes predominant. With respect to the "entanglement" mechanism steric and energetic effects have to be considered. Based on studies on the viscoelastic behavior of PAAm[10-13] the significance of the energetic interaction - possibly due to hydrogen bonding[6] - was emphasized.

As can be seen from Fig.1 the viscosity maximum of the AAm/sodiumacrylate-copolymers is even mor pronounced in the concentrated solution range. This effect will partly be due to the larger coil size of each chain, but a 3 to 4 fold increase of viscosity is an indication, that the above mentioned hydrogen bonds are not only formed along one chain but also to some degree between different chains.

The importance of the intermolecular interaction parameters becomes clear in the comparison of very different polymers, as shown in Fig.2. The very fast and stepp increase of the curve of Rhodopol is typical for the group of polysaccharides from bacterial synthesis. With regard to the viscosity levels reached at very small concentrations these polysaccharides are the most efficient polymers known to-day. It is evident, that another polysaccharide polymer HEC has a comparable steep viscosity function, while PAAm and PEO (in salt solution) show a comparably slow viscosity increase. The ability to form hydrogen bonded three dimensional gel structures is a well known property of biological polysaccharides. We believe that this high strength of intermolecular energetic interaction is the key to the interpretation of the very high viscosity levels of the biopolysaccharides at concentrations well below 1000 ppm. Incorporation of interacting side groups seems to be essential to obtain comparable effects with synthetic vinyltype polymers.

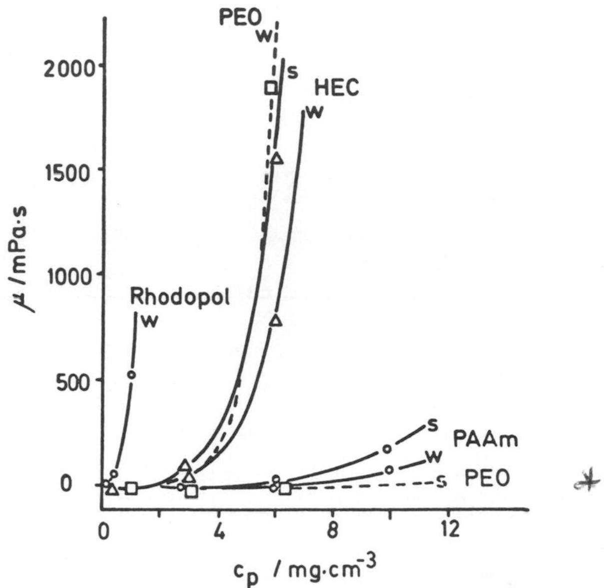

Fig.2: Concentration dependence of different polymer types (● Rhodopol: Polysaccharid; ■ PEO: Polyethylenoxid; ▲HEC: Hydroxyethylcellulosis; ● PAAm) in water (w) and salt solution (s)

CONCLUSIVE REMARKS

It is the intention of this contribution to emphasize the importance of chain extension and chain interaction as fundamental parameters in the development of efficient polymers for EOR application. While ionic polymers may well be used in low salinity deposits, there is certainly a need for nonionic polymers for the important group of high salinity deposits and in this latter case "sugar-polymers" of different structure might be advantageous.

The future of EOR techniques in general and of polymer related processes especially is not all clear[14]. But development of polymers and improved understanding of their solution behavior will be of value for other areas of application, e.g. drag reduction.

REFERENCES

1) United States Dep. Energy, Quaterly Progress Rev. "Enhanced oil and Gas Recovery", Bartesville Energy Technology Center, Bartesville, Okl., U.S.A.

2) J. Klein, Research Report to BMFT, Grant-No. ET 3159 A published as part of DGMK Research Rep. No. 165,Hamburg 1978

3) J. Klein and K.D. Conrad, Makromol. Chem. 181,(1980)227

4) J. Klein and R. Heitzmann, Makromol. Chem. 179,(1978) 1895

5) J. Klein and K.D. Conrad, Makromol. Chem. 179,(1978) 1635

6) J. Klein and W. Scholz, Makromol. Chem. 180(1979) 1477

7) V. Martin and J. Klein, Erdöl-Erdg. Z. 95(1979) 164

8) H.H. Naidlinger and L.McCormick, 26th JUPAC Symp. on Macromol. Mainz 1979, Preprint, p. 1584

9) J. Klein and F. Roesler, unpublished results

10) W.-M. Kulicke and J. Klein, Erdöl u. Kohle, Erdg. u. Petroch. 31(1978) 373

11) J. Klein and W.-M. Kulicke, Proceed. VII.Int. Congr. Rheol., Gothenburg 1976, p. 398

12) J. Klein and W.-M. Kulicke, Rheol. Acta 15(1976) 558

13) J. Klein and W.-M. Kulicke, Rheol. Acta 15(1976) 568

14) G.E. Weismantel, Chem. Engineering, November 1979, p. 102

RHEOLOGICAL PROPERTIES OF POLYMER SOLUTIONS USED FOR TERTIARY OIL RECOVERY

Werner Lange and Günther Rehage

Institute of Physical Chemistry
Technical University of Clausthal
FRG

INTRODUCTION

Our contribution deals with the rheological properties of polymer solutions used for polymer flooding, one method of tertiary oil recovery. The situation in most oil fields in the world is the following: only 20-40% of the oil in place has been yet drawn out. By means of tertiary oil recovery an additional oil winning of 8-16% of oil in place is expected.

The special problems occuring with the oil reservoirs in the Federal Republic of Germany (FRG) are the high salinity of the brine (up to 200 g/l salt content) and the high viscosity of the crude oil compared with oil deposits in the USA.

For an efficient polymer flooding it is necessary that the polymer solutions used for this process have to have a viscosity which is in the range of the viscosity of the crude oil under reservoir conditions. The polymer flooding will only be economic, if the polymer concentration necessary for this purpose can be kept very low; the value of 1000 ppm gives a suitable range.

EXPERIMENTAL

Measurements of shear viscosities were carried out by means of Low Shear rheometers with concentric cylinder geometry [+)]. These instruments allow measurements in the range of shear rates from 10^{-4} to $2.5 \cdot 10^{2}$ s^{-1} at viscosities of more than 1mPa·s.

+) LS2 and LS100; Contraves Company, Zürich

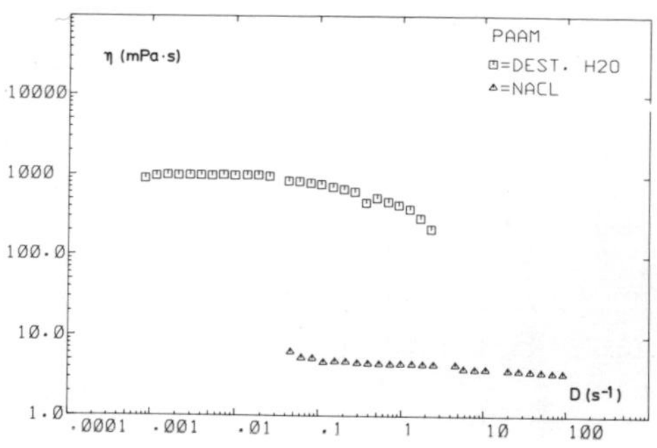

Fig.1: Viscosity η vs shear rate D for a solution of anionic PAAm in H_2O. T=25°C; □ =dist. H_2O; △ = 1% NaCl

Fig.2: Structure of the used homopolysaccharide

Measurements of the viscoelastic properties of the polymer solutions were performed by means of a rotational viscometer with an additional device for oscillatory measurements+). The frequency range is 10^{-3} to 1.6 cps. All three viscometers were automated.

The polymers used were commercial samples kindly supplied by the chemical industry. In our research many different polymers were tested. Only the rheological behaviour of the following polymers will here be discussed: 1) neutral and anionic polyacrylamides (PAAm), 2) synthetic homo- and heteropolysaccharides, 3) hydroxyethylcelluloses (HEC).

RESULTS AND DISCUSSION

In first experiments the influence of salt addition on the viscosity of polymer solutions in distilled water was tested. One demand of polymer flooding is , that the viscosity of the polymer solution should be higher than 20-30 mPa·s (shear rate $D=1-10\ s^{-1}$) at a polymer concentration of 1000 ppm. As mentioned above, oil reservoirs in the FRG have a very high salinity; therefore the viscosity should not decrease if relative large amounts of salts are added.

For reference purposes the polymer concentration is 1000 ppm for all experiments discussed in the following chapter.

For solutions of neutral polyacrylamides with high molecular weights in distilled water, a viscosity of only 4 mPa·s was found. This value does not change when salts are added. As there are no charged groups in the polymer, this is a normal behaviour.

+) LS30, Contraves Company, Zürich

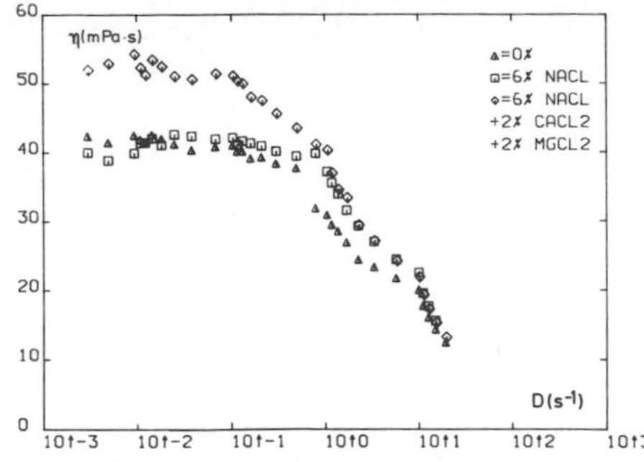

Fig.3: Viscosity η vs shear rate D for a solution of a homopolysaccharide in H_2O.T=25°C; polymer concentration: 1000 ppm; △ =dist. H_2O, □ =6% NaCl, ◇ =6% NaCl+2%$CaCl_2$+2%$MgCl_2$

Fig.4: Structure of the used heteropolysaccharide

Anionic polyacrylamides show quite different results (Fig.1). A solution of an anionic PAAm in distilled H_2O has a viscosity of 1000 mPa·s at 10^{-2} s^{-1}. The reason for the Non-Newtonian behaviour are the elastic properties of the solutions. In the region from 10^{-3} to 10^{-2} s^{-1} we assume a linear viscoelastic behaviour. The extreme high viscosity of several hundred mPa·s for all measured shear rates is caused by the pronounced polyelectrolyte character of this polymer. Negative charged groups at the polymer chain show an electrostatic repulsion and by this effect the volume of the polymer molecule is increased. This means an increase of the hydrodynamic volume, too.

On addition of simple salts (1% NaCl in Fig.1), the viscosity of the solution decreases from 1000 mPa·s to only 5 mPa·s. By the addition of NaCl counter-ions in excess are introduced into the system. The repulsion forces of the negative charged groups along the chain of the polymer are cancelled and the hydrodynamic volume of the polymer molecule decreases.

Solutions of a selected synthetic homopolysaccharide (Fig.2) show a different behaviour. The viscosity of solutions of this polymer in distilled water is not influenced if NaCl is added (Fig.3). If divalent cations, for example Ca^{++} are added, the viscosity increases in the range of low shear rates (10^{-3}-10^{-2} s^{-1}). In the case of NaCl-addition the viscosity remains constant, because there are no charged groups in this polymer. The divalent cations, however, may build up salt bridges between the OH-groups of the polymer molecules, which are stable only at low mechanical strains corresponding to shear rates of 10^{-2} s^{-1} or lower. At higher rates of shear these bridges are destroyed and the viscosity is equal to that of a solution without the presence of Ca^{++}-ions. The visco-

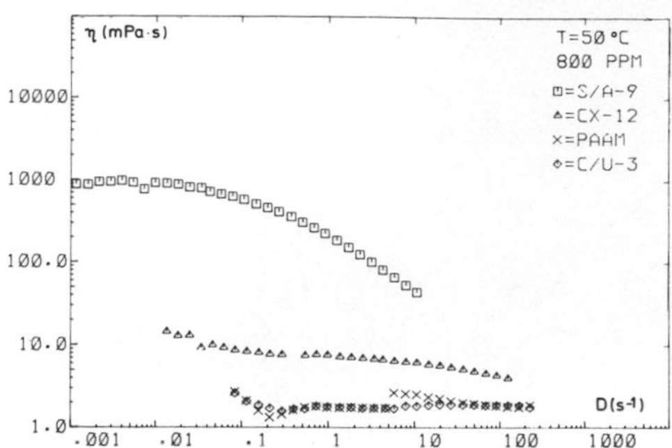

Fig.5: Viscosity η vs shear rate D for solutions of four selected polymers in the synthetic brine

sity of this solution is higher than 20 mPa·s, even if salts are added.

Addition of salts (NaCl, $CaCl_2$) to a solution of a synthetic heteropolysaccharide (Fig.4) results in a slight viscosity decrease at low shear rates, which is caused by the ionic groups of the molecule. The rigidity of the molecule due to the large side groups is responsible for the fact that there is only a small viscosity decrease. These solutions have a viscosity higher than 20 mPa·s, too.

In case of hydroxyethylcellulose there are no charged groups in the molecule; therefore the viscosity will not be altered, if salts are added to the solutions. Because there are no large side groups fixed to the chain, the viscosity does not reach the high values of polysaccharide solutions. At higher concentrations the viscosity values are sufficient ($>$20 mPa·s at 2500 ppm).

In order to gain a better relation to the practical demands the polymers were tested in a model system of high salinity. They were dissolved in a synthetic brine of a salt concentration of 200 g/l. The composition of the brine was representative for the oil deposits in the north of the FRG. In this brine there are not only simple salts like NaCl, $CaCl_2$ etc. but iron and boric salts, too, which may cause flocculation of the polymers. Flocculation of the polymer would plug the pores of the oil reservoir. Therefore the most important criterion is the direct and good solubility of the polymer in this model brine without flocculation.

Further criteria, which will not be discussed here, are the long term stability of viscosity under reservoir conditions and good filtration properties of the polymer solutions.

Within the great number of tested polymer samples, only a few were soluble without complications in the synthetic brine (200 g/l). In the group of the homopolysaccharides, samples received directly

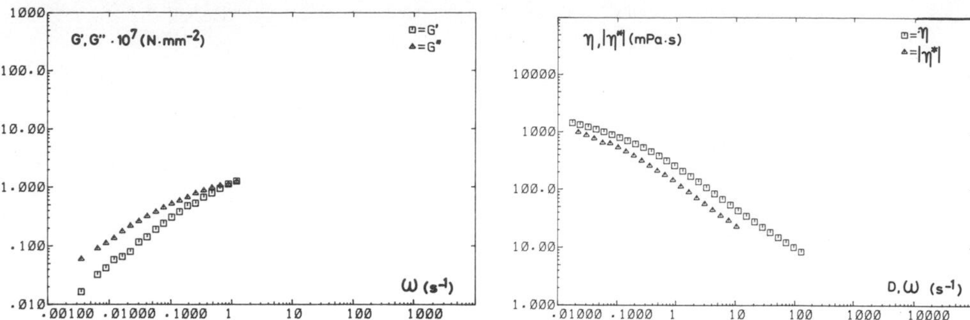

Fig. 6a: Frequency dependence of storage modulus G' and loss modulus G"

Fig. 6b: Complex dynamic viscosity |η*| and steady-flow viscosity η vs D,

for a solution of a homopolysaccharide ("liquid polymer") in H_2O. T=25 °C; polymer concentration = 800 ppm.

from the fermentation in the solved state ("liquid polymers"), showed good properties. For comparison the viscosity curves of four samples are presented in Fig. 5: (1) Homopolysaccharide (liquid polymer; Code No. S/A-9) (2) Heteropolysaccharide (Code No. CX-12), (3) Anionic polyacrylamide (Code No. PAAM), (4) Hydroxyethylcellulose (Code No. C/U-3). The solutions were obtained by mixing equal amounts of the polymer solution and the brine (salt content 200 g/l) to a final concentration of 800 ppm.

The sample S/A-9 has the highest viscosity values of the four samples and reaches a viscosity of 1000 mPa·s at shear rates of 10^{-3} s^{-1}. The high viscosity is firstly caused by the relatively high rigidity of the molecules, due to the voluminous side groups (Fig. 2) and secondly by complex formation of the divalent cations of the brine and the hydroxyl groups of the polymer. At shear rates of 10^{-3} to 10^{-2} s^{-1}, the viscosity is approximately constant. As already explained above in the case of the PAAm solution in distilled water, this does not mean that the solution is Newtonian. As will be shown below, the solution has pronounced elastic properties even at this relatively low polymer concentration. The viscosities of the other polymer samples are much lower.

If a polymer flows through a porous medium, the mechanical strain will not be uniform, but changes as the pore diameter changes. Therefore not only the viscous properties are important for the flow of the polymer solutions in the oil reservoir, but also the elastic properties. In Fig. 6a the dynamic moduli G' and G" of a solution of a homopolysaccharide ("liquid polymer") in distilled water are plotted vs radian frequency ω. (G' (ω) = storage modulus, G" (ω) = loss modulus). The figure shows that the storage modulus G' is not small compared with the loss modulus G", but is of the same magnitude even at this low concentration.

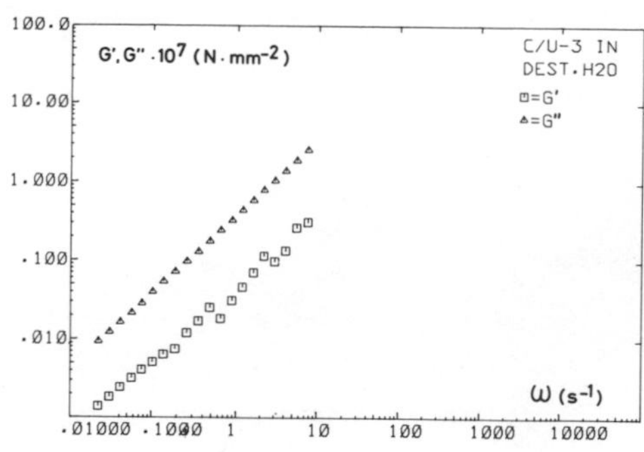

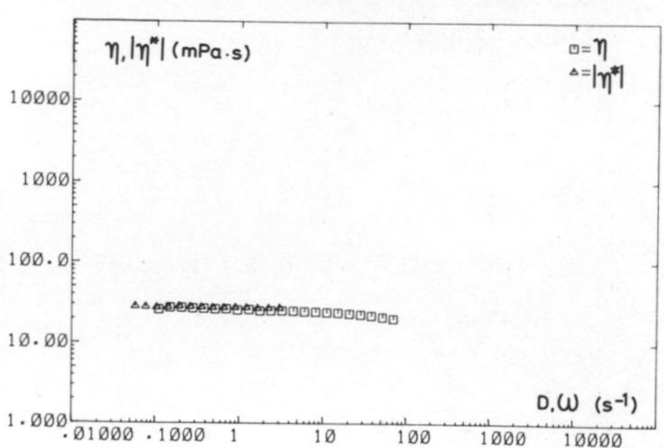

Fig. 7a: Frequency dependence of storage modulus G' and loss modulus G"

Fig. 7b: Complex dynamic viscosity $|\eta^*|$ and steady-flow viscosity η vs D

for a solution of HEC in H_2O. T= 25°C; polymer concentration = 2500 ppm

With the values of G' and G", the complex dynamic viscosity can be calculated according to the formula:

$$|\eta^*(\omega)| = (G'(\omega)^2 + G''(\omega)^2)^{\frac{1}{2}} / \omega$$

The formula holds in the range of linear viscoelastic behaviour. According to the empirical Cox-Merz rule[1], the steady-flow viscosity $\eta(D)$ and the absolute value of the complex dynamic viscosity $|\eta^*(\omega)|$ should be approximately equal at the same values of the shear rate D and the radian frequency ω.

In Fig. 6b the curves of the steady-flow viscosity η and the dynamic viscosity $|\eta^*|$ are presented for the same solution. The course of both curves is similar. In the measured frequency range from $\omega = 10^{-2}$ to 10, the complex dynamic viscosity $|\eta^*|$ is lower than the steady-flow viscosity η. A similar deviation from the Cox-Merz rule was found for a ethylene-methacrylic acid copolymer melt[2]. The lower values of $|\eta^*|$ may be explained by the concept that there is a quasi network structure in the solution of the homopolysaccharide measured in this work.

For a solution of a HEC-sample in distilled water, the dynamic moduli G' and G" are given in Fig. 7a. In Fig. 7b the viscosities η and $|\eta^*|$ are plotted as function of D and ω, respectively. The HEC-solution follows the Cox-Merz rule.

1) W.D. Cox, E.H. Merz, J. Polym. Sci. 28 (1958) p. 619

2) K. Sakamoto, W.J. Mac Knight and R.S. Porter, J. Polym. Sci. A2, 8, (1970) p. 277-287

DEPENDENCE OF DRAG REDUCTION ON POLYMER SIZE DISTRIBUTION AND CONFIGURATION

N. Charifi Sandjani, J.M. Sangster and H.P. Schreiber

Department of Chemical Engineering
Ecole Polytechnique
Box 6079, Station A
Montreal (Québec) H3C 3A7

The suppression of turbulence in fluids due to the addition of high molecular weight polymers has been the subject of much study (1). Recently, we postulated (2) that polymers function as drag reducing agents on a molecular level by inhibiting the coalescence of microvortexes present in the solvent, thereby delaying the onset of bulk turbulence. These arguments rationalize the known dependence of drag reduction effectiveness (DRE) on such variables as solute concentration, molecular weight and chain flexibility. Further, the concept indicates that in a polymer with broad molecular weight distribution, it should be the high molecular weight end of the distribution which dominates the DRE performance, much as the bulk viscosity is dominated by these long chains. One of the purposes of this communication is to test the concept experimentally.

A related problem of continuing interest is the known sensitivity of DRE for a given polymer on solvent quality (3), a problem which stresses the importance of macromolecular "architecture"(4) on the effective application of the D.R. principle. One possible approach to desensitizing polymer chain conformation to variations in solvent quality is to immobilize the chain in a preferred conformation state. Crosslinking the polymer in solution by γ-ray treatment may be a feasible approach (5-7). In this study we have briefly explored the value of this route toward stabilizing the DRE performance of polyacrylamides in aqueous solutions.

EXPERIMENTAL

i) Materials: Two samples of polyacrylamide, Dow Chemical Co. Separan AP-30 and Separan MGL, were used. The former is a partially hydrolyzed, the latter a more completely hydrolyzed specimen of the polymer. The molecular weight of sample MGL was not determined, but AP-30 was fractionated and characterized by gel permeation chromatography, thanks to the cooperation of Prof. A.E.Hamieleç Dept. of Chemical Engineering, McMaster University, Hamilton, Canada. For fractionation, powdered polymer was mixed with methanol (non-solvent) whereupon water was added to a concentration of 22.3 wt-%, previously determined as a critical concentration for the formation of a highly swollen gel. Additional water (∿ 5 wt %) was then added and the system refluxed with agitation for 10-14 hr. Upon centrifugation, the supernatant solution was removed and the polymer precipitated by evaporating the solvent/non-solvent mixture. The residual gel was then reprocessed as above, finally yielding a total of 6 fractions. The recovered fractions were reprecipitated prior to characterization. Their molecular weights and intrinsic viscosities [η], in tap water at 30°C, are compared with those of the parent polymer below.

	AP-30	f_1	f_2	f_3	f_4	f_5	f_6
$Mn.10^{-6}$	2.3	2.4	2.6	2.2	2.8	3.0	3.5
$Mw.10^{-6}$	7.2	5.8	5.2	5.0	5.5	8.1	7.6
$Mz.10^{-6}$	19.5	16.5	14.9	15.7	13.2	24.0	18.9
$[\eta].dl.g^{-1}$	12.0	12.6	11.3	11.9	9.9	10.9	10.7

Although the molecular weight distribution breadths of fractions remain appreciable, they were considered adequate for present purposes .

ii) Procedures: Drag reduction experiments were carried out in a 4 l.stainless-steel apparatus, containing a capillary holder, 35.5 cm in length. This could accomodate steel tubes ranging in diameter from 1.6-12.7 mm. A Moyno pump circulated the fluid through the tubes, via a buffer volume designed to minimize pressure pulses due to pumping. The apparatus was suitable for work up to Reynolds (Re) numbers in the range 7 x 10^4. The experiments were at 30°C, and pressure drops across the capillary tubes were read on calibrated precision gages. DRE was defined as usual (1) as the % reduction in the frictional coefficient in the flow of solvent due to the presence of polymer.

Polymer conformation was characterized by [η]. Only whole samples of AP-30 and MGL were used for studies of the interdependence between DRE and molecular architecture. The solvents involved were conductivity water, tap water, conductivity water to which was added 5 vol. % i-propanol in one case, and in another case NaCl to produce 0.55 M solution, intended to simulate a brine environment. The relevant [η], given elsewhere indicate that very pronounced changes in the conformation of AP-30 were produced. As expected, the hydrolyzed MGL polymer was much less sensitive to changes in solvent quality.

In an attempt to control the response of polymer conformation to changes in solvent quality, polymer solutions in pure water were exposed to γ-rays in a Gammacell 220, produced by Atomic Energy of

Canada Ltd, and used with permission of the Chemistry Dept., McGill University. Solutions of varying concentration were placed in 85 ml. cuvettes and exposed for periods from 3 min. to 24 hr. The specific viscosities of irradiated solutions were used to define critical gel concentrations(5,6) and thus to indicate regions of predominance for intra-molecular or intermolecular crosslinks introduced by the irradiation process. Additional irradiation experiments were performed at selected polymer concentrations, and the sensitivity of polymers to solvent changes was tested by comparing specific viscosities of irradiated and control solutions in pure water and in 0.55 NaCl solutions. The DRE of irradiated solutions was also measured.

RESULTS AND DISCUSSION

i) DRE and molecular size characteristics: Polymer AP-30 and its six fractions were used to measure the frictional coefficient as a function of polymer concentration and of Re. The magnitude of the DR effect is known (1) to increase systematically with Re, but varies in a more complex manner with concentration (3). We examined the concentration dependence of DR, and this was done at constant Re values 5 x 10^4, corresponding to the (approximate) maximum flow rate obtained for all systems. The performance of three of the fractions illustrate the results obtained in Fig. 1. As at all Re values where significant DR effects were observed the curves in Fig. 1 define a maximum, and the parameters $(DR)_m$ and $(C)_{opt}$. The $(DR)_m$ is Re dependent, but the polymer concentration $(C)_{opt}$ corresponding to that maximum was not significantly different at lower Re values.

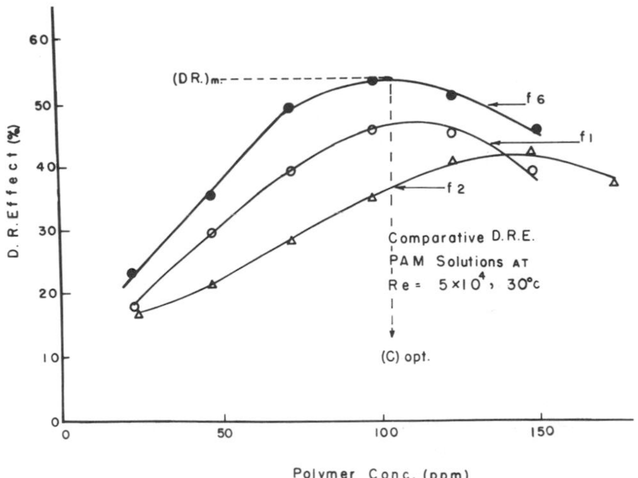

Fig. 1. D.R. performance of polyacrylamide fractions

The significant results in this portion of the study are summarized in Table I. This makes clear the molecular weight dependence of both $(DR)_m$ and $(C)_{opt}$, without immediately showing the importance of molecular weight distribution. It is clear however that the DR performance of the whole polymer corresponds approximately to that of f6; inspection of the molecular weight data shows these materials have very distinct M_n values, but similar higher moment values. In order to develop this point, the ratios $(DR)_m/M$ were calculated for the three known moments of the distribution and entered in Table I. The variation in quotient values decreases from ~60% for $(DR)_m/Mn$ to ~25% (M_w) and ~10% in the case of $(DR)_m/M_z$. The result confirms the postulate of our earlier work (2); the DR effect is dominated by the highest molecular weight members in the distribution of a polymer specimen. It seems reasonable that these members should contribute disproportionately to the immobilization of turbulence precursors. The observation establishes the rationality of considering DR phenomena from a molecular viewpoint and thereby links the DR phenomenon to other rheological manifestations, including the viscosity of polymer melts and solutions were similar effects of molecular weight distribution exist. Finally, the $[\eta]$ values (see above) were used to calculate the product, $C_{opt}^{1/3} \cdot [\eta]$ and entered in Table I. The near constancy of this product again confirms consequences drawn from earlier hypotheses (2), which equated this product with the total number of effective DR units in a volume of fluid and suggested that it be constant at a selected DRE (such as DR max) regardless of the chemistry of the drag reducing polymer.

Table 1: Comparison of (D.R.)m and (C)opt. (T = 30°C)

	(D.R.)m %	(C)opt. ppm	$(C)_{opt}^{1/3} \cdot [\eta]$	$\frac{DRm}{M_n}$	$\frac{DRm}{M_w}$	$\frac{DRm}{M_z}$
Parent:	55.5	110	2.68	24.1	7.70	2.84
f_1	46.0	120	2.90	19.2	7.93	2.79
f_2	40.5	145	2.75	15.6	7.79	2.72
f_3	43.0	125	2.77	19.5	8.60	2.74
f_4	39.5	150	2.45	14.1	7.18	2.99
f_5	58.0	170	2.80	19.3	7.16	2.42
f_6	53.0	105	2.65	15.1	7.00	2.80

ii) D.R. and molecular architecture: The conformation of AP-30 and MGL polymer responds to solvent quality, as was shown by $[\eta]$ results for the four solvents of this study. The presence of salts in tap water and in the NaCl solution produces a sharp drop in the hydrodynamic volume of the polymer coil, an effect of recognized importance to the drag reducing performance of polymers (2,3). Relatively slight changes were caused by the presence of i-propanol and in general the hydrolysed MGL sample is less sensitive to the changes in solvent quality. As expected, a close relationship was found between DRE and chain conformation. Very strong reductions in DRE were observed for AP-30, as the $[\eta]$ of this polymer decreased; the response of polymer MGL was less pronounced. In practical use, the DR performance of any water-soluble polymer will tend to vary markedly with the local quality of the solvent medium.

Our efforts to "desensitize" the macromolecule to solvent

effects by γ-ray treatment first called for the definition of a critical concentration for macrogel formation (5,6); in all irradiated solutions used for further study the polymer concentration was below that value. The effects of irradiation followed closely the behaviour reported for water-soluble polymers by Kiran and Rodriguez (6). Thus, with increasing radiation dose, the specific viscosity initially decreased to a shallow minimum, rising to a maximum at higher doses and finally decreasing rapidly. The position and magnitude of these features varied with polymer concentration, and their origins - though still controversial - may be ascribed to shifting balances between intra and intermolecular crosslinks and, at higher dosages, to chain scission (6,7). In our work, emphasis was placed on the first response region, ostensibly favoring intramolecular links. A summary of results is presented in Table II. The Table compares η_{sp}/C values for irradiated polymer in pure water and in 0.55 M NaCl solution, and reports a reference value of DRE.

γ-ray induced effects influence both absolute viscosities and their variation with solvent quality. The large decrease in $\eta sp/C$ with increasing dose for AP-30 is probably due partly to chain scission and partly to intramolecular linkages. Both factors would reduce the sensitivity of chain conformation to solvent interactions. The desensitization of chain conformation exacts a heavy penalty in drag reduction performance, which diminishes with rising irradiation dose. The AP-30 solution at 1.4 g/dl is near the critical gel concentration, hence in this case a more favorable environment for irradiation-induced intermolecular linkages exists. The drop in absolute viscosity is indeed less severe, but the DRE is even more drastically inhibited. Evidently, in contrast with Kiran's expectation (4), but in keeping with our earlier concepts (2), drag reduction efficiency varies directly with chain flexibility - a characteristic which is inhibited by the irradiation procedure. A similar pattern of diminishing solvent sensitivity and DRE is followed by sample MGL. Here, however, the maximum in η_{sp}/C near a dose of 0.3 M Rad. is clearly discernible. The predominance of intermolecular crosslinks in this response region apparently builds a larger macromolecule with somewhat restored flexibility, a feature also apparent in its enhanced DR performance.

The diminishing effectiveness of radiation - crosslinked polymers as drag reducing agents is more generally displayed in Fig. 2. The linear decrease in DR effectiveness for AP-30 indicates that drag reduction performance decreases exponentially with irradiation dose; the hydrolyzed MGL polymer, responds to irradiation in a characteristically different manner. We assume that the balance of crosslinking and scission events is significantly more heavily weighted toward the former mechanisms in this polymer.

We conclude that while γ-ray modification of high molecular

weight polymers in aqueous solution can greatly diminish the sensitivity of polymer conformation to subsequent changes in solvent quality, the chain stiffening and scission events occurring during irradiation are counterproductive to the polymer's function as drag reducer. This experimental approach may therefore enjoy only limited application.

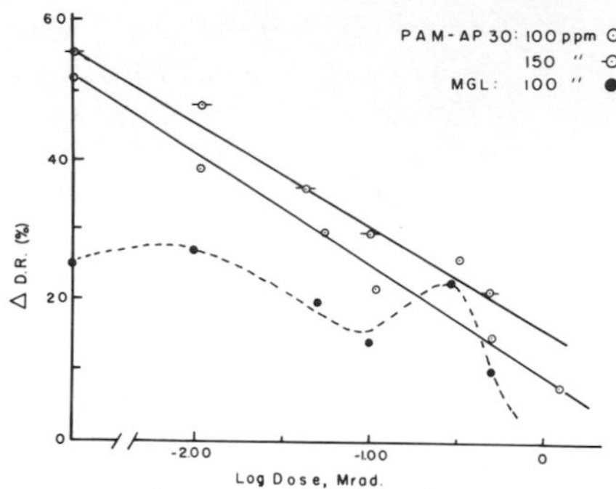

Fig. 2: D.R. performance versus irradiation dose

Table II
VISCOSITIES AND D.R. PERFORMANCE OF IRRADIATED POLYMERS

Irr. Dose M Rad.	AP-30, 0.5 g/dl η_{sp}/C H_2O	NaCl	% Chge	DRE*	AP-30, 1.4 g/dl η_{sp}/C H_2O	NaCl	% Chge	DRE*	MGL, 0.4 g/dl η_{sp}/C H_2O	NaCl	% Chge	DRE*
0	180	15	91.7	52	180	15	9.7	52	14.1	9.7	31.2	25
.04	130	28.7	78.5	39	153	19.3	87.3	40	8.0	4.5	43.7	27
.055	48	13.0	72.9	30	100	16.0	84.0	22	4.1	2.8	31.7	20
.110	32.5	8.8	66.7	22	79.5	11.7	85.3	17	4.5	3.3	26.7	14
.330	21.0	7.9	62.4	26	69.0	15.3	77.8	17	17.4	8.6	50.5	23
.550	13.9	7.0	49.6	15	55.0	24.7	55.1	5	9.5	7.4	22.1	10
1.40	4.0	3.6	21.7	8	28.2	14.3	50.0	--	----	---	----	--

* DRE measured at 30°C in H_2O at Re = 5 x 10^4 and polymer concentration = 100 ppm.

References

(1) For example: A. White, Drag Reduction by Additives, Review and Bibliography, BHRA fluid engineering, Cranfield, England (1976).

(2) B. Hlavacek, A.L. Rollin and H.P. Schreiber, Polymer, 17, 81 (1976).

(3) R.C. Little, Nature Phys. Sci. 242, 79 (1973) and R.C. Little J. Appl. Polym. Sci. 15, 3117 (1971).

(4) E. Kiran, Nature Phys. Sci. 238, 29 (1972).

(5) A. Charlesby, "Aspects of Radiation Reactions in Polymers" in Energetics and Mechanisms in Radiation Biology, G.O. Phillips, Ed. Academic Press, N.Y. (1968).

(6) E. Kiran and F. Rodriguez, XXIII Int. Congress of Pure and Appl. Chem. 8, 175 (1971).

(7) A. Rudin and T.C. Chan, J. Polym. Sci. Polym. Chem. Ed. 10, 3589 (1972).

APPLICATION OF THERMODYNAMICS TO

STABILITY OF FLOW AND DRAG REDUCTION

W. M. Jones

Department of Physics

U.C.W., Penglais
Aberystwyth, Dyfed, U.K.

INTRODUCTION

Dimensionless numbers exist to describe the onset of secondary flows and the onset of turbulence, in various geometries. For example, in flow in bent pipes and in porous materials secondary flows suddenly begin to contribute to the pressure gradient when the Dean number $L' = 276$; in Couette flow between coaxial cylinders (inner rotating) secondary flows suddenly contribute to the torque when the Taylor number $Ta \approx 3400$; in flow through straight pipes the flow becomes turbulent when the Reynolds number $Re > 1000$.

In the flow of CO_2 and of N_2 through porous materials, at high pressures, the values of L' can be different for the two gases at the onset of secondary flows. This difference may be attributed to the exchange of energy between the vibrational mode of the CO_2 molecule and the energy of bulk motion; no vibrational mode of the N_2 molecule is accessible at room temperature. Thermodynamics has been applied to discuss the processes governing the exchange of energy and its influence on the secondary flow.

Drag and stability of flow are also affected by small amounts of polymer additives and similar ideas to those used to interpret the flow of CO_2 have been applied to them. The object of this Paper is to summarise the concepts.

FUNDAMENTAL CONCEPTS

The fluid (the system) is in contact with the device which supplies energy to maintain flow (the surroundings). Suppose an instability has just about grown in the (non-equilibrium) flow.

Then in unit volume,

$$E_s = E_\mu + E_b + E_i, \qquad (1)$$

where E_s = energy supplied from the surroundings,

E_μ = viscous energy dissipated within the bulk fluid during the onset of the instability;

E_b = change of energy of bulk of fluid (potential & kinetic);

E_i = supply of energy to the disturbance to enable it be established.

Suppose now an additional energy mechanism at the molecular level is introduced due to peculiar circumstances, which exchanges energy with the disturbance and therefore competes for a supply of energy from the bulk flow, then

$$E'_s = E'_\mu + E'_b + E_a + E_i \qquad (2)$$

where E_a is the energy which is transferred as a consequence of the molecular mechanism; the disturbance requires the same energy to grow as in the absence of the molecular process. When E_a is positive (a 'sink' of energy) then the bulk energy of motion will have to increase to supply the extra energy and so the disturbance appears to be initiated at a higher Reynolds number (stabilization of the flow), whereas when E_a is negative (a source of energy) then the disturbance appears to be initiated at a lower Reynolds number (destabilization of the flow).

What are the peculiar circumstances which lead to the transfer of energy E_a? Firstly, the molecular energy of the system must consist of an active part E_A which is in instantaneous equilibrium with the bulk (macroscopic) energy of the system and an "inert" part E_O which is temporarily not in equilibrium with E_A such that

$$\dot{E}_A = -\dot{E}_O = (E_O - E_A) / \tau \qquad (3)$$

where τ is a characteristic time for the process. For the growth of the disturbance to be affected by the molecular process then two conditions must be fulfilled: (1) the molecular energy is transferred in that part of the system where the instability is growing, (2) the magnitude of E_a is a measurable fraction of E_i. The former of these conditions leads to

$$E_a = (E_o - E_A)(1 - e^{-1/L}) \qquad (4)$$

in which l is the length in the flow system over which the disturbance grows and L is the relaxation length for the molecular exchange of energy, that is $L = u\tau$ where u is the speed of flow in the region of the disturbance. Since most of the energy ($E_O - E_A$) is transferred when $l = L$ then we might expect one of the peculiar circumstances necessary for the molecular process to affect the flow is: $l \simeq L$. When $l \ll L$ or when $L \gg l$ there is no exchange of energy with the instability.

Feelings of doubt might arise at the idea that molecular energy can be transferred so as to actuate disturbances in bulk flow. These feelings might be put at rest by referring to the experiments of Betchkov (1957, 1961) which show strong evidence for the view that molecular fluctuations lead to energy being concentrated in a locality to grow into a disturbance in the flow system.

The problem in practice is to identify the molecular mechanism responsible, to quantify it, and then to identify and quantify l and L in equation (4). Some examples are given to clarify the ideas.

APPLICATION TO THE FLOW OF CO_2 and N_2 THROUGH POROUS MEDIA

The Phenomenon

In Figure 1, Δp_o is the observed pressure drop and Δp_e is that calculated on the assumption that Δp varies linearly with the rate of flow. $\Delta p_o/\Delta p_e > 1$ signifies the onset of a secondary flow within the porous material. R_e is the Reynolds number, Jones and Williams (1963). In the experiments $\Delta p \simeq 0.5$ atm.

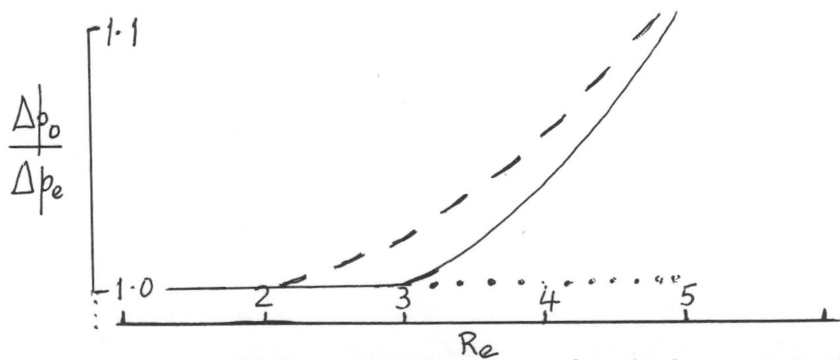

Figure 1. —— flow of N_2 through porous nickel sheet at various mean pressures ranging from 28 atm to 70 atm; ---- flow of CO_2 at 42 atm; .... flow of CO_2 at 14.5 atm. Results for CO_2 at 28 atm coincided with the results for N_2.

It is seen that destabilization of the flow of CO_2 occurs when $\bar{p}$ = 42 atm and stabilization when CO_2 = 14.5 atm.

The Mechanism

The energy of the N_2 molecule consists of translatory motion through space and of rotation as a rigid rod about the centre of gravity of the molecule. These energy modes are in instantaneous equilibrium with each other. The CO_2 molecule on the other hand has a vibratory mode of energy (bending of the molecule) as well as the translational and rotational modes. When the translational energy is suddenly changed, the vibrational mode requires 1.6 x 10^4 molecular collisions before it reaches equilibrium; at atmospheric pressure this corresponds to a relaxation time of 1.7 μs. (Because mean free path decreases as the pressure increases $\tau = 1.7/p$ μs where p is in atm). The mean linear speed of the gas through the porous material can be found and it varies with pressure at the onset of instability, and at 15 atm it is ≃ 3 ms so L ≃ 0.1 μm Arguments were put forward by Jones and Williams (1963) to suggest that $l = \bar{r}/5$ in which $\bar{r}$ is the mean pore radius. For the nickel sheet of Fig. 1 l/L varied from 0.4 to 3.1 in the range of pressure from 14.5 to 42.0 atm; for porous copper sheet which showed similar results it varied between 0.2 and 2.1; for a nickel sheet where CO_2 did not differ from N_2 l/L varied from 0.02 to 0.14.

What is the non-equilibrium condition in the flow that results in the separation of the energy levels? The nickel sheet was fabricated from nickel powder annealled and rolled. So the flow path would be a series of constrictions with relatively large openings between them. Expansion of the gas means work has to be done against inter molecular forces at the expense of thermal energy. Assume energy equilibrium is reached after each constriction, the non-equilibrium occurring in the constriction.

Consider the isothermal equilibrium states. They can be joined by paths consisting of an adiabatic (isenthalpic) expansion followed by isobaric work to restore the temperature to the isothermal condition of the experiment. In the isenthalpic process

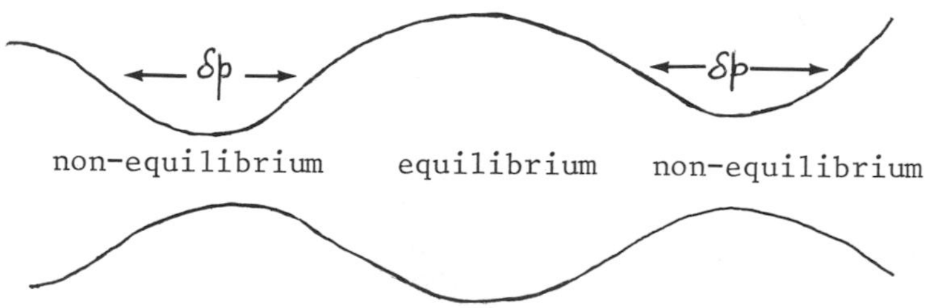

the energy gap ΔE_c, created in unit time in volume v is $v\rho c\mu'\delta p$ where v is volume rate of flow in the constriction, ρ = density, c = specific heat capacity, $\mu' = (\partial T/\partial p)_H$ where H is enthalpy and T temperature. Some of the work necessary to restore the isothermal condition has already been done against friction in the constriction; this is $\Delta E_f = v\,\delta p$ in unit time. $\Delta E_c/\Delta E_f = 10\rho c\mu'$ when ρ is in gcm^{-3}, and c in joules g^{-1}K^{-1} and μ' in deg atm^{-1}. $\mu' = (1/\rho c)(\beta T - 1)$ in which β is expansivity. Values of β are given in tables as a function of pressure. It is found that $\Delta E_c/\Delta E_f = 1$ when p = 15 atm, $\Delta E_c/\Delta E_f > 1$ when $p >$ 15 atm and $\Delta E_c/\Delta E_f < 1$ when $p <$ 15 atm. $\Delta E_c/\Delta E_f > 1$ implies E_a in equation 2 is negative (destabilization) and $\Delta E_c/\Delta E_f < 1$ implies stabilization. $\Delta E_c/\Delta E_f = 1$ implies no effect on stability. So the pattern of flow of Figure 1 can be obtained; that no effect occurs at 28 atm rather than 15 atm is attributed to the neglect of heat transfer to (or from) the metal of the porous material.

Jones and Williams (1963) found $\Delta E_f - \Delta E_c$ was 0.5 J mole^{-1} and the kinetic energy of bulk motion was about the same, so there is adequate energy in E_a to influence the flow pattern.

OTHER APPLICATIONS

Consider the flow of a dilute macromolecular solution. The Helmholtz function A is the appropriate thermodynamic function to apply to its study since it is isothermal and incompressible. Neglecting dissipative effects then for unit quantity of fluid flowing steadily through a geometry of uniform cross-section (no end-effects) dA = 0. However where there are changes in shear stress such as at the onset of instability or at changes in geometry, then dA = đW, where đW is the work done on unit quantity of the system during the change of stress. In equation (2), Ea = dA = v dP_e where dP_e is the extra pressure difference generated (neglecting dissipative effects) because of molecular interaction with the instability and v is the volume of unit quantity.

Now A = -NkTlnZ and dA = - NkT dZ/Z where Z is the partition function per molecule and N is the number of molecules. If it is assumed now that the energy of the molecule remains in the ground state (ε_0) in the shear field then $Z = \Omega_0 e^{-\varepsilon_0/kT}$ where Ω_0 is the weight (number of complexions) of that state

$dA = -NkTd\Omega_0/\Omega_0 = -TdS$, since $S = Nk\ln\Omega_0$, where S is the conformational entropy of the molecules in the ground state ('elastic entropy'). In a 'simple' macromolecule such as polyethylene or polyethylene oxide or even polyacylamide there are two gauche (g-) arrangements and one trans (t-) per bond. There is an energy gap between t- and g- but not between g- and g-.

dS will diminish (dA and E_a increase) if half the g- positions in the bonds become relatively inaccessible because of shear stresses. The relaxation time (when the stress is removed) is determined by the energy barrier between g- and g- and by the number of bonds affected by the shear field (i.e. the relaxation time is a function of shear stress).

It is now possible to identify some flows which are affected by these molecular processes.

In the steady shear field in the entry length of a pipe energy is stored in the molecule and E_a in equation 2 is negative (destabilization of the flow). When conditions are opportune for fully-developed turbulence to occur, the vortex stretching this entails would be restricted since E_a is then positive and that part of the spectrum of turbulence having a characteristic length similar to the relaxation length of the macromolecular processes will not occur at that rate of flow ('turbulence suppression' or 'drag reduction'). It would be expected therefore that destabilization and drag reduction would be quantitatively related and this has been shown experimentally.

Similarly, 'polymer effects' in flow through porous media, in flow between concentric cylinders and in bent pipes, can be correlated through comparison of characteristic lengths in the flowing fluid with those of the particular geometry containing the flow. The conformational relaxation time of a macromolecule is of order 1 ms (c.f. 1 μs in CO_2) so effects are observed in pipes and pores of correspondingly larger dimensions.

BIBLIOGRAPHY FOR REFERENCES AND FURTHER DETAILS OF WORK CITED

Astarita G and Sarti G C 1976 J. Non-Newt Fluid Mech 1, 39
Jarecki L 1979 Coll Polym Sci 257, 711
Jones W M and Williams I B 1963 Brit J Appl Phys 14, 877
Jones W M 1976 J Phys D Appl Phys 9, 721-70
Jones W M 1979 J Phys D Appl Phys 12, 369

CONTRIBUTION TO THE ELONGATIONAL VISCOSIMETRY OF HIGH-POLYMERIC SOLUTIONS

Ernst-Otto Reher and Reinhard Karmer

Merseburg Technological Institute
Department of Chemical Engineering
42 Merseburg, German Democratic Republic

INTRODUCTION

Elongational flows play a decisive role in many shaping processes of manufacturing technology. Specific flow phenomena, by which we mean a multitude of instabilities and breaks, may definitely limit the technological possibility of controlling elongation flows, above all in the spinning of artificial fibres, the coating of film materials, and film blowing. In these processes, the rheological characterization of the behaviour of the fluid materials to be processed under elongational load is an indispensable precondition. Work on these problems is carried out in order to define the material function of elongational flow by analogy with simple shear, and to ensure exact measurement in suitable rheometer flows. At present, the rheological characterization of the elongational behaviour of high-polymeric solutions presents difficulties greater than those found in other groups of materials.

MATERIAL FUNCTIONS OF ELONGATIONAL FLOW

Material functions can be defined in the so-called "rheometer flows", by which we mean uniform steady flows with constant and exactly determinable stress and rate of strain tensors. Apart from "steady simple shear", the following three elongational flows can be distinguished as basic steady flows:

Steady uniaxial elongation

with the rate of strain tensor

$$e_{ij} = \begin{pmatrix} \dot{\varepsilon} & 0 & 0 \\ 0 & -\frac{1}{2}\dot{\varepsilon} & 0 \\ 0 & 0 & -\frac{1}{2}\dot{\varepsilon} \end{pmatrix} , \quad \dot{\varepsilon} = \text{const}$$

and the so-called elongational viscosity (Trouton viscosity) as material function

$$\eta_T(\dot{\varepsilon}) = \frac{P_{11} - P_{22}}{\dot{\varepsilon}} ,$$

where P_{ii} are the components of the stress tensor.

Steady equal biaxial elongation

with the rate of strain tensor

$$e_{ij} = \begin{pmatrix} -2\dot{\gamma} & 0 & 0 \\ 0 & \dot{\gamma} & 0 \\ 0 & 0 & \dot{\gamma} \end{pmatrix} , \quad \dot{\gamma} = \text{const}$$

and the material function

$$\eta_{eb}(\dot{\gamma}) = \frac{P_{33} - P_{11}}{\dot{\gamma}}$$

Steady pure shear or strip biaxial extension

with the rate of strain tensor

$$e_{ij} = \begin{pmatrix} \dot{\Gamma} & 0 & 0 \\ 0 & -\dot{\Gamma} & 0 \\ 0 & 0 & 0 \end{pmatrix} , \quad \dot{\Gamma} = \text{const}$$

and the material functions

$$\eta^{1}_{sb}(\dot{\Gamma}) = \frac{P_{11} - P_{22}}{\dot{\Gamma}} , \quad \eta^{2}_{sb}(\dot{\Gamma}) = \frac{P_{33} - P_{22}}{\dot{\Gamma}}$$

In[1] the realization of this basic steady flow is shown to be practically impossible, since contrary to steady simple shear it is not possible, or only partially possible, to meet the demand for an alongational flow with temporally and locally constant stress and rate of strain tensors. The types of flow represented in Figure 1., i.e. uniaxial stretching, biaxial stretching, and spinning are the only ones which are used as basic elongational flows.

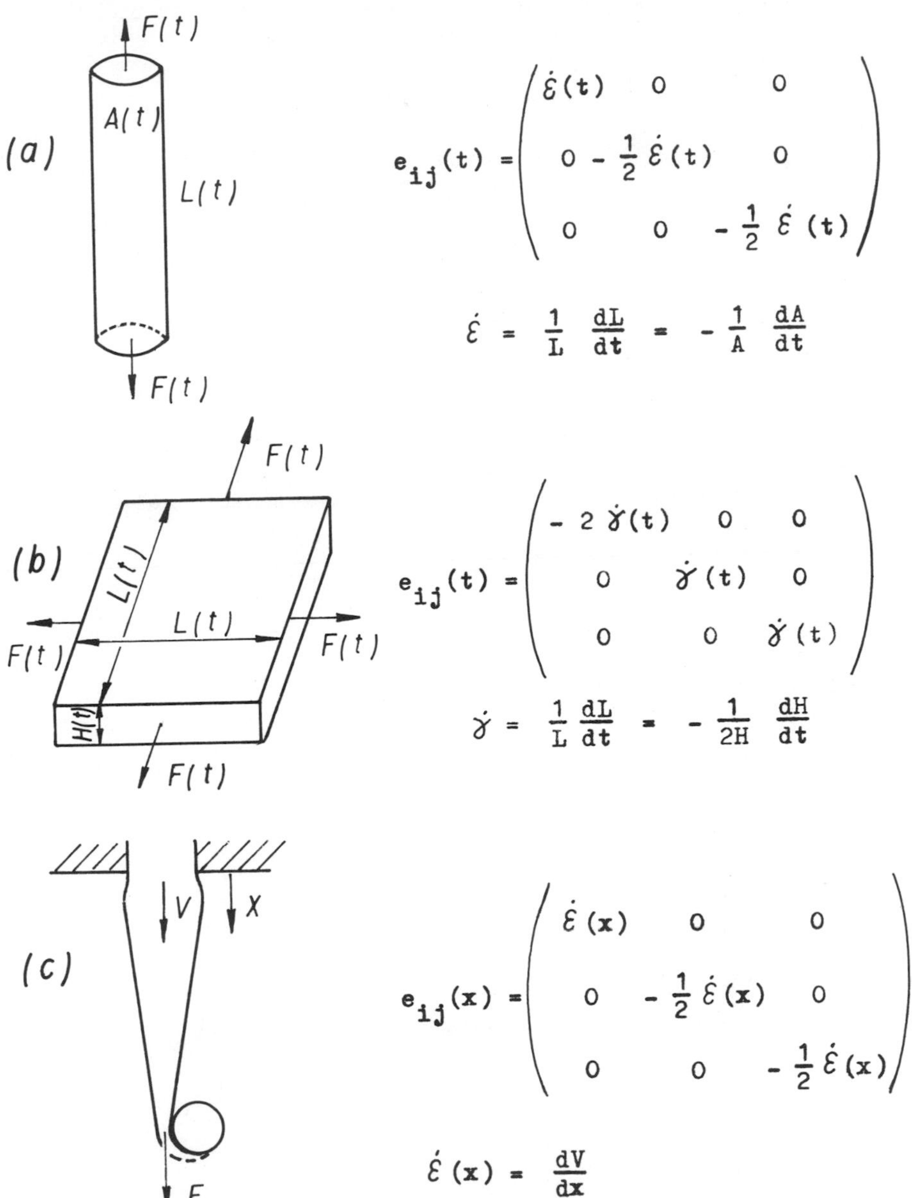

Fig. 1. Basic elongational flows. (a) uniaxial stretching; (b) biaxial stretching; (c) spinning.

It is important to statethat although the spinning flow worst of all meets the criteria of a rheometer flow with respect to steady deformation and a flow with constant prehistory, it presents the only relizable possibility of giving a rheological characterization of the elongation behaviour a high-polymeric solution.

EXPERIMENTAL DETERMINATION OF ELONGATIONAL VISCOSITIES

In order to characterize the elongational behaviour of high-polymeric solutions, the elongational viscosities of aqueous solutions of PEO of different concentrations were determined with the help of the spinning flow, using the spin balance represented in Fig. 2. which differs from the apparatus described in[2] by a modified power measurement by means of an elastic mounting of the nozzle and a rigid tubing for the experimental fluid. The measuring variables are the force F_o at the exit of the jet from the nozzle, and the photographically determined change of the area over the length of the fluid jet taken up by a roll. The elongational viscosity is obtained

$$\eta_T = \frac{P_{11} - P_{22}}{\dot{\varepsilon}} = \frac{T}{\dot{\varepsilon}} ,$$

from F_o and the respective cross section A of the jet at any point of the jet allowing for the influence of gravity force, air drag, inertia force, and net surface tension force.

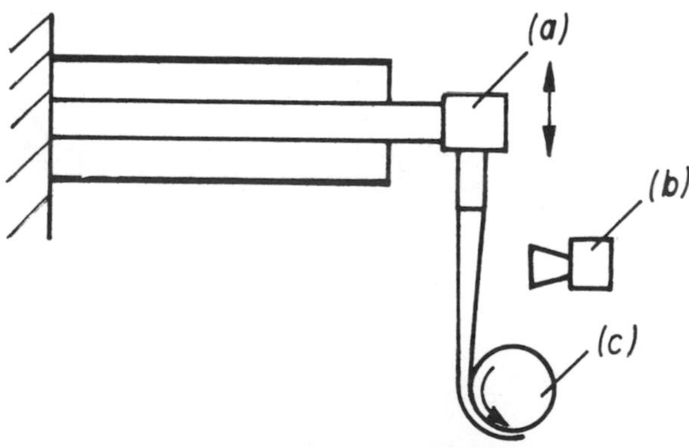

Fig. 2. Spin-balance. (a) elastic mounted nozzle; (b) camera; (c) take-up roll.

The elongational velocity $\dot{\varepsilon}$ is obtained from

$$\dot{\varepsilon} = \frac{dV}{dx} = \frac{dV}{dA}\frac{dA}{dx}$$

and $$v = \frac{Q}{A}$$

to $$\bar{\dot{\varepsilon}} = - v \frac{d\,(\ln A/A_o)}{dx}$$

and can thus be determined for each point x of the jet from the velocity v ($\dot{x}$) and the slope of the logarithmically plotted ratio of areas in dependence on the distance x.

Fig. 3. shows, as a characteristic example, the result of one spinning test, i.e. of one experiment with constant flow rate, take-up velocity, and jet length in the form of different plots of the elongational viscosity in dependence on the elongational stress, and the elongation rate $\dot{\varepsilon}$ (x). These figures clearly show the phenomenon of the increase of elongational viscosity with increasing load, which is typical of high-polymeric solutions. As has already been described in the literature[1-3], it is difficult to give an explicit representation of the results of different jet experiments with different flow rates, take-up velocities, and jet lengths. The authors interprete the differing courses of the curves of elongational viscosity shown in Fig. 4. as a consequence of the instationary flow conditions of the spinning flow. According to these conceptions, the state of stress at any place of the spinning jet is composed of a steady part which depends only on the elongation rate, and an insteady part

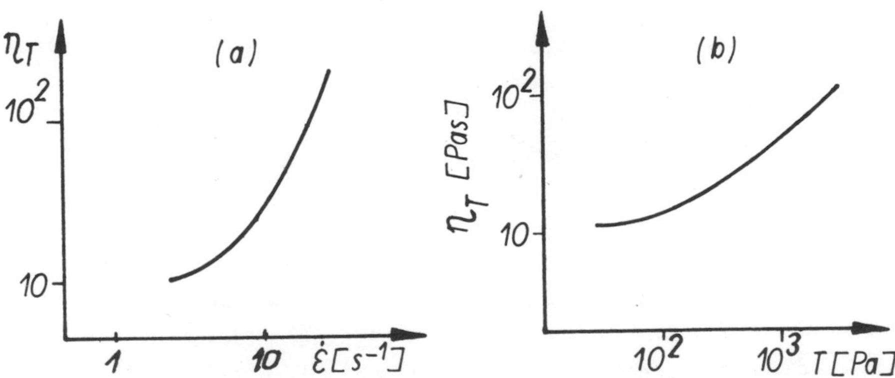

Fig. 3. Result of one spinning test. (a) elongational viscosity in dependence on elongation rate; (b) in dependence on elongational stress

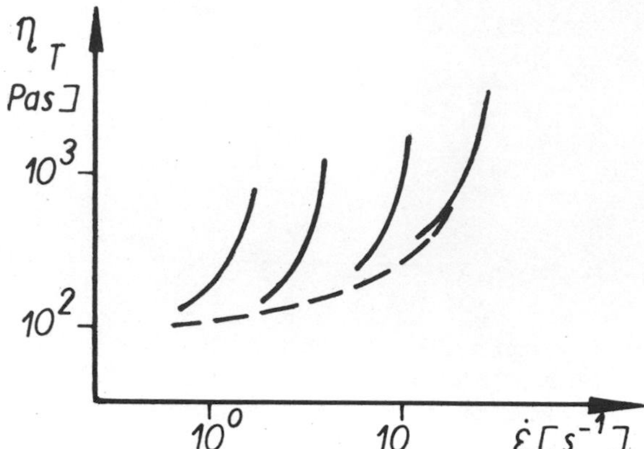

Fig. 4. Differing courses of curves of elongational viscosity (2 % PEO/H_2O-solution).

which can be attributed, as a relaxation effect, to the load changing in the direction of the flow. It is proposed to approximate only one value of the apparent course of the elongational viscosity resulting from one spinning experiment, without influence of relaxation or memory effects, the correction being based on a graphical plot of the elongational viscosities measured over the jet length in dependence upon flow time, and on the approximation of the abtained curves to straight lines with constant slopes. Thus, a region with constant elongational velocity is assumed, which is extended to a time t = 0 as is shown in Fig. 5. At this place, a stress may be supposed to exist which is characteristic of the elongation rate but is not yet influenced by elongation, which rises linearly. Fig. 4. represents the corrected viscosities as a dotted line.

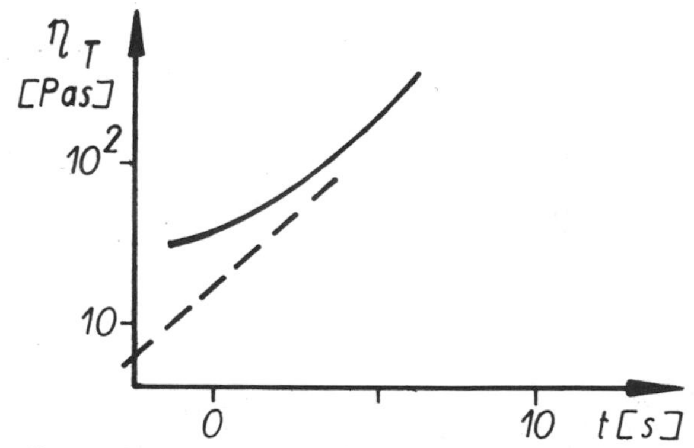

Fig. 5. Correction of elongational viscosity

1. C.I.S. Petrie, "Elongational flows", Pitman (1978)
2. M.Z.A. Zidan, Rheol. Acta 8 (1969) S. 89
3. R.K. Bayer et al., Rheol. Acta 17 (1978) S. 28

DYNAMIC OF FLEXIBLE AND LARGE MACROMOLECULES IN ELONGATIONAL FLOW USING FLOW REFRINGENCE

Robert Cressely and Roger Hocquart

Laboratoire d'Optique Moléculaire
Faculté des Sciences
Ile du Saulcy
57000 METZ (FRANCE)

INTRODUCTION

Several studies of the effect of a transverse shear stress on a macromolecular solution have been made by flow birefringence (F. B.) (e. g. (1)). For consequence of the unceasing rotation motion of the particles in this type of flow, the deformations of the flexible macromolecules remain generally small. But, in elongational flow and in the particular case where there is no rotational part, we can expect, in certain restrictive conditions, very strong molecular deformation such as the transition of the macromolecule from the coiled state to a streched form. Theoritical studies of this transition have been carried out (2-4), but there are only a few experiments comparable with proposed theories. The F.B. is a technique which lends itself perfectly to these studies. Localized flow birefringence (L.F.B.), appearing in experiments as thin luminous lines and showing the extention of the macromolecule in specific parts of the solution have first been observed with a system of opposed jets (5). But, in order to perform quantitative optical measurements, exploitable without ambiguity, we are limited to the use of two dimensional elongational flow. Such flows have already been realized in a system of two (6-9), four (10-11) or six rolls (12) and in the wake of an adequate obstacle in a uniform flow. Taking inspiration from the system of two opposed jets, we have made a simple cell, called "cross cell" (14-15), allowing a careful study of the transition "coil-stretch as well as the related relaxation phenomena. This is the aim of this study. In an other paper presented at this Congress, Lyazid et al (16) report the results of the study of the hydrodynamic field by laser anemometry in a similar cell.

EXPERIMENTAL

The cross cell, which has already been described in (14-15), is represented schematically on figure (1). The width e of the four channels and the thickness p of the liquid layer can be changed by using a set of different lids making thus the geometric of the cell suitable for the experimental conditions (concentration c, viscosity η, temperature, ...). In the central part (D) of this cell, we can postulate an hyperbolic structure of the hydrodynamical field. The velocity may be represented here by v = S(x, -y, O), S standing for the value of the elongational gradient, nearly constant in (D). All the measurements have been realized in laminar flow. The intensity of the L.F.B. has been determined by the classical method of Senarmont suitable for faint birefringence. In order to minimize the disturbing influence of the sharp edges at the four corners, we have built a cell with four fixed cylinders (fig (2)) and we found that the evolution of the phenomenon was the same. Thus we shall report here the results obtained with a "cross cell" where e = 2 mm and p = 15 mm to draw a parallel with the hydrodynamical measurements realized in such a cell. The aqueous solutions of polyethylene oxyde (PEO WSR 301 of $\overline{M} \simeq 4.10^6$), used in these experiments, were freshly prepared without strong mechanical stirring.

RESULTS

In the cell described, it has been possible to observe at concentration c superior to O,1 % but inferior to O,5 % a perfectly steady L.F.B. along the x x' axis of the cell, direction of outgoing flow. A minimum flow rate Q is necessary in order to obtain the L.F.B. line, but unsteady phenomena appears when Q becomes too great (the line is not straight anymore, it gets out of shape and widens out). The width i of the line, for small flows rate, is about 2/100 mm. When Q is increased, the line widens out, but, on the x x' axis, a deeper luminous line remains, the width of which is still of the same magnitude. As the variation of the F.B. on both sides of xx' is very rapid and as the width of this line is small, we found it difficult to draw accurately the cross section of the birefringence Δn. In order to caracterize the width i in terms of the flow rate Q, we have represented on fig (3), the variation of the width i of the birefringente zone up to $\Delta n > 4.10^{-7}$ (for a concentration of O,4 %). The birefringente medium being on the whole oriented along x x', the compensation by Senarmont method for a given Δn, causes two dark symetrical lines to appear on both sides of x x' for this compensation level and it is the distance separating these two dark lines that we have represented in terms of Q. We found out that the evolution is nearly linear.

On fig (4), we have represented the evolution of Δn in the centre O of the cell, against Q for three different concentrations. We were obliged to magnify the field a lot and to use a set of diaphragms in order to obtain satisfactorily accurate measurements.

Fig (5) represents the intensity Δn of the birefringence on x x', for different flow rates, in different points at a distance x from O, in a O,4 % solution.

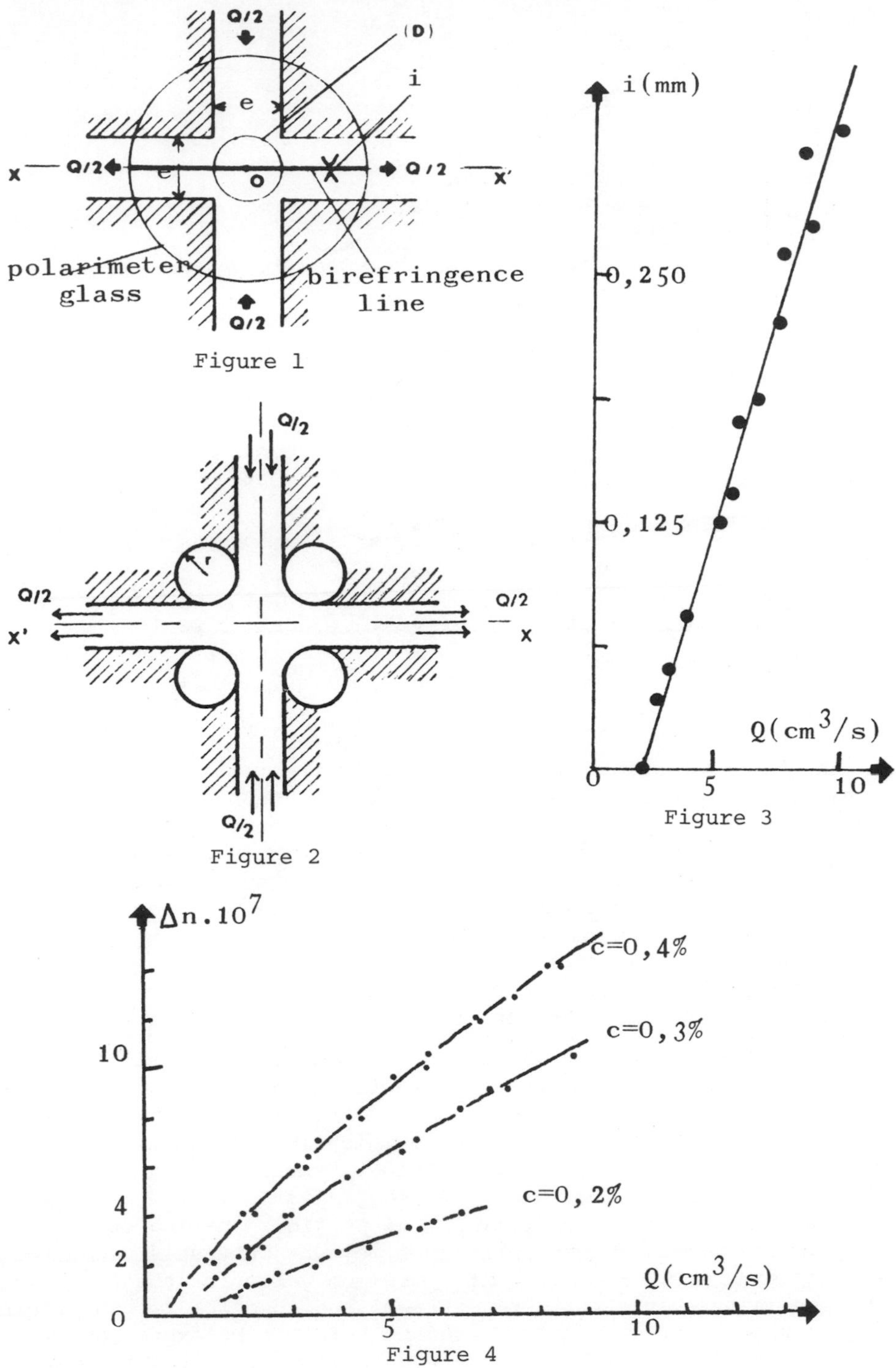

Figure 1

Figure 2

Figure 3

Figure 4

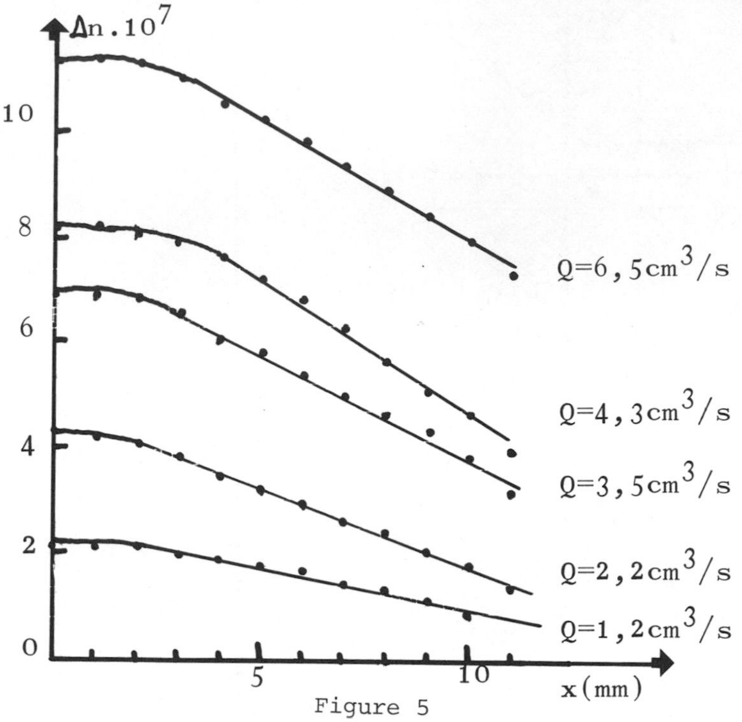

Figure 5

DISCUSSION AND ANALYSIS OF RESULTS

An important point in the analysis of the results we obtained is the really strong localization of the F.B. We notice that the only macromolecules which undergo the transition from the coiled state to the stretched state are the ones mowing near enough to the centre of the cell. This fact shows that a sufficiently strong gradient must act on the particle for a sufficiently long time. Indeed, the closer the macromolecules move to O, the longer they will be in the elongational field. The ones moving along the outmost edges of the birefringence line, that is a distance i/2 from xx', will stay in this gradient S for a time $t_m = \frac{1}{S} \log \frac{e}{i}$ and this is the minimum time the particle has to spend in the field for the extension to happen. From the experiments, we notice that i increases with Q, and consequently with S, and this means that the more important is S, shorter t_m has to be, which is physically quite logical (For monodisperse particles, this t_m should decrease more rapidly that 1/S when S increases). On the xx' axis of the cell, we can assume a time independant flow, the particles having stayed there, theoretically for an infinite time. The curves (4) show that Q and consequently S must present a minimum value for the uncoiling to happen (they show a non zero extrapolated abscissa at the origin).

The experimental results presented here can be explained (at least qualitatively) with the simple dumbell model valid in monodisperse solutions of non interacting particles. The concentrations used are such that our experiments are not performed in diluted solutions, but those realized in diluted solutions of PEO containing only a few tens of ppm (in water + glycerol) show similar phenomena of L.F.B. In addition, the polydispersity in our solutions is large, thus producing supplementary deformations of the curves drawn for monodisperse solutions. In ref (11), where the birefringence induced by a flow similar to the one we are interested in here, V = S(x, -y, O), has been calcualted, approximative forms suitable in the cases of small elongations have been used. This problem will be examined here, taking into account the finite size of the macromolecule undergoing strong deformations, thus allowing us to neglect the brownian motion (the problem is therefore solved from a purely hydrodynamical point of view).

The following approximation will be made :

a) The force Fe of entropic origin between the two ends of the macromolecule $Fe = \frac{kT}{a} L^{-1}(r)$ is taken equal to $Fe = \frac{3kT}{a} \frac{r}{1-r^2}$ where $r = l/L$ represents the ratio of the chain length in a given state to his entirely stretched state length.
$L = Na$ and L^{-1} is the inverse Langevin function.

b) We use a variable friction coefficient ζ from the coiled state to the stretched form (3-4)

$$\zeta = \frac{3kT}{L_o^2} \left| N^{1/2} r + 1 \right| \tau_o$$

where $L_o = N^{1/2} a$ and τ_o represents the first relaxation time of Zimm.

Neglecting the elongations in directions different from x x' we obtain on this axis ($t = \infty$), the relation

$$S\tau_o = \frac{1}{(N^{1/2} x+1)\ (1-x^2)} \quad \text{with the condition } S\tau_o > 1$$

We can then deduce, with our hypothesis the relation giving the intensity of the birefringence Δn on the x x' axis in the region (D) $\Delta n = \Delta n_o x^2$ where Δn_o represents the berefringence for an entirely strech chain. The figure (6) shows $\Delta n/\Delta n_o$ interms of $S' = S\tau_o$ for three types of macromolecules caracterized by N=100, N=1000 and N=10 000. For a given solution (τ_o and N determined), for $S'=S\tau_o<1$ we only get a faint birefringence resulting from small deformations and which will therefore be neglected here. For $S\tau_o=1=S'$ the macromolecule streches (with the corresponding x) and consequently a birefringence Δn_B appears. When S is still increased, the deformations of the paricle increases slowly thus producing a variation of the birefringence caracterized by the line BC. When S'(or S) decreases again, the berefringence also decreases, but now like the curve CBDEO, showing an hysteresis phenomenon. In a monodisperse solution, the theoretical birefringence curves versus the gradient should look like OABC because of the approximation made and should be compared to the experimental curves on fig(4). If we remember that our solutions are polydisperse we can easily unterstand the round shape of the transition AB and the abscissa at the origin, which is different from zero, would correspond to the transition

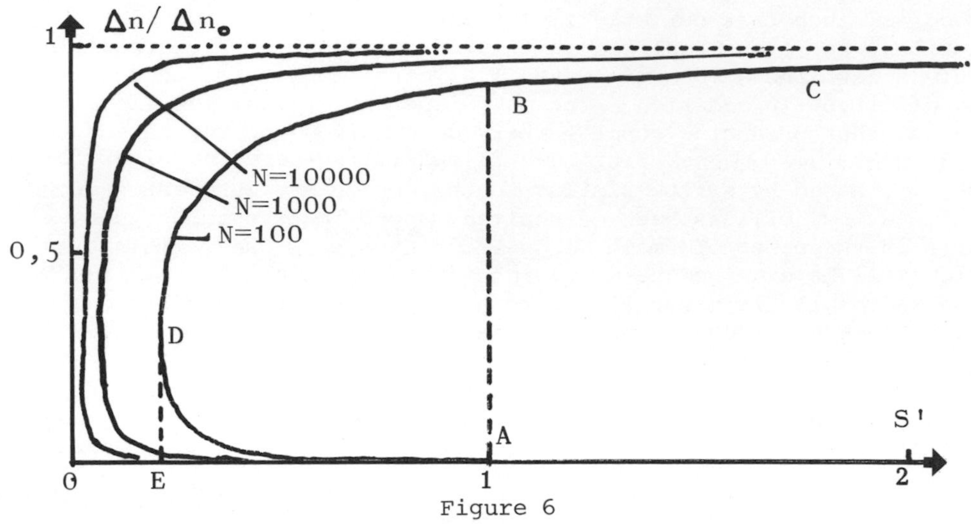

Figure 6

coil-stretch of the largest particles.

We have also performed a theoretical study of the decreasing F.B. along the axis x x' of outflowing in the cell with the same hypothesis. The results will not be presented here for lack of space, but, assuming that immediately after the region (D), S is equal to zero, we get an exponentially decreasing law of the F.B. On the other hand, if we take into account that disturbancs leading to a progressively decreasing gradient exist after (D) in the entry of the outflow channels (hypothesis confirmed by the study of the hydrodynamical field) we obtain with adequate value of τ_o a decrease of the same kind that the one observed in experiments.

CONCLUSION

This study shows the advantage that the F.B. offers for the study of strong deformations of macromolecules in elongational flow. If the simple theories agree qualitatively with the observed phenomena, a more precise comparing will require a) the use of a well definite monodisperse solution, b) the use of concentrations consistent with the theories suitable for diluted solutions (or to adapt these theories to more concentrated solutions). In these conditions we should be able to get, with our "cross cell", important characteristics and informations of the dynamic of large flexible macromolecules in elongational flow. This is the aim of our next study.

(1) V.N. TSVETKOV, "Never Methods of Polymer Caracterisation", Bacond Ke, Ed, New York Interscience (1964)
(2) A. PETERLIN, Pure Appl. Chem. 12, 563 (1966)
(3) P.G. DE GENNES, J. Chem. Phys. 60, 5030 (1974)
(4) E.J. HINCH, Phys. of Fluids, 20, 10, 523 (1977)
(5) M.R. MACKLEY, A. KELLER, Phil.Trans.Roy.Soc.Lond. 278,29(1975)
(6) F.C. FRANCK, M.R. MACKLEY, J. Polym.Sci. 14, 1121 (1976)
(7) R. CRESSELY, R. HOCQUART, O.SCRIVENER,Optica Acta, 25,559 (1978)

(8) C.J. FARREL, A. KELLER, J. Colloïd Polym. Sci.,256, 966 (1978)
(9) R. CRESSELY, R. HOCQUART, O. SCRIVENER, Optica Acta, 26, 1173 (1976)
(10) D.G. CROWLEY, F.C. FRANK, M.R. MACKLEY, R.G. STEPHENSON J. Polym. Sci. 14, 1111 (1976)
(11) D.P. POPE, A. KELLER, J. Colloïd Polym, Sci. 255, 633 (1977)
(12) M.U. BERRY, M.R. MACKLEY, Trans. Roy. Soc. Lond. 287,1,(1977)
(13) R. CRESSELY, R. HOCQUART, Optica Acta, to be published
(14) O. SCRIVENER, C. BERNER, R. CRESSELY, R. HOCQUART, R. SELLIN, N.S. VLACHOS, J. Non Newt. Fluid. Mech., 5, 475 (1979)
(15) R. CRESSELY, R. HOCQUART, A. LYAZID, O. SCRIVENER, I.U.P.A.C. Makro - Mainz, II, 999 (1979)
(16) A. LYAZID, O. SCRIVENER, TEITGEN, VIII Int. Congress of Rheologie, NAPLES (1980)

EXTENSIONAL FLOW OF DILUTE POLYMER SOLUTION*

S. T. J. Peng and R. F. Landel

Jet Propulsion Laboratory
California Institute of Technology
Pasadena, California USA 91103

INTRODUCTION

In the previous paper[1], the detailed experimental set-up and procedures were described for tubeless siphon or Fano flow, from which the tensile viscosity, η_T, is calculated. Here we investigate the simultaneous effects of temperature and concentration on tensile viscosity. A semi-empirical scaling law has now been established to relate the concentration, c, temperature, T, and deformation rate, $\dot{\varepsilon}$, by a reduced deformation rate variable to bring all η_T together. Furthermore, η_T has been compared with the steady and dynamic shear data. The onset of high η_T occurs in the vicinity of the onset of non-Newtonian viscosity observed in a simple shear flow experiment. Finally, the asymtotic solution of Hassager's kinetic theory of a bead-rod model for macromolecular solutions[2] has been compared with experimental data. It indicates that η_T depends strongly on the molecular weight, and the higher the molecular weight, the more sensitive the tensile viscosity in terms of $\dot{\varepsilon}$.

EXPERIMENTAL SET-UP AND DISCUSSION OF RESULTS

A. Experimental

The experimental set-up, experimental procedures and data reduction techniques are essentially the same as reported in the previous paper[1], where a liquid column is formed from the quiesent liquid reservoir by the action of a suction force at the tip of the capillary. Here we refine the experimental set-up by building a large temperature box encasing the whole apparatus

*This paper represents one phase of research carried out by the Jet Propulsion Laboratory, California Institute of Technology, Pasadena, CA, sponsored by NASA under Contract No. NAS7-100.

such that the temperature of solution can be maintained at a fixed value. The temperature range of the experiment is limited to 10^oF to 100^oF. A total of three concentrations and four temperatures were run. The material investigated was a very high-molecular weight hydrocarbon polymer, presumably linear, Conoco AM-1, in JP-8 solvent (military specification of jet aviation fuel). The data reduction scheme is the same as in Ref. (1).

B. Tensile Viscosity

Figures 1, 2 and 3 show for the first time the dependence of tensile viscosities on $\dot{\varepsilon}$, T, and c. The zero shear rate viscosities, η_{so}, at various c and T are indicated in the tables, along with the flow rates of each test condition. The value of n is the exponent of the following equation[1,3]

$$\frac{1}{R^n} = ax + b \qquad (1)$$

which is used to describe the profile of column.

Prior workers have calculated the results in terms of an average η_T or η_T which varies with the time or duration of deformation of a given liquid particle. They assume that $\dot{\varepsilon}$ is constant along the whole liquid column. However, in our experiment it was found that $\dot{\varepsilon}$ varied along the column from the tip to bottom. Then it must be stressed that the η_T reported in Figs. 1, 2 and 3 are <u>transient</u> in nature. The results of our experiments indicate that η_T has a very strong dependence on $\dot{\varepsilon}$, T and c. Also, the onset of higher tensile viscosity (i.e., $> 3\eta_{so}$) depends strongly on T and c. At the lowest c, i.e., c = 0.74%, a stable, symmetrical liquid column could be obtained only at lower T, i.e., 10^oF and 40^oF, but not at higher T, i.e., 70^oF or higher [Fig. 1].

Note that in the Fano flow configuration the fluid is in its quiescent state until the extensional flow began, hence it has the advantage over the spinning configuration used to study extensional flow, since in the spinning, the fluid had experienced certain shear history prior to exit from the orifice. The details of the inlet condition are not clearly defined, however, the extensional flow starts presently at some undetermined depth below the surface. Hence, it is important to investigate what happens when the fluid is presheared before Fano flow began, i.e., to investigate the consequence of "entrance" effects. In order to do that, various blockages of the inlet regime were investigated - plate beneath the surface with various depths and the guard ring of various diameters, but in no case was the liquid column markedly affected. However, if the liquid was through a cylindrical tube of 8 cm dia. and 15 cm height, packed with 2 mm glass bead as shown in the photo (Fig. 4), then η_T is markedly affected by the high preshear. The effect is seen that the onset of high tensile viscosity is shift to the left and the rising of

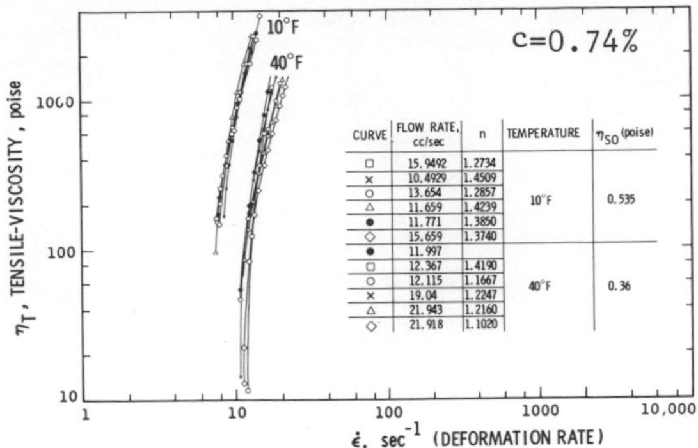

CURVE	FLOW RATE, cc/sec	n	TEMPERATURE	η_{SO} (poise)
□	15.9492	1.2734	10°F	0.535
×	10.4929	1.4509		
○	13.654	1.2857		
△	11.659	1.4239		
●	11.771	1.3850		
◇	15.659	1.3740		
●	11.997		40°F	0.36
□	12.367	1.4190		
○	12.115	1.1667		
×	19.04	1.2247		
△	21.943	1.2160		
◇	21.918	1.1020		

Figure 1 Tensile Viscosity vs Deformation Rate at Concentration c = .744% by Wt of AM-1 in JP-8 Aviation Fuel

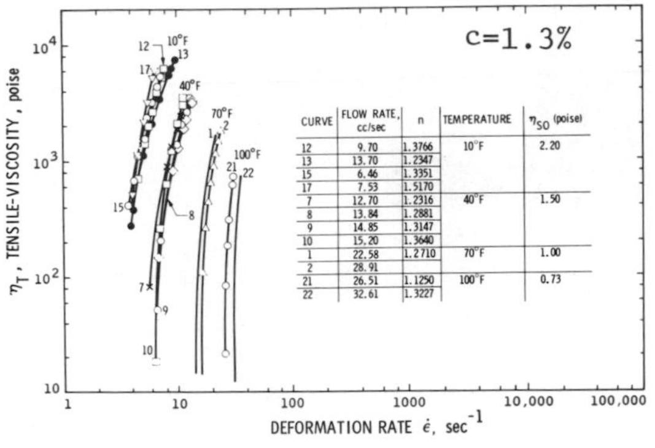

CURVE	FLOW RATE, cc/sec	n	TEMPERATURE	η_{SO} (poise)
12	9.70	1.3766	10°F	2.20
13	13.70	1.2347		
15	6.46	1.3351		
17	7.53	1.5170		
7	12.70	1.2316	40°F	1.50
8	13.84	1.2881		
9	14.85	1.3147		
10	15.20	1.3640		
1	22.58	1.2710	70°F	1.00
2	28.91			
21	26.51	1.1250	100°F	0.73
22	32.61	1.3227		

Figure 2
Tensile Viscosity η_T vs $\dot{\varepsilon}$ at c = 1.3 by Wt of AM-1 in JP-8 Aviation Fuel

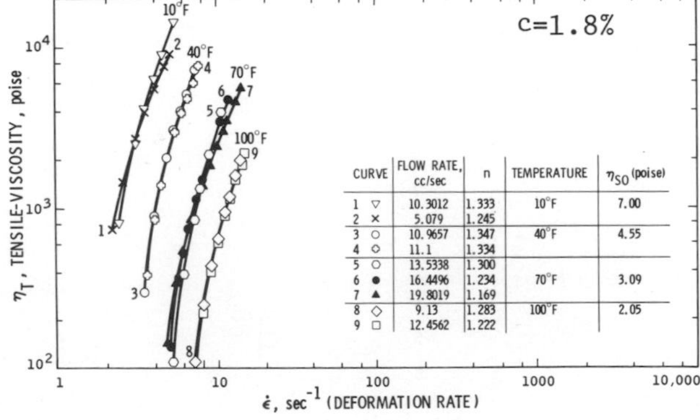

CURVE		FLOW RATE, cc/sec	n	TEMPERATURE	η_{SO} (poise)
1	▽	10.3012	1.333	10°F	7.00
2	×	5.079	1.245		
3	○	10.9657	1.347	40°F	4.55
4	○	11.1	1.334		
5	○	13.5338	1.300	70°F	3.09
6	●	16.4496	1.234		
7	▲	19.8019	1.169		
8	◇	9.13	1.283	100°F	2.05
9	□	12.4562	1.222		

Figure 3
Tensile Viscosity vs Deformation Rate at Concentration c = 1.8% by Wt of AM-1 in JP-8 Aviation Fuel

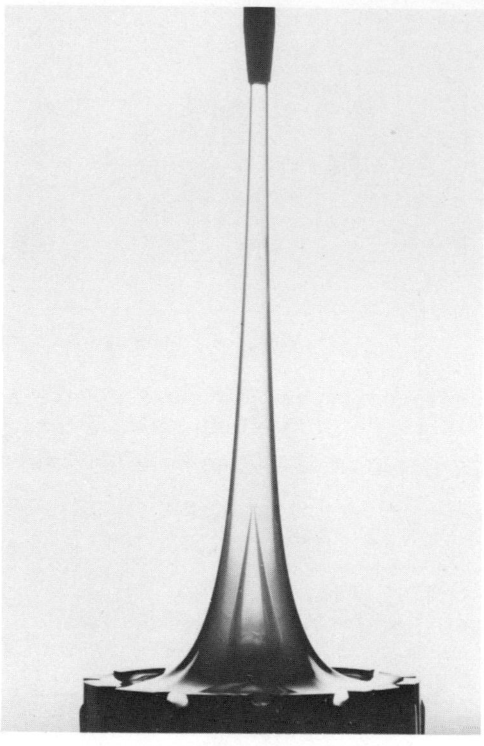

Figure 4 Photo of Tubeless Siphon Liquid Column from Presheared Solution Through Glass Bead Bed

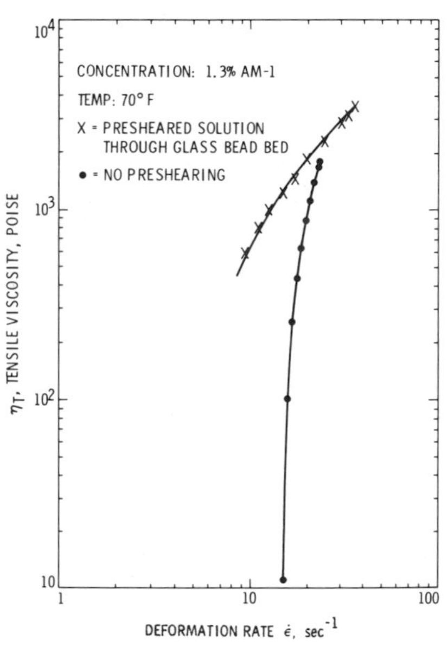

Figure 5 Effect of Preshear on Tensile Viscosity η_T

tensile viscosity is relatively mild and not so sharp as those without preshear (see Fig. 5). Since it is difficult to determine the histroy of preshear from the glass bead bed, a systematic study of the effect of preshear on tensile viscosity was not carried out. Also, it is interesting to note that a stable, symmetric void was formed inside the Fano column at the top of the glass bead bed (Fig. 4). We believe this phenomenon has never been reported.

C. Reduced Variables

From the principle of reduced variables and studies of c and T dependence of the dynamic properties of concentrated polymer solutions, it is known[4] that the dynamic viscosity scales along a 45° line, i.e., that the proper reduced variables are η'/η_{so} and $\omega\eta_{so}$, where η' is the real part of the dynamic viscosity and ω is the circular frequency. The dynamic modulus, or the elastic part of the response, on the other hand, scales principally by a horizontal shift along the time scale due to the change in viscosity, though there is a vertical shift as well, because of the change in c and T. That is, the reduced variables are $g'T./Tc$ and $\omega\eta_{so}T./Tc$ (taking unit c for the reference state).

Although η_T may combine both dissipation and storage of free energy, we believe the latter plays a much more dominant role. That is, in an alongational flow experiment the uncoiling of the molecules is the principal characteristic of the molecular response. Hence, η_T is essentially a manifestation of the elastic energy accumulated in the solution. This suggests by analogy with the dynamic response that a similar concentration reduction scheme should hold.

It turns out, however, that only the time scale is affected by the change of temperature. That is, plots of η_T vs $\dot{\varepsilon}\eta_{so}/c$ superposed to form a master curve. Finally, Figure 7 shows the complete superposition scheme such that the plots of η_T vs $\dot{\varepsilon}\eta_{so}c_oT_o/cT$ superpose all the curves to form a master curve in terms of temperature and concentration.

D. Discussion

It is important to compare the onset of the tensile viscosity and the onset of the non-Newtonian effects observed under both steady shear and dynamic experiments with a cone and plate geometry, as shown in Figure 8, by putting the deformation rate $\dot{\varepsilon}$, shear rate $\dot{\gamma}$ and circular frequency ω together in the same abscissa coordinate. Both G' and N_1 (the first normal stress difference) are approaching their limiting slope of 2 at the point on the abscissa where η' is approaching η_{so}, as should be expected. The onset of high tensile viscosity is in the vicinity of the onset of non-Newtonian shear viscosity, which indicates that the time scales necessary to induce the elastic ef-

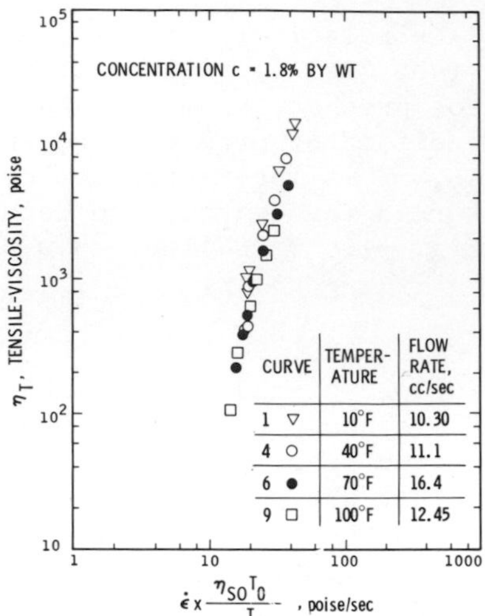

Figure 6 Temperature-Reduced Tensile-Viscosity Curve for AM-1 at Concentration c= 1.8% by Wt

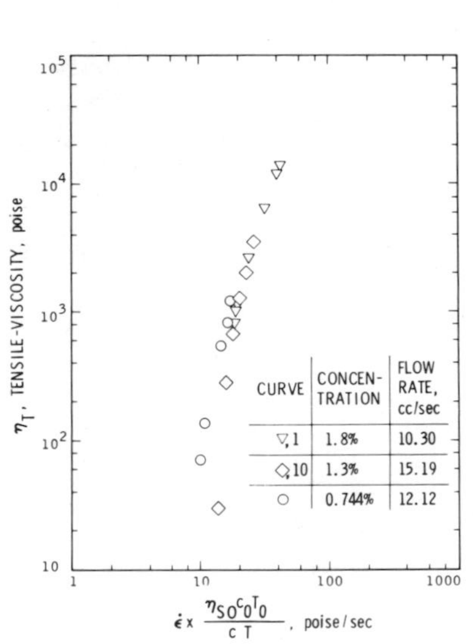

Figure 7 Concentration-Temperature-Reduced Tensile Viscosity Curve for AM-1

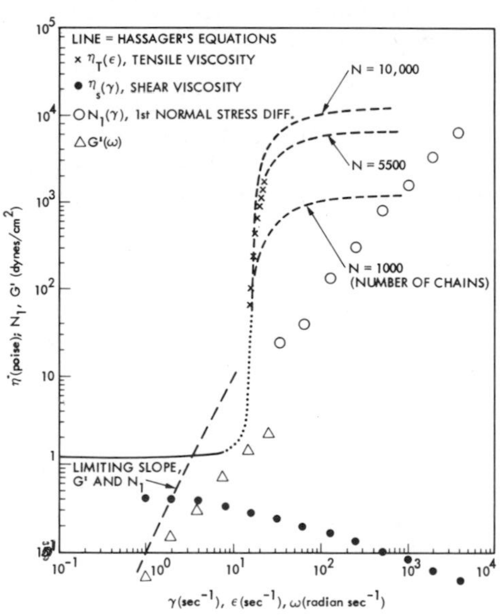

Figure 8 Comparison of Tensile Viscosity, Steady Shear Viscosity η_{so}, 1st Normal Stress Difference, G' (ω) and Hassager Hassager's Equation (AM -1, c= 1.3% by Wt, 70°F)

fect for both types of deformation rate are the same or very close.

Finally, Hassager's kinetic theory for a bead-rod model of macromolecular solutions[2] is considered. Because of mathmatical complexity, he developed expressions for only two regions of deformation rates for an arbitrary number of segments of molecules N_1, i.e., (1) at very slow steady flow regions where the coefficient of expansion up to the secondary order is considered, and (2) at large deformation rate region, where an asymptotic value is obtained. Knowing the requisite parameters for the model, e.g. steady flow viscosity and number of chain segments, the two expressions for the low and high $\dot{\varepsilon}$ regions can be plotted on the same graph. We find that the central region can then be easily interpolated, with results as shown in Fig. 8. Here the solid and dashed lines represent the response as calculated with his two expressions (using three values of N in the second expression), and the dotted lines represent the interpolation. Although his analysis neglects the entanglement effect (intermolecular forces), it indicates a strong dependence of η_T on the $\dot{\varepsilon}$ and the number of chain segments N.

The results can be compared with experiment by a horizontal translation of the theoretical curve. The results match the trend of the date very well, even though the theory is for steady and the experiments are for transient behavior, giving confidence in the strikingly large increase in η_T observed with increasing $\dot{\varepsilon}$.

[Acknowledgment]

Authors would like to express their appreciation to Professor Mike Williams for his kindness to help measure the steady and dynamic shear data.

References

1. S. T. J. Peng and R. F. Landel, J. Appl. Phys. Vol. 47, No. 10, p. 4255, 1976.

2. D. Hassager, J. Chem. Phys., Vol. 60, No. 5, p. 2111, 1974.

3. F. A. Kanel, Ph.D. Thesis (University of Delaware).

4. J. D. Ferry (Viscoelastic Properties of Polymer," p. 209, John Wiley, N.Y. 1961.

EFFECT OF MOLECULAR WEIGHT AND FLOW TYPE ON FLOW BIREFRINGENCE OF DILUTE POLYMER SOLUTIONS

Gerald G. Fuller and L. Gary Leal

Department of Chemical Engineering
California Institute of Technology
Pasadean, California 91125 U.S.A.

INTRODUCTION

Birefringence measurements can be used to study conformation changes in a polymer solution due to the presence of flow. In the dilute concentration range, such studies have been carried out for simple shear flow[1] and uniaxial extensional flow,[2] but only for a single molecular weight in each case and for concentrations above 300 ppm. Neither the effect of flow type nor molecular weight has been adequately investigated, though each may be extremely important in "applications" such as the well-known "drag reduction" phenomenon. The present paper summarizes the results of flow birefringence experiments for 50 and 100 ppm solutions of three different MW samples of polystyrene dissolved in a viscous polychlorinated biphenyl solvent. A range of two-dimensional flow types was generated using a four roll mill.

EXPERIMENTAL

The apparatus used in the experiments has been described in detail elsewhere[3] and consisted of a laser light source which was first passed through a prism polarizer and then passed through a four roll mill which was mounted on a cross slide rotary table. The beam was then sent through a second polarizer which was rotated to extinction with respect to the first. The intensity of the exiting beam was measured with a laser power meter.

The four roll mill consists of four independently rotating cylinders which are situated at the corners of a square. It is designed to allow simulation of the general two-dimensional velocity

$$\underline{v} = \gamma \begin{pmatrix} 0 & \lambda & 0 \\ 1 & 0 & 0 \\ 0 & 0 & 0 \end{pmatrix} \cdot \underline{r} \tag{1}$$

field locally near the center of the device.

Figure 1 shows the manner in which this simulation is achieved. The magnitude of the velocity gradient, specified by γ is determined by the rate of rotation of the fastest pair of rollers (2 and 4). The flow field is a linear superposition of purely extensional flow (λ=1) and purely rotational flow (λ=-1) with simple shear flow corresponding to λ=0. The ratio of vorticity to the rate of strain in the flow is $w = (1-\lambda)/(1+\lambda)$.

Flow visualization studies[3] have demonstrated that the device is capable of accurate simulation of this class of two dimensional flows as long as $|\lambda|>0.1$. The technique of homodyne light scattering spectroscopy was used to directly measure the local velocity gradients at the center stagnation point.[4,5] It was found that the velocity gradient was a linear function of the roller velocity over the entire accessible range of velocity gradients.

The PCB solvent had a viscosity of 454 cp and refractive index of 1.63 at 20° C (all the experiments were run at 20° C). The characteristics of the three polymer samples are listed in table 1.

If the intensity of the incident beam is I_0 and if the principal axes of the refractive index tensor of the deformed polymer solution

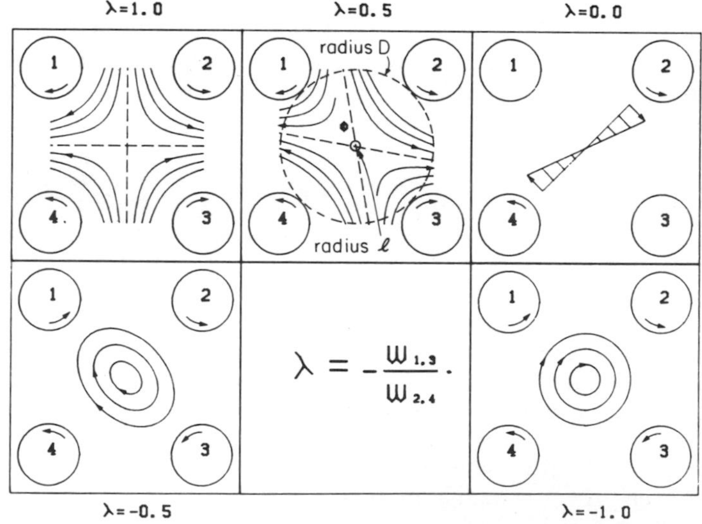

Fig. 1. Simulation of two dimensional flows with the four roll mill.

Table 1. Characterization of the Polystyrene Samples

Sample	Molecular Weight	M_w/M_n	n	$(\mu_1-\mu_2)^6$	N
1	2.00×10^6	1.3	1.6	-145×10^{-25} cm^3	594
2	4.48×10^6	1.14	"	"	1330
3	8.42×10^6	1.17	"	"	2500

are oriented at an angle θ with respect to the polarization vector of the incident light, it can be shown that the light intensity measured at the power meter is $I = I_0 \sin^2\theta \sin^2(\delta/2)$ where $\delta = 2\pi\Delta n d/\lambda$, d is the thickness of the sample path, λ is the wavelength of the light and Δn is the birefringence. The experiment was performed by rotating the flow device in order to deterimine the orientation where $\Theta = \pm 45^\circ$ (at which I was a maximum). At this point the intensity was measured and the birefringence Δn was calculated. It must be pointed out that the PCB solvent also contributed to the birefringence. This contribution was subracted out using the method of Philippoff[1] which assumes that the polarizibility tensors of the solvent and the polymer are additive.

Using the procedure outlined above, the birefringence from the polystyrene in solution was determined for each molecular weight over a wide range of the parameter λ. In all cases, the measurements were taken by passing the laser beam through the center stagnation point of the flow field. The results are shown in figure 2 where the birefringence is plotted against the quantity $\sqrt{\lambda}\gamma$ which is the eigenvalue of the velocity gradient tensor of equation (1). It is evident that the birefringence is very well correlated against $\sqrt{\lambda}\gamma$. Indeed, use of this variable causes the data for all three molecular weights to collapse roughly onto single curves. This result is qualitatively consistent with the theoretical investigations of Tanner[7] and Olbricht et al.[8] which show that the criterion for a flow to be "strong" and therefore to induce significant deformation is $\tau\xi > 1/2$ where τ is the rest state relaxation time of the polymer and ξ is the largest, positive eigenvalue of the velocity gradient tensor of the flow. The quantity $\tau\xi$ in fact controls the degree of deformation (and birefringence) of the polymer for any given steady flow field.

Besides demonstrating the effect of flow type (as characterized by λ), figure 2 also illustrates the role of molecular weight. As the molecular weight increases, the relaxation time of the polymer τ, increases proportionally to M^α where the exponent α is a function of the quality of the solvent and has a value of 1.5 for solvents at theta conditions where the polymer chains obey Gaussian statistics. Therefore, as the molecular weight is increased, the quantity $\tau\xi$ for any given value of the eigenvalue of the velocity gradient tensor will increase and the corresponding degree of deformation and birefringence will be higher.

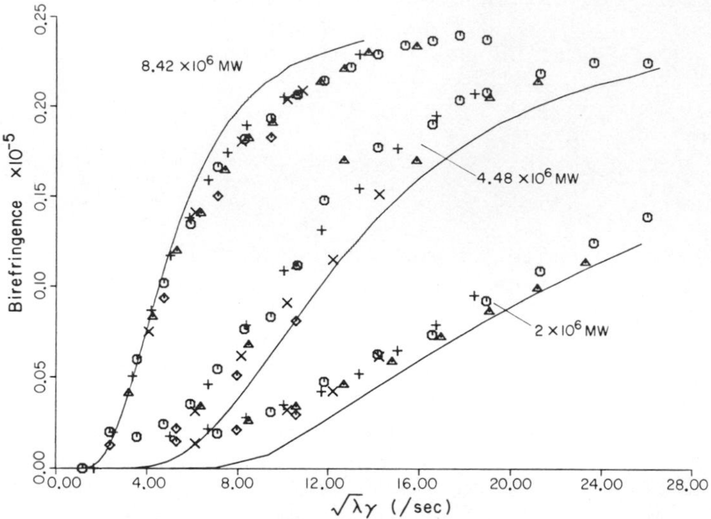

Fig. 2. Birefringence vs $\sqrt{\lambda}\gamma$. ʘ λ=1., $\Delta\lambda$=.8, + λ=.5, x λ=.33, ◇λ=.2 (except for MW=8.42x10^6 where λ=.25). The solid curves were generated using the dumbbell model for λ=1. and ℓ/D=0.003.

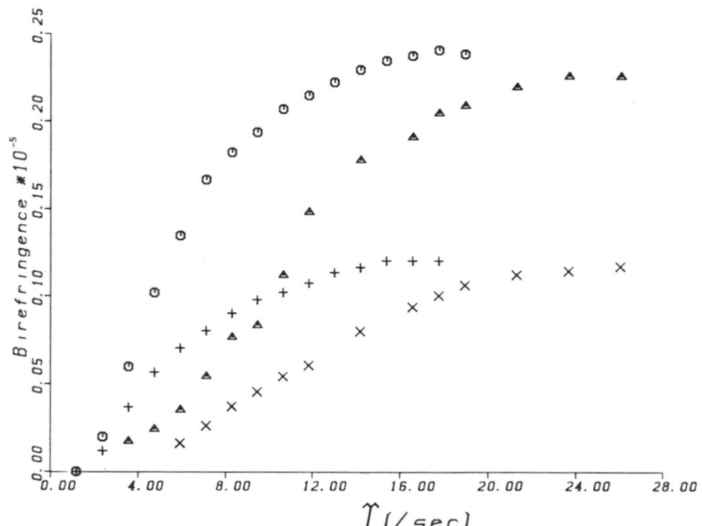

Fig. 3. Birefringence vs. γ. (ʘ,+) MW=8.42x10^6 , (Δ,x) MW=4.48x10^6, (ʘ,Δ) c=100 ppm, (+,x) c=50 ppm.

Figure 3 shows the effect of polymer concentration on the birefringence for 50 and 100 ppm solutions of the 4.48 and 8.42×10^6 MW samples. It can be seen that the birefringence is a linear function of the concentration in this range, which suggests that the polmer molecules are responding in a largely independent manner to the flow.

Another important feature illustrated by figures 2 and 3 is that the quantity ($\Delta n/nc$) approaches an asymptotic value at high velocity gradients which is independent of both the concentration and the molecular weight. This is an important result and is in agreement with the theoretical prediction of Peterlin[9] using the elastic dumbbell model, where it was found that this saturation value should be

$$(\Delta n/nc)_\infty = \{(n+2)^2 \ /3n\}^2 \ (\mu_1-\mu_2)(N_A M)N \qquad (2)$$

where c is the concentration in g/cm^3, N_A is Avogadro's number, M is the molecular weight and ($\mu_1-\mu_2$) is the polarizibility difference between the parallel and normal dirctions along the polymer chain. The parameter N is proportional to the molecular weight and represents the number of statistical subunits which make up the polymer chain. Since the saturation value of $(\Delta n/nc)_\infty$ occurs when the polymer chains are fully extended, an estimate of the parameter N can be obtained by measurement of $(\Delta n/nc)_\infty$. From figure 3, a value of $.015 \pm .001$ was determined for $(\Delta n/nc)_\infty$ which is in the same neighborhood as the value reported by Pope and Keller[2] of .04 who studied 2 million MW polystyrene/xylene solutions of 300-5000 ppm subjected to uniaxial extensional flow. Using the value of $(\Delta n/nc)_\infty = .015$ in equation (2), the parameter N for each molecular weight sample was calculated and the results are tabulated in table 1. If Gaussian statistics are assumed for the polymer chains, the ratio of the size of the chains at maximum extension to the rest state size is $\sqrt{N}$ which would suggest that extension ratios of 50 and 36 were achieved in the experiments for the 8.42 and 4.48×10^6 MW samples respectively.

DISCUSSION

The four roll mill makes it possible to study the effect of flow type on flow induced deformation of polymer molecules and it has been demonstrated that the eigenvalue of the velocity gradient tensor determines the strength of the flow and therefore the degree of polymer deformation. We have also carried out calculations using the dumbbell model in order to analyze the effects of molecular weight distribution and finite transit times in the flow device. A detailed discussion of the model and the methods of solution can be found elsewhere[5]. The dumbbell model incorporated both a conformation dependent friction factor[10,11] and nonlinear entropic spring function (the "Warner" spring[12] was used). The solid lines appearing in figure 2 are the model predictions for the respective values of M_w/M_n and N listed in table 1. The log-normal distribution was used to model the distribution of molecular weights. The effect of finite transit times in the flow field was taken into account by calculating the level of birefringence from dumbbells starting from the rest state on the perimeter of the circle of radius D and entering the sample volume at the point $(\ell/\sqrt{2},\ell/\sqrt{2})$ (see figure 1). Since there is actually a distribution of transit times inside the sample volume, this will provide only a rough approximation of the transit time effect. In order to generate the curves in figure 2 it was also necessary to input the relaxation time τ into the model. For this purpose we used the dynamic light scattering data of Sankur[13] on 4.1×10^6 MW polystyrene in cyclohexane at the theta condition where τ was reported to be 97×10^{-6} seconds. This value was then adjusted for the molecular weight, viscosity and temperature. In adjusting for the molecular weight, we have assumed that $\tau \sim M^{1.5}$. This simple model appears to be able to simulate the experimental data very well and illustrates the importance of accounting for both molecular weight distribution and finite transit times in the flow. For the large values of the parameter N which were used in fitting the data, the precise form of the nonlinear spring constant and the presence of a variable hydrodynamic friction factor were found to not change the qualitative trends of the model predictions. Use of the variable hydrodynamic friction factor did, however, improve the quantitative comparison of the data and the model curves.

REFERENCES

1. Philippoff, W., in: E. H. Lee (ed.), Proc. IV Intern. Congr. Rheol., Providence 1963, Vol. 2, p 343, Interscience Publ. (N. Y. 1965).
2. Pope, D. P. and Keller, A., Colloid & Polym. Sci. 255,633(1978).
3. Fuller, G. G. and Leal, L. G., Flow Birefringence of Concentrated Polymer Solutions in Two-Dimensional Flows, forthcoming publication (1980).
4. Fuller, G. G., Rallison, J. M., Scmidt, R. L. and Leal, L. G., The Measurement of Velocity Gradients by Homodyne Light Scattering Spectroscopy, forthcoming publication (1980).
5. Fuller, G. G. and Leal, L. G., Flow Birefringence of Dilute Polymer Solutions in Two-Dimensional Flows, forthcoming publication (1980).
6. Polymer Handbook, Eds. Brandrup, J. and Immergut, E. H., John Wiley and Sons Publ. (1965).
7. Tanner, R. I., AICHE J. 22,910(1976).
8. Olbricht, W. L., Rallison, J. M. and Leal, L. G., A Criterion for Strong Flow Based on Microstructure Deformation, forthcoming publication (1980).
9. Peterlin, A., Polymer 2,257(1961).
10. Hinch, E. J., Proc. Symp. Polym. Lubrification, Brest (1974).
11. de Gennes, P. G., J. Chem. Phys. 60,5030(1974).
12. Warner, H. R., Ind. Eng. Chem. Fund. 11, 379 (1972).
13. Sankur, V. D., PhD. Thesis, California Institute of Technology (1977).

VISCOELASTIC PROPERTIES OF MIXTURES OF OPTICAL ISOMERS OF POLY-BENZYLGLUTAMATE IN LIQUID CRYSTAL SOLUTION IN TETRAHYDROFURAN

Donald B. DuPré and Dahyabhai L. Patel

Department of Chemistry
University of Louisville
Louisville, Kentucky 40208

Liquid crystal solutions of the polypeptide enantiomers poly-benzyl-L-glutamate (PBLG) and polybenzyl-D-glutamate (PBDG), prepared so as to keep the volume fraction of polymer constant, provide a unique opportunity to examine separately the effect of intermolecular helical twisting power on rheological properties with other molecular factors held constant. The viscoelastic behavior of solutions of the individual polymers[1-4] and the racemic mixture[2,4] have been studied at concentrations both below and above the critical point for liquid crystal formation. One might expect that the viscous component of the liquid crystal is in some way related to the degree of cholesteric twist superposed on the nematic ordering of the essentially rod shaped[5,6] macromolecules. This twist accessory must to some extent be disrupted by the flow field and the impedence may be overcome by either a tilting of the spiral axis or a partial unwinding of the molecular superstructure. The cumulative pitch, P, of the cholesteric liquid crystal is given by[7] $P = 2\pi d/\langle\theta\rangle$ where $\langle\theta\rangle$ is the average angular displacement between adjacent molecular planes separated by a distance d. The reciprocal of P is consequently a measure of the twisting power[8] of the medium. $\langle\theta\rangle$ may be either positive or negative depending on the chirality of the polypeptide and the nature of the solvent.[9] P may be altered in the solutions of this study by the progressive substitution of the D isomer for the L as the two antipodes support opposite liquid crystal twist senses in the same solvent. In fact an equimolar mixture of PBLG and PBDG results in complete compensation of the twist torque and a nematic liquid crystal occurs with an infinite pitch value.[9] As the L to D ratio of the isomers is adjusted to increase P, it is expected that the viscosity would drop as rod orientation in flow can take place more readily without the twist encumbrance.

It is easier for the molecules to flow past one another in the absence of the spontaneous twist. It is for this reason that the viscosity of nematic liquid crystals is generally lower than the isotrope of the same molecule, while cholesterics generally have higher viscosities than the corresponding isotropic fluids. (The viscosity of thermotropic nematic liquid crystals is also Newtonian whereas cholesterics exhibit non-Newtonian effects.[10]) According to this reasoning as the cholesteric twist is reduced the resistance to flow should also decrease in a continuous manner. We have examined this thesis for a series of variable pitch mixtures of PBLG and PBDG in the solvent tetrahydrofuran (THF). We have measured the elastic constant K_{22} for the twist deformation by a nonmechanical means that relies on the monitoring of the field dependence of the pitch as the cholesteric liquid crystal passes into an aligned nematic phase under the influence of a strong magnetic field. We find that K_{22} is essentially constant for all proportions of the polymeric isomers. The elasticity of the medium at least for this distortion thus remains unchanged despite the variation of the overall helical twist in the preparations. Crude bulk viscosity measurements, performed as a function of the fraction of L isomer in PBLG/PBDG liquid crystals, show that the viscosity indeed decreases as the pitch goes up except at the racemic ratio where the viscosity is anomolously large. The latter observation is not explained by any current theory or molecular model.

EXPERIMENTAL SECTION

Samples of PBLG and PBDG both of molecular weight 150,000 were used in this study. Six lyotropic liquid crystals all containing the same total concentration of polymer (24.7% wt/v) in purified tetrahydrofuran were prepared as previously described[11,12] in standard 2 mm pathlength spectrophotometric cells. Examination in the polarization microscope revealed the presence of a cholesteric texture for all solutions except the racemic mixture. A micrometer eyepiece was employed to measure the striation separation which is one half the pitch, P, of the macroscopic twist.[11] Samples were introduced into the field of a 4" electromagnet and the resultant pitch dilation[11] was followed as a function of slowly increasing field strength. Field intensities were determined using a rotating coil fluxmeter and at least 24 hours was allowed between pitch measurements at each field increment to assure equilibration.[13] Flow rates discussed below were obtained from the measurement of the time for the meniscus of a solution to traverse a distance of 2 cm inside the cell at room temperature. Each measurement was repeated at least 10 times and averaged with a reproducibility of ±0.1 sec.

Pitch and Twist Modulus

Table I lists values of the zero field pitch, P_o, as a function of the weight fraction, $\phi_{L/D}$, of L isomer in the PBLG/PBDG liquid crystals. ($\phi_{L/D}$=1.0 indicates a solution of the L isomer only.) It is seen that as D isomer is substituted for L in the liquid crystal the pitch increases due to the compensatory effect of the optical antipodes. The remnants of striations in the racemic mixture ($\phi_{L/D}$ = 0.5) were too large to measure, *i.e.*, were outside the field of view of the microscope. $P \sim \phi_{L/D}^{1.5}$ over the range where accurate pitch measurements could be obtained.

It is now well established[14] that the application of a sufficiently strong electric or magnetic field perpendicular to the spiral axis of a cholesteric liquid crystal will act to unwind the helical superstructure and convert the liquid crystal to an aligned nematic organization. This occurs when the susceptibility of the material is positive which is the case for these polypeptide macromolecules.[15] The process can be followed as a function of field strength by the change in the pitch or striation separation. For these preparations the pitch is large enough to be monitored in the optical microscope.[11,12,15,16] The helical unwinding, which is found to be slowly varying at lower field strengths, proceeds rapidly as a critical field, H_c, is approached. It has been shown[17,18] theoretically that H_c is inversely proportional to P_o and involves the twist modulus through:

$$H_c = (\pi^2/2) \cdot (K_{22}/\Delta\chi_m)^{\frac{1}{2}} \cdot (1/P_o) \qquad (1)$$

where $\Delta\chi_m$ is the diamagnetic susceptibility. The value of the twist modulus can thus be obtained from a measurement of the critical field and P_o using Eqn. (1). The results are listed in Table I where the modulus appears as the ratio $K_{22}/\Delta\chi_m$. The diamagnetic susceptibility of PBLG/PBDG liquid crystals in THF has not been measured. It has been shown[15] however that $\Delta\chi_m$ is the same for left and right handed conformations of this polypeptide and is independent of molecular

Table I: Parameters of PBLG/PBDG Liquid Crystals in THF

$\phi_{L/D}$	P_o (μ)	H_c (kG)	$K_{22}/\Delta\chi_m$ (emu cgs)	K_{22} (dyne)
1.00	29.3	>14	----	----
0.89	39.0	10.7	70.6	45.2×10^{-8}
0.78	47.8	9.20	79.5	50.9×10^{-8}
0.71	58.7	7.40	77.4	49.5×10^{-8}
0.59	85.3	4.75	67.4	43.2×10^{-8}

weight. There is no reason to believe therefore that $\Delta\chi_m$ would change in the preparations studied here. If we take the value of $\Delta\chi_m = 0.64 \pm 0.07 \times 10^{-8}$ emu/cc measured[12,15] in a similar solvent (dioxane) we may convert the ratio to our best estimate of K_{22} in this solvent. It is seen that either the ratio $K_{22}/\Delta\chi_m$ or K_{22} so calculated is constant, indicating that the elasticity of the medium is invariant to the isomer composition. Theoretical calculations[19] for K_{22} based on the elasticity of a suspension of long hard rods gives $K_{22} = 14.0 \times 10^{-8}$ dyne for the racemic mixture of PBLG and PBDG. This result is in good agreement with the experimental values of Table I.

Flow Properties

Our crude measurements of the flow rate of these liquid crystals demonstrates substantial changes in the bulk viscosity that are a result of changes in the polypeptide composition. As shown in Figure 1 progressive substitution of the D isomer for the L is accompanied by a decrease in viscosity, presumably a result of the gradual elimination of the twist encumbrance to ordering of the macromolecules in flow. At weight fractions above 0.50 a linear decrease is found as $\phi_{L/D}$ drops. A linear correlation of the pitch (inverse twisting power) with flow rate is also found and is presented in Figure 2. These observations are in accord with the thesis of macromolecular reorientation outlined above. An anomoly, not explainable on the basis of this model of the lyotropic phase, occurs however at the racemic condition ($\phi_{L/D} = 0.50$). At this point we observe

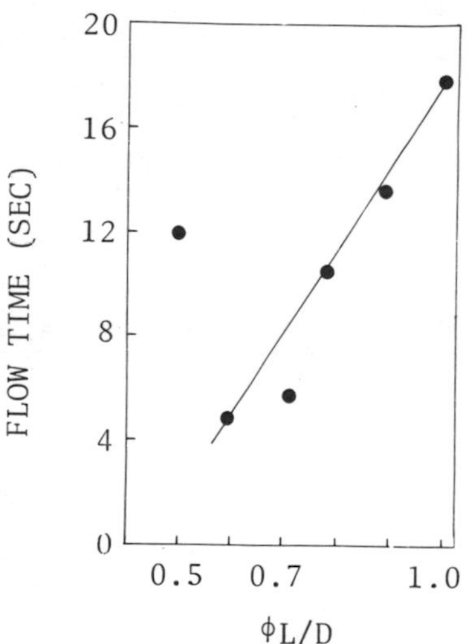

Fig. 1 Flow time in seconds versus $\phi_{L/D}$ for PBLG/PBDG liquid crystals. An anomoly appears at the racemic point $\phi_{L/D} = 0.50$ in the otherwise linear decrease of viscosity.

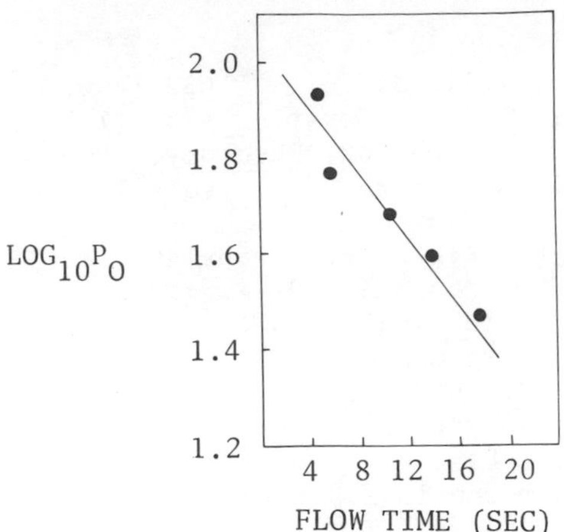

Fig. 2 Correlation of the $\log_{10}P_O$ with flow time for the preparation of Fig. 1.

an increase in the viscosity that was otherwise in decline. In fact the viscosity of the racemic mixture where the twist hindrance should be the minimum approaches that of the solution of the pure L isomer. Such a rheological peculiarity for the racemic mixtures of polypeptides in the liquid crystal condition has been noted previously. Kiss and Porter[4] found that, contrary to expectation, the viscosity of their racemic mixtures of polybenzylglutamate in m-cresol was very similar to that reported earlier by Hermans[1] who studied only the L isomer in the same solvent. Iizuki[2] has also compared polyethyl-L-glutamate (PELG) with the equimolar mixture of PELG and PEDG in CH_2Br_2 and finds similar viscosity behavior. Both these studies were carried out over a range of shear rates.

We also performed a separate flow rate experiment on a liquid crystalline solution of PBLG (in another solvent, pyridene) held both below and above the critical magnetic field strength for the cholesteric to nematic transition. The flow time of the same liquid crystal was found to average 39±1 seconds for the 2 cm travel in the cholesteric phase and 34±1 seconds in the nematic condition. The above suggests that the rheological oddity is with the racemic mixture and not the nematic phase otherwise obtained. At this time a theory of the rheology of liquid crystalline solutions that includes more than just the heuristic elements of the sort discussed in this introduction is apparently lacking. Whatever rigorous theory eventually emerges must include this anomoly of the racemate in its exposition.

Further details of this experiment are found in reference (20).

References

1. J. Hermans, Jr., J. Colloid. Sci. 17 638 (1962).
2. E. Iizuka, Mol. Cryst. Liq. Cryst. 25 287 (1974).
3. R. S. Porter and J. F. Johnson, Rheology, Theory and Applications, F. R. Eirich (ed.), Vol. IV, Academic Press, NY, pp. 317-345.
4. G. Kiss and R. S. Porter, Polym. Preprints 18 185 (1977); J. Poly. Sci. Symp. 65 193 (1978).
5. M. F. Perutz, Nature 167 1053 (1951).
6. L. Pauling, R. B. Corey and H. R. Branson, Proc. Natl. Acad. Sci. U.S. 37 205 (1951).
7. P. N. Keating, Mol. Cryst. Liq. Cryst. 8 315 (1969).
8. H. Baessler and M. Labes, J. Chem. Phys. 52 631 (1970).
9. C. Robinson, Tetrahedron 13 219 (1961).
10. R. S. Porter, E. M. Barrall II, J. F. Johnson, J. Chem. Phys. 45 1452 (1966).
11. R. W. Duke and D. B. DuPré, J. Chem. Phys. 60 2759 (1974).
12. D. B. DuPré and R. W. Duke, J. Chem. Phys. 63 143 (1975).
13. R. W. Duke and D. B. DuPré, Macromolecules 7 374 (1974).
14. P. G. deGennes, The Physics of Liquid Crystals, Oxford University Press, London, 1974, Ch. 6.
15. C. Guha-Sridhar, W. A. Hines, and E. T. Samulski, J. Chem. Phys. 61 947 (1974).
16. D. B. DuPré, R. W. Duke, W. A. Hines and E. T. Samulski, Mol. Cryst. Liq. Cryst. 40 247 (1977).
17. R. B. Meyer, Appl. Phys. Lett. 14 208 (1968).
18. P. G. deGennes, Solid State Comm. 6 163 (1968).
19. J. P. Straley, Phys. Rev. A8 2181 (1973).
20. D. L. Patel and D. B. DuPré, Rheol. Acta 18 662 (1979).

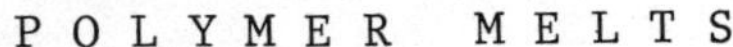

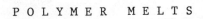

POLYMER MELTS

ELONGATIONAL VISCOSITY OF POLYETHYLENE MELTS

G. Attalla, G. Corrieri° and D. Romanini°

Istituto Guido Donegani S.p.A., Montedison Group
28100 Novara, Italy
°Montedison S.p.A., Plastics Division, Research Centre
44100 Ferrara, Italy

INTRODUCTION

The measurement of elongational viscosity of polymer melts was extensively carried out by several authors (1-6) using different techniques, as reviewed by Cogswell (7) and Hill & Cuculo (8). Han (cf. (9) p. 205) has collected results obtained by several authors on different polymers using various techniques; the data reported appear to be discordant in some cases.
A general theoretical model for the interpretation of the elongational behaviour has not yet been found. That is mainly due to the fact that the molecular mechanism in extensional flow is not clearly understood.
Object of this work is the measurement of the elongational viscosity of some polyethylene samples, both high-and low-density, whose molecular and shear viscosity characteristics are known from previous work (10-12). The aim is to evaluate the influence of molecular weight and long chain branching (LCB) on the elongational viscosity.

EXPERIMENTAL

Measurements were carried out on some polyethylene samples, obtained with different processes, using the method of isothermal spinning (e.g. cf. (2)).
The molecular characteristics of the samples tested are shown in Tab. 1. Samples 1-4 are low-density polyethylenes produced by high pressure processes (10,11), samples 5, 6 are linear low-density polyethylenes and samples 7,8 are high-density polyethylenes.
The experimental procedure was as following: the polymer pellets were fed into a 15 mm diameter single-screw extruder, where plastification occurred; the melt was forced through a cylindrical die

Table 1 - Molecular characteristics of the samples investigated

Nr.	Sample	Density [a] g cm^{-3}	MFR [b] g/10'	$[\eta]$ [c] dl g^{-1}	Mw$\cdot 10^{-3}$ [d]	β [e]	g_B [f]
1	T-1	0.9176	2.14	0.88	280	5.4	0.29
2	TM-2	0.9234	2.13	0.83	135	4.1	0.45
3	V-3	0.9156	6.58	0.96	420	9.6	0.25
4	VM-2	0.9204	1.69	0.89	185	5.5	0.40
5	G-1	0.9310	2.50	1.34	105	4.4	0.85
6	G-2	0.9260	0.60	1.78	160	4.0	0.86
7	A-1	0.9505	1.76	1.26	180	4.6	0.56
8	A-2	0.9539	1.46	1.19	220	5.9	0.47

a ASTM D 1505/63T, B
b ASTM D 1238/73, B
c Orthodichlorobenzene, 135°C
d Light scattering, α-chloronaphthalene, 135°C
e Dimensional polydispersity from G.P.C.
f Ratio between $[\eta]$ and intrisic viscosity of a linear polyethylene having same molecular weight and polydispersity

(2 mm diameter) by a gear pump, mounted on the head of the extruder, that guaranteed a constant flow rate. The extrudate passed through an isothermal chamber, kept at the same temperature of the melt (180°C), and was drawn by the rollers of the Rheotens apparatus (manufacted by Göttfert). Experiments were carried out at constant flow rate (2 g/min) and different draw ratios (3.6; 7.2; 14.4).
The profile of thread diameter was measured photographically and the tensile force was recorded on a strip-chart y-t recorder. The photos were taken when the tensile force F had reached a constant value.
The elongational viscosity η_E was calculated from the well known equations (cf. (9), chapt. 8):

$$v(z)=\frac{W}{\rho A(z)}; \quad \sigma(z)=\frac{F_T}{A(z)}; \quad \dot{\varepsilon}(z)=\frac{dv}{dz}; \quad \eta_E=\frac{\sigma(z)}{\dot{\varepsilon}(z)} \qquad [1]$$

where v is the thread velocity in z-direction, W is the extrudate flow rate, ρ is the extrudate density, A is the cross sectional area of the thread, σ is the tensile stress, F_T is the total tensile force needed for the deformation of the thread, $\dot{\varepsilon}$ is the elongational rate and z is the distance from die along the thread.
In the calculations F_T has been considered equal to F; this is justified by the short distance between die and rollers (about 30 cm.).

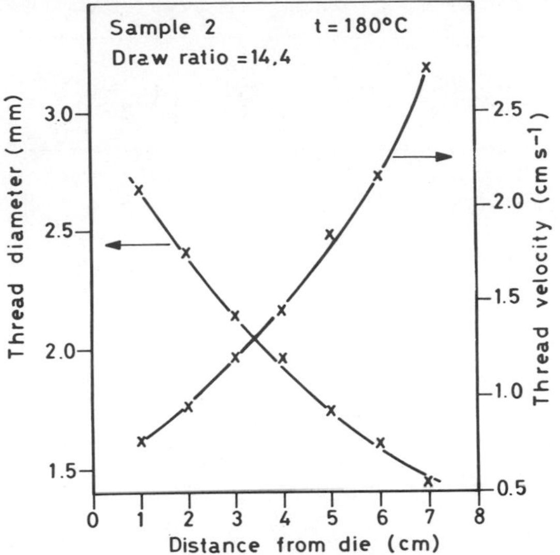

Fig. 1 - Profiles of thread diameter and thread velocity for sample 2

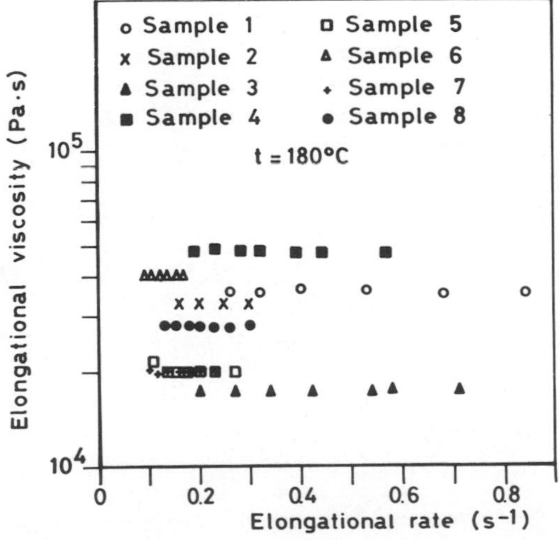

Fig. 2 - Elongational viscosity versus elongational rate for all the samples

ANALYSIS OF RESULTS

The typical extensional behaviour of the samples investigated is shown in Fig.1, where the profiles of thread diameter and thread velocity for sample 2 are plotted versus the distance from die. If one plots the thread velocity on logarithmic scale versus the distance from die on linear scale, one finds a straight line for all the samples. That means that the thread velocity is an exponential function of the spinning way:

$$v = v_0 \exp(kz) \qquad [2]$$

where v_0 and k are two constants. In this case one easily obtains from eqs. [1] that the elongational rate is given by:

$$\dot{\varepsilon} = kv(z) \qquad [3]$$

and that the elongational viscosity is independent of $\dot{\varepsilon}$. The values of elongational viscosities of all the samples investigated versus the elongational rate are shown in Fig. 2.

In order to determine the influence of LCB on elongational viscosity we have chosen as rheological index giving the LCB degree of each sample, the ratio $(\eta_0/\eta_{0\ell})^{\frac{1}{2.84}}$ where η_0 is its zero-shear viscosity and $\eta_{0\ell}$ is the zero-shear viscosity of a linear sample having the same βMw product (11).

The values of the rheological LCB indexes for all the samples are shown in Tab.2.

Table 2 - Rheological characteristics of the samples investigated

Nr.	Sample	η_0 [a] Pa s	η_E Pa s	$\eta_E/3\eta_0$	MS [b] g	BSR [c]	$(\frac{\eta_0}{\eta_\ell})^{\frac{1}{2.84}}$
1	T-1	13,000	36,000	0.9	2.4	70	0.30
2	TM-2	12,000	33,000	0.9	1.7	253	0.80
3	V-3	4,000	17,500	1.4	2.3	63	0.07
4	VM-2	16,000	48,000	1.0	3.5	119	0.48
5	G-1	6,000	20,000	1.1	0.9	759	0.75
6	G-2	26,000	40,000	0.5	2.2	305	0.90
7	A-1	8,500	20,000	0.8	0.8	332	0.47
8	A-2	17,000	28,000	0.5	1.2	419	0.38

a From Weissenberg Rheogoniometer
b,c Melt strength and breakage stretch ratio from Melt Tension Tester

It should be noted that for samples 2 and 3 they differ sensibly from those obtained from solution measurements (g_B), shown in Tab.1. That is probably due to the fact that these samples have types of branching affecting viscosities of diluite solutions in a different way from viscosities of melts, where interactions between molecules are determinant.

$(\eta_0/\eta_{0\ell})^{\frac{1}{2.84}}$ is the best rheological LCB index we were able to find to correlate the elongational viscosity with the molecular weight. The correlation is shown in Fig. 3 where an approximately linear relationship appears to hold between the elongational viscosity and the molecular parameter $(\eta_0/\eta_{0\ell})^{\frac{1}{2.84}}\,\beta$ Mw.

If the elongational viscosity is plotted versus the product β Mw (Fig. 4), samples having approximately the same branching degrees appear to be on straight lines, parallel to that drawn in Fig. 3. That means that LCB lowers the elongational viscosity for the same β Mw product.

The influence of LCB on elongational viscosity is the same as on shear viscosity, as reported in (11).

In Fig. 5 a plot is shown of elongational viscosity, determined with the method of isothermal spinning, versus melt strength (MS), measured in non isothermal condition with the Melt Tension Tester apparatus, described in (11). A linear relationship appears to hold for all the samples but one. That seems to show that measurement of melt strength, technically very quick to carry out, can be reasonably utilized to test the extensional behaviour of polyethylene melts.

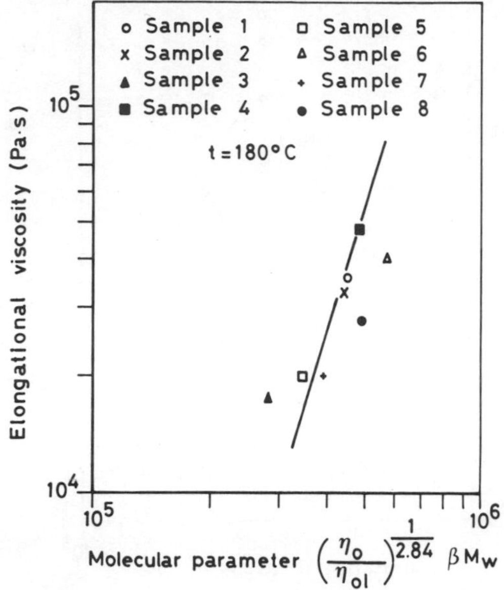

Fig. 3 - Elongational viscosity versus $(\eta_o/\eta_{ol})^{\frac{1}{2.84}}$ β Mw product for all the samples

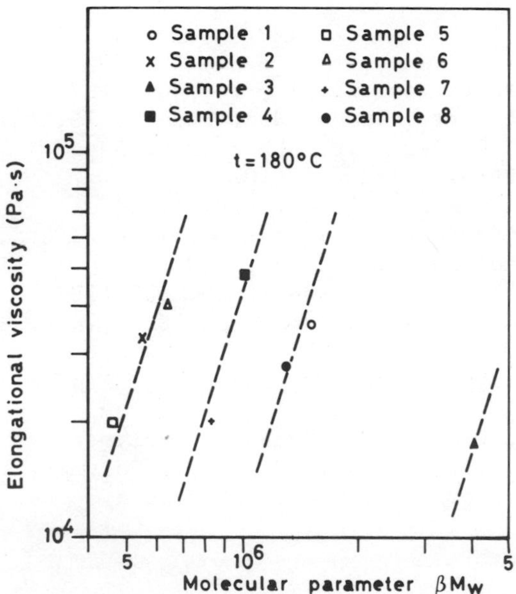

Fig. 4 - Elongational viscosity versus β Mw product for all the samples

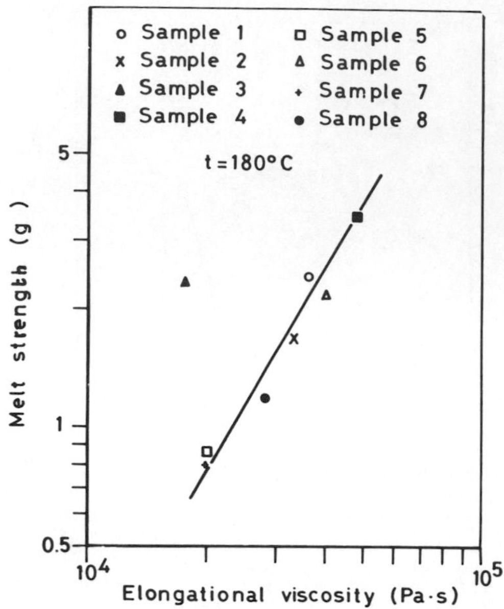

Fig. 5 - Melt strength versus elongational viscosity for all the samples

REFERENCES

1) J. Meissner, Trans. Soc. Rheol., 16, 405 (1972).
2) C.D. Han & R. Lamonte, Trans. Soc. Rheol., 16, 447 (1972).
3) I-J. Chen, G.E. Hagler, L.E. Abbot, D.C. Bogue & J.L. White, Trans. Soc. Rheol., 16, 473 (1972).
4) H. Münstedt, J. Rheol., 23, 421 (1979).
5) R.K. Bayer, H. Schreiner & W. Ruland, Rheol. Acta, 17, 28 (1978).
6) R.K. Bayer, Rheol. Acta, 18, 25 (1979).
7) F.N. Cogswell, Trans. Soc. Rheol. 16, 383 (1972).
8) J.W. Hill & J.A. Cuculo, J. Macromol. Sci., Rev. Macromol. Chem., C14, 107 (1976).
9) C.D. Han, Rheology in Polymer Processing, Academic Press, New York, 1976.
10) G. Gianotti, A. Cicuta & D. Romanini, Polymer, in press.
11) D. Romanini, A. Savadori & G. Gianotti, Polymer, in press.
12) G. Attalla & D. Romanini, unpublished data.

THE ELONGATIONAL BEHAVIOUR OF VARIOUS POLYMER MELTS

Helmut Münstedt

Mess- und Prüflaboratorium
BASF Aktiengesellschaft
D-6700 Ludwigshafen/Rhein, Germany

INTRODUCTION

It has been shown for a low density polyethylene [1,2,3,4] and some polystyrenes[5,6] that the elongational behaviour of their melts is remarkably different from the shear properties in the nonlinear range of deformation. Having a reliable experimental method for measuring the elongational behaviour of polymer melts in hand now[5] which uses only a small amount of material, it seems challenging to investigate the influence of the molecular structure on the elongational properties in order to obtain a more thorough understanding of the flow behaviour of polymer melts. From experiments in shear it is well-known how decisively the flow behaviour of a polymer melt is influenced by the molecular weight and the molecular weight distribution.

ELONGATIONAL BEHAVIOUR OF LOW DENSITY POLYETHYLENES AND POLYSTYRENES

Fig. 1 represents the stress-strain curves of three low density polyethylenes of a similar weight average molecular weight M_w, but different molecular weight distributions which are given in Fig. 2*. The sample with the high molecular weight tail shows a

*Some data on the rheological behaviour and the film blowing performance of these three samples are given elsewhere[7].

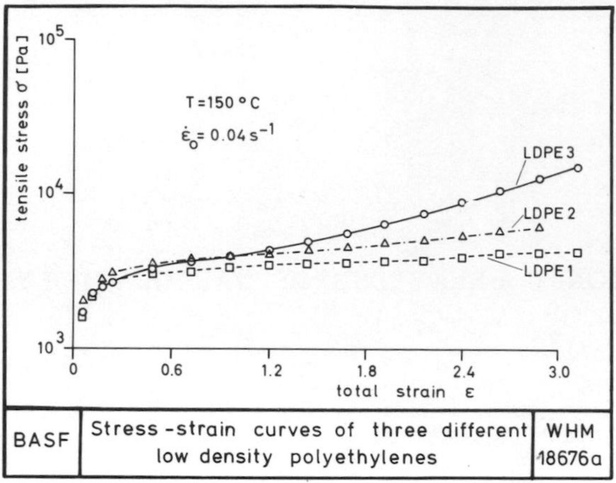

Stress-strain curves of three different low density polyethylenes

Fig. 1

pronounced strain-hardening behaviour. Whereas in the constant strain-rate experiment a steady state of flow is not reached at the highest applied total strain, in a creep test, i.e. in an experiment at constant tensile stress, the strain rate and the recoverable strain attain their equilibrium values (cf. Fig. 3). These results confirm the findings on a LDPE[2] that the steady state is reached at smaller total strain in a tensile creep than in a constant strain-rate test.

Similar to LDPE, a strong influence of a high molecular weight component on the stress-strain curve is found for polystyrene, too, as Fig. 4 clearly demonstrates. More detailed measurements of the elongational properties of polystyrene are given elsewhere[6].

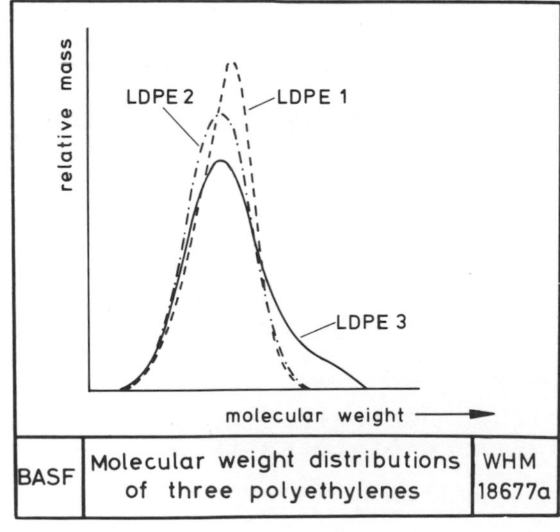

Molecular weight distributions of three polyethylenes

Fig. 2

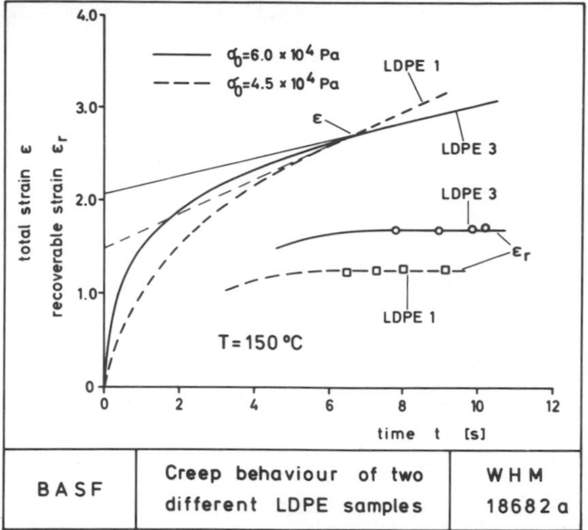

Fig. 3

From the equilibrium values a steady-state elongational viscosity defined as $\mu_s = \sigma_0/\dot{\varepsilon}_0$ can be calculated. This quantity is plotted as a function of tensile stress in Fig. 5. The viscosity maximum which has been found for the LDPE IUPAC A[3] occurs for the three other LDPE-samples, too, and comes out to be the higher the more pronounced the high molecular weight tail is.

The narrow-distributed polystyrene PS 1 shows only a weak maximum of the elongation viscosity, whereas for PS 2 with the distinct high molecular weight

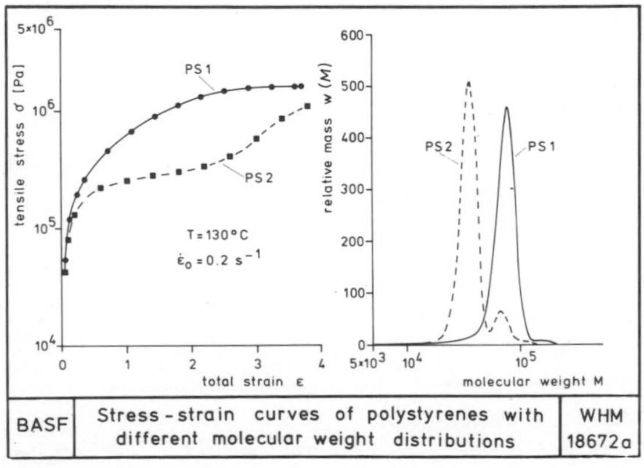

Fig. 4

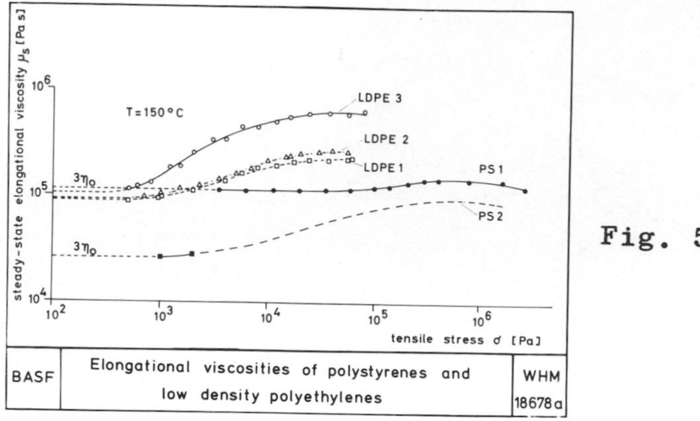

Fig. 5

component, μ_s runs through a pronounced maximum*.

Another quantity which can be determined by stretching experiments is the steady-state recoverable strain $\epsilon_{r,s}$. In this quantity the molecular weight distribution is reflected, too, as becomes obvious from Fig. 6. PS 1 with the narrowest distribution shows the lowest recoverable strain, the high molecular weight component of PS 2 increases $\epsilon_{r,s}$ remarkably. Qualitatively the same effect is found for the three LDPE samples in Fig. 6. Whereas at low stresses the recoverable strains of the various samples differ

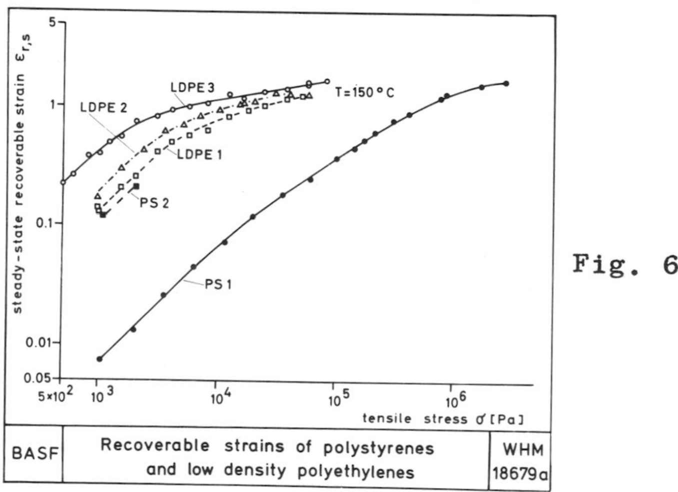

Fig. 6

*In the case of PS 2 a steady state of elongational flow could only be attained at low tensile stresses. The steady-state elongational viscosities at larger stresses are still higher than the values represented by the broken line.

by more than one decade, at higher stresses they seem to approach a very similar value which lies at around $\epsilon_{r,s} = 2$.

ELONGATIONAL BEHAVIOUR OF A LOW DENSITY POLYETHYLENE FILLED WITH GLASS BEADS

The maximum which seems to be the most significant feature of the steady-state elongational viscosity can be influenced not only by a change of the molecular weight distribution but by the addition of fillers, too. Fig. 7 demonstrates that the viscosity maximum becomes the smaller the higher the filler content is. This effect leads to the somewhat peculiar feature that in the maximum region the elongational viscosity of the filled system becomes smaller than the viscosity of the matrix material although $3\,\eta_o$ (η_o is the zero shear viscosity) steadily increases with the concentration of glass beads.

DISCUSSION

The occurrence of a maximum in the steady-state elongational viscosity is not a function of branching alone as can be concluded from a comparison of the results on the linear polystyrenes and the branched polyethylenes. The high molecular weight components show up to be the decisive factor which determines the height of the maximum. For low density polyethylenes and polystyrenes as well, a high molecular weight component leads to an increase of the maximum of the steady-state elongational viscosity and a growth of

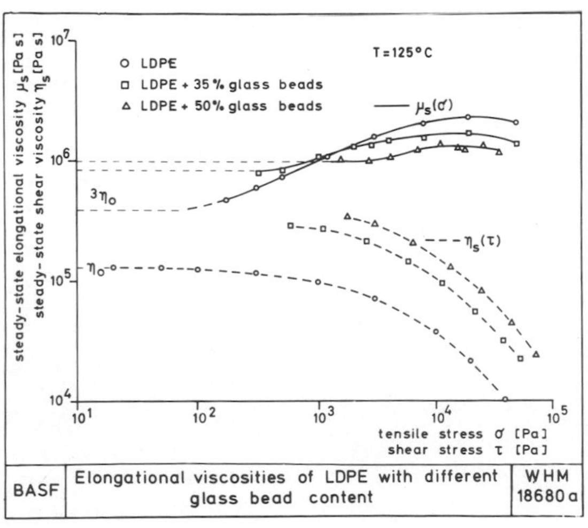

Elongational viscosities of LDPE with different glass bead content

Fig. 7

the recoverable strain. The weight average molecular weight does not influence the shape of the viscosity curve but only its level according to the molecular weight dependence of the zero-shear viscosity.

A different molecular weight distribution may explain the findings of various authors who have published controversial curves for the elongational viscosities of samples of one class of polymers like polystyrene or low density polyethylene, for example[8,9].

The decrease of the viscosity maximum with an increase of the glass bead content may be due to a prevailing shear deformation between the fillers during the elongation of the sample[10].

The strain-hardening behaviour has a distinct effect on the homogeneity of the sample deformation. All the specimens with a pronounced strain-hardening range show a better homogeneity than those where such a behaviour is not found. This fact can easily be understood by a kind of self-healing mechanism following from the existence of strain hardening, i.e. a locally occurring higher strain results in an increase of the tensile stress necessary for elongation.

REFERENCES

1. J. Meissner, J. Appl. Polym. Sci. 16, 2877 (1972)
2. H. M. Laun, H. Münstedt, Rheol. Acta 15, 517 (1976)
3. H. M. Laun, H. Münstedt, Rheol. Acta 17, 415 (1978)
4. H. Münstedt, H. M. Laun, Rheol. Acta 18, 492 (1979)
5. H. Münstedt, J. Rheol. 23(4), 421 (1979)
6. H. Münstedt, J. Rheol., in press
7. R. N. Shroff, C. W. Macosko, H. Münstedt, Paper given at the Golden Jubilee Meeting of the American Society of Rheology, Boston, 1979
8. F. N. Cogswell, Appl. Polym Symp. 27, 1 (1975)
9. J. L. White, Y. Ide, Appl. Polym. Symp. 27, 61 (1975)
10. J. D. Goddard, personal communication

STRESSES AND RECOVERABLE STRAINS OF STRETCHED POLYMER MELTS AND THEIR PREDICTION BY MEANS OF A SINGLE INTEGRAL CONSTITUTIVE EQUATION

Hans Martin Laun

Mess- und Prüflaboratorium
BASF Aktiengesellschaft
D-6700 Ludwigshafen/Rhein, Germany

EXPERIMENTAL

The elongational behaviour of various polymer melts was measured by means of an elongational rheometer using a rotating clamp[1,2]. The servo control makes it possible to apply varying or constant stresses or strain rates (Fig. 1). The recoverable strain and the quality of the test is determined from three cut-offs.

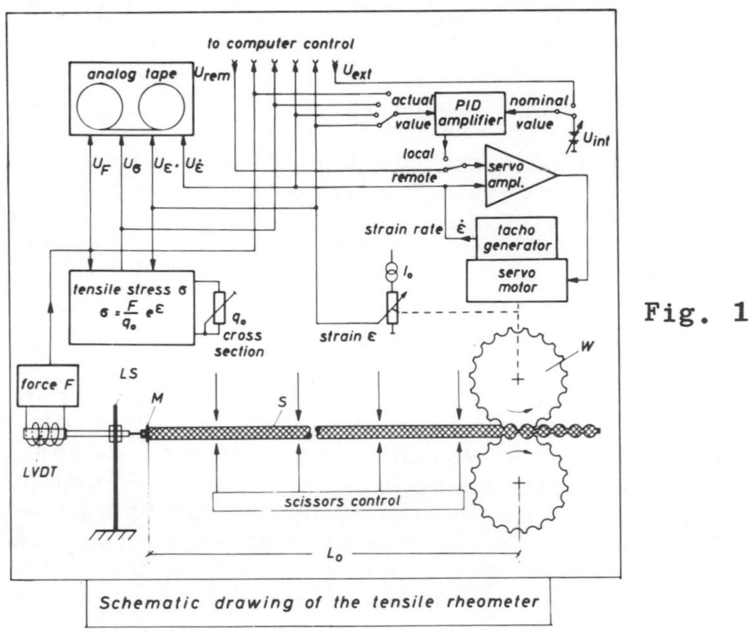

Schematic drawing of the tensile rheometer

Fig. 1

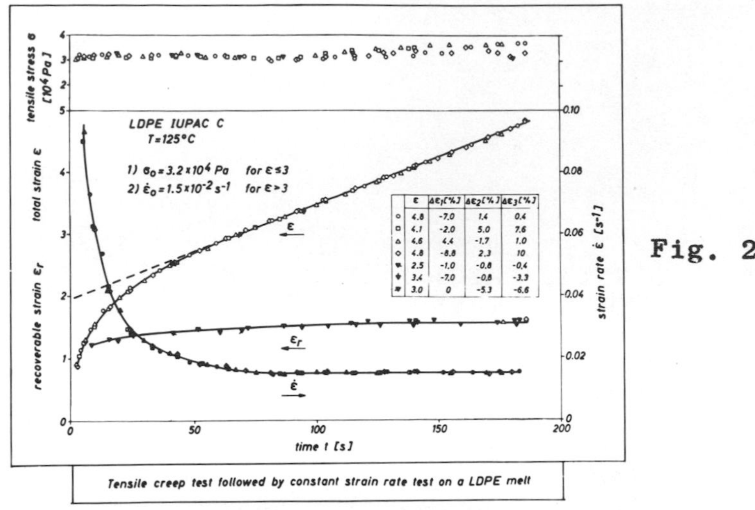

	ε	$\Delta\varepsilon_1$ [%]	$\Delta\varepsilon_2$ [%]	$\Delta\varepsilon_3$ [%]
○	4.8	-7.0	1.4	0.4
□	4.1	-2.0	5.0	7.6
△	4.6	4.4	-1.7	1.0
◇	4.8	-8.8	2.3	10
▼	2.5	-1.0	-0.8	-0.4
♦	3.4	-7.0	-0.8	-3.3
▼	3.0	0	-5.3	-6.6

Tensile creep test followed by constant strain rate test on a LDPE melt

Fig. 2

Fig. 2 gives the results of tensile creep tests followed by constant strain rate tests - the strain rate of which had been determined from the steady strain rate of preceding tensile creep tests. Within experimental accuracy it appears that a steady-state of elongational flow is achieved over about two strain units. The Table in Fig. 2 gives the deviations of the cut-off strains from the recorded ε-signal (compare Fig. 1) for several tests. Unfilled symbols show the excellent reproducibility for four tests. Filled symbols represent final values of other tests. Those contained in the Table are indicated by a tic.

THEORY

An extension of the Lodge-type single integral constitutive equation[3] with strain dependent and factorized memory function $\mu(t-t')h(\varepsilon)$ as proposed by Wagner[4] is used for the calculation of material functions.

Linear Viscoelastic Behaviour

For computational simplicity a spectrum of nine discrete relaxation times τ_i and relaxation strengths g_i is chosen, the values of which are determined by fitting the dynamic moduli $G'(\omega)$ and $G''(\omega)$ (Fig. 3)[5]. The linear viscoelastic memory function $\mu(t-t')$ has the form

$$\mu(t-t') = \sum_i \frac{g_i}{\tau_i} \exp\left[-\frac{t-t'}{\tau_i}\right] . \tag{1}$$

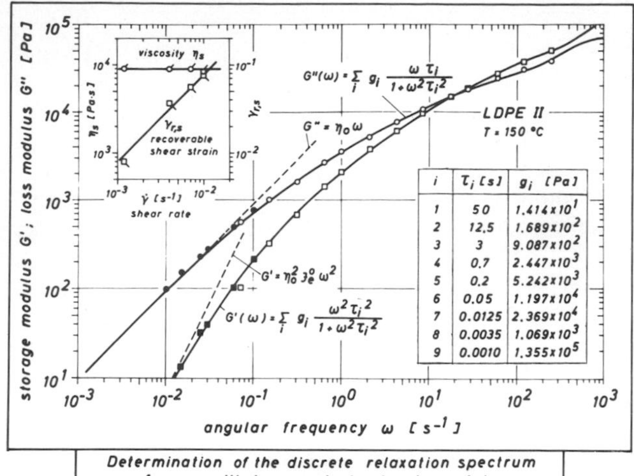

i	τ_i [s]	g_i [Pa]
1	50	1.414×10^1
2	12.5	1.689×10^2
3	3	9.087×10^2
4	0.7	2.447×10^3
5	0.2	5.242×10^3
6	0.05	1.197×10^4
7	0.0125	2.369×10^4
8	0.0035	1.069×10^3
9	0.0010	1.355×10^5

Determination of the discrete relaxation spectrum from oscillatory and steady shear data

Fig. 3

In order to correctly introduce the long relaxation times - which strongly influence the elastic properties of the melt - the limiting course of the moduli (broken lines in Fig. 3) is obtained from the zero shear viscosity η_o and the steady-state shear compliance J_e^o

$$J_e^o = \frac{1}{\eta_o} \lim_{\dot{\gamma} \to 0} \frac{\gamma_{r,s}}{\dot{\gamma}} \qquad (2)$$

determined in steady shear experiments[6] as indicated by the inserted plot.

The experimentally determined moduli are represented by N data sets ω_k, G_k', G_k''. Approximately half a decade spacing of the discrete relaxation times is used, the longest relaxation time τ_1 being the reciprocal value of the angular frequency at which the storage modulus begins to significantly deviate from the limiting course (broken line). The relaxation strengths g_i are obtained by linear regression from the condition

$$\sum_{k=1}^{N} \left[\left(\frac{G'(\omega_k)}{G_k'} - 1 \right)^2 + \left(\frac{G''(\omega_k)}{G_k''} - 1 \right)^2 \right] = \text{Min} \,. \qquad (3)$$

For a linear viscoelastic liquid the recoverable strain ϵ_r, the tensile stress σ, and the strain rate $\dot{\epsilon}$ are interrelated by the equation

$$\frac{d\epsilon_r}{dt} = \dot{\epsilon} - \frac{\sigma}{\mu_o} \qquad (\mu_o = 3\,\eta_o \text{ Trouton viscosity}) \qquad (4)$$

which for a constant strain rate test yields the formula for $\epsilon_r(t)$ given in Fig. 4. The only time dependent ratio $\epsilon_r/\dot{\epsilon}_o$ is more sensitive to long relaxation times than the time dependent viscosity $\sigma/\dot{\epsilon}_o$. The expression for $\epsilon_r(t)$ can be used for an approximate prediction of the recoverable strain in the nonlinear range up to limited total strains as demonstrated in Fig. 4. It is interesting to note that the highly elastic deformation at $\dot{\epsilon}_o = 1\ s^{-1}$ is correctly predicted up to relatively high strains.

Nonlinear Viscoelastic Behaviour

The nonlinearity of the elongational behaviour is governed by the 'damping' function $h(\epsilon)$. The symbols in Fig. 5 represent the strain dependence numerically evaluated from stress strain curves. These data are expressed analytically by the function given in Fig. 5. The parameters m and ϵ_o have been introduced by Wagner[7]. q influences the transition between the limiting courses $\exp[-m\epsilon]$ for $\epsilon \approx 0$ and $\exp[-2\epsilon]$ for $\epsilon \gg \epsilon_o$.

RESULTS

Fig. 6 gives the measured time dependent elongational viscosities for three polyethylene and a polystyrene melt. The full lines represent the calculated behaviour. It is clearly seen from Figs. 5 and 6, that the parameter ϵ_o essentially determines the shape of the stress-strain curves and the (within the limits of

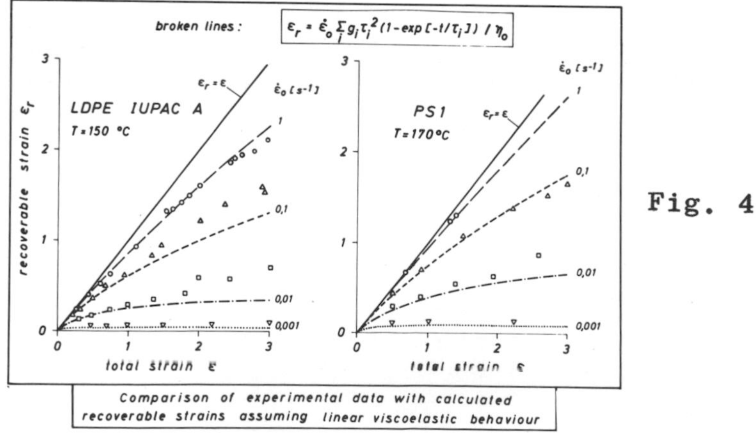

Fig. 4

Comparison of experimental data with calculated recoverable strains assuming linear viscoelastic behaviour

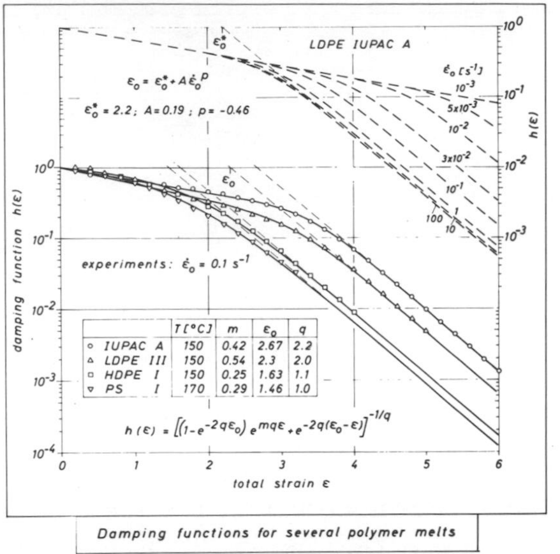

Damping functions for several polymer melts

Fig. 5

our experiments steady-state) elongational viscosities at high strains. ϵ_0 depends on molecular quantities like the degree and kind of branching as well as the molecular weight distribution.

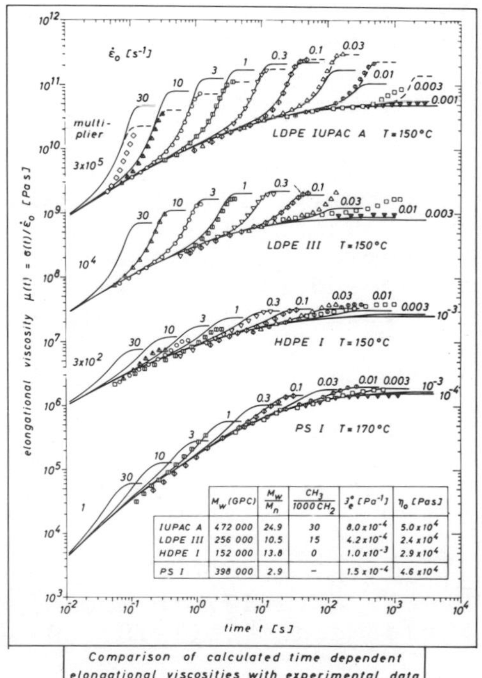

Comparison of calculated time dependent elongational viscosities with experimental data

The discrepancies between theory and experiment in Fig. 6 may only partly be attributed to experimental errors like a time delay in the stress response at high strain rates or an unsufficient thermal stability of the melt at the low strain rate measurements. A significant difference between the measured and calculated strain rate dependence at high elongations is observed (compare IUPAC A in Fig. 6) in ranges where there is no doubt about the reliability of the experimental data.

Fig. 6

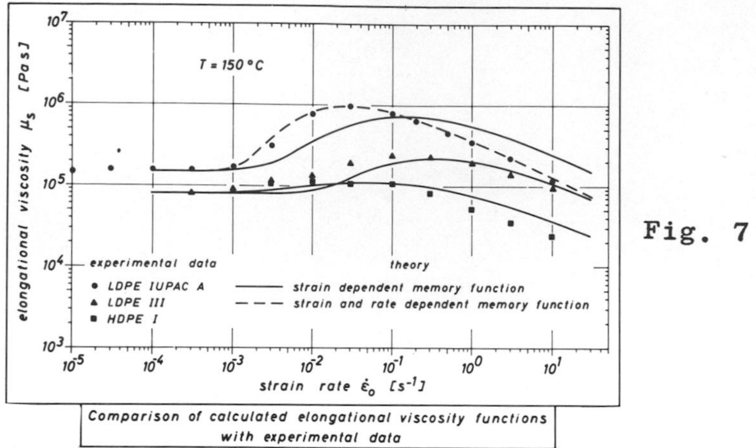

Comparison of calculated elongational viscosity functions with experimental data

Fig. 7

The kind of deviation comes out more clearly in plots of the elongational viscosity as a function of the strain rate (Fig. 7).

The heights of the maxima of the measured viscosity functions are reasonably described by the strain dependent memory function. However, the calculated viscosity functions are shifted to higher strain rates. A correct description of the observed behaviour can only be obtained by introducing a strain rate dependence of the parameter ϵ_o. By using

$$\epsilon_o = \epsilon_o^* + A\,\dot{\epsilon}P \approx \epsilon_o^* + A/\sqrt{\dot{\epsilon}} \qquad (5)$$

the agreement is improved considerably. This is demonstrated in Figs. 5-7 for LDPE IUPAC A by the broken lines. In this respect it is important to note that the limitations of a purely strain dependent 'damping' function become visible only at relatively high strains.

REFERENCES

1. J. Meissner, Trans. Soc. Rheol. 16, (1972) 405
2. H. M. Laun, H. Muenstedt, Rheol. Acta 17, (1978) 415
3. A. S. Lodge, "Body Tensor Fields in Continuum Mechanics", Academic Press, New York (1974)
4. M. H. Wagner, Rheol. Acta 15, (1976) 136
5. H. M. Laun, Rheol. Acta 17, (1978) 1
6. H. M. Laun, J. Meissner, Rheol. Acta 19, in press
7. M. H. Wagner, J. Non.-J. Fluid Mech. 4, (1978) 39

UNIAXIAL EXTENSIONAL EXPERIMENTS WITH LARGE STRAINS PERFORMED WITH LOW DENSITY POLYETHYLENE (LDPE)

T. Raible and J. Meissner

Department of Industrial and Engineering Chemistry
Swiss Federal Institute of Technology (ETHZ)
CH - 8092 Zürich, Switzerland

In this paper results of three different types of extensional experiments are presented, which were performed with a new version of the Uniaxial Extensional Rheometer with rotary clamps: (A) extension with constant strain rate, (B) recovery after cessation of extension with constant strain rate, and (C) extension with constant tensile force. The LDPE used had a density at 20°C of ρ_{20} = 0.919 g/cm^3 and a melt flow index of MFI = 1.37 g/10 min /1/. The test temperature was 150°C.

The details of the apparatus used were reported already /2/. As an illustration, the clamp connected with the force measuring device is shown in fig. 1. Because of the excellent temperature homogeneity, large strains up to 7 units in the HENCKY strain measure $\varepsilon = \ln \lambda$ could be achieved, i.e. a maximum stretch $\lambda = 1097$ /2,3/. All strains are expressed as HENCKY strains throughout this paper.

Each tensile test is started at the instant t = 0 and terminated at $t = t_1$ by cutting the sample into small pieces. From the mass m_i of each piece and the incompressibility assumption for the melt, the cross-sectional areas $q_i(t_1)$ follow with the average $\bar{q}$. The true total strain $\varepsilon_1 = \ln \lambda_1$ at the end of each test is equal to $\varepsilon_1 = \ln (q_o/\bar{q})$, where $q_o = q(t=0)$. ε_1 can be compared with the total nominal strain $\varepsilon_{nom}(t_1)$, which follows from the total angles of the clamp rotations.

It has been shown previously /2,3/ that for large total strains it is vital to judge the quality of the test performance by means of the following two parameters:

$$\Delta\varepsilon/\varepsilon_{nom}(t_1) = \left[\varepsilon_1 - \varepsilon_{nom}(t_1)\right]/\varepsilon_{nom}(t_1) \qquad (1)$$

$$\Delta q/\bar{q} = (q_{i,max} - q_{j,min})/\bar{q} \tag{2}$$

The first parameter characterizes the deviation of the true strain from the nominal strain, the second one characterizes the homogeneity of the deformation at t_1 because Δq is the difference in the largest and the smallest cross-sectional area.

Deformation with Constant Extensional Strain Rates

Preliminary results of tests performed with the new rheometer /2,3/ are complemented in this paper by results obtained at additional constant nominal strain rates $\dot{\varepsilon}_o$. For this mode of operation the nominal strain at t_1 is $\varepsilon_{nom}(t_1) = \dot{\varepsilon}_o t_1$. The true strain rate $\bar{\dot{\varepsilon}}$ is defined by the relation $\varepsilon_1 = \bar{\dot{\varepsilon}} t_1$ and is assumed to be constant during the duration of the test considered. With $\Delta\dot{\varepsilon} = \bar{\dot{\varepsilon}} - \dot{\varepsilon}_o$, the quality parameter eq. (1) follows:

$$\Delta\dot{\varepsilon}/\dot{\varepsilon}_o = \Delta\varepsilon/\varepsilon_{nom}(t_1) \tag{3}$$

Only tests of sufficient performance, i.e. tests with quality parameters $\Delta q/\bar{q} < 0.12$ and $|\Delta\dot{\varepsilon}/\dot{\varepsilon}_o| < 0.02$, were further evaluated. From the recorded tensile force $\bar{F}(t)$ the tensile stress $\sigma(t) = F(t)/q_o\exp(-\bar{\dot{\varepsilon}}t)$ is calculated. As final result the true extensional viscosity $\mu(t) = \sigma(t)/\bar{\dot{\varepsilon}}$ is given in fig. 2 for different nominal strain rates $\dot{\varepsilon}_o$. At the lowest strain rate $\dot{\varepsilon}_o = 0.001\ s^{-1}$, the linear viscoelastic limit $\mathring{\mu}(t) = 3\mathring{\eta}(t)$ is obtained where $\mathring{\eta}(t)$ is the linear limit measured in shear*). At this low $\dot{\varepsilon}_o$, one stabilized sample (symbol +) was used yielding the same result as the unstabilized sample (symbol I). Therefore, unstabilized samples were used for all other tests.

For $\dot{\varepsilon}_o > 0.001\ s^{-1}$ the curves in fig. 2 coincide for short deformation times, but start to grow above the linear viscoelastic limiting curve at times which are shorter at higher $\dot{\varepsilon}_o$ and correspond to a strain $\varepsilon \approx 0.5$. This strain hardening behaviour of LDPE has been reported previously /2-7/. At higher strains $\varepsilon > 4$ the curves of fig. 2 show a transition to a maximum followed by a decrease.

The existence of this maximum means that, in the range of ε investigated, for $\dot{\varepsilon}_o > 0.001\ s^{-1}$ no rheologically steady state of flow exists. The maximum could not be found at $\dot{\varepsilon}_o = 1\ s^{-1}$ because at this strain rate rupture of the sample occured at strains $3.6 < \varepsilon < 4$. For $\dot{\varepsilon}_o = 0.001\ s^{-1}$ it is still open whether the viscosity remains constant or shows the strain hardening at still longer periods of deformation.

---

*) The shear measurements were kindly performed for us by Dr. H.M. LAUN at a shear rate of $\dot{\gamma}_o = 0.001\ s^{-1}$.

Recovery after Cessation of Extensional Flow with Constant Strain Rate

In order to investigate the time dependent recovery, the copper cover (CC of fig. 1) of the rheometer was opened before cutting the sample at t_1, and the shrinking cut-offs were photographed through the glass cover GC. A special cutting length L_A = 250 mm was used. With the recovery time scale $t' = t - t_1$, the shrinkage expressed in the HENCKY measure is

$$s(t') = \ln(L_A/L(t')) \tag{4}$$

There are two reasons for the shrinkage: The elastically stored recoverable strain $\varepsilon_r(t')$ which decreases with increasing t', and the interfacial tension α. With the notations of fig. 3 it follows that

$$s(t') = \varepsilon_s(t') + \varepsilon_\alpha(t') \tag{5}$$

$\varepsilon_s(t')$ is the recovered elastic strain which, for $t' \to \infty$, becomes constant and equals the total recovery ε_R at t_1: $\varepsilon_R(t_1,\dot{\varepsilon}_o) = \lim_{t'\to\infty} \varepsilon_s(t'; t_1, \dot{\varepsilon}_o)$.

From the conservation of energy during recovery /8/ the following relation results

$$\varepsilon_s(t'') = s(t'') - (\alpha/\mu_r R_1) \int_{t'=0}^{t''} \exp(-s/2)dt' \tag{6}$$

where μ_r is the effective viscosity during recovery and R_1 the radius of the sample at t_1. For $t' \to \infty$, the interface tension is the only force deforming the cut-off, and therefore the following relation can be derived

$$\alpha/\mu_r R_1 = 2 \frac{d}{dt'} (\exp(s/2)) \qquad (t' \to \infty) \tag{7}$$

If $\alpha/\mu_r R_1$ is constant, a linear relation should follow for exp (s/2) as a function of t'. As shown in fig. 3 for t' > 600 s, a straight line results from the slope of which $\alpha/\mu_r R_1$ can be calculated. With this expression and eq. (6), $\varepsilon_s(t')$ can be determined. This function is also shown in fig. 3 for $\dot{\varepsilon}_o = 0.03\ s^{-1}$ and t_1 = 200 s.

Fig. 4 shows the recovered strain $\varepsilon_s(t')$ obtained after deforming the sample at $\dot{\varepsilon}_o = 0.03\ s^{-1}$ and different total strains $\dot{\varepsilon}_o t_1$. For t' > 600 s, ε_s shows the equilibrium value representing the total recovery $\varepsilon_R = \varepsilon_s(t' \to \infty)$. From the ε_R-values for the different total strains $\dot{\varepsilon}_o t_1$, it follows that not only the stress or the extensional viscosity, but also the total recovery as a function of the total strain, shows a maximum as stated already /2,3/.

From the results given it is obvious that the influence of the interface tension α must be taken into account in evaluating recovery experiments. It should be added that the tests used for figs. 3 and 4 had the quality parameters $\Delta q/\bar{q} < 0.15$ and $|\Delta\dot{\varepsilon}/\dot{\varepsilon}_o| < 0.02$.

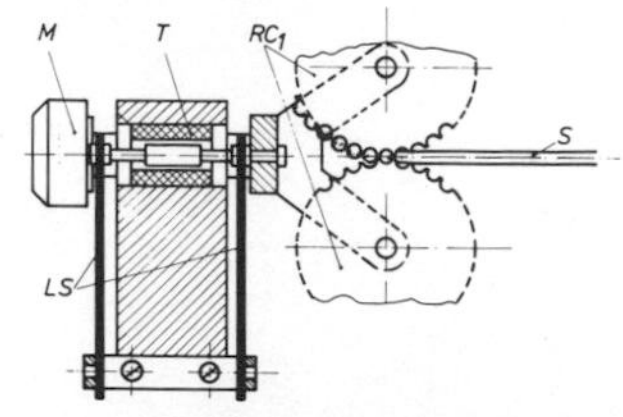

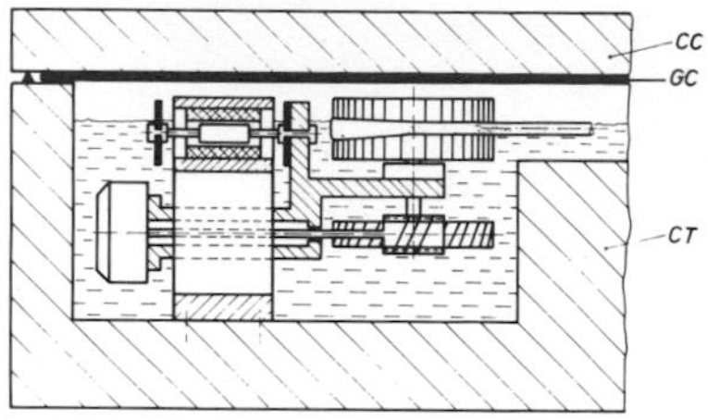

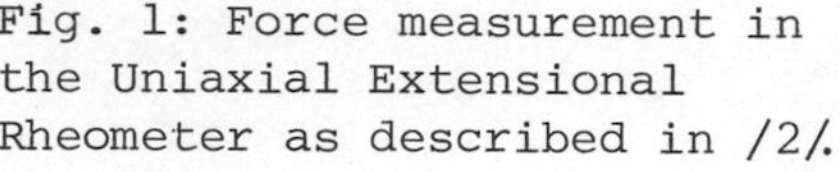
Fig. 1: Force measurement in the Uniaxial Extensional Rheometer as described in /2/.

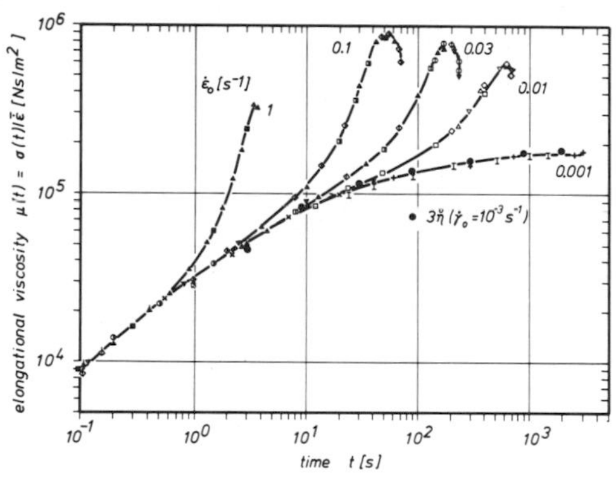

Fig. 2: Elongational viscosity $\mu(t)$ for different strain rates $\dot{\varepsilon}_o$ and comparison with the (linear viscoelastic) shear viscosity $\mathring{\eta}(t)$.

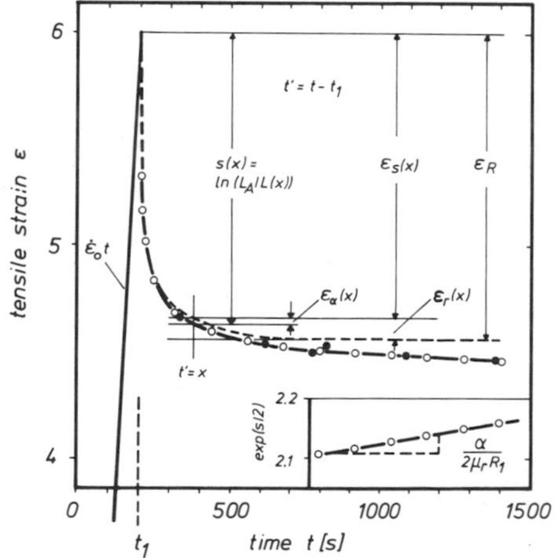

Fig. 3: Recovery and influence of the interface tension α after extension with $\dot{\varepsilon}_o$ = 0.03 s^{-1} and t_1 = 200 s. Length measurement optically (o) or mechanically after quenching (●).

$s(t')$: measured shrinkage
ε_R : total recovery, $\varepsilon_R = \varepsilon_s(t' \to \infty) = \varepsilon_r(t' = 0)$
$\varepsilon_s(t')$: recovered strain
$\varepsilon_r(t')$: recoverable strain
$\varepsilon_\alpha(t')$: shrinkage due to interfacial tension

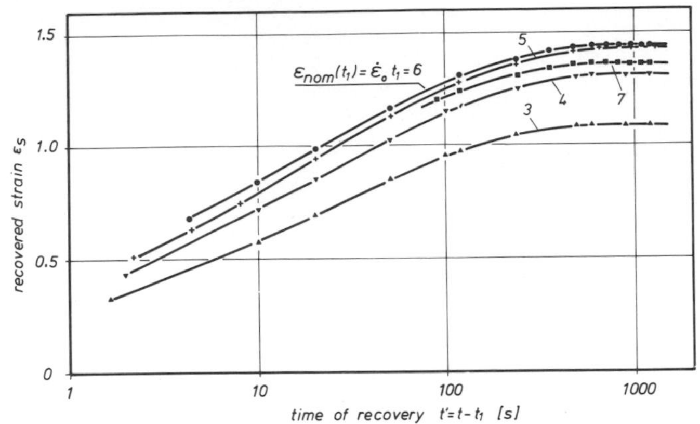

Fig. 4: Recovered strain ε_s as a function of the recovery time t' for different total strains $\varepsilon_{nom}(t_1)$; $\dot{\varepsilon}_o = 0.03\ s^{-1}$.

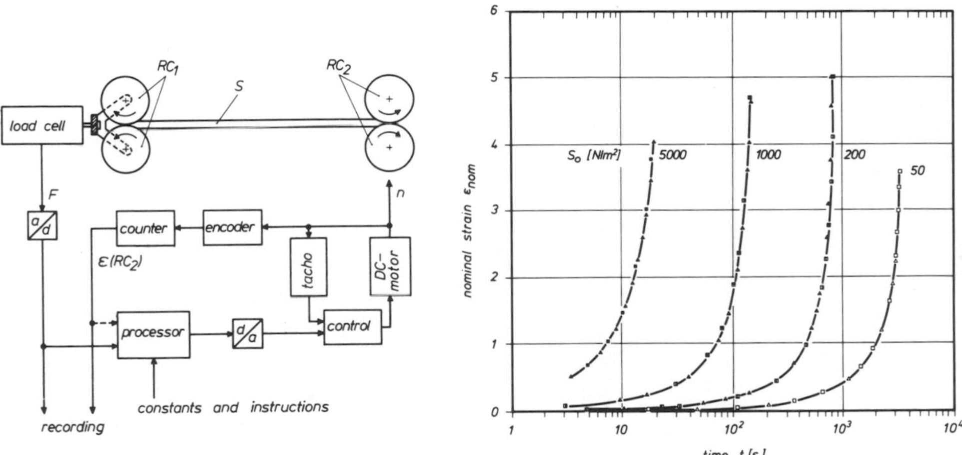

Fig. 5: Schematic diagram of the rheometer drive control for general test modes.

Fig. 6: Extension at constant nominal stress S_o, i.e. constant reduced tensile force, $S_o = F_o/q_o$.

Extension with Constant Tensile Force

This test mode is of practical as well as of theoretical interest, e.g. WAGNER /9/ predicted a "catastrophic" extension at a finite time at which the filament is drawn down to zero cross-sectional area. Fig. 5 shows schematically the digital control of the rheometer drive for the performance of general test modes.

For extensional tests with constant tensile force F_o, the "nominal stress" S_o is defined as force per initial cross-sectional area, $S_o = F_o/q_o$. The tests were performed with different S_o between 50 and 5000 N/m^2. All the samples were cut at different times t_1 in order to determine ε_1. The values of the quality parameters for these tests were $\Delta q/\bar{q} < 0.12$ and $|\Delta\varepsilon/\varepsilon_{nom}(t_1)| < 0.03$. Because of the necessary good test performance the maximum strain achieved at present in this test mode is $\varepsilon_{nom} = 5$.*) The result of these measurements is given in fig. 6 which shows the time dependent increase of the strain $\varepsilon_{nom}(t)$ for different values S_o. All curves end with a steep growth supporting the prediction of the catastrophic extension. In fact, for $S_o = 5000$ and 1000 N/m^2, the data of fig. 6 follow closely the curves predicted for these two S_o values and a very similar melt ("melt I") from the constitutive equation of the modified rubberlike liquid by WAGNER /9/.

Acknowledgements

We thank Dr. M.H. WAGNER for helpful discussions and Messrs. W. Achermann, A. Demarmels, R. Frischknecht, H. Hürlimann, N. Marugg, and A. Reichling for their contributions in developing the rheometer and performing the measurements. The help from BASF Aktiengesellschaft Ludwigshafen/Germany (delivery of the LDPE used for this investigation) and the financial support from the Swiss National Foundation for Scientific Research is gratefully acknowledged.

References

1/ US and German standards: ASTM-D1238-73 and DIN 53735.
2/ J. Meissner, T. Raible, and S.E. Stephenson, paper presented at the Golden Jubilee Meeting of the Society of Rheology, Boston, Mass. 1979. J. of Rheol., to be published.
3/ T. Raible, A. Demarmels, and J. Meissner, Polymer Bulletin 1, 397 (1979)
4/ J. Meissner, Rheol. Acta 8, 78 (1969)
5/ J. Meissner, Rheol. Acta 10, 230 (1971)
6/ J. Meissner, Trans. Soc. Rheol. 16, 405 (1972)
7/ H.M. Laun and H. Münstedt, Rheol. Acta 15, 517 (1976)
8/ H.M. Laun and H. Münstedt, Rheol. Acta 17, 415 (1978)
9/ M.H. Wagner, Rheol. Acta 18, 681 (1979)

*) This requires additionally that the true tensile force F is kept within 3% during $0 < t < t_1$.

LARGE HOMOGENEOUS BIAXIAL EXTENSION OF POLYISOBUTYLENE AND COMPARISON WITH UNIAXIAL BEHAVIOUR

S.E. Stephenson and J. Meissner

Department of Industrial and Engineering Chemistry
Swiss Federal Institute of Technology (ETHZ)
CH - 8092 Zürich, Switzerland

INTRODUCTION

Compared with the activity during the last decade in uniaxial extension of polymer melts, the investigation of their biaxial behaviour has been rather limited because of experimental difficulties. In a previous paper /1/, a new type of biaxial rheometer has been described, which has distinct advantages over the conventional bubble inflation technique. Using this new rheometer, equibiaxial extensional experiments on polyisobutylene at room temperature (22°C) and constant strain rates have been made. The results of these experiments are compared with the response of the same material in uniaxial extension and simple shear.

EXPERIMENTAL PROCEDURE

Eight rotary clamps are used in the rheometer. A typical clamp, whose function is both to produce the deformation and to measure the corresponding force required, is shown in fig. 1. The clamp described in /1/ has been modified to the form shown here to enable more accurate measurement of force. The clamps are positioned alternately around a circle with pneumatically operated scissors as shown schematically from above in fig. 2. All rollers are driven at the same speed by an internal stepping motor and gear system, extending the disc shaped sample radially. The scissors intermittently cut the rim of the sample forming strips which are wound up on the rollers WU of fig. 1.

All experiments presented here were performed on polyisobutylene at room temperature (22°C). The polyisobutylene used (BASF-Oppanol B15), had a number average molecular weight of 13 000 /2/. The samples of thickness h_o = 5 mm, were extrusion moulded using a devolati-

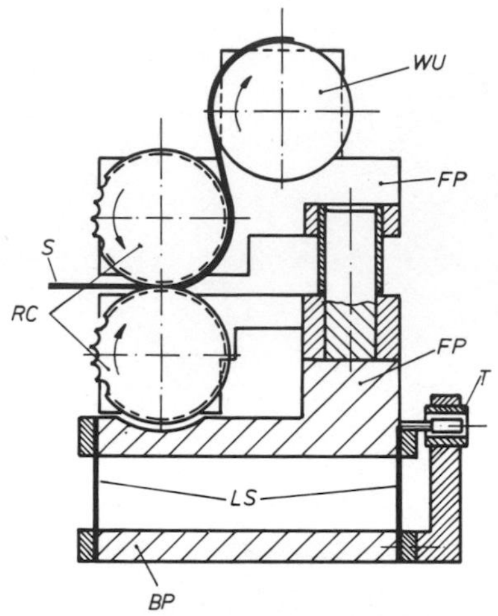

Fig. 1 The Rotary Clamp

BP Base Plate
FP Frame Plate
LS Leaf Springs
RC grooved Rotating Cylinders (77 mm long x 74 mm dia.)
S Sample
T Displacement Transducer used to measure force F_i
WU Wind-Up roller

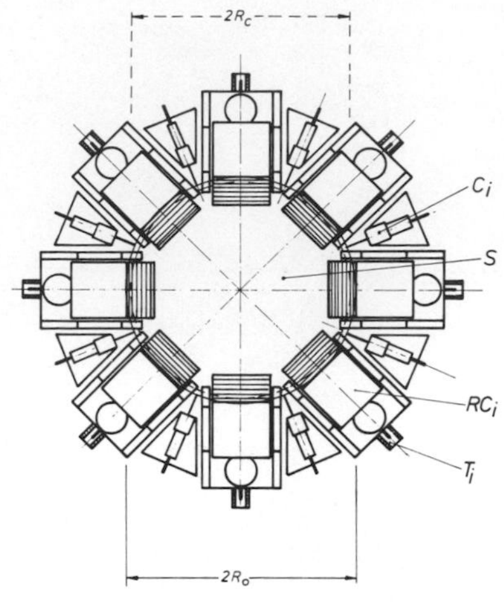

Fig. 2 The Biaxial Rheometer

C_i Cutter i
RC_i Rotary Clamp i
R_c Radius of circle through tips of cutters (150 mm)
R_o Radius of circle through centre of clamping cylinders (156 mm)
S Sample
T_i Transducer for F_i

zing screw at $100^{\circ}C$. Before introduction into the rheometer each sample was marked with a grid of black ink on its upper surface. The samples were initially supported by a layer of talcum powder resting on a table in the centre of the rheometer.

The deformation of the grid on the sample, photographed at regular intervals, was used to determine the radial HENCKY strain $\varepsilon_{rr} = \ln \lambda_{rr}$ (λ_{rr} = radial stretch), in each of two perpendicular directions. Fig. 3 shows four such pictures selected out of a series taken every four seconds, and demonstrates the homogeneity of the deformation /1/.

As a first approximation the radial stress σ_{rr} at the centre has been calculated assuming

$$\sigma_{rr}(t) = \left[\sum_{i=1}^{8} F_i(t)\right] / 2\pi\, R(t) h(t)$$

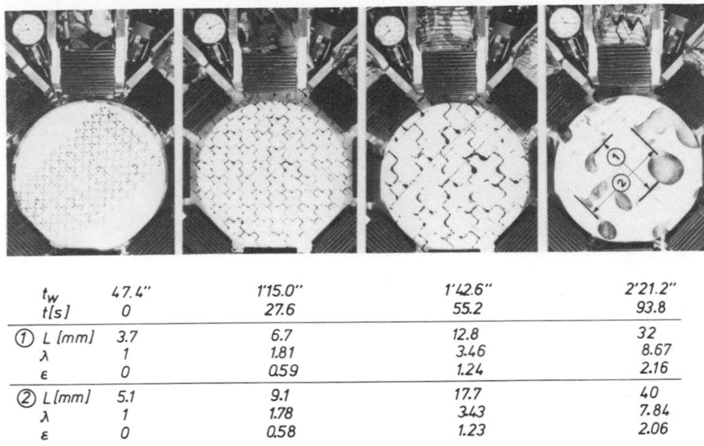

t_W	47.4"	1'15.0"	1'42.6"	2'21.2"
$t[s]$	0	27.6	55.2	93.8
① L [mm]	3.7	6.7	12.8	32
λ	1	1.81	3.46	8.67
ε	0	0.59	1.24	2.16
② L [mm]	5.1	9.1	17.7	40
λ	1	1.78	3.43	7.84
ε	0	0.58	1.23	2.06

Fig. 3 Photographic determination of strain. t_W = time indicated by watch, t = absolute time of deformation

where $F_i(t)$ are the forces measured by the clamps, R(t) is the radius of the circle through the tips of the cuts, and h(t) is the thickness of the sample at time t. h(t) follows from h_o, $\varepsilon_{rr}(t)$ and the incompressibility assumption for the melt. A detailed consideration of the distribution of stress within the sample, and a more accurate method of determining the strain than that indicated in fig. 3, will be dealt with in a forthcoming paper.

RESULTS

Typical results are shown in fig. 4. As can be seen from the strain $\varepsilon_{rr}(t)$, the strain rate $\dot{\varepsilon}_{rr}(t)$ remains constant during most of this test, falling off slightly towards the end. The effective strain rate $\bar{\dot{\varepsilon}}_{rr}$ is determined by the slope of the linear portion of this strain-time curve. The steps in the force curve, due to cutting at the rim, are no longer to be seen in the biaxial viscosity $\mu_b = \sigma_{rr}/\bar{\dot{\varepsilon}}_{rr}$. The large fluctuations in the μ_b-curve arise as a result of vibrations induced in the measurement system by the cutting action and do not represent actual material response.

Further experimental results for a nominal strain rate of 0.021 s^{-1} are shown in fig. 5. Both $\dot{\varepsilon}_{rr}(t)$ and $\varepsilon_{rr}(t)$ are shown for one experiment in which a total radial strain in excess of 3.2 was reached. However, $\dot{\varepsilon}_{rr}$ is not ideally constant throughout the test, the deviation towards the end is probably the cause of the maximum shown by μ_b. Nevertheless, the total radial strain achieved, and the total radial strain achieved at constant strain rate, are greater than

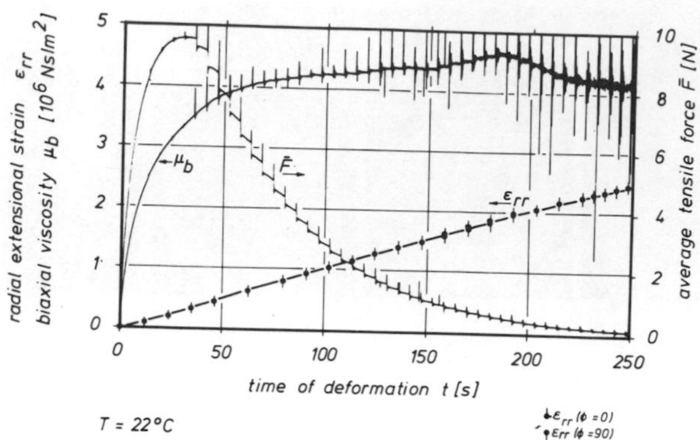

Fig. 4 Test results for the effective radial strain rate $\bar{\dot{\varepsilon}}_{rr} = 0.0102\ s^{-1}$

those which can be achieved with bubble inflation. At large strains, the sample is very thin. e.g. at $\varepsilon_{rr} = 3.2$ the thickness h is 8 μm. Therefore, $\dot{\varepsilon}_{rr}$ might deviate from the intended value due to insufficient clamping. An improved clamp design should overcome this problem.

The results of equibiaxial extension measured at nominal strain rates of 0.01 and 0.021 s^{-1} are shown in fig. 6, where they are compared with results in uniaxial extension and shear. The uniaxial extensional viscosity μ_u was measured using two opposed rotary clamps

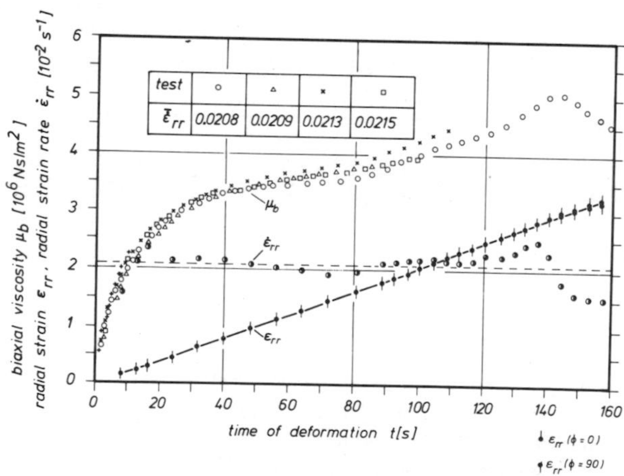

Fig. 5 Results for different tests with a nominal radial strain rate of 0.021 s^{-1}.

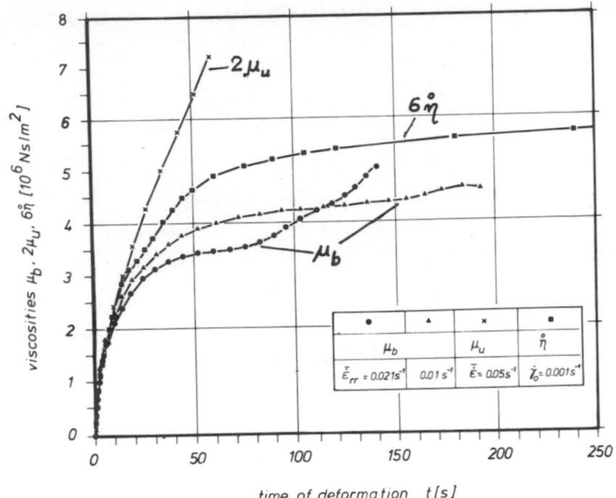

Fig. 6 Comparison of the viscosities for equibiaxial extension, uniaxial extension, and shear

312 mm apart. The sample was a flat strip, 5 mm thick and 35 mm wide. The shear viscosity curve $\mathring{\eta}(t)$ was kindly measured for us by Dr. H.M. LAUN.

For both radial strain rates $\bar{\dot{\varepsilon}}_{rr}$, the biaxial viscosities $\mu_b(t)$ rise sharply at first, flatten out in a plateau region and rise again at a total strain of between 1.4 and 1.6. The curves indicate that the apparent steady-state values measured previously /3/ with the bubble inflation technique actually correspond to the plateau region of the μ_b-curves of fig. 6. Even for the low strain rates applied here, a steady state in biaxial flow is not achieved.

There are two more interesting results: (i) In the plateau region, μ_b decreases with the increase in $\dot{\varepsilon}_{rr}$ from 0.01 to $0.021 s^{-1}$, and (ii) in this region $\mu_b(t)$ is less than the linear viscoelastic limit $6\mathring{\eta}(t)$. Maerker and Schowalter /3/, using bubble inflation, have found the same result for similar polyisobutylenes. In contrast to the biaxial behaviour, the uniaxial viscosity $\mu_u(t)$ is never less than the linear viscoelastic limiting curve $3\mathring{\eta}(t)$.

Although it is not the purpose of this paper to compare these results with predictions of various theories, a qualitative comparison is made with the predictions of the LODGE rubberlike liquid model. Beyond a critical strain rate, this model predicts a steady state neither in uniaxial nor in equibiaxial extension. The uniaxial extensional viscosity is predicted to rise above the linear viscoelastic limit and to continue to rise indefinitely. The prediction for the biaxial viscosity is a decrease below its linear viscoelastic limit before a similar indefinite rise. The LODGE model qualitatively describes the important features of the behaviour observed here in uniaxial and biaxial extension.

CONCLUSION

With the new biaxial rheometer it has been possible to produce large homogeneous biaxial strains in polyisobutylene at constant strain rates. Total strains as large as $\varepsilon_{rr} = 3.2$ have been reached, corresponding to a compression in the axial direction of 6.4. That such large strains can be achieved is due on one side to the construction of the rheometer and on the other to the inherent stability of biaxial extension discussed already by White /4/.

Even with the very large total strains applied here, no rheologically steady state could be found in extending polyisobutylene, in spite of the fact that the extensional strain rates were rather low. In biaxial extension, an intermediate plateau of the viscosity function is found which is below the linear viscoelastic limit. The plateau is terminated by a further increase at high extension. In contrast to the biaxial case, the viscosity for uniaxial extension always lies above the linear viscoelastic limit. The LODGE rubberlike liquid model explains the general features of the observed behaviour.

ACKNOWLEDGEMENT

We express our thanks to all coworkers who helped in the design and construction of the new rheometer, and to Dr. H.M. LAUN for measurement of the shear viscosity of our sample. The polyisobutylene polymer was kindly donated by BASF Aktiengesellschaft, Ludwigshafen, Germany. The financial support of the Charles Kolling Travelling Scholarship of the University of Sydney, Australia, is gratefully acknowledged.

REFERENCES

1/ J. Meissner, T. Raible and S.E. Stephenson, paper presented at the Golden Jubilee Meeting of the Society of Rheology, Boston, Mass. 1979. J. of Rheol., to be published.

2/ Technical Information Sheet, BASF Aktiengesellschaft

3/ J.M. Maerker and W.R. Schowalter, Rheol. Acta 13, 627-638 (1974)

4/ J.L. White, Rheol. Acta 14, 600-611 (1975)

NON-ISOTHERMAL EFFECTS IN THE ELONGATIONAL FLOW OF POLYMER MELTS

D. Acierno, L. Dieli, F. P. La Mantia, G. Titomanlio

Istituto di Ingegneria Chimica, Università di Palermo

Viale delle Scienze, 90128 Palermo, Italy

Non-isothermal histories occur in almost all polymer processing, and however only a few experimental and theoretical studies have been devoted to the rheological response of the polymeric fluids under non-isothermal conditions. A systematic study, which cannot however be considered exhaustive, has been recently carried out in this field by Bogue and coworkers[1-3].

In the present work some stress relaxation experiments have been performed under both small and large elongational strains on a commercial polyisobutylene, while temperature increased almost linearly during the tests. Various heating rates were of course considered. Also some stress-strain data have been collected in the same temperature conditions and at various elongation rates.

The experimental results are analyzed on the basis of a generalized Maxwell model already presented for the isothermal case[4], accounting for the temperature rise through free volume variations.

EXPERIMENTAL

All tests were performed in tension by an Instron testing machine mod. 1115, equipped with the temperature cabinet, on a polyisobutylene commecially known as Oppanol B50, manufactured by Basf.

Strip samples 1 cm wide were cut from 0.3 cm thick sheets obtained by compression moulding at 200°C. Pneumatic jaws were adopted and the pressure was held at a minimum value. The distance between the jaws at beginning of each test was 4.5 cm.

The temperature was measured by a thermocouple placed in a polyisobutylene piece having dimensions equal to those of the test specimen. In order to simplify theoretical analysis, the temperature controller was regulated so as to maintain a temperature growth, r , as constant as possible during the tests. The imposed heating rates were 15.6°C/min, 7.3, 2.5, 0.96 and 0 (isothermal); for the largest rate additional heating sources were placed inside the temperature cabinet. The initial temperature was 24°C for all runs.

Stress relaxation tests were performed under small and large deformations; in particular for the elongation ratio $\alpha = l/lo$ the values 1.1, 1.75 and 2.15 were imposed. In all cases the samples were loaded with the maximum velocity allowed by the machine, i.e. 50 cm/min.

Constant velocity stress-strain measurements were performed with several values of the initial elongation rate, $\dot{\Gamma}_o$, in the range $2.10^{-3} \div 2.10^{-2}$ sec^{-1}.

In order to account for the volume changes (expansion) due to temperature rise, runs were carried out with stress free samples; the sample strain $\alpha-1 = \Delta l/lo$, recorded during these runs, are plotted vs time in Fig. 1.

EQUATIONS OF THE MODEL

The model considered here is based on the idea that free volume changes according to a differential equation by effect of stress or temperature variations. An isothermal formulation has been al-

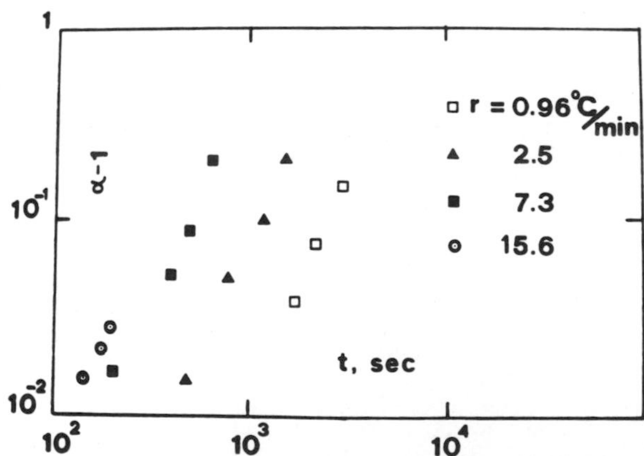

Fig.1 Elongational strain vs time for stress free samples subjected to various heating rates.

ready presented[4] . The following set of equations describe in detail the model

$$\tau = \Sigma_i \tau_i \quad (1)$$

$$\tau_i + \lambda_i \frac{\delta\tau_i}{\delta t} = 2\lambda_i G_i D \quad (2)$$

$$\lambda_i = \lambda_{oi} \exp\left(\frac{1}{f_i} - \frac{1}{f_{To}} \right) \quad (3)$$

$$\frac{df_i}{dt} = \{ bH_i \frac{E_i}{G_i} - (f_i - f_T)\} \frac{1}{\lambda_i} \quad (4)$$

Eq. 1 gives the stress tensor as a sum of contributions τ_i arising from a discretized spectrum of material modes. Each material mode is defined by an equilibrium relaxation time λ_{oi} and a modulus G_i. Eq. 2 is the well known equation of a Maxwell model, written with the contravariant convected derivative of the stress tensor. Eq. 3, following Doolittle, relates the relaxation time λ_i to free volume f_i ; f_{To} is the rest equilibrium free volume at a reference temperature T_o , where the relaxation time is λ_{oi}. Eq. 4 gives the rate of change of free volume for each complexity i . The first term on the right hand side is the rate of the free volume formation by effect of stress: H_i is the value of the relaxation spectrum related to λ_{oi} , E_i is the elastic energy associated to the ith mode, given by $\frac{1}{2}\mathrm{trace}(\tau_i)$ and b is the single parameter of the model. The second term represents the tendency of free volume to reach its rest equilibrium value at temperature T . f_T is simply related to f_{To} by the relation

$$f_T = f_{To} + (T - T_o)\,\Delta\alpha \quad (5)$$

where $\Delta\alpha$ is the incremental expansion coefficient of the material.

COMPARISON AND DISCUSSION

Theoretical predictions have been obtained numerically on the basis of the equilibrium relaxation spectrum, already reported[5] . As for model parameter b , the value $1.5\ 10^{-7}$ cm^2/dynes has been held as previously determined in considering isothermal data on the same material[4] . Account was taken of the data collected during stress free runs. In particular the values of $\dot{\Gamma}$ as function of time, obtained for each heating rate from the data of Fig. 1, have been considered as negative contribution to the deformation rate.

The stress relaxation data under small deformations are considered in Fig. 2. These are compared with both the model predictions (obtained by means of Eqs 1-5) and a family of curves calculated after substitution of Eqs 4 and 5 with

$$f_i = f_{To} + (T - T_o)\ \Delta\alpha \tag{6}$$

Eq. 6 may be considered the steady state solution of Eq. 4 in the limit of small stresses when the effect of E_i is negligible.

The fact that the curves evaluated making use of Eq. 6 show a relaxation much faster than that observed experimentally indicates that free volume, and thus relaxation times, does not follow instantaneously temperature changes. On the other hand the need of a differential equation in order to properly describe the relation between free volume and temperature changes was already evidentiated in many other cases[6]. The agreement between the model predictions and the data of Fig. 2, which follows a favorable comparison with isothermal data taken under very non linear conditions[4], indicates that Eq. 4 accounts correctly for free volume variations generated separately by stress and temperature changes.

When temperature changes take place in presence of large stresses, a more complicated situation was found. A good agreement between the data and the model predictions is shown in Fig. 3, where the stress relaxation behavior following an elongation with α =1.75 is considered; less satisfactory is the comparison of Fig. 4 which

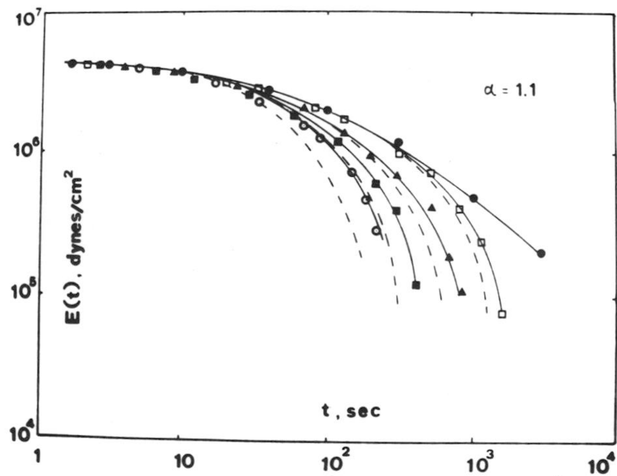

Fig.2 Relaxation moduli in the linear range. Symbols as in Fig.1; ●, isothermal run. —— model predictions; --- curves obtained after substitution of Eqs 4-5 with Eq. 6.

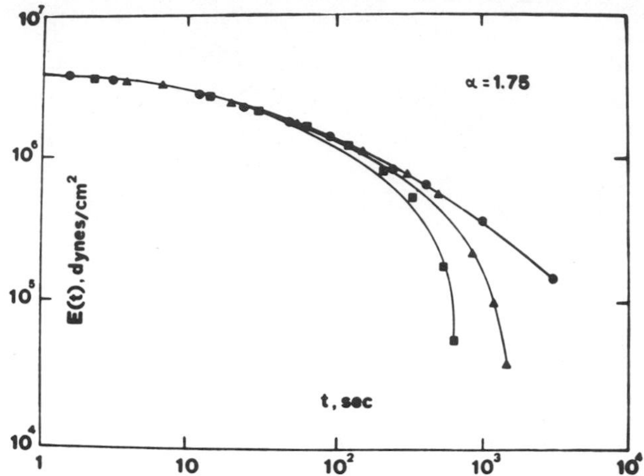

Fig.3 Relaxation moduli at large deformation. Symbols as in Figs 1 and 2.

refers to data relative to α = 2.15. As for the stress strain behavior considered in Fig. 5, the model predictions for the stress σ reproduce the experimental results for small values of the heating rate; as the heating rate increases, when the elongation ratio becomes larger than about 3, the data start to remain significantly below the model predictions. In these cases however, at the end of the test a considerable unhomogeneity in the sample deformation was

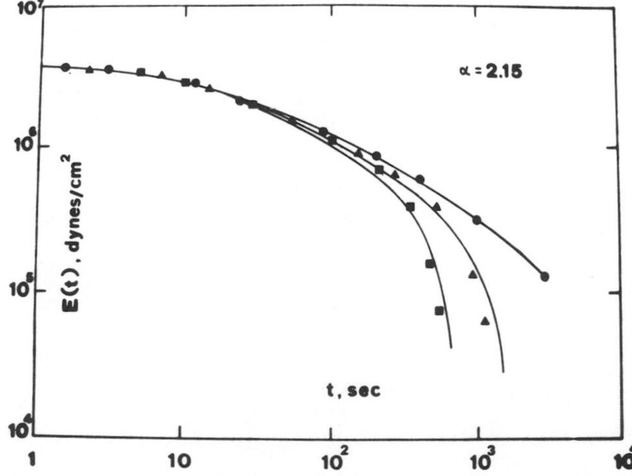

Fig.4 Relaxation moduli at large deformation. Symbols as in Figs 1 and 2.

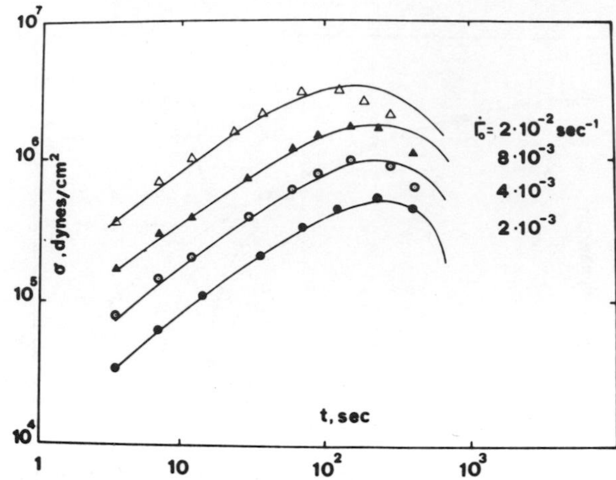

Fig.5 Constant velocity stress growth in elongation. The heating rate was 7.3 °C/min.

observed. Furthermore larger temperature differences along the specimen can be expected when the length of the test piece increases. Both aspects deserve further investigation.

REFERENCES

1. M. Matsui and D. C. Bogue, Non - Isothermal Rheological Response in Melt Spinning and Idealized Elongational Flow, Polym. Eng. Sci. 16:735 (1976)
2. M. Matsui and D. C. Bogue, Studies in Non - Isothermal Rheology, Trans. Soc. Rheol. 21:133 (1977)
3. T. Matsumoto and D. C. Bogue, Non - Isothermal Rheological Response During Elongational Flow, Trans. Soc. Rheol. 21:453 (1977)
4. F. P. La Mantia and G. Titomanlio, Testing of a Constitutive Equation with Free Volume Dependent Relaxation Spectrum, Rheol. Acta 18:469 (1979)
5. D. Acierno, F. P. La Mantia, B. de Cindio and L. Nicodemo, Transient Shear and Elongational Data for Polyisobutylene Melts, Trans. Soc. Rheol. 21:261 (1977)
6. L. C. E. Struik, "Physical Aging in Amorphous Polymers and Other Materials", Elsevier, Amsterdam (1978).

DOUBLE STEP STRAIN AND RELAXATION OF POLYMER MELTS

G.Marrucci, F.P.La Mantia, G.Spadaro, G.Titomanlio

Istituto di Ingegneria Chimica, V.delle Scienze, Palermo

Istituto di Principi di Ingegneria Chimica
Piazzale Tecchio, 80125 Napoli

INTRODUCTION

A possible description of the viscoelastic behaviour of polymer melts in the non-linear range which has been recently considered makes use of stress equations of the form

$$\underset{\sim}{\sigma}(t) = \int_{-\infty}^{t} \dot{\mu}(t-t') \, \underset{\sim}{Q}_t(t') \, dt' \qquad (1)$$

where the tensor $\underset{\sim}{Q}_t(t')$ is determined by the deformation occurred between t' and t and the memory kernel $\dot{\mu}$ is a function of time only.

Experiments of stress relaxation support Eq.1 insofar as, to a good approximation, the relaxing stress can be factorized as

$$\underset{\sim}{\sigma}(t) = \mu(t) \, \underset{\sim}{Q} \qquad (2)$$

where $\underset{\sim}{Q}$ only depends on deformation and μ on time. Eq.1 applies this separability between time and deformation to an arbitrary deformation history thus introducing a superposition principle which extends to the non-linear range the well known Boltzmann principle of linear viscoelasticity.

From the viewpoint of molecular theories, Eq.1 has been derived by Doi and Edwards [1] by using a "reptation" model. The same equation with a modified form for $\underset{\sim}{Q}$ has been recently advanced by Marrucci and Hermans [2].

In this paper, Eq.1 is tested against experiments of stress relaxation following a deformation which is applied in two steps. Since the experiments are made in simple extension, we use the extension ratio λ to characterize the deformation. In single step relaxation, Eqs 1 or 2 would predict

$$\sigma(t,\lambda) = \mu(t)\, Q(\lambda) \tag{3}$$

In double step relaxation, if λ_1 and λ_2 are the first and second deformation respectively, $\hat{t}$ the time interval between the deformation steps and t the time after the second deformation has been applied, Eq.1 would give

$$\begin{aligned}\sigma(t) &= \mu(\hat{t}+t)\, Q(\lambda_1\lambda_2) + \{\mu(t) - \mu(\hat{t}+t)\}\, Q(\lambda_2) = \\ &= \sigma(\hat{t}+t,\, \lambda_1\lambda_2) + \sigma(t,\lambda_2) - \sigma(\hat{t}+t,\, \lambda_2)\end{aligned} \tag{4}$$

In Eqs 3 and 4, $\sigma(t,\lambda)$ represents the set of curves obtained for the stress in single step relaxation experiments at various λ's. The relaxing stress after a double step deformation has been indicated simply as $\sigma(t)$. As shown by Eq.4, $\sigma(t)$ should be obtained by suitably combining the results of single step experiments.

Equations similar to Eq.4 were derived by Zapas and Craft [3] using a BKZ model and by Osaki et al.[4] for the case of shear deformations. The tensile experiments by Zapas and Craft support Eq.4 but the deformations used by them are relatively small ($\lambda_1\lambda_2 < 2$). In the case of the large shear deformations used by Osaki et al., significant deviations from Eq.4 were observed.

EXPERIMENTAL RESULTS

The tensile tests were run on a polyisobutylene known as Oppanol B 50 by using an Instron machine. The temperature was held constant at 24°C. Strip samples 1 cm wide were cut from a 3 mm thick sheet obtained by compression moulding at 200°C. The initial distance between the machine jaws was 5 cm. These dimensions assured a good uniformity of the deformation up to the maximum elongation before rupture, which occurred at values of λ somewhere between 3.5 and 4.

Fig.1 shows the results of single step relaxation experiments which provide the function $\sigma(t,\lambda)$. As indicated by Fig.1, the curves for the different λ-values run approximately parallel to one another in agreement with Eq.3 .

Fig.2 shows a typical result of a double step experiment (λ_1 =

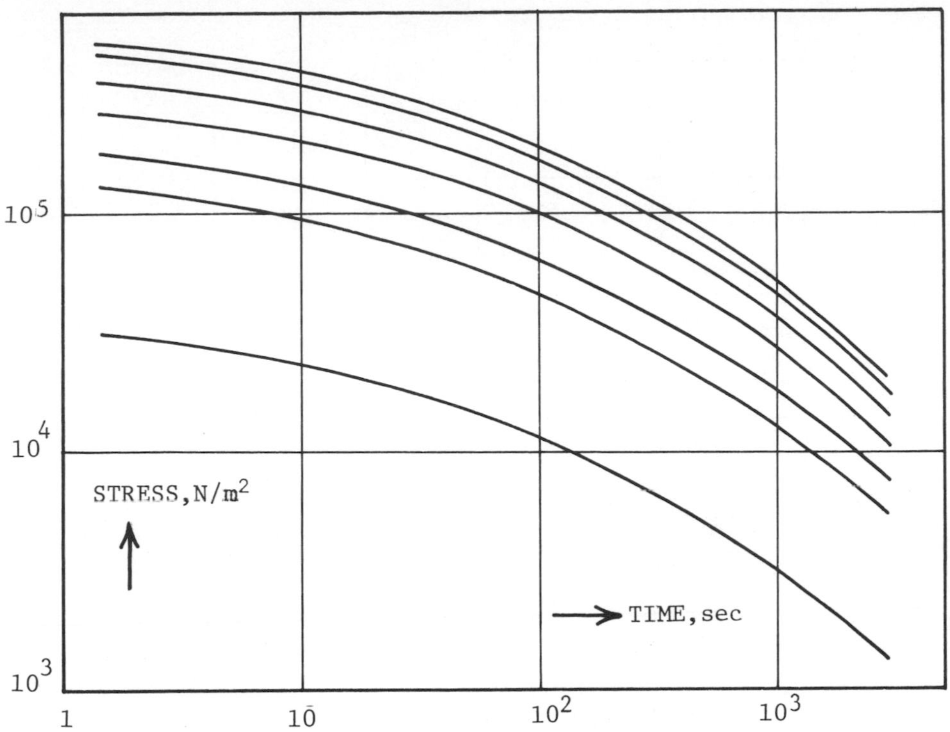

Fig.1. Single step relaxation. The values of λ are (starting from the lowest curve) 1.07, 1.26, 1.40, 1.75, 2.15, 2.75, 3.05 .

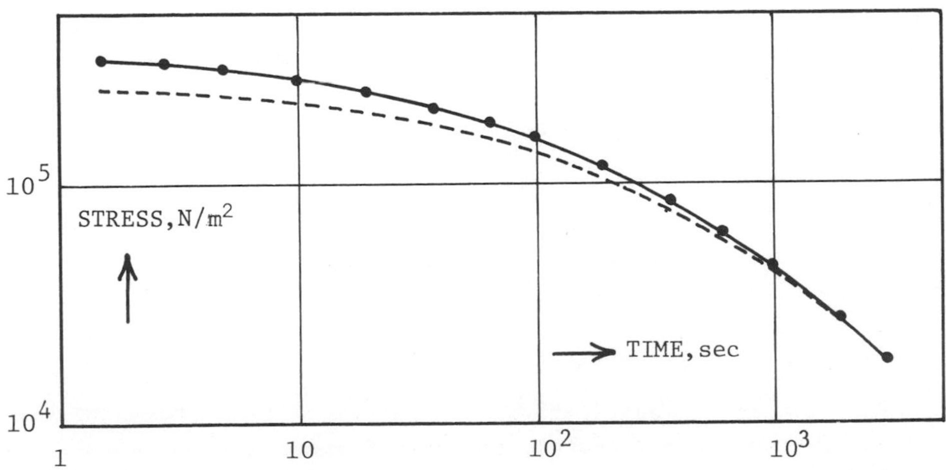

Fig.2. Double step relaxation. See text for explanations.

Table 1

λ_1	λ_2	λ	$\hat{t}$, sec	t",sec	$\frac{\mu(t'')}{\mu(\hat{t})}$	$\frac{\lambda^2+2/\lambda}{\lambda_1^2+2/\lambda_1}$
1.4	1.26	1.76	100	50	1.25	1.25
"	"	"	1200	600	1.55	"
"	"	"	3000	1500	1.7	"
2.15	1.26	2.71	100	35	1.4	1.45
"	"	"	1200	500	1.7	"
2.15	1.4	3.01	100	20	1.75	1.75
"	"	"	1200	300	2.25	"
"	"	"	3000	400	3.9	"
1.26	2.15	2.71	1200	100	4.0	2.55

2.15, λ_2 = 1.26, $\hat{t}$ = 100 sec). The circles in Fig.2 are the experimental values of σ whilst the dotted curve is obtained by using Eq.4 and the results of Fig.1 . As shown by Fig.2, the data are higher than the predictions of Eq.4 especially at low values of t . Similar results were obtained for all double step experiments. The conditions of the runs are summarized in Table 1 .

The full curve drawn in Fig.2, which fits the data, was obtained by the following procedure. By trial and error, a fictitious value of time was determined, called t", to be used in Eq.4 in place of $\hat{t}$. In view of the form of Eq.4, the fit is always obtained at a value of t" which is smaller than $\hat{t}$. The values of t" for the various runs are also reported in Table 1 . The last two entries of the same table are discussed in the following section.

DISCUSSION

If the fit of the double step data obtained by using Eq.4 with t" in place of $\hat{t}$ is not accidental, there follows the concept that by applying the second deformation the system which was relaxing the first one somehow "rejuvenates".

This concept is qualitatively in agreement with the model by Doi and Edwards. In this model, the "tube" which surrounds each macromolecule gets deformed by the applied deformation and then, during the relaxation, an increasingly large part of it recovers its equilibrium structure as a consequence of the macromolecule diffusing back and forth along its length. Indeed, if in Eqs 2 or 4, $\mu(t)$ is normalized so as to be 1 at the beginning of the relaxation, $\mu(t)$ at any other time represents the average fractional length of tube which has kept the original deformation up to time t.

Eq.4 can be directly interpreted in terms of the Doi-Edwards model. The first term is the contribution to the stress at time t of the tube section which was still λ_1-deformed when the second deformation, λ_2, was applied. Its contribution is therefore proportional to $Q(\lambda_1\lambda_2)$. The second term gives the contribution of the remaining part, already at equilibrium at that time, which is thus proportional to $Q(\lambda_2)$ only.

Using a value of t" smaller than $\hat{t}$ in Eq.4 is justified by the following argument. When the second deformation is applied, the length of tube which was still deformed according to the first deformation gets abruptly enlarged. In simple extension, it is easily shown that the enlargement factor should be

$$F \equiv \sqrt{\frac{\lambda^2 + 2/\lambda}{\lambda_1^2 + 2/\lambda_1}} = \frac{\mu(t'')}{\mu(\hat{t})} \qquad (5)$$

where $\lambda = \lambda_1\lambda_2$ is the overall extension ratio. Eq.5 equates the enlargement factor to the ratio $\mu(t'')/\mu(\hat{t})$. In fact, the abrupt length increase of the tube section which had kept the original deformation would correspond to a rejuvenation. In deriving Eq.1 from the Doi-Edwards model, this rejuvenation effect was lost by introducing the so called "independent alignement approximation"[1]

Notice that when small deformations are considered, by letting $\varepsilon = \lambda - 1$ and $\varepsilon_1 = \lambda_1 - 1$, we obtain

$$F = 1 + O(\varepsilon^2) + O(\varepsilon_1^2) \qquad (6)$$

so that the influence of the enlargement factor becomes a second order effect. In this case, $t'' = \hat{t}$ and the superposition principle of linear viscoelasticity is recovered.

The last two entries of Table 1 are the values of the ratio $\mu(t'')/\mu(\hat{t})$ and of the group $(\lambda^2 + 2/\lambda)/(\lambda_1^2 + 2/\lambda_1) = F^2$. The ratio of the μ's was obtained from the values of t" and $\hat{t}$ by using the lowest curve ($\lambda = 1.07$) of Fig.1 . Should the time-deformation separability expressed by Eq.3 be completely satisfied, this ratio could be calculated by any other of the curves of Fig.1 with the same result. In practice, slight differences are found and the values obtained by using the curve for $\lambda = 1.07$ are the smallest.

Comparing the values in the last two columns of Table 1 shows that Eq.5 is not satisfied quantitatively. Agreement is found only in that, by changing λ and λ_1, F and $\mu(t'')/\mu(\hat{t})$ both increase or decrease together. However, $\mu(t'')/\mu(\hat{t})$ is always larger than F and increases as $\hat{t}$ increases. It may be noticed that at low values of $\hat{t}$, $\mu(t'')/\mu(\hat{t})$ is approximatively equal to F^2 (not F). No interpretation of these results is advanced here. It seems, however, that relaxation mechanisms other than the reptation process should be considered.

It is finally noted that the deviations from Eq.4 observed by Osaki et al.[4], albeit apparently in the opposite direction than found here, are still consistent with a rejuvenation concept, i.e. with a value of t" smaller than $\hat{t}$. In fact, for the large shear deformations used in their work the stress becomes a decreasing function of the strain so that the difference $\sigma(\hat{t}+t, \lambda_1\lambda_2) - \sigma(\hat{t}+t, \lambda_2)$ appearing in Eq.4 is negative in their case.

REFERENCES

1. M. Doi and S. F. Edwards, Dynamics of Concentrated Polymer Systems, Parts 1, 2, 3, J.Chem.Soc.Faraday Trans.II, 74:1789, 1802, 1818 (1978)
2. G. Marrucci and J. J. Hermans, Non-linear Viscoelasticity of Concentrated Polymeric Liquids, Macromolecules, in press
3. L. J. Zapas and T. Craft, Correlation of Large Longitudinal Deformations with Different Strain Histories, J.Res.Nat. Bur.Stand., A 69:541 (1965)
4. K. Osaki, Y. Einaga, M. Kurata, N. Yamada and M. Tamura, Stress Relaxation of Polymer Solutions under Large Strains: Application of Double-Step Strain, Polymer J., 5:283 (1973)

CONSTITUTIVE EQUATIONS, DEFORMATION RATE SOFTENING AND ELONGATIONAL FLOW CHARACTERISTICS OF POLYMER MELTS

J.L. White, H. Tanaka, W. Minoshima

Polymer Engineering, The University of Tennessee, USA

(Abstract)

Single integral constitutive equations representing the flow behavior of polymer melts and particle reinforced polymer melts are contrasted. The model for the melt is a traditional non-linear viscoelastic formulation. The constitutive equation representing the particle reinforced melt contain a yield value and is a plastic-viscoelastic material. This is expressed through a von Mises yield criterion beyond the integral form.

The predictions of these constitutive equations are compared with emphasis on the response in elongational flow. In the case of the viscoelastic fluid, the elongational flow is dominated by the deformation rate softening character. The magnitude of the yield value determines the elongational flow response in the case of the plastic viscoelastic fluid. The implications of this behavior are described and compared with experiments on elongational flow of initially virgin filaments and melt spinning.

RHEO-OPTICAL STUDY OF THE FLOW OF MOLTEN POLYMERS

Claude Dehennau

Solvay & Cie, S.A.
Laboratoire Central
310, rue de Ransbeek
B - 1120 Bruxelles

ABSTRACT

The flow of molten polymers in converging and diverging channels has been studied by rheo-optical techniques including flow birefringence and flow velocity measurements. The flow patterns of HDPE, LDPE, PS, PP, PVC and copolymers of PAN were analysed. The results show that, unlike the stress distribution, the velocity profile is unaffected by the viscoelastic properties of the melt when the converging flow is forced or when the flow occurs in a divergent geometry. The size of the secondary flow which can be observed in a sharp-edged slit die entrance, varies with the flow rate and the rheological parameters of the melt. On the contrary, when the flow is developed in a 180° slit die exit, the secondary flow becomes independant of these parameters and is volumetrically constant. These concluding remarks and the study of the stress distribution in the converging flow permit to correlate the shape of the secondary flow, and particularly the natural converging angle, to the viscoelastic properties of the melt. Indeed the natural converging angle can be related not to the Weissenberg number, but to the ratio of tensile stress to shear stress.

INTRODUCTION

The development of a new grade of thermoplastic for extrusion processing can be greatly facilitated if the influence of the viscoelastic properties of the molten polymer on its flow in a die is known. Particularly, the knowledge of the relationship between die geometry, viscoelastic properties of the melt and flow stagnation or flow instability at a fixed output extrusion rate is important.

A first study[1] allowed us to develop experimental techniques and numerical simulation programmes which can be used to predict the effect of the flow in a channel of constant geometry on thermal degradation and sticking of a thermal sensitive molten polymer. We are able to adjust the output rates, the temperature or the geometrical parameters or to modify the raw material itself to obtain a product of constant quality during extrusion.

These experimental and theoretical techniques are not able to predict the stagnation or secondary flow occurring in a more complex die having convergent or divergent channels or some discontinuity in the cross section variation[2].

A better understanding of converging and diverging flows of viscoelastic polymeric melts has become important in order to caracterise completely the effect of viscoelastic properties and die geometry on stagnation[3,4] and on flow instabilities[5].

The goal of this study is to understand and explain the effect of the viscoelastic properties of the molten polymers on the shape and the stability of its flow in a complex die geometry.

EXPERIMENTAL

The flow of the molten polymer was observed by rheo-optical methods in a special die fed by a BM 30 Maillefer extruder. The cross-section of the die was rectangular and its thickness was continuously varied between 8 and 2 mm. Convergence half angle of 90, 67.5, 45 and 10 degrees could be used. As the die is symetric, diverging and converging flow could be studied with the same apparatus. The flow velocity distribution was measured through Pyrex windows by following the displacement of glass microspheres acting as optical tracers. A high speed cine camera Hitachi 16 HM, provided with Nikon 50 objectives was used to record the motion of the optical tracers. The analysis of the data was made on a scanning table. As a consequence of the techniques used to study the flow, the calculated velocities are mean values. The stress distribution in the flow was measured by means of the flow birefringence technique. Colour photography was used to record the stress birefringence patterns.

A system similar to an extruder screenchanger enables the die to be changed without dismantling and cleaning the equipment.

By adjusting a by-pass opening, it was possible to vary the flow rate through the die from 0 to 0,7 ml s without any modification of the extrusion conditions. All the rheo-optical results were completed by flow studies on a capillary Instron viscosimeter. Special capillaries having 1 mm diameter, different lengths and half angles of 90, 67,5, 45, 22,5 and 10 degrees at the entrance were used to calculate the shear and tensile viscoelastic properties. Shear viscosity and Weissenberg number were measured in Poiseuille flow. The tensile properties were evaluated by means of the loss pressure and the die swell measured during the flows through a short die as proposed by Cogswell[6].

Low- and high-pressure polyethylene, polypropylene, PVC and copolymers of PAN were tested. From a rheological point of view, the selection was made to provide polymers of different chemical compositions, long chain branching and copolymerisation levels and shear properties. For instance, the viscosity and Weissenberg number varied respectively from 300 to 20000 Pa s and from 0,25 to 2,5 for a shear rate of 10 s^{-1}.

RESULTS AND DISCUSSION

The cross section of a convergent is decreasing from the entrance till the exit and the fluid is accelerated and subjected to a complex flow which involves shearing and stretching. On the contrary, the cross section of a divergent is increasing and only shearing stresses can be developed in the flow.

Flow and velocity profiles

The secondary flow which can be observed in a sharp-edged slit die entrance disappears when the half angle at the die entrance becomes lower than a critical value depending on the melt's viscoelastic properties. For instance, most of the polymers do not give secondary flow in the convergent die of 67,5 degrees while a convergent die of 22,5 degrees must be used at high shear rates for high pressure polyethylene to prevent it. When the half angle of the convergent becomes lower than 90°, the velocity field of the melt converges as more a sink flow as the half converging angle is decreasing so that any flow line can be extrapolated to the angle formed by the two walls of the tapered-slit die.

If the flow is developed in a 90° slit die exit, or if the tensile deformation of the flow becomes weak, the size of the secondary flow is independant of the flow rate and of the visco-elastic properties of the molten polymer. The same conclusion can be reached when the velocity flow distribution in divergent and convergent dies is analysed.

For instance, at a given flow rate the flow velocity profiles of high and low pressure polyethylenes through a 90 or a 45 degrees tapered slit die are different ; on the contrary, for a converging forced flow without secondary flows, the velocity profiles are similar and only weakly influenced by the viscoelastic properties of the melt. However, the measured velocities being mean values, it could be possible that some weak effect of the melt's viscoelastic properties could be unobserved.

These results show that simple flow theories such as the sink flow theory developed for newtonian and non-elastic melts, can be used to calculate the experimental velocity distribution found in a forced converging or diverging flow, when the convergence half angle becomes small. The flow velocity V_R calculated at a radial position R by means of the sink flow theory is given by the equation :

$$V_R = \frac{Q}{Rb} \frac{\cos 2\alpha - \cos 2\theta}{\sin 2\alpha - 2\alpha \cos 2\alpha}$$

where Q is the flow rate,
θ an angular parameter,
b the width of the die and
α the half angle of the convergent.

The ratio of the difference $R_1 - R_2$ and the transit time t between the 2 points gives the mean velocity between the radial position R_1 and R_2

$$t = \int_{R_1}^{R_2} \frac{dr}{V_R}$$

At a given flow rate, the theoretical and experimental velocity profiles in a 10 degrees divergent or convergent die are similar.

Stress distribution

The flow stress distribution is measured by means of the flow birefringence technique. To simplify the theoretical analysis we calculate only the principal shear stress given by the classical relationship :

$$\frac{\sigma_1 - \sigma_2}{2} = \frac{\Delta n}{2c}$$

where σ_1 and σ_2 are the principal stresses
c is the stress optical coefficient
Δn is the birefringence

For each flow rate and die geometry, the value of the principal shear stress is a function of the viscoelastic properties of the melt. As expected for viscoelastic fluids, the distribution of the stresses is different in converging and diverging flows.

A comparative study of the isostress patterns shows that in a diverging flow, the isostress curves are parallel to the wall of the die and that the axial principal shear rapidly reaches a low value. These different isostress patterns must be related to the flow deformation which the melt undergoes in a convergent or a divergent channel. A complex tensile and shear flow develops in a convergent die and only a shear flow in a divergent die. In particular, the axial principal shear stress in a converging flow can be described as a function of mean tensile strain and tensile rate in such a way that the relationship becomes independent of the geometry of the convergent.

The stresses measured in a secondary flow occuring in a sharp-edged slit die entrance are small and this part of the flow seems black or grey in a flow birefringence observation with white light. This property allows us to evaluate the natural converging angle. For the output rates explorated and a stable flow, the values found with HDPE and PP are practically independent of the flow rate and are respectively close to 85 and 75 degrees. Polystyrene and LDPE give lower values which vary versus the flow rate. On the contrary when the flow becomes divergent, the value of the natural angle is now independent of the flow rate and of the viscoelastic properties of the melt.

Capillary viscosimeter experiments

The results show that below a critical value of the half converging angle of the capillary entrance, the die swell and the critical shear rate above which instabilities occur, are functions of the entrance geometry. The presence of this critical value must be related to the natural converging angle seen during rheo-optical analysis of the flow.

Capillaries having a diameter of 1 mm, L/R ratios of 1 and 15, and a half angle of 22,5 degrees at the entrance have been used to measure the tensile and shear properties of the molten polymers. For all the polymers tested a relationship has been found between the ratio of tensile stress to the shear stress measured at the shear rate at the wall of the capillary and the natural converging angle. This influence of this ratio on die entrance phenomena, was first proposed theoretically by Cogswell[6]. It must be used to analyse the flow difference found between converging and diverging velocity and stress flow patterns.

The author wish to express his gratitude to the Institut pour l'Encouragement de la Recherche Scientifique dans l'Industrie et l'Agriculture (I.R.S.I.A.) for its financial support of this work.

1. J.C. Chauffoureaux, C. Dehennau and J. van Ryckevorsel, Journal of Rheology, 23 1 (1979)
2. R.F. Ballenger and J.L. White, Journal of Applied Polymers Science, 15 1949 (1971)
3. H. Nguyen and D.V. Boger, IUTAH Symposium on Non Newtonian Fluid Mechanics, Louvain-la-Neuve, August 29-September 1 (1978)
4. Kimimara Ritoh, Japan Plastics 12 16 (1978)
5. C.D. Han, Journal of Applied Polymers Science 17 1403 (1973)
6. F.N. Cogswell, Polymer Engineering Science 12, 64 (1972)

TWO SIMPLE TIME-SHEAR RATE RELATIONS COMBINING VISCOSITY AND FIRST NORMAL STRESS COEFFICIENT IN THE LINEAR AND NON-LINEAR FLOW RANGE

Wolfgang Gleissle

Institut für Mechanische Verfahrenstechnik
Universität Karlsruhe (TH)
7500 Karlsruhe, West-Germany

GENERAL

The theory of linear viscoelasticity is able to describe the time dependent flow behaviour of viscoelastic liquids in the range of small shear rates κ. Using a memory function, defined for example in [1], with a suitable set of constants the transient zero viscosity $\eta_0(t,\kappa\to 0)$ can be calculated. The theory of linear viscoelasticity however, predicts no shear rate dependency of the viscosity as aobserved for nearly all polymers at higher shear rates. The steady-state viscosity η_{st} at constant shear rate κ_{st} is defined by the equation

$$\eta_{st} = \tau_{st} / \kappa_{st} \tag{1}$$

To define the transient viscosity $\eta(t)$ at the beginning of a shear flow the following equation is used

$$\eta(t,\kappa(t)) = \tau(t) / \kappa(t) \tag{2}$$

Making a "stressing test"[2] with constant shear rate, you connot neglect the acceleration time to constant shear rate, in particular at high shear rates and for fluids with a fast response. In such cases you have to measure $\kappa(t)$ during each experiment [3] to avoid great errors in calculating the transient viscosity from shear stress $\tau(t)$. In the linear viscoelastic range the transient viscosity becomes independent of the shear rate.

$$\eta(t) = \tau(t) / \kappa(t) = \eta_0(t) \neq f(\kappa) \tag{3}$$

$$\Theta_{st}(\kappa) = (\sigma_1-\sigma_2)_{st}/\kappa_{st}^2 \tag{4}$$

The first steady-state normal stress coefficient Θ_{st} is defined by eq.(4). σ_1 means the stress in the direction of the shear flow and σ_2 the stress in the direction of the shear gradient. Analogous to the transient viscosity $\eta(t)$, the transient normal stress coefficient $\Theta(t)$ is defined as

$$\Theta(t,\kappa(t)) = (\sigma_1-\sigma_2)\ (t)\ /\ \kappa^2(t) \tag{5}$$

Because of the square of the shear rate, bad assumptions about the actual shear rate at time t lead to still greater errors in the calculation of $\Theta(t)$ from $(\sigma_1-\sigma_2)(t)$, than for the viscosity.

For investigating the transient stress behaviour of polymers undergoing shear flow with high shear rate, we have developed a special cone and plate rheometer. To get a good time resolution of the signals all transducers have a natural frequency of more than 2 kHz. With its very dynamic drive we can realise extremely short acceleration times to stationary shear flow. The shear rate κ=10,000 s^{-1} for example can be reached within 5 ms.

VISCOSITY-RELATION

The diagram on the left side of Fig. 1 shows a double logarithmic plot of the transient viscosity $\eta(t)$ of a high polymer silicone oil at 23°C. For experiments with different shear rates the viscosities calculated from transient shear stresses $\tau(t)$ are plotted versus the shear time from 1 ms up to 10 s. The shear rate has been varied in the range of $1 < \kappa < 4500\ s^{-1}$.

Because of the pseudoplastic behaviour of the silicone oil you get the lowest steady-state viscosity for the highest shear rate of 4500 s^{-1}. In this case the response of the fluid is very rapid. The overshoot, a typical behaviour in the non-linear viscoelastic flow range, is reached after about 3 milliseconds. After a shear time of 10 ms the

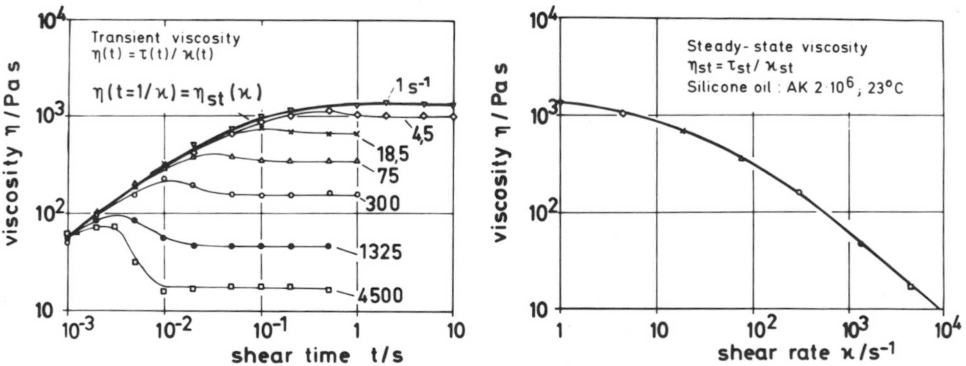

Fig.1 Transient and steady-state viscosity of a high polymer silicone oil [3] (Parameter: shear rate κ_{st})

viscosity has reached its steady-state value. For experiments with lower shear rates the viscosity becomes higher and the maximum value of the viscosity appears later after starting. The maximum values become smaller in respect to the steady-state value. For the experiment with $\kappa=1\ s^{-1}$ no significant overshoot can be observed. For still lower shear rates the viscosity η_{st} does not increase remarkably. Therefore η_{st} $(\kappa=1\ s^{-1})$ ist nearly the Newtonian zero shear viscosity η_o for this silicone oil and the transient viscosity $\eta(t)$ for $\kappa=1\ s^{-1}$ is the linear viscoelastic transient zero shear viscosity $\eta_o(t)$.

From the steady state values the viscosity function $\eta_{st}(\kappa)$ can be computed. The result is shown in the right diagram of Fig.1. AK 2.10[6] is a significant pseudoplastic fluid and its viscosity $\eta_{st}(\kappa)$ becomes smaller in the tested range from more than thousand down to ten Pa s. Comparing $\eta_o(t)$ on the left side and $\eta_{st}(\kappa)$ on the right side of Fig.1 you get a surprising result: Both curves have the same shape and the absolute values of transient viscosity $\eta_o(t)$ and the viscosity function $\eta_{st}(\kappa)$ have the same value if the product of shear rate κ and shear time is equal to one, respectively if the shear time is the reciprocal value of the shear rate or vice versa. The same consistency of measuring values we have found in all our experiments on polymers: different silicone oils, polyisobutylene and a polyethylene melt [3]. To be sure that this is not a result of our measuring equipment only, we have looked for testing series from other investigators in which $\eta_o(t)$ and $\eta_{st}(\kappa)$ has been measured. They all show the same correlation [4,5,6]. The results of measurements at a LDPE-melt, published in [6], are shown in Fig.2. For this example we have transferred the viscosity function from the right diagram to the times $t=1/\kappa$ in the left diagram. This curve is identical with $\eta_o(t)$.

It seems that there ecists a class of fluids for which the pseudoplastic behaviour is the mirror function of the linear viscoelastic behaviour. The viscosity function is given with the time behaviour in the linear viscoelastic range. The values of the viscosity function

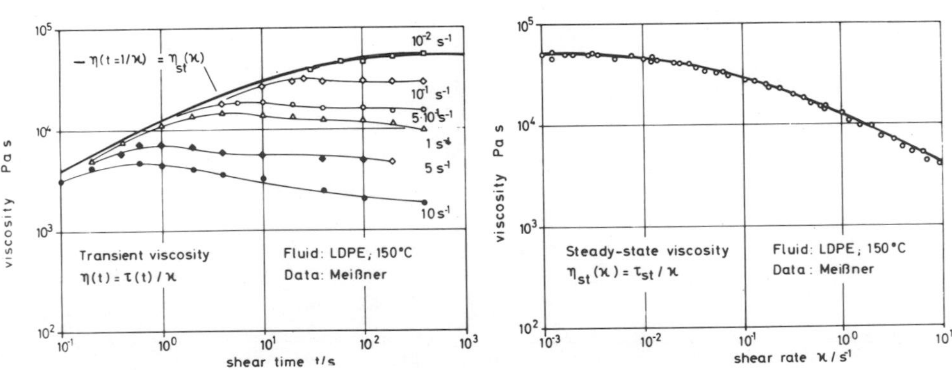

Fig.2 Transient and steady-state viscosity of a LDPE-melt[6] (Parameter: shear rate κ)

and the transient zero shear viscosity are equal if the product of shear rate κ and shear time t is equal to one.

$$\eta_{st}(\kappa) = \eta_o(t) \text{ if } \kappa t = 1 \qquad \text{rel.(1)}$$

If rel. (1) holds, you can transform the well-known exponential equation (eg.(6))for the linear viscoelastic transient zero shear viscosity into the viscosity function by substituting current time t and discrete times t_i by shear rate κ and κ_i. Transient zero shear viscosity and the shear dependent viscosity function have the same form (eqs. (6,7)).

$$\eta_o(t) = \Sigma_i \, \eta_i \, \left[1 - \exp \, (-t/t_i)\right] \qquad (6)$$

$$\eta_{st}(\kappa) = \Sigma_i \, \eta_i \, \left[1 - \exp \, (-\kappa_i/\kappa)\right] \qquad (7)$$

For practice rel. (1) means: if you have measured the viscosity function $\eta_{st}(\kappa)$ you can directly conclude on the time behaviour of the fluid. If you have equipmentwhich give the transient zero shear viscosity with good time resolution you get the viscosity function by replacing the inverse value of time t with the shear rate κ. The Newtonian fluid is a limiting case of rel. (1). Because it has no shear rate dependency you cannot expect any time effects.

NORMAL STRESS COEFFICIENT

The direct correlation between $\eta_o(t)$ and $\eta_{st}(\kappa)$ leads to the question if there exists a similar correlation for the normal stess coeffi-

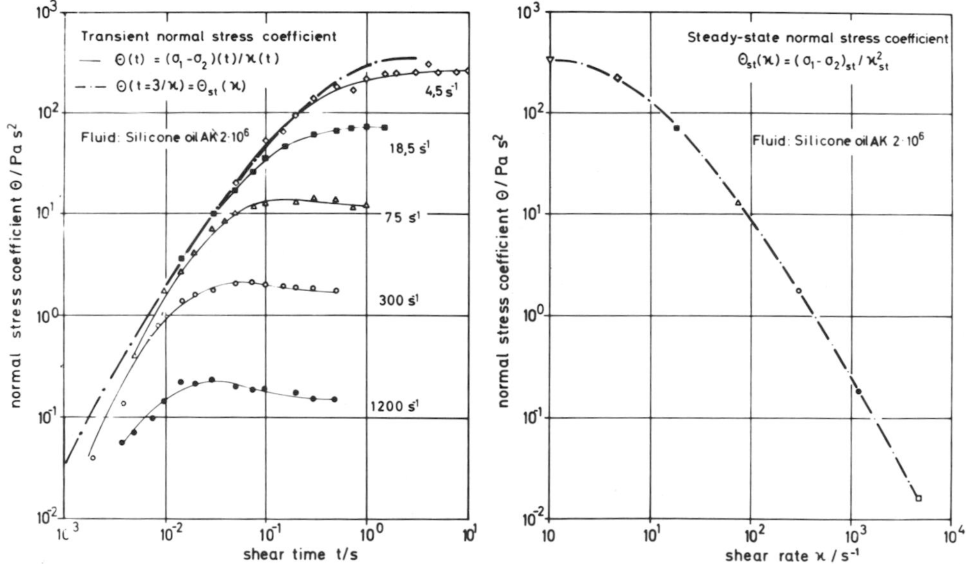

Fig.3 Transient and steady-state normal stress coefficient

cients. Steady-state normal stress coefficient $\Theta_{st}(\kappa)$ and the envelope of the transient normal stress coefficients $\Theta(t)$ have the same shape as demonstrated in Fig.3 for AK 2.10^6. This behaviour is similar to that of the viscosities. Comparing the values of $\Theta_{st}(\kappa)$ and $\Theta_o(t)$, according to the condition that $\kappa t=1$, you find no agreement. There exists no immediate relation between κ and t as found for $\eta_o(t)$ and $\eta_{st}(\kappa)$. But transferring the steady-state normal stress coefficient $\Theta_{st}(\kappa)$ in that way, that the times **t** are equal to $1/\kappa$ you get a curve which goes parallel to $\Theta_o(t)$. Shifting this curve along the time axis towards greater times, $\Theta_{st}(\kappa)$ is equal to $\Theta_o(t)$. For our measurements with silicone oil, polyisobutelene and LDPE you get, independent from the kind of fluid, the best fit of datas, if you chose k=3 as shifting facotor. The dashed curve in the time diagram of Fig.3 is $\Theta_{st}(\kappa)$ in the transformed manner. Fig.4 shows $\Theta_{st}(\kappa)$ and $\Theta(t)$ of the LDPE-melt, measured from Meißner. The lined and dotted curve in the time-diagram is $\Theta_{st}(\kappa)$ shifted with factot k=3. The correlation with $\Theta_o(t)$ is very good.

You get the same relation between $\Theta_{st}(\kappa)$ and $\Theta_o(t)$ as well as for the viscosities. Different to rel. (1) however, for identity the product κt is not unity.

$$\Theta_{st}(\kappa) = \Theta_o(t) \quad \text{if} \quad \kappa t = k \qquad \text{rel. (2)}$$

For all data of $\Theta_{st}(\kappa)$ and $\Theta_o(t)$ of very different fluids we have measured and found in publications the faktor k is in the range of $2,5 < k < 3$. This means that the transformation of the time behaviour directly into the shear dependency or vice versa using rel. (2) gives

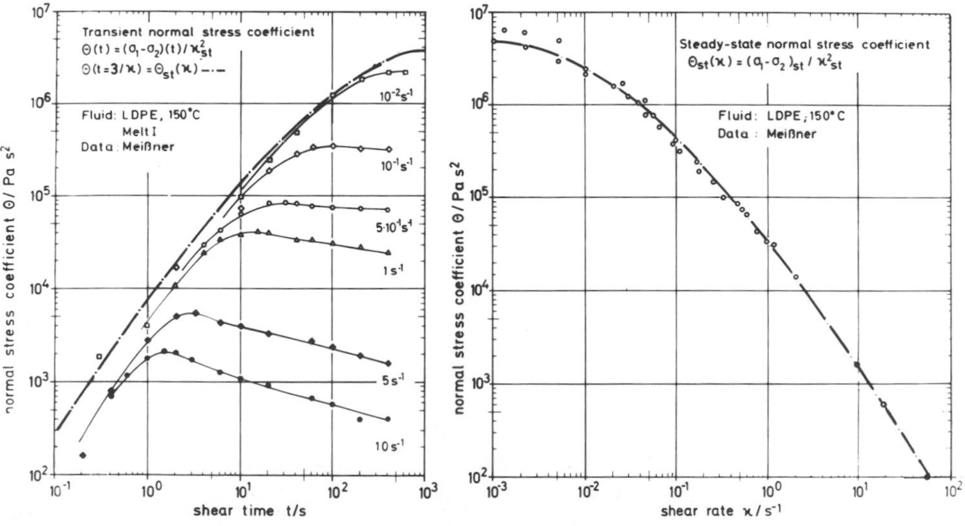

Fig.4 Transient and steady-state normal stress coefficient

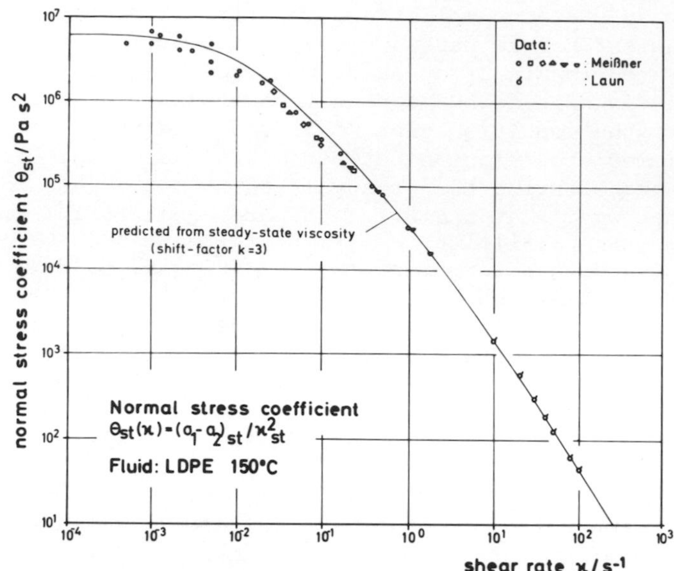

Fig.5 Comparison between calculated and measured normal stress

good agreement of $\Theta_{st}(\kappa)$ and $\Theta_o(t)$ for a lot of fluids. Derived from the theory of rubberlike liquid you are able to calculate the transient zero shear normal stress coefficient $\Theta_o(t)$ from $\eta_o(t)$ with a set of suitable viscosities η_i and times t_i.

$$\Theta_o(t) = 2\sum_i \eta_i t_i \left[1-(1+t/t_i)\exp(-t/t_i)\right] \tag{8}$$

Substituting t_i and t in eq. (8) by suitable values which you get from the condition of rel. (2), you obtain $\Theta_{st}(\kappa)$.

$$\Theta_{st}(\kappa)=2\sum_i(\eta_i/\kappa_i)\left[1-(1+k\kappa_i/\kappa)\;\exp(-k\kappa_i/\kappa)\right] \tag{9}$$

If rel.(2) holds, you can calculate with eq.(9) the normal stress coefficient, if you have measured nothing else than $\eta_{st}(\kappa)$. An example of these calculations is given in Fig. 5 for the LDPE-melt.

1. A.S. Lodge "Elastic Liquids", Academic Press, New York(1964)
2. H. Giesekus, Proc. Forth Int. Congress on Rheology,Providence (1963) Part 3, Interscience New York (1965)
3. W. Gleissle, Diss. Universität Karlsruhe (1978)
4. F.H. Gortemaker, M.G. Hansen, B. de Cindio, H.M. Laun, H. Janeschitz-Kriegl, Rheol. Acta 15 (1976) 256-267
5. K. Osaki, S.Ohta, M. Fukuda, M. Kurata, J. of Polym. Sci. 14 (1976) 1701-1715
6. M.H. Wagner, Rheol. Acta 16 (1977) 43-50

APPLICATION OF THE LASER DOPPLER VELOCIMETRY TO POLYMER MELT FLOW STUDIES

H. Kramer and J. Meissner

Department of Industrial and Engineering Chemistry
Swiss Federal Institute of Technology (ETHZ)
CH - 8092 Zürich, Switzerland

INTRODUCTION

Laser Doppler velocimetry (LDV) is a new tool for fluid dynamics studies /1/. By LDV the velocity vector field can be measured without disturbing the flow itself. However, in polymer melt fluid dynamics, difficulties arise because of temperature problems, the relatively high pressure involved, and the often extremely slow flows. In this paper a LDV system is described which was set up especially for polymer melt flow investigations.

EXPERIMENTAL PROCEDURE

The optical system used for this investigation is schematically shown in fig. 1. The light beam from the argon laser (AL)* is split by the optical unit (OU) into two parallel beams, which are focussed by the lens L and form at their intersection the measuring volume (MV), of the dimension 0.04 x 0.04 x 0.2 mm^3 and with an interference pattern of bright and dark planes. These planes are perpendicular to the optical plane formed by the two beams and parallel to the optical axis, i.e. the median line between the two beams.

A Bragg cell within (OU) shifts the interference planes with constant velocity in the direction of their normal. A particle passing through (MV) produces scattered light of changing intensity, the frequency of which is proportional to the difference between the velocity of the interference planes and the particle velocity component v_m normal to these planes. Hence, the Bragg cell allows one to measure the magnitude and the sign of v_m. The perpendicular component follows from an additional measurement after a rotation of the optical plane around the optical axis by $\pi/2$.

---

*SPECTRA PHYSICS argon laser; we use the green line (514.5 nm)

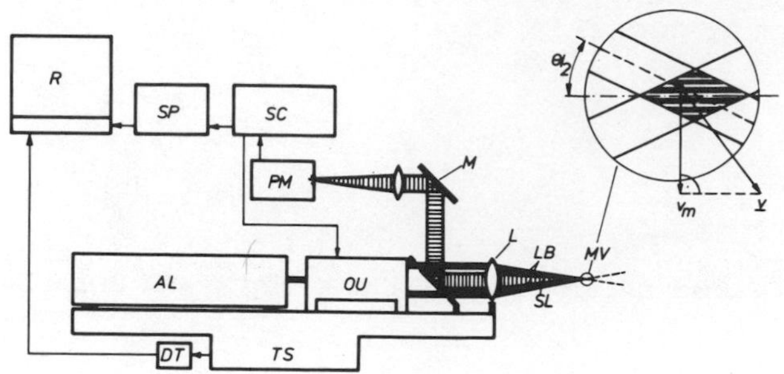

Fig. 1 Schematic diagram of the Laser-Doppler system.

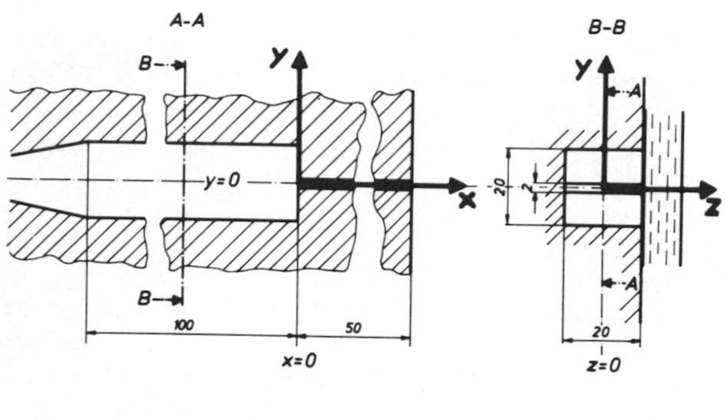

Fig. 2 Flow channel, slit die, and coordinate system.

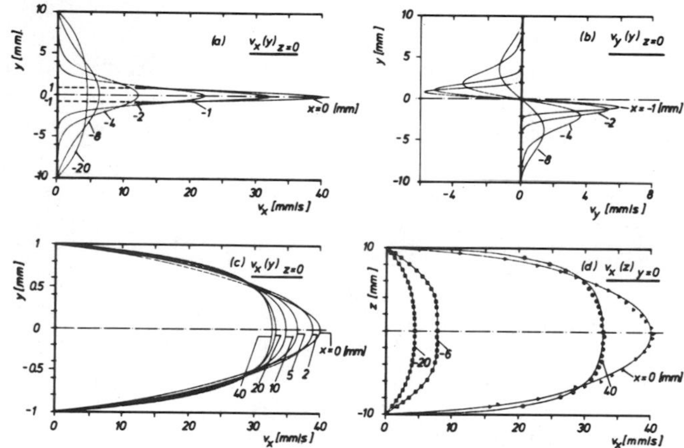

Fig. 3 Velocity profiles for different planes x = const. in front of and within the die. LDPE at 150°C, $D = 65\ s^{-1}$.

As illustrated in fig. 1, the light which is scattered backward from a particle in (MV) on the central part of lens (L), is collimated and focussed on the photomultiplier (PM). The frequency within each random-burst signal, of the intensity of scattered light, is measured by a counter type signal processor (SP) and recorded as velocity by the recorder (R). Because of the Bragg cell, the high standards in noise free amplification, and the high frequency stability, very small velocities should be measurable from one single burst, i.e. for one particle passing through (MV). However, because of externally induced mechanical oscillations, the lowest limit of the velocity measurable at present is 0.1 mm/s.

The optical system of fig. 1 was used to determine the velocity field of a LDPE melt streaming through the flow channel and the slit die shown in fig. 2 together with the cartesian coordinates (X,Y,Z) used in this paper. One side of the slit and the channel was covered by a polished PYREX glass plate through which the light could pass. The mechanical vibrations could be reduced by mounting the flow channel and die together with the optical system on a heavy granite table. The flow channel and the extruder (20 mm screw) providing the melt flow, were interconnected by a flexible, steel armed PTFE tube. By means of a translation stage (TS of fig. 1) the optical system could be moved in order to position the measuring volume (MV) at any place within the channel and the die of fig. 2.

The polymer melt used for this investigation was low density polyethylene (LDPE) with a density at 20°C of ρ_{20} = 0.918 g/cm^3 and a melt flow index MFI (190/2.16) = 1.3 g/10 min /2/. The density of light scattering particles in the material as received was too small, therefore TiO_2 powder was added (grain size < 1 µm, 1 mg TiO_2/ 1 kg LDPE). The temperature of the melt within channel and die was 150°C.

VELOCITY PROFILES

Two output rates were applied at which the melt flow was always stable, i.e. no melt fracture occured. The corresponding apparent shear rates in the slit were D = 13.7 and 65 s^{-1}. The velocity profiles measured at D = 65 s^{-1} are shown in fig. 3, where the curves are symmetric polynomials fitted to the measured values. Only in fig. 3(d) a difference can be detected between the measured values indicated as points and the best fit symmetric polynomials.

Fig. 3(a) and (b) show the development of the entrance flow field in front of the die for the plane of symmetry z = 0 (see the coordinate system of fig. 2). x = 0 denotes the entrance of the slit die. At x = -20 mm, the fully developed flow profile in the flow channel is not yet disturbed by the slit, hence v_y(x = -20 mm) is zero within the accuracy of our measurements. Because of geometrical reasons, v_y cannot be measured within the die presently. Therefore, in fig. 3(c) and (d) only v_x is shown for x > 0. v_x can be measured as

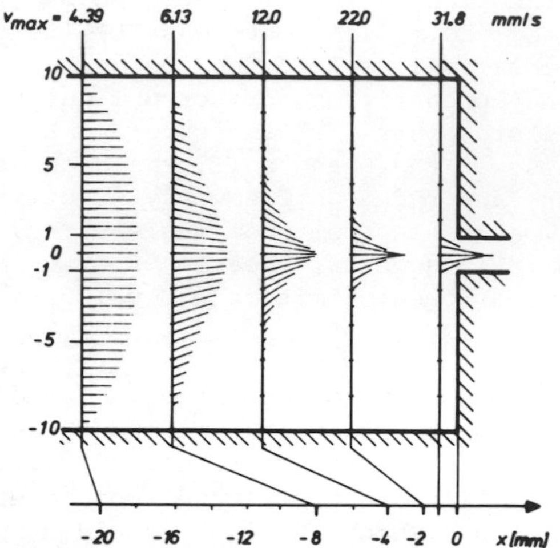

Fig. 4 Velocity vector field at the slit entrance (x = const., D=65 s^{-1}).

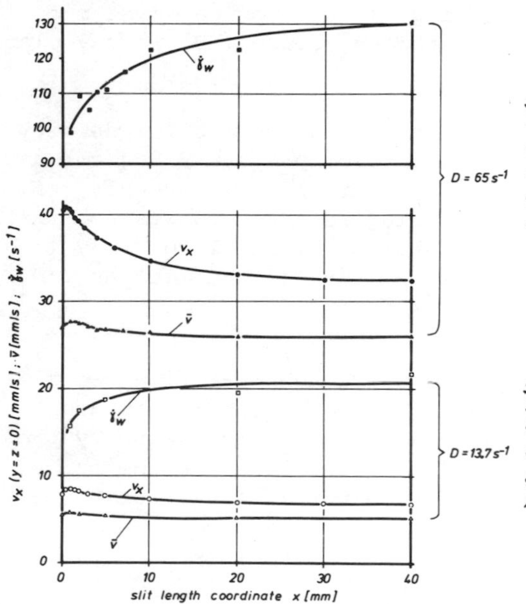

Fig. 5 Development of the equilibrium values for $v_x(y = z = 0)$, $\bar{v}$, and $\dot{\gamma}_w$ within the slit die.

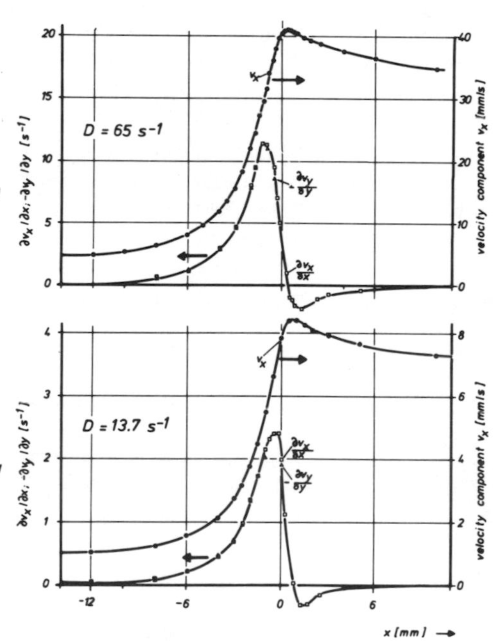

Fig. 6 Velocity and deformation rate components along the centerline of the slit entrance.

close as 0.03 mm to the slit wall ($y = \pm l$). For the best fitted symmetric polynomials no wall slip was assumed: $v(y = \pm l) = 0$. We notice from fig. 3(c) that the velocity profile $v_x(y)$ changes remarkably within the slit die.

In order to investigate the width of the planar flow region, the profile $v_x(z)$ was measured in front of and within the die for the plane $y = 0$. Small deviations from the symmetric profile, to be seen in fig. 3(d), are due to temperature problems connected with the glass cover. Fig. 3(d) also shows that the profile $v_x(z)$ and the width of the planar flow region change with the length coordinate x. Assuming that the profiles of fig. 3(c) also have the same shape for planes $z \neq 0$, but scaled down by the appropriate factor taken from fig. 3(d), the flux through a plane x = const. can be calculated. For x = 40 mm, the result of this calculation deviates only by 3.4 % from the output rate q measured directly by cutting the strand at the die exit.

For $z = 0$, the velocity vectors in the entrance channel are shown as line segments in fig. 4 for different planes x = const. The scale factor for each plane is indicated by the maximum v_{max}. Obviously, the velocity field in the secondary flow regions at the sides of the central melt stream are too small to be drawn in this graph.

DISCUSSION

It follows from fig. 3 that a considerable length within the die is needed for full development of the velocity profile (if in view of heat production within the melt, such a profile exists at all). Hence, under the conditions investigated, a flow entrance region exists not only in front of but also within the slit die, and this result is interesting for melt viscometry when slit dies of flat entrance are used. For the plane $z = 0$ and as functions of the coordinate x within the die, fig. 5 shows the maximum velocity $v_x = v_{x,max}$, the average velocity $\bar{v}$ (calculated from the areas of the profiles $v_x(y)$, and the true shear rate at the slit wall, $\dot{\gamma}_w = \partial v_x/\partial y$ for $y = \pm l$.

For both shear rates D applied, v_x shows a maximum at the entrance of the slit die. A steady decrease follows up to the equilibrium value at x = 30 mm for D = 13.7 s^{-1}. For D = 65 s^{-1}, the equilibrium value of v_x is not achieved within the slit length used. The decrease of v_x is accompanied by an increase of the velocity gradient near the wall and, correspondingly, of $\dot{\gamma}_w$. Even $\bar{v}$ is not constant along x, but shows a small maximum and a slight decrease at the entrance. This behaviour of $\bar{v}$ indicates that in this region of the die $\partial v_z/\partial z \neq 0$.

From the velocity measurements, the deformation history can easily be derived for a material particle moving along the centerline (y = 0, z = 0) of the channel and die. For the two apparent shear rates applied, fig. 6 gives the v_x-component of this particle as a function of x, its derivative $\partial v_x/\partial x$, and from the profiles of fig. 3(b) the negative derivative $-\partial v_y/\partial y$. The latter curve can be determined presently only in the flow channel in front of the die. Fig.6 shows that the two derivatives give a sum of zero, i.e. $\partial v_x/\partial x = -\partial v_y/\partial y$. In orther words, for the space near the central axis of the channel the condition of "pure shear" deformation exists within the accuracy of our measurements and with the incompressibility assumption for the melt. Notice that for the particle considered, the maximum of the deformation rate component $\dot{\varepsilon}_{11} = \partial v_x/\partial x = -\dot{\varepsilon}_{22}$ is very high and located in front of the die entrance. All other components of the deformation rate tensor are zero. Within the die there is a region where $\dot{\varepsilon}_{11} < 0$, probably due to the rapid recovery of the elastic tensile deformation at the die entrance.

CONCLUSION

The present investigation demonstrates that the LDV method can be applied for the measurement of velocity vector fields of polymer melt flows even at temperatures higher than room temperature. From these measurements, interesting results follow (a) for the fluid dynamics situation chosen and (b) for the deformation history of the melt under test.

ACKNOWLEDGEMENT

We are very much indebted and express our thanks to Dr. P.D. Iten for his immense help in setting up the Laser-Doppler system. The LDPE polymer used for this investigation was kindly donated to us by BASF Aktiengesellschaft Ludwigshafen/Germany.

REFERENCES

1/ F. Durst, A. Melling and J.H. Whitelaw: Principles and Practice of Laser-Doppler Anemometry, Academic Press, 1976

2/ US and German standards: ASTM - D 1238-73 and DIN 53 735

EFFECT OF CARBON BLACK ON THE RHEOLOGICAL PROPERTIES OF STYRENE n-BUTYL METHACRYLATE COPOLYMER MELT

S. K. Ahuja

Joseph C. Wilson Center of Technology

Rochester, N.Y. 14644

INTRODUCTION

Carbon black reinforcement in elastomers, as well as glassy polymers, has been the subject of an intense study for almost fifty years.[1] Grades of carbon black are distinguished by surface area and bulkiness (dibutyl phthalate absorption), the latter being a function of the number and arrangement of particles within an aggregate. Carbon black surface area increase results in an increase in the elastic modulus at low amplitude of deformation because of the presence of interaggregate network.[2] At higher amplitudes of deformation, there is breakage and reformation of this network. Aggregate bulkiness increases both the elastic and the viscous modulus without changing their ratios.[3] Lobe and White[4] have recently observed that rheological behavior of a molten filled polymer at 10-20% loading is that of a gel. They also found that viscosity does not level off at decreasing rates of deformation but continues to increase.

Our study is concentrated on the effects of carbon black, its structure, surface area, and concentration on elastic moduli and viscosities of a copolymer. Its implication to networks and pertinent existing models (Graessley, Bingham-Maxwell) is discussed.

EXPERIMENTS

Carbon blacks from Cities Service, Raven types, varying in surface acidic groups and surface area, were dispersed in styrene/n-butyl methacrylate of 65/35 composition at 0, 10, 20, and 30% concentrations. Surface acidic groups (carboxylic groups) on the surface of carbon black are increased by an oxidation process of surface treatment. The Banbury and roll mill were used to disperse the carbon blacks. The technique of using Banbury and roll mill for mixing carbon blacks in polymers recently has been described by Lobe and White.[4] Compression-molded disks were prepared and their rheological properties were determined using the ERD device of a Rheometrics Mechanical Spectrometer.[5] Experiments were conducted from 90-180°C at rotational frequencies of 10^{-2} to 10^{2} sec $^{-1}$. The elastic moduli and viscous moduli at varying temperatures and frequencies were shifted to a reference temperature, say 120°C, using time-temperature superposition. Figures 1 and 3 show the master curves in elastic moduli. From elastic and viscous moduli, dynamic complex viscosities at different temperatures and frequencies are obtained. These too are shifted to a reference temperature and are shown in Figure 4. The master curves of viscous moduli were used to construct relaxation moduli according to the iteration technique discussed previously.[6] Figure 2 shows relaxation moduli as functions of time at 120°C.

RESULTS AND DISCUSSION

Figure 1 shows the effects of surface treated carbon black concentration on the copolymer elastic shear moduli at 120°C. The copolymers containing 0% and 10% carbon blacks of 580 m^2/g (surface area) show a terminal region, a plateau region, and a high frequency region, characteristic regions of a linear viscoelastic spectrum.[5,6] The copolymers containing 20% and 30% levels of carbon black show second plateaus at lower frequencies, $5 \times 10^{-4} sec^{-1}$ and $5 \times 10^{-3} sec^{-1}$, respectively, with corresponding elastic moduli plateaus at 3×10^{3} dynes/cm^2 and 1.3×10^{4} dynes/cm^2.

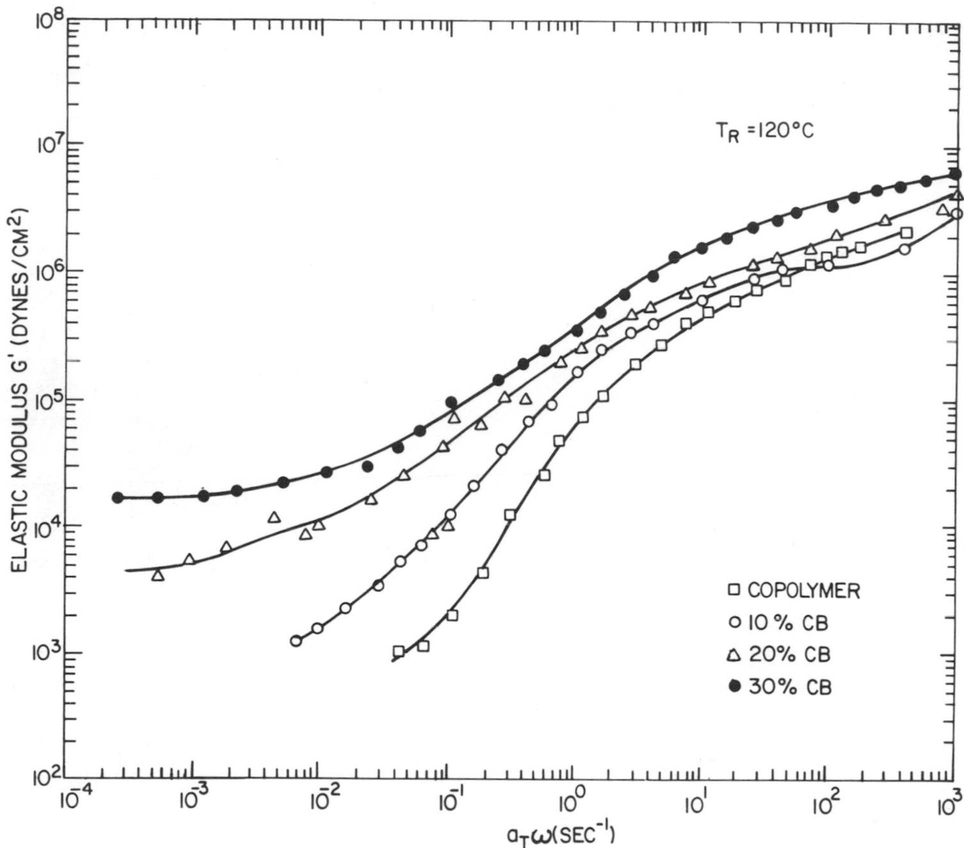

FIGURE 1. Effect of carbon black concentration on elastic moduli, $G'(a_T\omega)$.

The elastic modulus increases nonlinearly with increasing magnitude of strain beyond 60% strain in the polymer. The threshold in nonlinearity diminshes with increasing carbon black loading. The percentage of strain in the polymer for a nonlinear viscous modulus also decreases with carbon black loading.

Figure 2 shows the relaxation moduli of polymers at varying surface treated carbon black concentrations. The relaxation modulus curves as functions of time flatten with increasing carbon black concentrations. There are significant differences in relaxation moduli between 0% and 10% carbon black and between 20% and 30% levels of carbon black. The relaxation moduli do not increase linearly with increase in carbon black concentration.

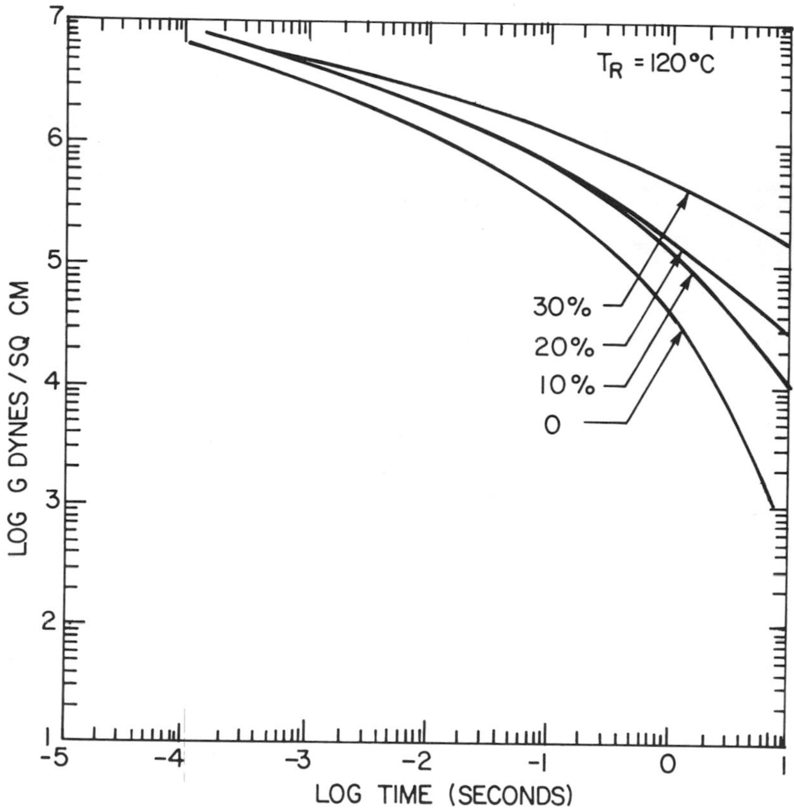

FIGURE 2. Effect of carbon black concentration on relaxation moduli, G(t).

Figure 3 shows the effect of surface areas of untreated and treated carbon blacks at 10% concentration in a copolymer of wider distribution than represented in Figure 1.

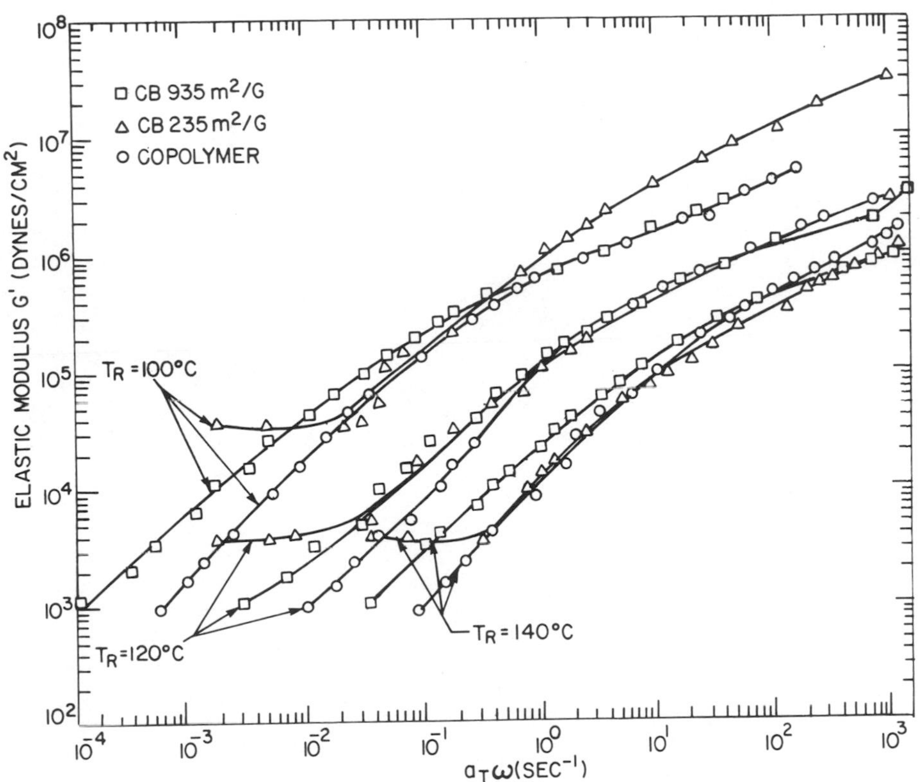

FIGURE 3. Effect of carbon black surface treatment and surface area on elastic moduli, $G'(a_T\omega)$.

Both the copolymers with untreated carbon black of 235 m^2/g surface area and surface-treated carbon black of 930 m^2/g at 10% concentration give higher elastic moduli than base polymer at lower frequencies. The copolymer with untreated carbon blacks shows a distinct second plateau. The differences between the elastic moduli of untreated and treated carbon blacks diminish with increasing temperature. This would indicate that the polymer carbon black networks involve temperature-dependent bonding. This bonding may be of several types, i.e., Van der Waal's forces, hydrogen bonding, etc.

Figure 4 shows the effect of dynamic complex viscosities at different frequencies for copolymers with varying concentrations of carbon blacks. In the copolymer without any carbon black, there is a typical flat region and a power law region. The power index of viscosity with frequency is in agreement with the entanglement model of Graessley.[7] At 10% and 20% concentrations of treated carbon black, there are two power law regions separated by a flat region. At 30% concentration of carbon black, the flat region disappears. Lobe and White[4] observed that viscosities of polystyrene at 20% and 30% concentration of carbon black did not show any flat region. A Bingham-Maxwell model has been recently proposed[4] to explain the behavior of the polymer-carbon black network. The network behaves as a Bingham body at low rates of deformation and as a Maxwell fluid at high rates of deformation. The Bingham body of the network has a yield stress. The Maxwell body of the network would behave as a Newtonian fluid at low shear rates (the flat region in the viscosity shear rate curve) and show power law shear thinning at high shear rates. Our experiments show that the flat region present between two power law regions decreases with increasing concentrations of carbon black. The Bingham-Maxwell model thus can not explain the phenomena that we have observed. This is contrary to predictions one would make on the basis of a Bingham-Maxwell model or the entanglement model of Graessley. Both models appear inappropriate to explain the behavior of filled polymers.

In conclusion, the results observed by Lobe and White have been confirmed in the styrene n-butyl methacrylate system. We also find that the molten polymer-carbon black network is a function of carbon black structure, surface area, and concentration. The network can be influenced by carbon black treatment. The network also is highly temperature-dependent and cannot be described by existing models.

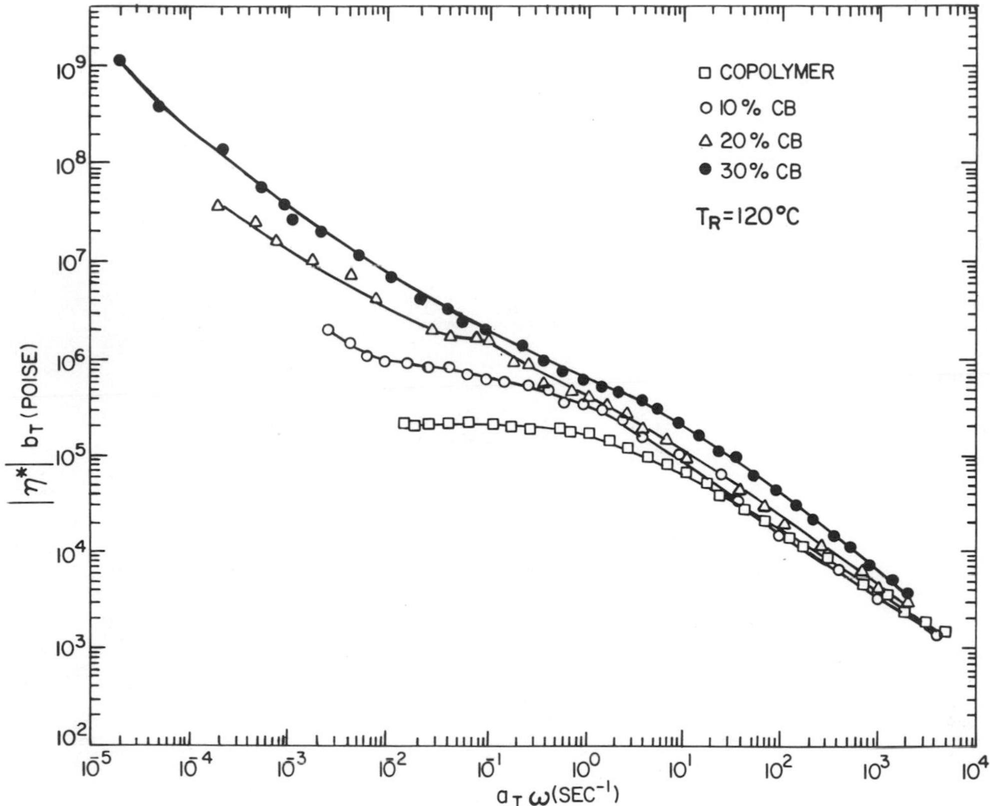

FIGURE 4. Effect of carbon black concentration on complex dynamic viscosity $|\eta^*|b_T(a_T\omega)$.

REFERENCES

1. A.I. Medalia, Rubber Chem. Technol. 47:411 (1974).

2. F.R. Graziano, R.E. Cohen, and A.I. Medalia, Rheol. Act. 18:640 (1979).

3. A.R. Payne and R.E. Whittaker, Rubber Chem Techol. 44:440 (1970).

4. V.M. Lobe and J.L. White, IUTAM Symposium on Non-Newtonian Fluid Mechanics 1:1 (1978).

5. G. Martin and W.W. Graessley, Rheological Acta 16:527 (1977).

6. S.K. Ahuja, Rheological Acta, in press.

7. W.W. Graessley, Advances in Polymer Science, 16 (1974).

LARGE ELASTIC DEFORMATIONS OF

FLOWING POLYMERS IN SHEAR AND EXTENSION

A.Ya. Malkin, O.Yu. Sabsai

Research Institute of Plastics
Moscow, USSR

V.P. Begishev

Department of Polymer Physics
Urals Scientific Center
Academy Sciences of the USSR
Perm, USSR

The stress-strain relationships for various loading conditions cannot be arbitrary since they must satisfy the requirement of invariance. The general method for the fulfillment of this requirement is the use of the elastic potential W, which is represented in the form of a certain function of the invariants of the strain tensor.

Of special interest in this respect is the two-constant potential proposed by Tschoegel [1], which is written in the following manner:

$$W = \frac{2G}{n}\left(\lambda_1^n + \lambda_2^n + \lambda_3^n - 3\right) \tag{1}$$

where G and n are empirical constants and λ_i are the principal values of the extension ratio.

We made an attempt to use formula (1) to describe the rubber-like properties of polymer melts and solutions.

The elastic deformations of polymer solutions and melts are invariably associated with dissipation losses in flow. In this case, the behavior of the liquid under steady-state flow conditions is described by two "equilibrium" functions - the flow curve, which characterizes the flow properties, and the elastic potential, which characterizes its rubberlike properties.

If the function $W(\lambda_i)$ is known, we can calculate the stress-strain relations for any geometric scheme of loading.

Of interest for the purposes of the present discussion are two principal loading schemes: uniaxial extension and simple shear. The first is characterized by the magnitude of the degree of equilibrium reversible elongation κ; the second is characterized by the magnitude of the equilibrium elastic shear deformation ε.

For uniaxial extension of an incompressible liquid, the expressions for the principal elongations have the following form:

$$\lambda_1 = \kappa; \quad \lambda_2 = \lambda_3 = \kappa^{-1/2} \tag{2}$$

And for simple shear the principal values are expressed in terms of ε as follows [2]:

$$\lambda_1 = \tan\chi; \quad \lambda_2 = \cot\chi; \quad \lambda_3 = 1 \tag{3}$$

where the angle χ is related to the shear deformation by the following formula:

$$\chi = \tfrac{1}{2}\arctan(2/\varepsilon)$$

Note that the shear stress τ, which is the basic force characteristic of the state of stress in simple shear, is expressed in terms of W in the following manner:

$$\tau = \frac{dw}{d\varepsilon} \tag{4}$$

For the elastic potential (1) the following formulas hold for the dependence $\sigma(\kappa)$ for uniaxial extension and the dependence $\tau(\varepsilon)$ for simple shear:

$$\sigma = 2G(\kappa^n - \kappa^{-n/2}) \tag{5}$$

$$\tau = \frac{2G}{2^n\sqrt{(4+\varepsilon^2)}}\left[(\sqrt{(4+\varepsilon^2)}+\varepsilon)^n - (\sqrt{(4+\varepsilon^2)}-\varepsilon)^n\right] \tag{6}$$

Now let us find the "initial" values of the elastic modulus in extension, E_0, and in simple shear, G_0. The appropriate calculations show that

$$E_0 = 3nG; \quad G_0 = nG \tag{7}$$

that is, in the limiting case of infinitesimal strains there is fulfilled the standard relationship, which is common to all incompressible bodies: E_0 $3G_0$. Here it should be emphasized that the quantity G figuring in formula (1) does not have the meaning of a "modulus" (though it has the same dimensions as a modulus).

Using expressions (7), we can write the final formulas in the following manner:

$$\frac{3\sigma n}{2G_0} = \kappa^n - \kappa^{-n/2} \tag{8}$$

$$\frac{\tau\varepsilon^n}{2G_0} = \frac{\varepsilon}{2^n\sqrt{(4+\varepsilon^2)}}\left[(\sqrt{(4+\varepsilon^2)}+\varepsilon)^n - (\sqrt{(4+\varepsilon^2)}-\varepsilon)^n\right] \tag{9}$$

If we now denote the left-hand sides of formulas (8) and (9) by Y and the right-hand sides by X, then the dependence Y(X) must be invariant to the form of the state of stress.

For experiments we used a commercial sample of atactic polystyrene with a wide molecular-mass distribution because this material displays clearly pronounced nonlinear properties both in shear and in extension. The molecular mass of the sample is 3×10^5. The experiments were carried out at 130 and 150°C.

The measurement error in all cases did not exceed $\pm10\%$. The results of measurements of the functions $\sigma(\kappa)$ and $\tau(\varepsilon)$ are given in Figs. 1 and 2, respectively.

For the experimental data to be represented in invariant form it is necessary to know in advance the value of the exponent n. But if $\varepsilon > 1.15$ and $\kappa > 2.25$, formulas (8) and (9) may be given in the following approximate form:

$$\log\left(\frac{3\sigma}{2E_0}\right) = n\log\kappa - \log n \tag{10}$$

$$\log\frac{\tau\sqrt{(4+\varepsilon^2)}}{G_0} = n\log(\sqrt{(4+\varepsilon^2)}+\varepsilon) - [(n-1)\log 2 + \log n] \tag{11}$$

The treatment of the corresponding experimental data on the dependences $\sigma(\kappa)$ and $\tau(\varepsilon)$ by means of these formulas gives the value of $n = 2.29 \pm 0.05$.

Now, knowing the value of n, we can make use of the general formulas (8) and (9); the experimental data can now be given in invariant form in the coordinates Y and X. This is done in Fig. 3.

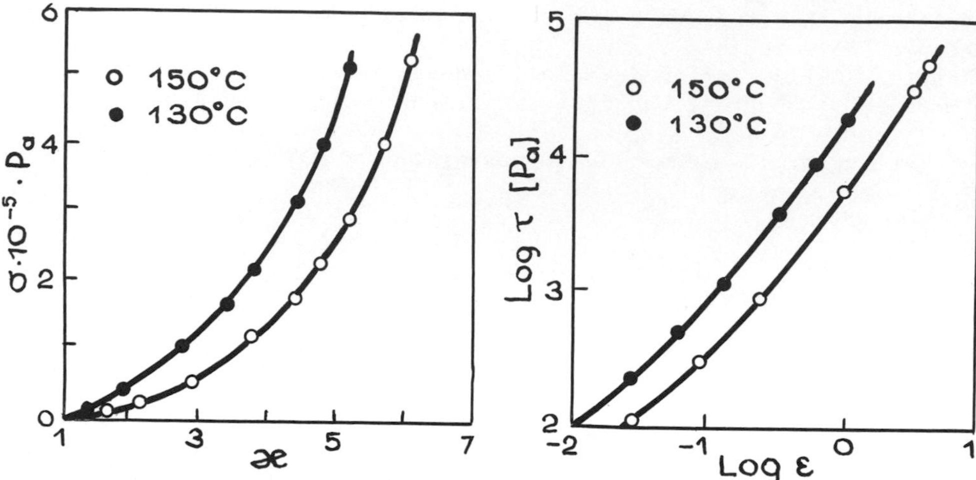

Fig. 1. Dependence of stress on the extension ratio in uniaxial extension.

Fig. 2. Dependence of shear stress on elastic deformation strain in simple shear.

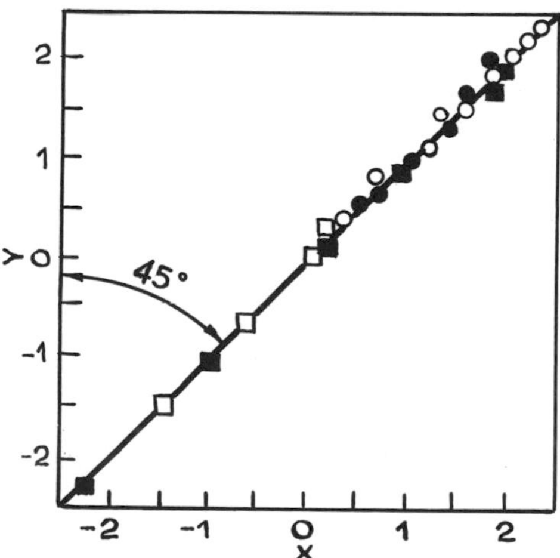

Fig. 3. Invariant representation of experimental data on the dependence of stress on strain in uniaxial extension and in simple shear. Y represents the left-hand sides and X the right-hand sides of formulas (8) and (9). The numerical data are the same as in Figs. 1 and 2.

As we have seen, the use of the potential (1) enables one to attain excellent agreement between the invariant representation of experimental data on shear and extension over the entire range of deformation rates and elastic deformations. This proves the applicability of the Tschoegl elastic potential for a quantitative description of elastic deformations stored during the flow of polymers.

References

1. N. Tschoegl, lecture given at the Institute of Petrochemical Synthesis, Academy of Sciences of the USSR, November, 1973; P.J. Blatz, S.C. Sharda, and N.W. Tschoegl, Trans. Soc. Rheol., 18, 145 (1974).
2. G.V. Vinogradov and A.Ya. Malkin, Rheology of Polymers, Khimiya, Moscow, 1977 (in Russian; English translation in press).

MICROVISCOELASTICITY AND VISCOELASTIC PROPERTIES OF LINEAR FLEXIBLE-CHAIN POLYMERS

G.V.Vinogradov, Yu.G.Yanovsky, L.I.Ivanova, V.S.Volkov

Institute of Petrochemical Synthesis of the USSR Academy of Sciences, Moscow, GSP1, USSR

(Abstract)

A concept of microviscoelasticity is introduced, according to which the motion of a macromolecule among similar ones is assumed to be equivalent to its motion in a viscoelastic medium whose characteristics are not known in advance. Theoretical equations have been derived, expressing the relationship between the components of the complex modulus of a linear viscoelastic liquid with an arbitrary set of relaxation times and a low-concentration polymer solution in it.

With a view to establishing a quantitative relation between the micro-and observed macroviscoelasticity, binary mixtures of 1,4-polybutadienes with a narrow molecular-weight distribution have been investigated, their molecular weights being 10 to 100 times different. The content of the high-molecular component varied from 0.1 to 10%.

These polymer mixtures have been experimentally found to exhibit different viscoelastic behaviors in different regions of concentrations of the high-molecular component, and a narrow region of composition with a corresponding "critical" concentration has been established. The calculations of the critical concentration values corresponding to the beginning of polymer coil overlapping, agree well with the experiment.

THE EFFECT OF ULTRASONIC VIBRATION ON POLYMERS IN FLUID STATE

M.L.Fridman, S.L.Peshkovsky, A.I.Tukachinski,
V.I.Brizitsky, G.V.Vinogradov

Institute of Petrochemical Synthesis of the USSR
Academy of Sciences, and Scientific and Industrial
Association "Plastik", Moscow, USSR

(Abstract)

This paper reviews the advances in studies of the effect of ultrasonic vibration on fluid polymer systems. It is shown that at definite vibration rates the hydraulic resistance of molding channels substantially decreases, while their flow capacity increases, and the superimposition of ultrasonic vibration permits controlling not only viscous but also highly elastic characteristics of polymers. The latter manifests itself, for example, in a considerable reduction of swelling and greater critical values of stresses and shear rates, corresponding to the onset of unsteady flow ("degradation" of the melt).

The paper is also concerned with the power efficiency of using ultrasonic vibration to intensify the flow of unfilled and composite polymer materials. In addition to the theoretical aspects, practical recommendations are given as regards the application of low-amplitude ultrasonic vibration in the extrusion of polymers.

Emphasis is placed on the new rheological effects revealed during exposure of fluid polymers to ultrasound, such as formation of drops from the liquid film on the surface of the acoustic radiator and abnormally sharp decrease in the viscosity ("superfluidity" of melts of high-molecular compounds when ultrasonic vibration is applied to them.

CAPILLARY RHEOMETRY OF POLYPROPYLENE: INFLUENCE OF MOLECULAR WEIGHT ON DIE SWELLING

D. Romanini and G. Pezzin°

Montedison S.p.A., Research Centre of Plastics Division, Ferrara (Italy)
° C.N.R., Macromolecular Physics Centre, Bologna (Italy)

INTRODUCTION

Although there is a relatively large number of papers in the literature concerning the phenomenon of "die swell", the knowledge of the molecular parameters that influence the expansion of a polymer at the die exit is still scarce, and most of the data collected up to now relate to polystyrene and polyethylene (1-7). To the best of the authors' knowledge, only Coen and Petraglia (8), Han and coworkers (9), and Fujiyama and Awaya (10) have investigated polypropylene. This polymer is essentially linear and it is well suited for investigations in which a better understanding of the relationship between average molecular weight and rheological properties is required.
Recently, we studied a series of polypropylene samples carefully characterized with respect to their molecular properties, and the results are reported in the present work.

EXPERIMENTAL

The molecular characteristics of the six samples of polypropylene (all produced by Montedison) investigated in this work are collected in Table 1. Whereas the average molecular weight $\overline{M}w$ changes within a relatively large range (2.7 to 6.8 x 10^5) the molecular weight distribution, as measured by the ratio $\beta = \overline{M}w/\overline{M}n$ obtained by G.P.C., can be considered substantially constant ($7 \leqslant \beta \leqslant 9.7$).
The Instron capillary rheometer was used to carry out the measurements of die swelling and of shear stress $\tau_w = PR/2L$ and apparent Newtonian shear rate at the wall $\dot{\gamma}_a = 4Q/\pi R^3$ (where the letters used have the well known meanings). The capillary diameter D_o was 1.25 mm, the L/D_o ratio was 40, and the capillary entrance angle 90°.

Table 1 - Sample characteristics

Sample	$[\eta]$ [a] dl/g	MFR [b] g/10'	$\bar{M}w \cdot 10^{-3}$ [c]	β [d]	$\eta_o \cdot 10^{-3}$ [e] Pa·s
PP-1	1.50	12.4	270	7	4.7
PP-2	1.75	6.5	340	8.2	10
PP-3	2.00	3.0	390	8.6	23
PP-4	2.20	2.0	440	9.1	28
PP-5	2.50	0.72	570	9.3	60
PP-6	2.65	0.39	680	9.7	110

a Tetrahydronaphthalene, 135°C
b ASTM D 1238-73, B
c Light scattering, α-chloronaphthalene, 150°C
d G.P.C., orthodichlorobenzene, 135°C
e Zero shear viscosity measured at 200°C by a Weissenberg Rheogoniometer

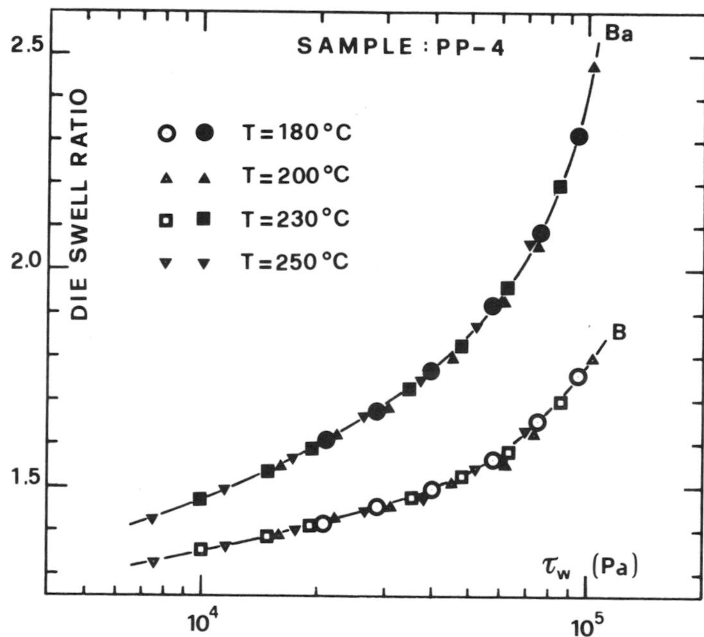

Figure 1 - Die swell ratio before (B) and after (Ba) annealing versus shear stress τ_w for several extrusion temperatures.

For the collection and handling of the extrudates a standard procedure, similar to that described by Mendelson (11) was followed. The extrudate's diameter D, from which the die swelling ratio D/D_o is obtained, was measured with a micrometer at room temperature before and after annealing for 10 minutes at 190°C in a silicon oil bath. From preliminary measurements, it was found that equilibrium values of D/D_o and very small (less than 1%) swelling of the polymer with silicon oil are obtained with these conditions.
The correction of the diameter measured at room temperature to account for the density change was made using the equation:

$$B = D/D_o \, (\varrho \, / \, \varrho_m)^{1/3} \qquad 1$$

where ϱ is the density at 23°C (0.904 g/ml) and ϱ_m is the density of polypropylene at the extrusion temperature (12).
Die swelling data are reported both for non-annealed extrudates (B) and for annealed extrudates (B_a).

RESULTS AND DISCUSSION

Die-swell data for the materials listed in Table 1 were obtained usually in the range of shear rate from 3 to 300 s^{-1} and at four selected temperatures (180°, 200°, 230° and 250°C).

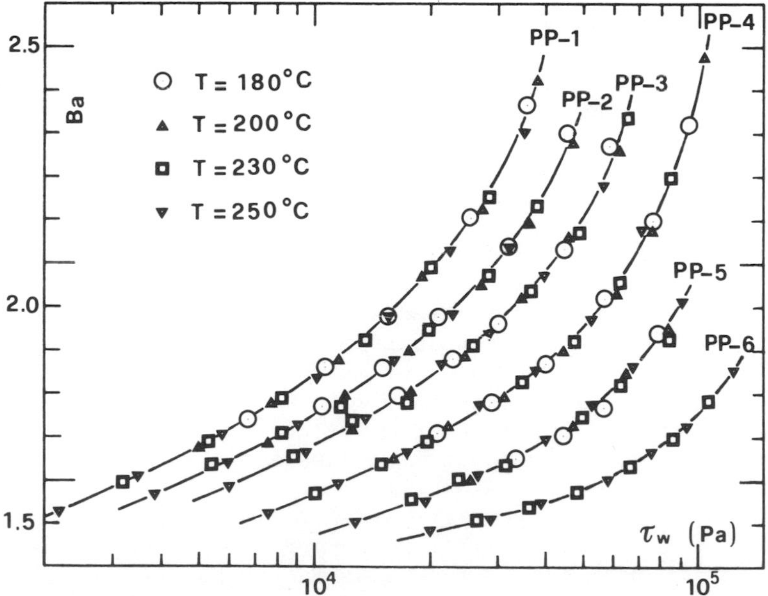

Figure 2 - Master curves of die swell after annealing (B_a) versus shear stress τ_w for all samples extruded at several temperatures.

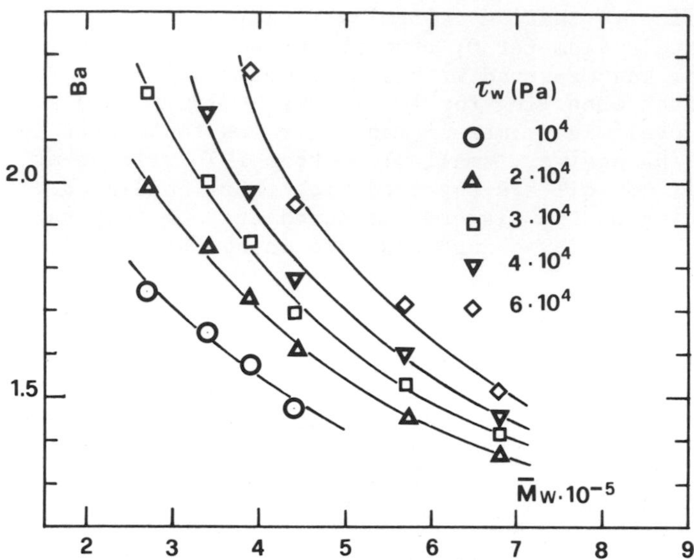

Figure 3 - Molecular weight dependence of die swell ratio at fixed values of shear stress τ_w for the polypropylene samples

As found by others (3,7) the temperature dependence is eliminated if die-swell is plotted against shear stress, as shown in Figure 1 for a sample of intermediate $\bar{M}w$. Also shown in the Figure is the marked effect of the annealing of the extrudates on the swelling ratio.

The master curves of die swell versus shear stress for all the samples investigated are shown in Figure 2. It is seen that the influence of molecular weight on the curves is a marked one; for example at $\tau_w = 3x10^4$ Pa the Ba value of sample PP-1 ($\bar{M}w$ = 270,000) is 2.2, whereas that of sample PP-6 is 1.4 ($\bar{M}w$ = 680,000).

When the shear stress τ_w is used as a parameter, the molecular weight dependence of Ba can be better visualized by plotting Ba versus $\bar{M}w$ (Figure 3). Although any extrapolation has to be regarded with caution, it is interesting to note that the limiting value of Ba at "both" low values of τ_w (Figure 2) and high values of $\bar{M}w$ (Figure 3) appears to be about 1.3, substantially larger than predicted for non-elastic liquids (7).

Since the curves of Figure 2 are essentially parallel, they can be superposed by horizontal translation along the log τ_w axis to give a master "master curve", i.e. a plot of Ba versus the log of the product $\alpha_M \tau_w$ (Figure 4). The values of $\log \alpha_M$, the logarithmic shifts needed to obtain superposition with respect to the curve belonging to sample PP-4, depend of course on the average molecular weight $\bar{M}w$ according to the least-square equation

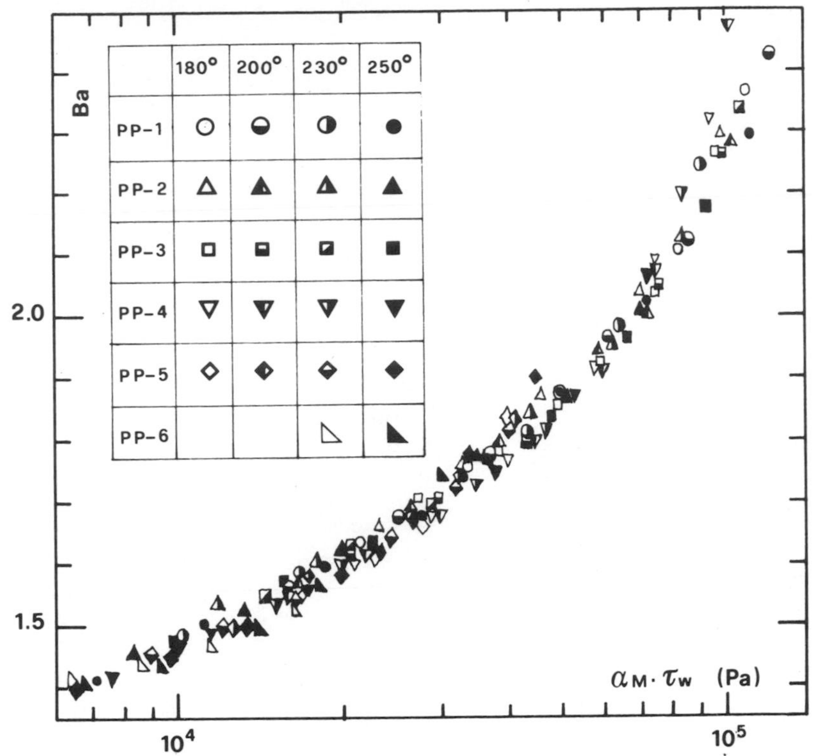

Figure 4 - Master "master curve" Ba vs. log ($\alpha_M \tau_w$) obtained by translation along the log τ_w axis of the master curves of Figure 2

$$\log \alpha_M = -2.77 \times 10^{-6} \, \bar{M}w + 1.27 \qquad 2$$

whose correlation coefficient is 0.998.
The master curve of Figure 4, giving the shear stress dependence of Ba for polypropylene independently of $\bar{M}$w, is similar in shape to those reported for other polymers (3,7).
From a log-log plot, one finds that at the upper limit of the τ_w range explored:

$$Ba \propto \tau_w^{\,0.25} \qquad 3$$

According to the analysis of Tanner (7,13) Ba depends on the first normal stress difference N_1 and shear stress τ, which are related by the shear modulus G

$$N_1 = 2\,\tau^2/G \qquad 4$$

With suitable assumption (7) one can write

$$B = \left[1 + 1/2(N_{1w}/2\ \tau_w)^2 \right]^{1/6} \qquad 5$$

If G is taken as constant, and $1/2\ (N_{1w}/2\ \tau_w)^2$ as much larger than 1, one obtains

$$B \propto \tau^{0.33} \qquad 6$$

where the exponent is not too different from that found experimentally.

The molecular weight dependence of die swelling found in the present work for polypropylene is the opposite to that reported for polystyrene (7) and low density polyethylene (14), and similar to that found for high density polyethylene (15).

Further development of theoretical models to understand this peculiar behaviour is clearly needed.

REFERENCES

1) C.D. Han,Rheology in Polymer Processing, Academic Press, N.Y. (1976).
2) N. Nakajima and M. Shida, Trans. Soc. Rheol., 10, 299 (1966).
3) W.W. Graessley, S.D. Glasscock and R.L. Crawley, Trans. Soc. Rheol., 14, 519 (1970).
4) J. Vlachopoulos, M. Horie and S. Lidorikis, Trans. Soc. Rheol., 16, 669 (1972).
5) M.G. Rogers, J. Appl. Polym. Sci., 14, 1679 (1970).
6) G. Locati, Rheol. Acta, 15, 525 (1976).
7) R. Racin and D.C. Bogue, J. Rheology, 23, 263 (1979).
8) A. Coen and G. Petraglia, Polym. Eng. Sci., 10, 79 (1970).
9) C.D. Han, M. Charles and W. Philippoff, Trans. Soc. Rheol., 14, 393 (1970).
10) M. Fujiyama and Awaya, J. Appl. Polym. Sci., 16, 275 (1972).
11) R.A. Mendelson, F.L. Finger and E.B. Bagley, J. Polym. Sci. C, 35, 177 (1971).
12) F. Danusso, G. Moraglio, W. Ghiglia, L. Motta and G. Talamini, La Chimica e l'Industria, 41 (8), 748 (1950).
13) R.I. Tanner, J. Polym. Sci., A-2, 8, 2067 (1970).
14) D. Romanini, A. Savadori and G. Gianotti, Rheology of branched LDPE, Polymer, in press.
15) R.A. Mendelson and F.L. Finger, J. Appl. Polym. Sci., 19, 1061 (1975).

THE CREEP BEHAVIOR OF A HIGH MOLECULAR WEIGHT POLYSTYRENE

D.J. Plazek, N. Raghupathi* and V.M. O'Rourke

Metallurgical and Materials Engineering
University of Pittsburgh
Pittsburgh, PA 15261

INTRODUCTION

In 1932 Warren Busse noted that high molecular weight linear polymers at temperatures above their glass tempearature, T_g, responded for a time as if they were crosslinked: i.e. they exhibited a rubberlike molulus before viscous flow dominated the deformation[1]. He proposed that entanglements involving neighboring threadlike molecules behaved as temporary crosslinks. The appearance of what is now referred to as the rubbery plateau in the creep compliance, J(t), and the stress relaxation modulus, G(t), curves plotted as a function of logarithmic time is the result of the diminution of viscous deformation by the entanglements. The enhanced molecular weight dependence of the viscosity at high molecular weights, $M^{3.4}$ over the first power dependence observed at low molecular weights at constant monomeric friction coefficient[2,3] is a related phenomenon attributed to the existence and influence of the molecular entanglements. The decrease of the entanglement concentration at high rates of shear is believed to be the principal mechanism responsible for the strong decrease in the shear viscosity with increasing shear rates. Graessley has developed a theory of non-Newtonian flow based on this concept[4,5].

The measurements of Stratton and Butcher[6] on polymer solutions have clearly shown a shearing and rest-time dependence of the stress-overshoot effect. These time dependences have been interpreted in terms of the entanglement and disentanglement processes.

The recent interpenetrating network studies of Ferry, et al[7-9]

*Present address: Ford Motor Co., Detroit, MI 48239

where one of the two networks is the molecular entanglement network, altered by various degrees of relaxation, add immensely to the evidence for the reality and the value of the entanglement concept.

The enormous effect of solvent on the local mobility of polymer segments evidenced by the sharp drop of T_g with the introduction of small amounts of plasticizers, coupled with the slight effect of solvent on the shape and extent of the distribution of retardation times, i.e. the retardation spectrum, up to concentrations of 50% solvent[10] should dispel arguments that the strong enhancement of drag forces and retardation times with the increase of molecular weight of an amorphous polymer can be explained in terms of secondary bond forces between neighboring chain molecules. The increased average distance between neighboring molecular segments would cause drastic effects on the extent of the retardation spectrum and its shape if secondary forces were determining it. The functionality of the molecular weight dependence of the viscosity might also be expected to change from the 3.4 power dependence.

Several years ago we looked for an increased rubbery plateau compliance in samples that had been freeze-dried from very dilute solutions of a polystyrene with a molecular weight of approximately 2 million. It was believed that a lower than equilibrium concentration of entanglements would persist in such a processed material if it were not heated much above T_g. This test of the entanglement concept proved to be inconclusive. However, the creep measurements on a polystyrene with a molecular weight of 44 million have shown that such an effect of altered entanglement concentration was unavoidable.

EXPERIMENTAL

Materials and Techniques

The ultra-high molecular weight polystyrene studied was graciously provided to us by Professor Lew J. Fetters of Akron University. It was anionically polymerized in tetrahydrofuran with low molecular weight polystyrllithium as the initiator[11-13]. Its molecular weight averages, M_w and M_z, were determined to be 4.37 x 10^7 and 4.88 x 10^7 respectively[14]. Freeze-drying was carried out with reagent grade benzene, certified ACS specifications, low thiophene, from Fisher Chemical Co., Pittsburgh PA 15238, cat. no. ΔB-245. The polystyrene is designated LJF-13.

Torsional creep and recovery measurements were carried out in a magnetic bearing creep apparatus[15] *in vacuo*, ca 10^{-2} Torr after flushing with nitrogen. Details of construction and operation are presented elsewhere[15].

Results and Analysis

The recoverable compliance, $J_r(t)$, cm^2 $dyne^{-1}$ or $Paschals^{-1}$, measurements being reported were preceded by a study of a 56 weight percent solution of the same high molecular weight polystyrene. When the first measurements were made on the bulk polystyrene at 109.6, 122.4, 141.2, 172.8 and 208.3°C a comparison of the reduced curves indicated that the rubbery plateau of the solution was more extensive than that for the undiluted polymer. This anomolous result was rationalized by hypothesizing that the undiluted polymer had never achieved its equilibrium level of entanglement after being recovered from solution.

In spite of fruitless attempts to enhance the state of entanglement, it was felt that a decrease would be possible by exaggerating the conditions that were hypothesized to cause the presumed lower than equilibrium level. According to the ordinary concept of randomly entangled threadlike polymer molecules, the entanglement concentration is proportional to the concentration of polymer. In a dilute solution therefore the entanglement concentration is greatly reduced. We therefore prepared a 0.1% solution of the LJF-13 in benzene. In an attempt to maintain the low level of chain entanglement this dilute solution was freeze-dried. The molecular chains are separated in the dilute solution, immobilized in the frozen benzene matrix and remain so after sublimation because they are 75°C below their T_g. It was thus concluded that very little reentangling could take place.

Creep measurements were carried out at 109 and 119°C on the material recovered from dilute solution. It was clear that at these temperatures no viscous deformation could be measured; therefore $J(t)$ and $J_r(t)$ are indistinguishable. These results along with the recoverable creep compliances obtained on the original material which is identified as being freeze-dried from concentrated solution are presented in Figure 1. The reduced compliance curve of a lower molecular weight narrow distribution polystyrene (A-19; $M=6.0 \times 10^5$) is shown as a dashed line for orientation purposes. It was surprising to find that the softening dispersion of the sample freeze dried from dilute solution was found at shorter times at 100°C so that a time scale shift slightly greater than 2 decades was necessary to achieve superposition with the softening dispersions of the other samples. The nearly two fold increase observed in compliance level accomplished by separating the entangled molecular coils and immobilizing them must be considered as clear evidence for a lower state of entanglement. The converging of the curves in the softening dispersion gives assurance that sample coefficient errors are not serious. In Figure 2 it can be seen how the higher level of compliance is built up and reflected in a much broader peak in L. The approaches of the two L curves

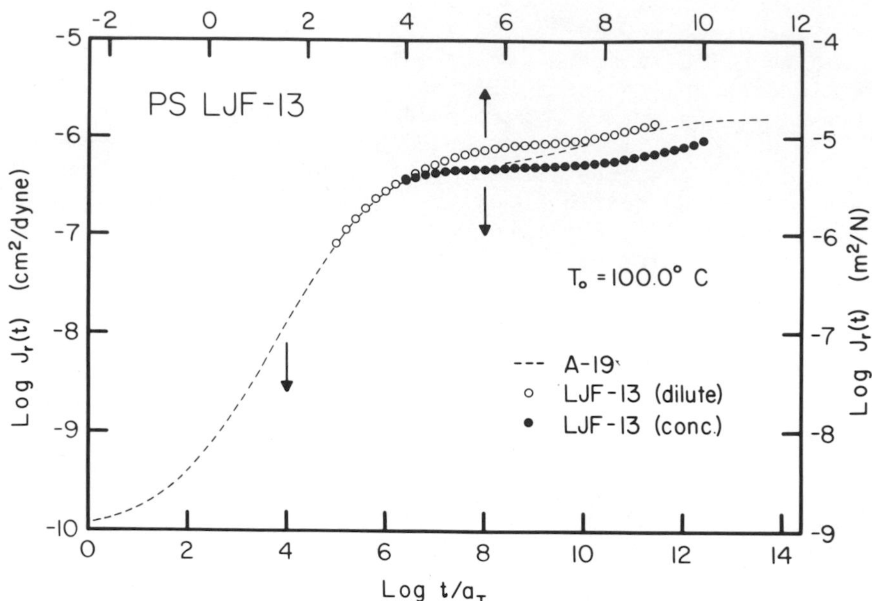

Figure 1. Logarithmic plot of $J_r(t)$ for the bulk LJF-13 polystyrene against the logarithm of the reduced time scale at a reference temperature of 100°C. Measurements follow freeze-drying from dilute and concentrated solutions. The dashed line represents the behavior of a lower molecular weight polystyrene, A-19 ($M=6.0 \times 10^5$).

to the terminal maximum support the argument for the smaller concentration of entanglements in the LJF-13 (dilute material since its rubbery plateau is indicated to be shorter. The rubbery plateau compliance, J_N, was calculated by integrating under the first peak in L. The dotted extrapolations of the first peaks in Figure 2 were used in the calculation of the integrals. The values found for log J_N were -6.07 and -6.31 for the "dilute" and "concentrated" specimens respectively. These values correspond to molecular weights per entangled unit of 2.7×10^4 and 1.6×10^4, when T = 373°K, ρ(100°C) = 1.029g cm^{-3} and $M_e = J_N \rho RT$.

CONCLUSIONS

An anomalously short rubbery plateau in the recoverable creep compliance $J_r(t)$ was encountered for an anionically polymerized polystyrene with an extremely high molecular weight, 4.4×10^7. Making the widely accepted assumption that the origin for this plateau is the presence of a transient network of physically entangled molecules and that the entanglement concentration was below

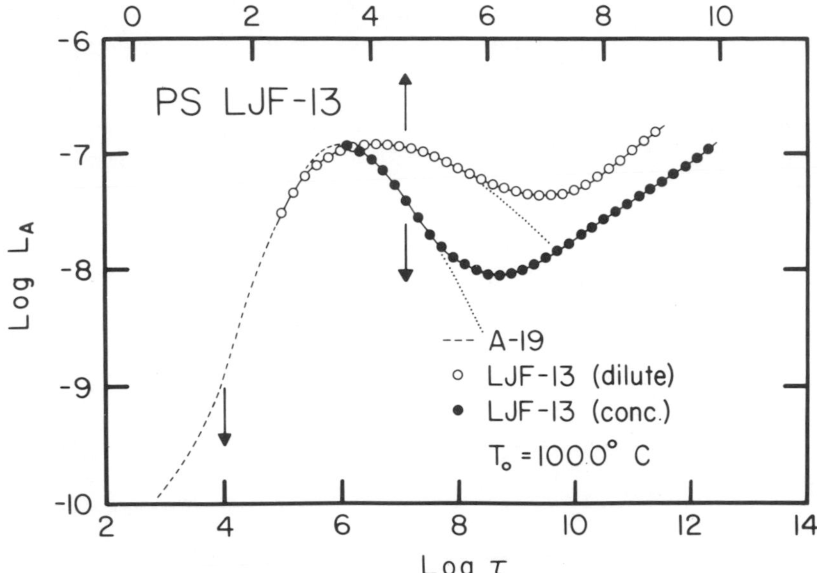

Figure 2. Double logarithmic presentation of the retardation spectra corresponding to the recoverable compliance curves of Figure 1. The dotted lines indiate the extrapolations utilized in the integrations to obtain J_N values. Dashed line from lower molecular weight polystyrene, A-19, showing the spectrum in the molecular weight insensitive region. The reference temperature, T_o, is 100°C. Arrows indicate pertinent retardation time scales.

the equilibrium level because of sluggish diffusion processes it was possible to predict a procedure to further curtail the length of the plateau and to nearly double its magnitude. It is believed that this is the first documentation of a persistent non-equilibrium conformational state in a liquid at temperatures appreciably above its T_g.

ACKNOWLEDGEMENTS

The authors gratefully acknowledge support of this research by the Polymer Program of the Division of Materials Research of the National Science Foundation through Grant No. DMR 76-09804 and the Chemical Process Program of the Engineering Division through Grant No. Eng. 78-18838.

REFERENCES

1. W.F. Busse, J. Phys. Chem., 36, 2862 (1932).
2. T. G Fox and V.R. Allen, J. Chem. Phys., 41, 337 (1964).
3. G.C. Berry and T. G Fox, Adv. Polymer Sci., 5, 261 (1968).
4. W.W. Graessley, J. Chem. Phys., 43, 2696 (1965).
5. W.W. Graessley, J. Chem. Phys., 47, 1942 (1967).
6. R.A. Stratton and A.F. Butcher, J. Polymer Sci. Phys., 11, 1747 (1973).
7. R.L. Carpenter, O. Kramer and J.D. Ferry, Macromolecules, 10, 117 (1977).
8. O. Kramer and J.D. Ferry, J. Polymer Sci. Phys. Ed., 15, 761 (1977).
9. J.D. Ferry, Polymer, 20, 1343 (1979); R.L. Carpenter, H.C. Kan and J.D. Ferry, Polym. Eng. Sci., 19, 267 (1979); H.C. Kan and J.D. Ferry, Macromolecules, 11, 1049 (1978); 12, 494 (1979).
10. D.J. Plazek, E. Riande, H. Markovitz and N. Raghupathi, J. Polymer Sci., Phys., 17, 2189 (1979).
11. E. Slagowski, L.J. Fetters and D. McIntyre, Polymer Prepr., Amer. Chem. Soc., Div. Polymer Chem., 12, 753 (1971).
12. D. McIntyre, L.J. Fetters and E. Slagowski, Science, 176, 1041 (1972).
13. E.L. Slagowski, L.J. Fetters and D. McIntyre, Macromolecules, 7, 394 (1974).
14. E.L. Slagowski, Ph.D. Thesis, University of Akron, 1972.
15. D.J. Plazek, J. Polymer Sci. A-2, 6, 621 (1968).

FLOW CHARACTERISTICS OF EPOXIDE PREPOLYMERS

J.V. Aleman

Instituto de Plasticos y Caucho, Madrid, Spain

(Abstract)

Rheological behaviour of four epoxide prepolymers and some flow pecularities are described: flow curve irregularities, shift factor and temperature independent flow activation energy, shear stress and shear rate effects, and their graphical and analytical description.

LINEAR AND NON-LINEAR VISCOELASTIC BEHAVIOUR OF ABS MELTS

H.C. Booij and J.H.M. Palmen

DSM, Central Laboratories

Geleen, Netherlands

THEORETICAL BACKGROUND

Recent investigations make it more and more clear that the non-linear viscoelastic behaviour of polymer solutions and melts can be described by constitutive equations of the type proposed by M.C. Phillips[1]

$$p(t) = -pI + \int_{-\infty}^{t} \mu(t-t')\, J(t-t')\, dt' \qquad (1)$$

where J(t-t') is a generalised tensorial measure, not further specified, of the strain between time t and t'. Special cases of this equation are those using the relative Finger tensor F(t-t') as a strain measure, e.g. that derived by Lodge[2] and by Doi and Edwards[3], and the one used by Wagner[4] and Laun[5] in which

$$J(t-t') = h\,(I_1, I_2)\, F(t-t') \qquad (2)$$

Here the so-called damping functional $h\,(I_1, I_2)$ is an adjustable material functional[6] of the first and second invariants, I_1 and I_2 resp., of the Finger strain tensor.

Phillips[1] obtained good results with a strain-independent material memory function $\mu(t-t')$, in conformity with findings of, e.g., Wagner[4], Laun[5] and Doi and Edwards[3] for some polymer solutions and melts. This function satisfies the relation

$$\mu(t-t') = \frac{\partial G(t-t')}{\partial t'} = \int_{-\infty}^{+\infty} \frac{H(\tau)}{\tau} \exp.\left[-\frac{t-t'}{\tau}\right] d\ln\tau \qquad (3)$$

where G(t) and $H(\tau)$ represent the linear viscoelastic shear relaxation modulus and relaxation spectrum, resp.

Table 1. Molecular Characteristics of the Samples

Sample	Rubber content % vol.	Molecular Mass of Extractable SAN		
		M_w^* (kg/mol)	M_w^*/M_n^*	M_z^*/M_w^*
Ronfalin MST	24	140	2.1	1.7
Ronfalin MT	16.8	-	-	-
Ronfalin Q	12	-	-	-
Ronfalin S (SAN)	0	197	2.5	1.7
Ronfalin F	16.8	118	2.2	1.8
Cevian N (SAN)	0	115	1.9	1.6

It is the aim of this paper to show that this approach is also useful for melts of a special type of rubber-modified plastics.

MATERIALS AND EXPERIMENTAL

Some characterization quantities of the investigated materials are given in Table 1. The molecular mass distributions were obtained by ALC-GPC, calibrated with polystyrene. All materials were dried for two hours at 80 °C and compression moulded at 180 °C. The experiments were performed with the cone-and-plate or parallel-plate system of the Rheometrics Mechanical Spectrometer, instrumented with a Solartron Frequency Response Analyser 1172, a Hewlett Packard 9835A calculator, an HP 2613A printer and an HP 9872A plotter.

EXPERIMENTAL RESULTS

Linear Oscillatory Experiments

Dynamic mechanical experiments over the frequency range from 0.02 up to 12.6 Hz, at temperatures of 140, 159, 179 and 200 °C, and at strain amplitude smaller than 0.06, provided the data on the frequency dependence of the phase angle δ and of the absolute value of the complex shear modulus G^* for all six materials. All exhibit perfectly the behaviour of thermorheologically simple materials: the δ vs. low ω curves can be superimposed by a mere shift along the log ω axis. The temperature dependence of the horizontal shift factor a_T appears not to be modified, within experimental accuracy, by the molecular mass of the SAN or by the rubber content[7,8]. Time-temperature reduction of the modulus curves was performable by applying the horizontal shift factors obtained for the phase angle, together with a slight shift along the vertical modulus axis.

The storage and loss moduli, $G'(\omega a_T)$ and $G''(\omega a_T)$ resp., are calculated from the master curves, and the relaxation spectra are estimated by means of an iteration procedure. The results are represented in Fig. 1. The intensity of the spectrum at long τ is higher according as the rubber content is higher[7-11] and a lower M_w^*(SAN) shifts the spectrum to shorter times[10].

The approach employed denies to the rubber-modified samples something like a yield stress, which is adhered to by some authors[11]. Instead, extra relaxation mechanisms with very long 'network' relaxation times are introduced, which make a large, nearly frequency-independent, contribution to G', and cause only a small rise of G".

Non-linear Stress Relaxation Experiments

Lodge and Meissner[12] have shown that for rubberlike liquids subjected to an instantaneous shear strain of magnitude γ at time t = 0 the relaxation of the stresses satisfies the relation

$$p_{11}(t) - p_{22}(t) = \gamma\, p_{12}(t) \qquad \text{for } t > 0.$$

The experimental proof of this relation is rather difficult. Measurements[13] of the time-dependent flow birefringence of polystyrene solutions in an apparatus without movable transducers confirm the relation, as do stress relaxation measurements Laun[5] performed on a low-density polyethylene melt. However, the Weissenberg Rheogoniometer is obviously too compliant for transient experiments[14-16]. The normal stress appears to relax more slowly than the shear stress, especially at low cone angles[17,18], or the quotient of these stresses is much larger than γ[17,18]. The Mechanical Spectrometer, although a factor of 5 stiffer in the vertical direction[18], is deficient in transient measurements in the same way[18]. With this instrument we found that for large plate distances and large cone angles the Lodge-Meissner relation is approximated or satisfied, but for small angles and narrow gaps the normal stress relaxation falls behind that of the shear stress, to an extent depending on the viscoelasticity of the melt. No transients in the shear stress were noticed for ABS melts at 159 °C after some 5 or 10 times the time needed for appli-

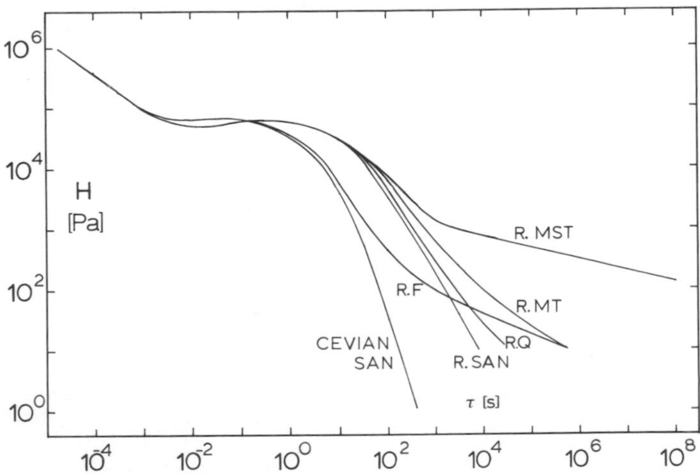

Fig. 1. Relaxation spectra at 159 °C for the various products.

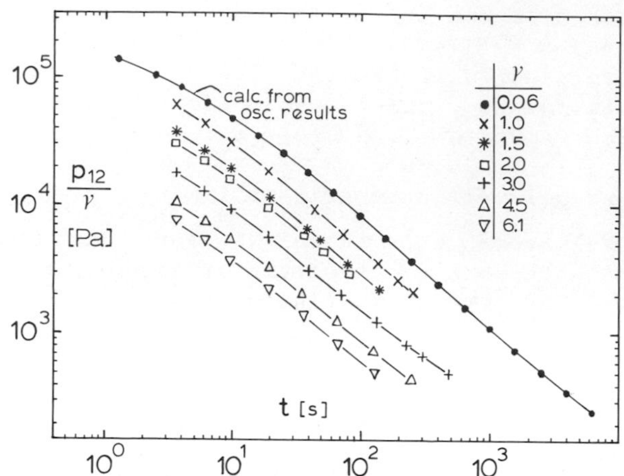

Fig. 2. The relaxation of p_{12} after different shear strain steps γ for Ronfalin MT at 159 °C. The top curve represents the calculated linear G(t) function.

cation of the strain if a cone angle of 0.1 or 0.2 rad and a plate radius of 2.5 cm were used.

According to Eqs. (1) to (3), after a step strain the relaxation of the stresses proceeds as expressed by[5]:

$$p_{12}(t)/\gamma = p_{11}(t) - p_{22}(t)/\gamma^2 = G(t)\, h(\gamma) \qquad (4)$$

which states that the time dependence is not affected by the magni-

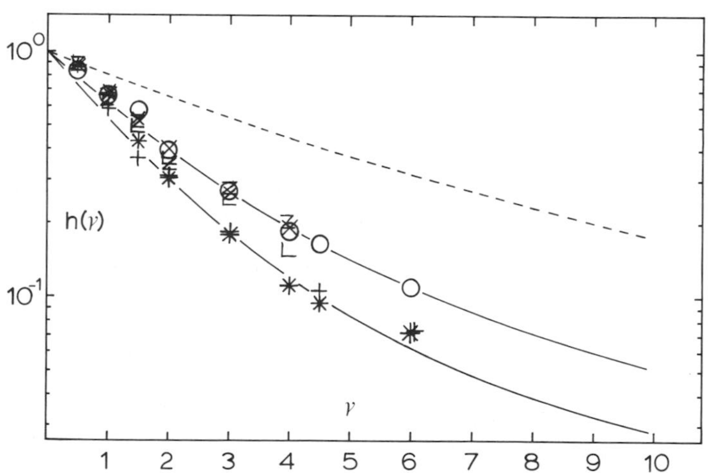

Fig. 3. Damping function $h(\gamma)$ for the various melts. Points indicated by Z, X, O, L, + and ✳ pertain to Cevian N, Ronfalin F, S, Q, MT and MST samples, resp.

tude of γ. This has indeed been found for the shear stress for all ABS melts, as shown e.g. in Fig. 2 for Ronfalin MT. The function G(t) was calculated from $H(\tau)$, after which $h(\gamma)$ could be derived from the relaxation stress at various strain levels. The values of this quantity are for the various products given in Fig. 3; they reveal that the non-linearity of ABS melts is considerably stronger than that of a low-density polyethylene melt examined by Phillips[1] and Laun[5]. No effect of M_w^*(SAN) can be discovered, but the non-linearity increases somewhat as the rubber content becomes higher. With a view to facilitating further calculations the damping functions are fitted with a sum of exponential functions

$$h(\gamma) = \sum_{i=1}^{N} f_i \exp\left[-n_i\gamma\right] \quad \text{with} \quad \sum_{i=1}^{N} f_i = 1 \tag{5}$$

In most cases two terms are sufficient. For Ronfalin S we found $f_1 = 0.78$, $n_1 = 0.6$, $f_2 = 0.22$ and $n_2 = 0.15$, and for Ronfalin MST $f_1 = 0.88$, $n_1 = 0.7$, $f_2 = 0.12$ and $n_2 = 0.15$. These functions are shown by full lines in Fig. 3.

Stress Growth Experiments

For the growth of the stresses after the sudden imposition of a constant shear rate $\dot{\gamma}$ Eqs. (1) to (5) yield[5]:

$$p_{12}(t) = \sum_{i=1}^{N} f_i \int_{-\infty}^{+\infty} H(\tau)\dot{\gamma}\,\tau_{r,i}^2/\tau \left[1 - \left\{1 - n_i\dot{\gamma}t\tau/\tau_{r,i}\right\} * \right.$$
$$\left. * \exp\left(-t/\tau_{r,i}\right)\right] d\ln\tau \tag{6}$$

and

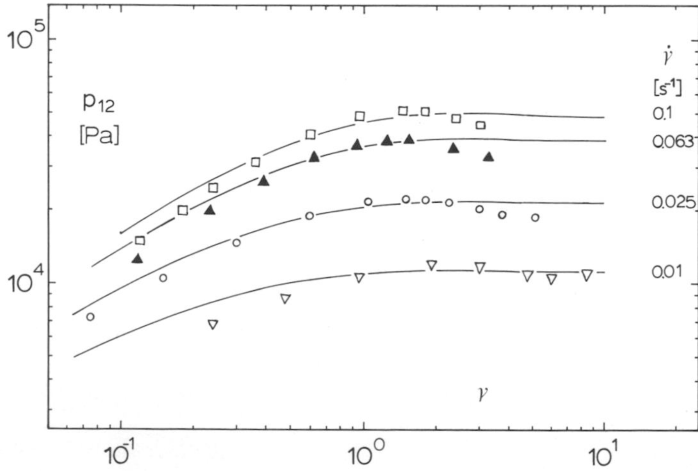

Fig. 4. The growth of the shear stress at various shear rates $\dot{\gamma}$ at 159 °C for Ronfalin S. (———) prediction using Eq. (6).

$$p_{11}(t) - p_{22}(t) = \sum_{i=1}^{N} f_i \int_{-\infty}^{+\infty} H(\tau) 2\dot{\gamma}^2 \tau_{r,i}^3/\tau \left[1 - \left\{ 1 + t/\tau_{r,i} - \tfrac{1}{2} n \dot{\gamma} t^2 \tau / \tau_{r,i}^2 \right\} \exp(-t/\tau_{r,i}) \right] d \ln \tau \quad (7)$$

with $\tau_{r,i} = \tau/(1 + n_i \dot{\gamma} \tau)$.

A comparison of $p_{12}(t)$ calculated for various shear rates with Eq. (6), the known function $H(\tau)$ and the parameters f_i and n_i, with experimental data is given in Fig. 4 for Ronfalin S. Qualitative agreement is obtained. For the strongly elastic MST melts the agreement is less satisfactory. However, experimental transient data in the non-linear viscoelastic range are for ABS melts so far not sufficiently reliable.

ACKNOWLEDGEMENTS

The authors wish to thank Mr. J.W.A. Sleijpen and Mr. Ch.C.M. Fabrie for their kind assistance in measurements and calculations.

REFERENCES

1. M.C. Phillips, J. Non-Newtonian Fluid Mech. 2:109 (1977).
2. A.S. Lodge, "Elastic Liquids", Academic Press, London (1964).
3. M. Doi and S.F. Edwards, J.C.S. Faraday Trans.II 74:1789, 1802, 1818 (1978); 75:38 (1979).
4. M.H. Wagner, Rheol. Acta 15:136 (1976); 16:43 (1977); 18:681 (1979).
5. H.M. Laun, Rheol. Acta 17:1 (1978).
6. M.H. Wagner and S.E. Stephenson, J. Rheol. 23:489 (1979).
7. R.A. Mendelson, Polym. Eng. Sci. 16:690 (1976).
8. K. Itoyama and A. Soda, J. Appl. Polym. Sci. 23:1723 (1979).
9. A. Zosel, Rheol. Acta 11:229 (1972).
10. H. Kubota, J. Appl. Polym. Sci. 19:2299 (1975).
11. H. Münstedt, in "Toughening of Plastics", Conf. RPI, London, pp. 9.1-9.8 (1978).
12. A.S. Lodge and J. Meissner, Rheol. Acta 11:351 (1972); 14:664 (1975).
13. K. Osaki, N. Bessho, T. Kojimoto and M. Kurata, J. Rheol. 23:617 (1979).
14. K.I. Chang, S.S. Yoo and J.P. Hartnett, Trans. Soc. Rheol. 19:155 (1975).
15. J. Meissner, J. Appl. Polym. Sci. 16:2877 (1972).
16. P. Le Roy and J.M. Pierrard, in "Proceedings of the VIIth International Congress on Rheology", C. Klason and J. Kubát, eds., Gothenburg, pp. 530-531 (1976).
17. N.J. Mills, Europ. Polym. J. 5:675 (1969).
18. S. Pedersen and L.L. Chapoy, J. Non-Newtonian Fluid Mech. 3:379 (1978).

THE INFLUENCE OF MOLECULAR WEIGHT ON THE RHEOLOGICAL PARAMETERS AND RELAXATION SPECTRUM OF POLY(2,6-DIMETHYL-1,4-PHENYLENE OXIDE)

L.R. Schmidt and J.A.O. Emmanuel

General Electric, Corporate Research & Development
Chemical Laboratory, P.O. Box 8
Schenectady, New York 12301 U.S.A.

INTRODUCTION

Poly(2,6-dimethyl-1,4-phenylene oxide), PPO* resin, exhibits thermodynamic compatibility with polystyrene over the entire composition range. A single glass transition temperature (T_g) is observed for a blend of these polymers when measured by differential scanning calorimetry.[1] This behavior is somewhat surprising considering the large differences between physical properties of these two homopolymers.

In an earlier paper, Schmidt[2] reported on the rheological responses of PPO resin, high impact polystyrene, HIPS, and a 35-65 blend of these two resins. The measured rheological parameters and the computed relaxation spectra indicated a high degree of segmental mixing of the PPO resin and the polystyrene in the HIPS. Hence, it is possible to tailor the properties of such a blend by carefully choosing the PPO resin and the polystyrene. In particular, the molecular weight (MW) and the molecular weight distribution (MWD) of each resin can be varied to achieve the proper level of viscous and elastic response for a specific application.

This paper reports on a series of PPO resins of increasing MW with nearly the same polydispersity. The melt properties were measured over a broad range of temperature and frequency. The relaxation spectra, computed from the melt data, provide a general picture of PPO resin rheology which has been summarized in a single, analytical expression. The basic features of these spectra and the

*PPO is a registered trademark of the General Electric Company.

fitting parameters of the general equations have been correlated with the molecular weight data.

EXPERIMENTAL

Eight different PPO resins were obtained from the General Electric Plastics Division in Selkirk, New York. These resins were produced by a process similar to the commercial process used to make PPO resin. Molecular weight data were obtained in chloroform at 25°C by gel permeation chromatography and laser light scattering. Intrinsic viscosities (IV) were also measured in chloroform at 25°C. Table 1 summarizes the MW data for the as-received powders. Samples A through G were all polymerized with the same catalyst system and with nearly the same reactor conditions. Sample H was polymerized using a different catalyst system from the other resins. Reactor conditions for Sample H were adjusted to compensate for different catalyst activity.

All PPO samples were compression molded into sheets measuring 10.2 cm x 10.2 cm x 0.2 cm with the platten temperature set at 293°C. The thermal history imposed during molding increased the MW which did not change further during the melt testing. The MW data on these sheets are also summarized in Table 1.

The dynamic viscoelastic properties of the melts were measured with a Rheometrics Mechanical Spectrometer Model RMS-7200, using parallel disks (eccentric rotating disk (ERD) test mode). Earlier test results[2] clearly demonstrated the advantages of this test mode over the plate-and-cone test mode which required critical spacing between the two fixtures.

The PPO sheets were cut into 25 mm squares and preconditioned in a vacuum oven at 60°C and 10 mm Hg for about 12 hours. This procedure was required to remove trapped air and/or residual solvent from the polymerization even though the compression molded sheets appeared to be free of trapped gases. The specimens were loaded between the hot test fixtures and sufficient time allowed to reach the set temperature. Excess material, extending beyond the disks, was next removed with a blade, leaving a very smooth polymer-air interface. The distance between disks was then adjusted to remove residual stresses associated with sample loading. All tests were run in a hot nitrogen environment which was controlled by an environmental chamber surrounding the test fixtures.

The tests were run with both positive and negative eccentricity, a. This procedure automatically accounted for any base line shift during the run. The shear strain γ for ERD is computed from the eccentricity and the spacing, h, between the parallel disks:

$$\gamma = a/h \qquad (1)$$

All tests were run at 5% strain, which was verified to be well within the linear region of material response. The storage modulus G' is computed as

$$G' = F_y h/\pi R^2 a \qquad (2)$$

where F_y is the force generated colinearly with the eccentricity and R is the disk radius. The loss modulus G" is computed as

$$G'' = F_x h/\pi R^2 a \qquad (3)$$

where F_x is the force in the shear plane, orthogonal to the eccentricity.

RESULTS AND DISCUSSION

Figure 1 shows typical loss modulus data over three decades of frequency for four different temperatures. At 300°C a different sample was used for each decade of frequency. Similar data were obtained for the storage modulus but these data show considerable scatter at the lower frequencies and higher temperatures due to a relatively small response which could not be accurately followed by our strain transducers. Both types of moduli data were empirically shifted onto master curves using the time-temperature superposition principle which assumes that all relaxation processes for a particular material have the same temperature dependence. A reference temperature of 260°C was arbitrarily chosen for each material. Figure 2 shows the loss moduli master curve for the data plotted in Figure 1. This technique extends the characterization to six decades of frequency and provides the required data for the computation of the relaxation spectrum for each material. The shift factors, a_T, follow the form of the well-known WLF equation[3]. All eight samples exhibited the same rubbery plateau modulus (G" = 1.5 x 10^6 dynes/cm^2), and thus the same entanglement molecular weight, Me, which was computed from an expression derived by Marvin[4]:

$$Me = 0.32\rho RT/G'' \qquad (4)$$

where ρ is the melt density, R is the universal gas constant and T is absolute temperature. The calculated entanglement molecular weight for all of the PPO resins studied is 9,100. This result indicates that up to this molecular weight, the chain formation is probably independent of catalyst and reactor conditions. The catalyst system, reaction temperature and oxygen up-take become important factors as this critical molecular weight is exceeded.

The steady shear viscosity was computed from the dynamic data using the Cox-Merz empirical relationship which was verified in the earlier work[2]. Flow activation energies were then calculated from the Arrhenius equation at zero shear rate and a shear rate of 1.0 sec^{-1} which is in the shear-sensitive portion of the viscosity function. Table 2 summarizes the activation energies for each resin.

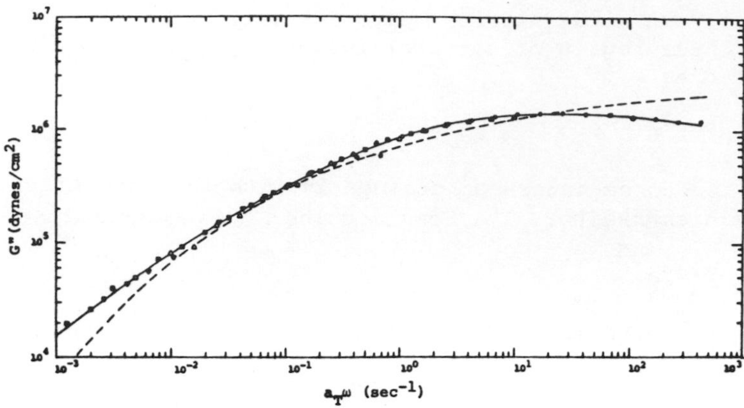

Figure 2

Master curve of loss moduli reduced to 260°C for Sample C: —— experimental data; --- computed from integral expression using Eq. 5.

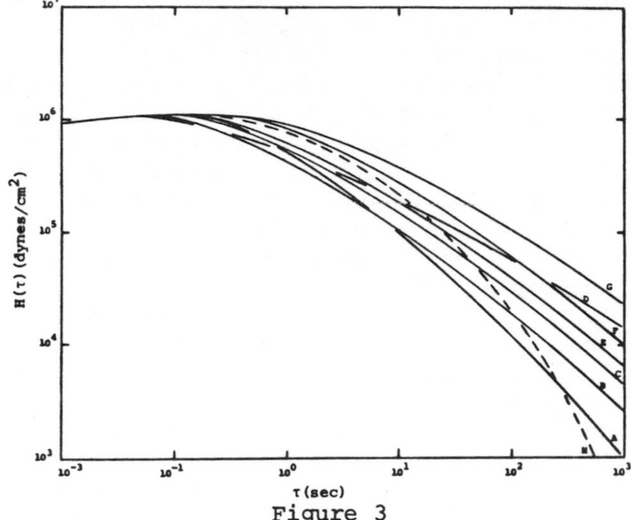

Figure 3

Relaxation time spectra for Samples A through H.

TABLE 2

ACTIVATION ENERGIES AND FITTING PARAMETERS

RESIN	E,kcal/mole		Fitting Parameters	
	γ=0 sec^{-1}	γ=1.0 sec^{-1}	α	β
A	47.3	40.0	.793	.208
B	49.5	41.7	.674	.208
C	42.0	34.1	.634	.208
D	53.7	43.9	.490	.208
E	47.4	30.5	.611	.208
F	49.1	35.2	.579	.208
G	32.2	22.2	.460	.208
H	46.9	34.2	.263	.373

4. R.S. Marvin, "In Viscoelasticity," J.T. Bergen, Ed., Academic, New York (1960).

TABLE 1

MOLECULAR WEIGHT DATA

	POWDER	SHEET			
RESIN	IV(dl/g)	IV(dl/g)	$\overline{M}n (x\ 10^{-3})$	$\overline{M}w (x\ 10^{-3})$	$\overline{M}w/\overline{M}n$
A	.42	.51	22	57	2.6
B	.49	.58	25	62	2.5
C	.45	.60	25	65	2.6
D	.54	.59	26	77	3.0
E	.50	.63	34	79	2.3
F	.54	.66	27	80	3.0
G	.49	.69	27	81	3.0
H	.51	.64	28	71	2.5

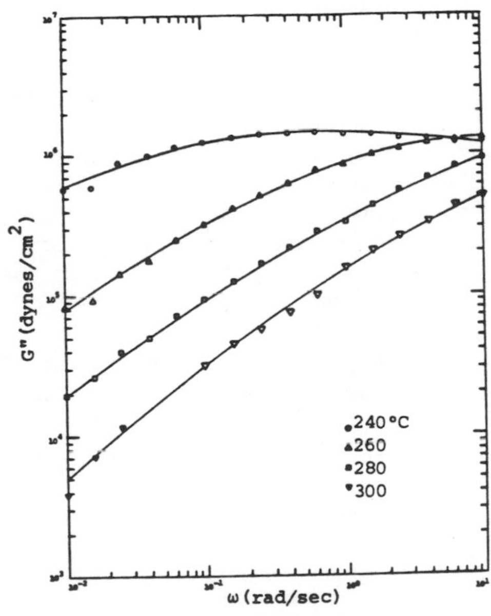

Figure 1

Loss modulus vs. frequency for Sample C

Relaxation spectra were calculated using the numerical method of Ninomiya and Ferry[3]. All of the relaxation spectra are shown in Figure 3. A broadening effect is observed with increasing MW, producing a family of curves with the same basic shape. Two exceptions to the general trends are noted. First, Sample H, which was polymerized by a different catalyst system exhibits very different curvature. Starting from the common plateau value, $H(\tau)$ drops off much more rapidly with increasing τ.

The second exception is Sample D which exhibits a relaxation spectrum which initially drops off from the plateau value much faster than expected and then shows a greater broadening of relaxation responses when compared to the other resins in this family. The reason for this behavior is still not clear but appears to be related to molecular structure.

A linear regression analysis was used to fit the data in Figure 3 to an analytical expression. The form of the relaxation function is

$$\log H = 6.32 - \alpha\tau^{\beta} \tag{5}$$

The values of the fitting parameters, α and β, are also listed in Table 2. A linear dependence was found between α and IV. However, the relationship between α and $\overline{Mw}$ or polydispersity, $\overline{Mw}/\overline{Mn}$ are not as simple. This is also the case for other relationships with α, such as activation flow energy.

The basic form of a three constant expression (Eq. 5) appears to be adequate to describe all of the rheological responses of PPO molecules over this extended portion of the relaxation spectrum. Eq. 5 can be used in the integral expressions to predict various rheological parameters. The approximate integration technique of Smith[3] was used with the analytical expression for the slope of $H(\tau)$ to generate a G" function for Sample C. The computed result is shown in Figure 2 as the dashed line. The constant 6.32 reflects the entanglement MW but the physical significance of the parameters α and β is still not completely understood. The exponent, β, seems to indicate a basic molecular structural form since it is different for the two catalyst systems. The coefficient α apparently incorporates several different molecular weight effects and quite possibly the higher moments of the molecular weight distribution.

REFERENCES

1. A.R. Shultz and B.M. Gendron, J. Appl. Polym. Sci., 16:461 (1972).
2. L.R. Schmidt, J. Appl. Polym. Sci., 23:246 (1979).
3. J.D. Ferry, "Viscoelastic Properties of Polymers," Wiley, New York (1970).

OSCILLATORY RHEOLOGY: POLYCARBONATE

C. J. Aloisio

Bell Laboratories
2000 Northeast Expressway
Norcross, Georgia 30071

V. W. Boehm

Western Electric
2000 Northeast Expressway
Norcross, Georgia 30071

INTRODUCTION

The oscillatory shear response of polymers offers a convenient measure of the viscoelasticity of these materials. Coupling well known analytical approaches with one of the many commercial devices available, one may determine polymer structural parameters of scientific and engineering importance. Such dynamic mechanical properties may be used to characterize the temperature and molecular weight dependence of the linear viscoelastic response of rubbery amorphous polymers. Recognition of the importance of these properties in polymer flow behavior clarifies many flow related phenomena.

EXPERIMENTAL

A Rheometrics eccentric rotating disc rheometer[1] was used to obtain the components of the complex shear modulus,

$$G^*(\omega) = G' + iG'' \tag{1}$$

The device, shown schematically in Figure 1, subjects a disc shaped sample to an oscillating shear deformation as a consequence of the axial offset, a, between the upper and lower disc. From the measured forces F_x and F_y in Figure 1

$$G' = \frac{F_y/\Pi r^2}{a/h} \tag{2}$$

and

$$G'' = \frac{F_x/\Pi r^2}{a/h} \tag{3}$$

where r is the sample radius.

Four bisphenol A polycarbonate samples were investigated. The molecular parameters determined using a Waters gel permeation chromatograph[2] are given in Table I.

TABLE I

Sample	Mn	Mw	Mw/Mn
A	11,700	26,900	2.30
B	13,000	32,400	2.50
C	14,100	36,300	2.58
D	16,700	40,400	2.42

The components of G* were obtained over a temperature range of 185°C to 260°C and a frequency range of 0.1 to 100 sec^{-1}. The resulting storage modulus (G') and loss modulus (G") master curves are shown in Figure 2. The isotherms are presented elsewhere.[3]

RESULTS

The master curves of Figure 2 were formed according to

$$G^*(\omega,T) = G^*(\omega a_T,T_o) \qquad (4)$$

using a reference temperature of 200°C. The data points in Figure 3A, representing all molecular weights, were adequately fit by the following WLF equation[4]

$$\log a_T = \frac{-5.23(T-To)}{98.4+T-To} \qquad (5)$$

Referencing the constants to Tg = 150°C one obtains $C_1{}^g = 10.6$ and $C_2{}^g = 48.4$. These yield a free volume fraction at Tg of 0.041 and a free volume thermal expansion coefficient of $8.4 \times 10^{-4}\,°C^{-1}$.

Using an approach similar to the development of a_T[4] the molecular weight dependence may be incorporated in equation (4),

$$G^*(\omega,T,M) = G^*(\omega a_T a_M,T_o,M_o) \qquad (6)$$

The shear modulus curves shifted readily with respect to molecular weight to form the master curves of Figure 4. The molecular weight dependence of a_M, Figure 3B is

$$a_M = \left(\frac{M_w}{M_o}\right)^{3.83} \qquad (7)$$

with $M_o = 32,400$.

COMPLEX VISCOSITY

The complex viscosity is, by definition,

$$|\eta^*(\omega,T)| = \frac{|G^*(\omega,T)|}{\omega} \tag{8}$$

Using equation (6) we obtained

$$|\eta^*(\omega,T)| = \frac{|G^*(\omega a_T a_M, T_o, M_o)|}{\omega} \tag{9}$$

Applying a horizontal shift in equation (8) gives

$$\eta^*(\omega a_T a_M, T_o, M_o) = \frac{|G^*(\omega a_T a_M, T_o, M_o)|}{\omega a_T a_M} \tag{10}$$

Substituting equation (9) into equation (10), and rearranging yields

$$|\eta^*(\omega,T,M)| = a_T a_M |\eta^*(\omega a_T a_M, T_o, M_o)| \tag{11}$$

Using the results of equation (11) the curves of complex viscosity were formed analytically using equations (5) and (7) (Figure 3).

CONCLUSIONS

The analytical results of equation (11) would seem to require equal viscosity and frequency scaling in any molecular theories. Using the plateau value of log G' = 7.43 as G_e in the expression for the entanglement molecular weight

$$M_e = \frac{\rho RT}{G_N^o}$$

one obtains M_e = 1470 very close to the 2000 obtained by Mercier etal.[5] From the terminal zone in G' (Figure 2) the decade breadth of the plateau zone, Δ, was found to be

$$\Delta = 4 \log \frac{M_W}{2M_e} \tag{13}$$

This parameter is essential for prediction of certain shear rate dependent flow properties.

Finally we note that the apparent activation energy values calculated from equation (5) shown in Table II are remarkably close to the 26-30 kcal/mole reported for polycarbonate flow.[6] Constant activation energy for flow must be a consequence of limited experimental data obtained on the high end of the WLF temperature dependence.

TABLE II

T, °C	Kcal/mole
200	54.3
225	38.2
250	29.2
275	23.4
300	19.6
325	16.8
350	14.8

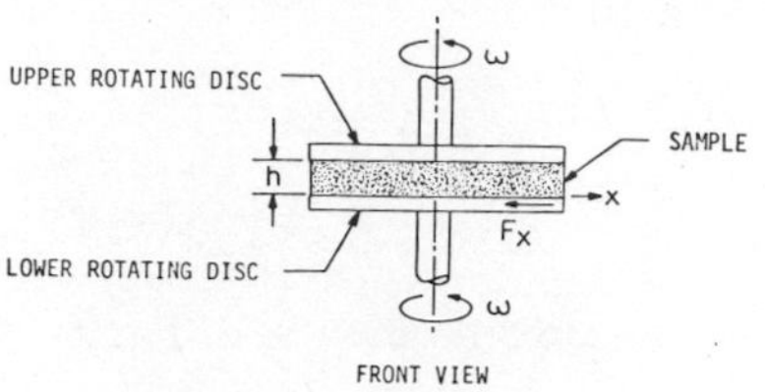

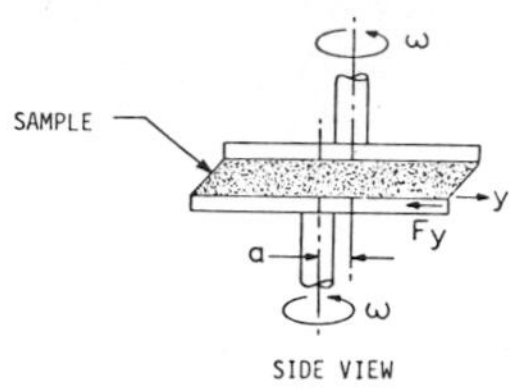

Figure 1. Schematic of Test Apparatus Illustrating the Forcing of the Sample Through a Shear Cycle with each Revolution

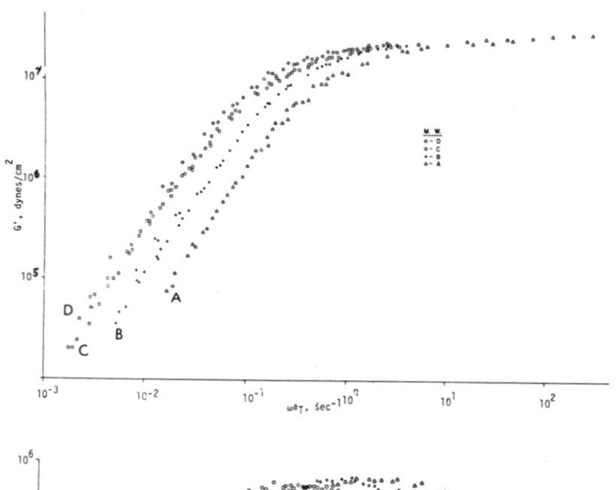

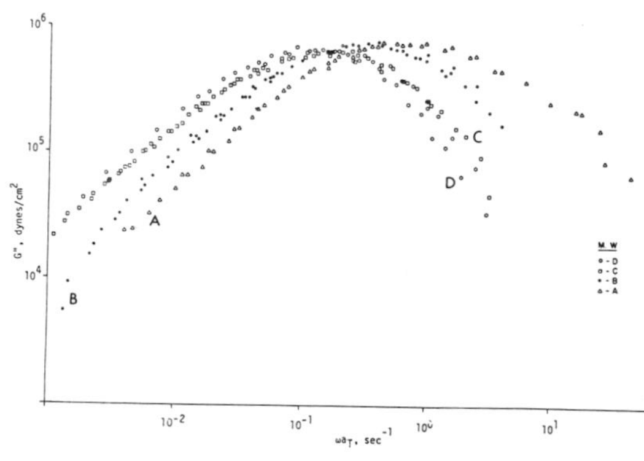

Figure 2. Loss Modulus Master Curves vs. Reduced Frequency at Reference Temperature, T_o = 473°K

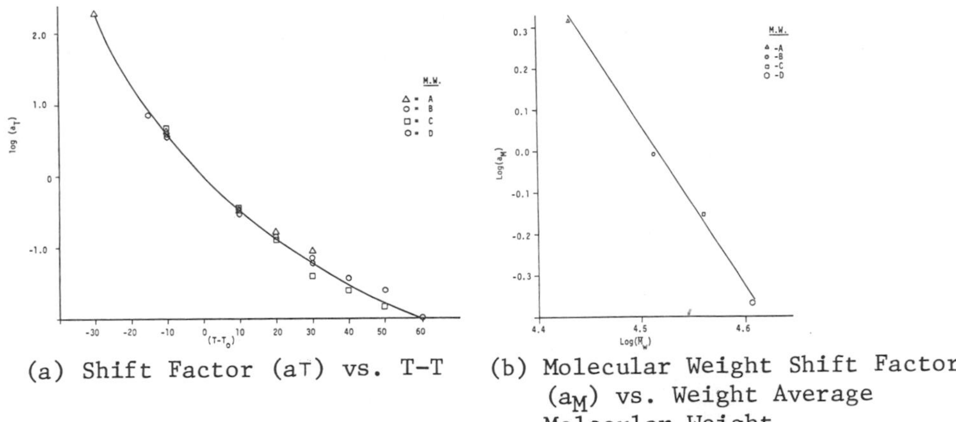

(a) Shift Factor (a_T) vs. T-T

(b) Molecular Weight Shift Factor (a_M) vs. Weight Average Molecular Weight

Figure 3

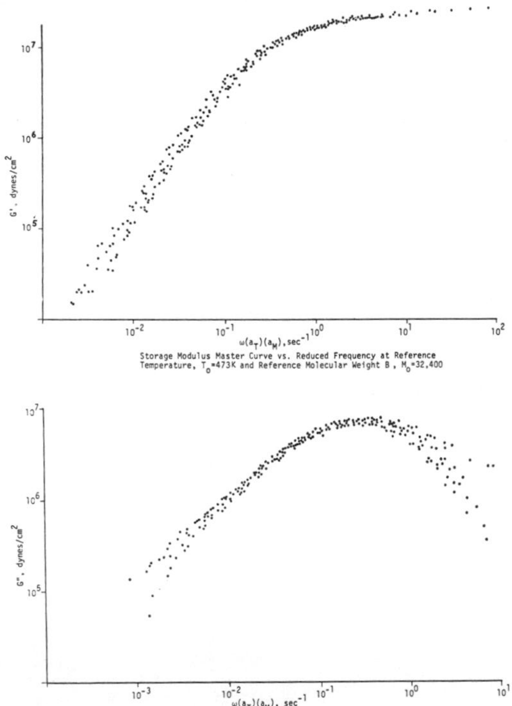

Figure 4. Master Curve vs. Reduced Frequency at Reference Temperature T_o = 473K and Reference Molecular Weight B, M_o = 32,400

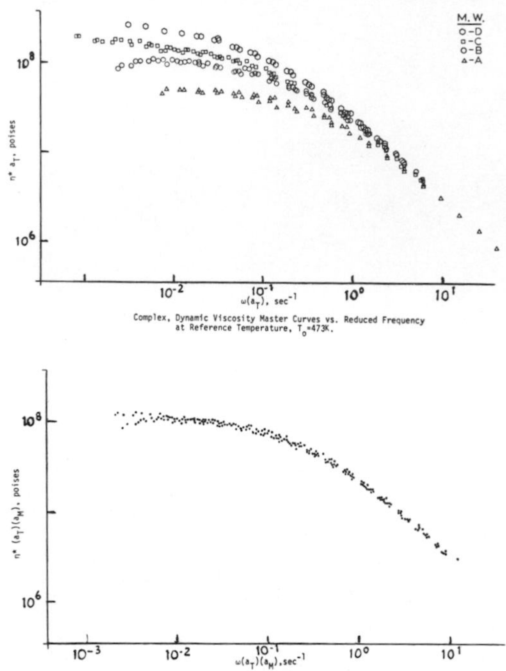

Figure 5. Complex Dynamic Viscosity Master Curve vs. Reduced Frequency at Reference Temperature T_O = 473K and Reference Molecular Weight B, M_O = 32,400

1. C. W. Macosko and J. M. Starita, Society of Plastics Engineers Journal 27, 38, (November 1971).
2. Dr. F. Kama, Mobay Chemical Corporation, Private Communication.
3. V. W. Boehm, "A Dynamic Mechanical Characterization of Polycarbonate" Thesis, March 1978, Georgia Institute of Technology.
4. J. D. Ferry, Viscoelastic Properties of Polymers, (John Wiley and Sons, New York, 1970).
5. Mercier, J. P., J. J. Aklonis, M. Litt and A. V. Tobolsky, J. Appl. Polymer Sci. 9, 447-459 (1965).
6. L. E. Nielsen, Polymer Rheology, Marcel Dekker, Inc., New York, 1977.

THE VISCOELASTIC PROPERTIES OF MELTS OF POLYPROPYLENE AND ITS BLENDS WITH HIGH-DENSITY POLYETHYLENE

J.S. Anand, I.S. Bhardwaj

Indian Petrochemical Corporation Limited
Gujarat, India

(Abstract)

Using an indigenous rheometer set up for the lower range and the Brabender Plasticorder for the higher range, the rheological studies for polymer melts through circular dies of varying (Length/ Diameter) ratio were made in the shear rate range of 20-4000 sec^{-1}. The samples used were three of polypropylene (MFI = 2.9, 0.62 and 0.3) and one of high-density polyethylene (MFI = 0.3) and blends of PP with hdPE. The blending characteristics of PP with hdPE were studied by means of a Brabender roller mixer where the concentration of hdPE in PP was varied from 0-20%. The torque absorption passes through a maximum as the concentration of hdPE varies in the blends. Viscosity average molecular weight and molecular weight distribution were determined by means of solution viscometry and melt flow indexer, respectively. The rheological measurements made were stress-strain behaviour and die swell characteristics of the virgin polymers, polymer blends and degraded polymers obtained after repeated extrusion. These were the interpreted to identify the effect of mol.wt. and mol.wt.distribution on the rheological properties of the polymer melts in the light of existing molecular viscoelastic theories. It was found that the polymer melt elasticity is greater for polymers having a broad mol.wt. distribution than the samples having higher average mol.wt. but lower mol.wt. distribution. The polymer degradation resulted in reduced average mol.wt. and mol.wt. distribution and also reduced melt elasticity.

RHEOLOGY AND MORPHOLOGY OF DISPERSED

TWO-PHASE POLYMER BLENDS IN CAPILLARY FLOW

Narasaiah Alle and J. Lyngaae-Jørgensen

Instituttet for Kemiindustri
Building 227, The Technical University of Denmark
DK-2800 Lyngby, Denmark

INTRODUCTION

In recent years there has been a great deal of interest in the studies of the structure and properties of polymer mixtures. However, there is little information in the literature concerning the morphology-rheology relationship and the principles which govern the morphology developments for the dispersed two-phase polymer blends in the molten state under continuous shear flow[1-5]. In previous papers[6-7] a blending law based on model morphologies proposed earlier by Vinogradow[8,9] was derived in order to evaluate the morphology of dispersed two-phase polymer blends during capillary flow. The purpose of the present study is to elucidate the morphology in the molten state of polymer mixtures under shear flow conditions.

EXPERIMENTAL

Sample materials investigated in this work are commercial grades of polystyrene (Hoechst, Hostyrene N7000, $\bar{M}_w = 3.67 \times 10^5$, $\bar{M}_n = 1.17 \times 10^5$) and polymethylmethacrylate (ICI, Diakon L0951, $\bar{M}_w = 8 \times 10^4$, $\bar{M}_n = 3.76 \times 10^4$). The molecular weight distribution was determined by GPC, as described elsewhere[10].

Blend Preparation

All the blends were prepared by melt mixing in a Brabender Plasticorder at 200°C. The speed of the mixing blades was set at 20 rpm and mixing continued till a constant torque was reached which took about 12-15 minutes.

Rheometry

Melt flor properties at shear rates in the range of 0.3 to 1000 sec^{-1} were determined by using an Instron constant speed rheometer at four temperatures 180°, 200°, 220° and 240°C respectively. The capillary

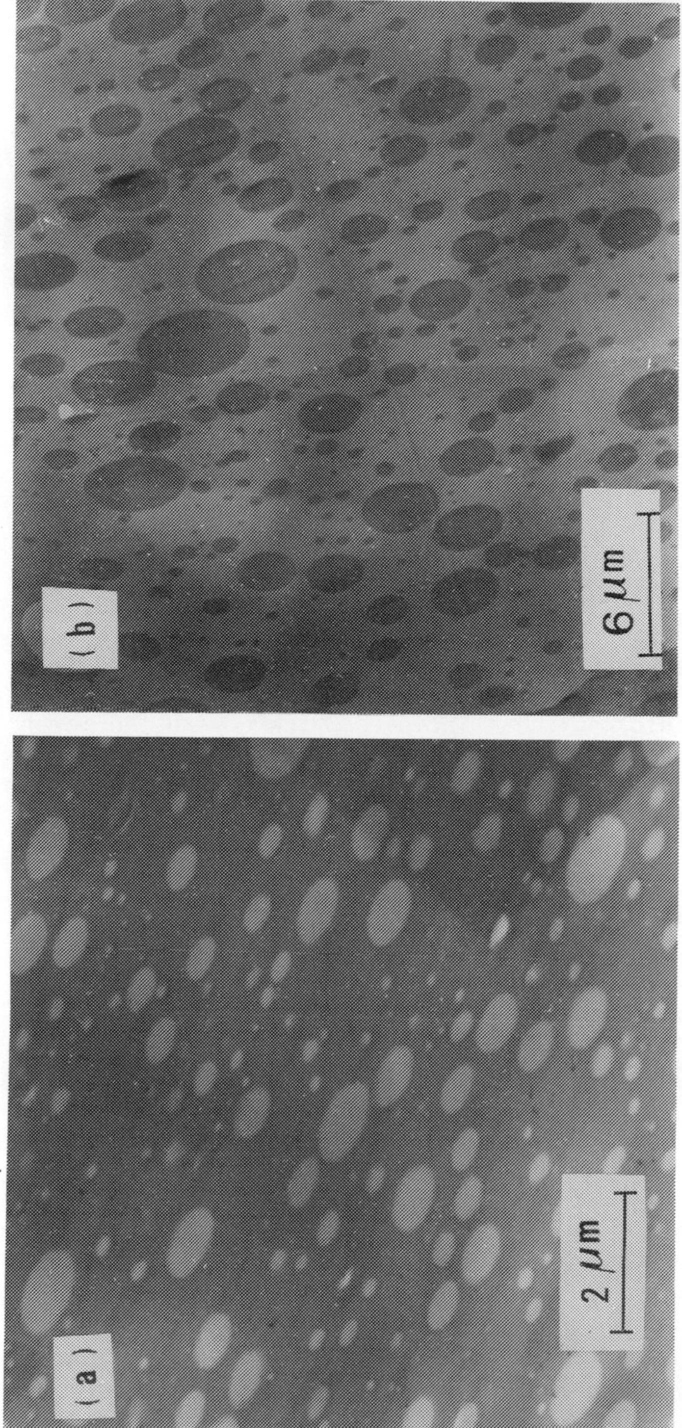

Figure 1. Morphology of PS/PMMA blends obtained after melt mixing in Brabender (a) PS/PMMA = 75/25, (b) PS/PMMA = 25/75

used has length/diameter ratio 81.79 with a diameter of 1.245 mm and 90° entrance angle. The Rabinowitsch correction[11] was applied to all data in calculating the true wall shear rate ($\dot{\gamma}$).

Morphology

Transmission electron micrographs of the well mixed blends obtained from Brabender were taken on a JEOL electron microscope (JEOL, Model JEM-100B-TR). Samples were sectioned by using an Ultramicrotome (Reicht "OM U3", Austria) at room temperature. The dark and light regions corresponding to the PS and PMMA phases were sufficient to distinguish the two-phases.

RESULTS AND DISCUSSION

Morphological Characterization

Figure 1 shows the electron micrographs of PS-PMMA blends obtained after melt mixing in the Brabender Plasticorder at 200°C at different blending ratios. It is seen in Fig. 1a that PS (black) forms the continuous phase and PMMA (white) forms the discrete phase, whereas the phase inversion takes place in Fig. 1b. The scale of phase separation estimated from Fig. 1 shows that the size of the domains of PS/PMMA = 75/25 and PS/PMMA = 25/75 are in the range of 0.2-0.6 μm and 0.5-2.0 μm respectively. Further, it is noted in Fig. 1 that the domains have a tendency to become spherical.

As shown by Fig. 1, different phase continuities and domain sizes are obtained for PS-PMMA blends by changing the composition ratio at the same mixing conditions. These results confirm the previous findings[2, 12]. That is, for a blend system when the minor component has a lower viscosity than the major one, the minor component is finely dispersed in the matrix of the major component (Fig. 1a). On the other hand when the minor component has a higher viscosity than the major component the minor component is coarsely dispersed in the continuous phase of low viscous major component (Fig. 1b).

Shear Rate - Temperature Superposition

For many polymer systems, the shear rate dependence of viscosity can be expressed in reduced variable form[13-15]:

$$\frac{\eta}{\eta_o} = V \cdot (\dot{\gamma}\lambda_o) \tag{1}$$

On the basis of various molecular theories[15-18] λ_o for the same polymer melt at different temperature can be expressed as

$$\lambda_o = \frac{\eta_o F}{\rho T} \tag{2}$$

$$\eta/\eta_o = V \cdot \left(F \cdot \frac{\eta_o \dot{\gamma}}{\rho T}\right) \tag{3}$$

where λ_o is a relaxation time constant, V is the viscosity master function, ρ is the density, T is the absolute temperature and F is a property which depends on molecular structure but is independent of temperature. Based on Equation (3), one would expect for a given polymer that the reduced curves of η/η_o versus $\eta_o\dot{\gamma}/\rho T$ should superimpose.

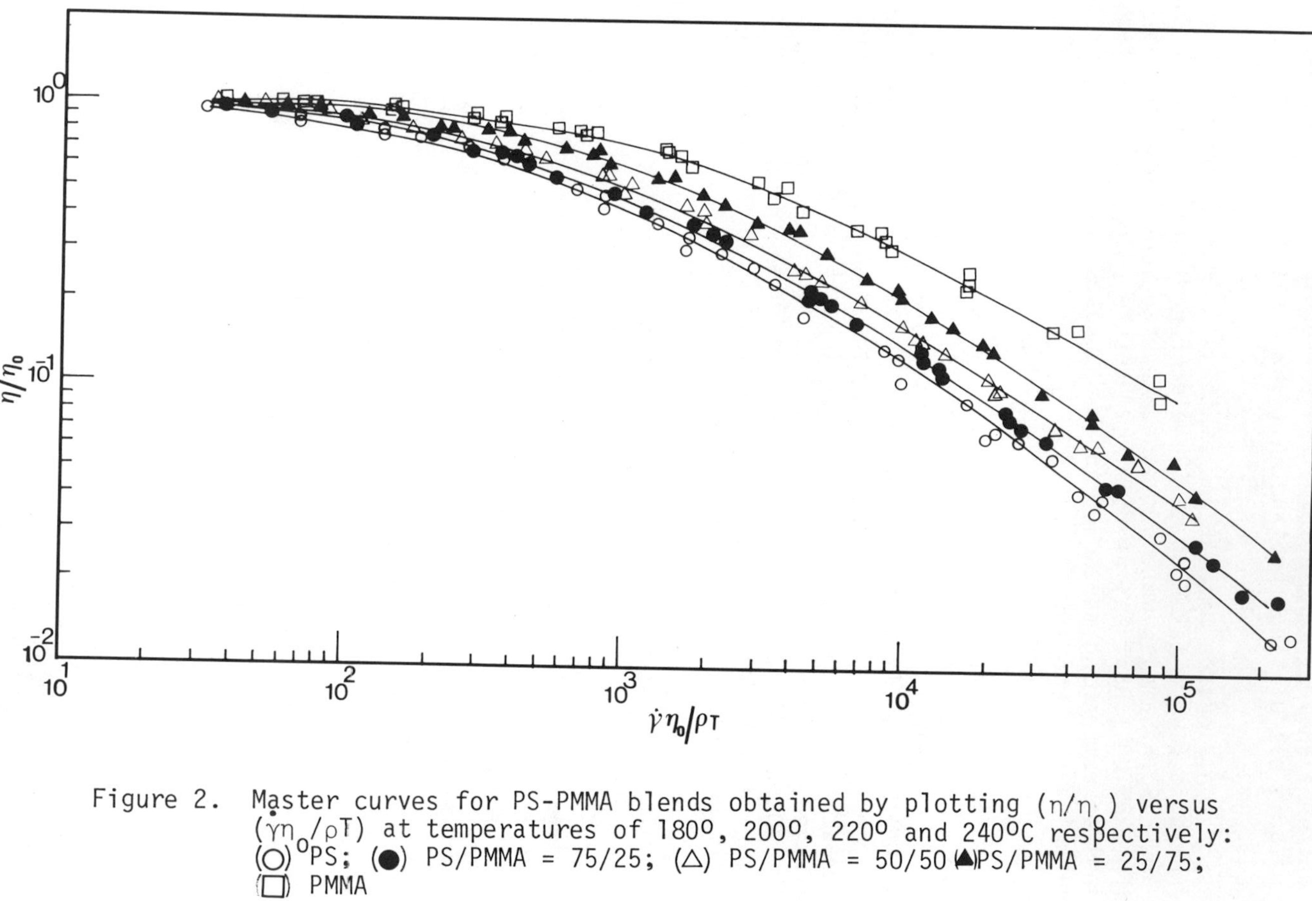

Figure 2. Master curves for PS-PMMA blends obtained by plotting (η/η_0) versus $(\dot{\gamma}\eta_0/\rho T)$ at temperatures of 180°, 200°, 220° and 240°C respectively: (○) PS; (●) PS/PMMA = 75/25; (△) PS/PMMA = 50/50 (▲) PS/PMMA = 25/75; (□) PMMA

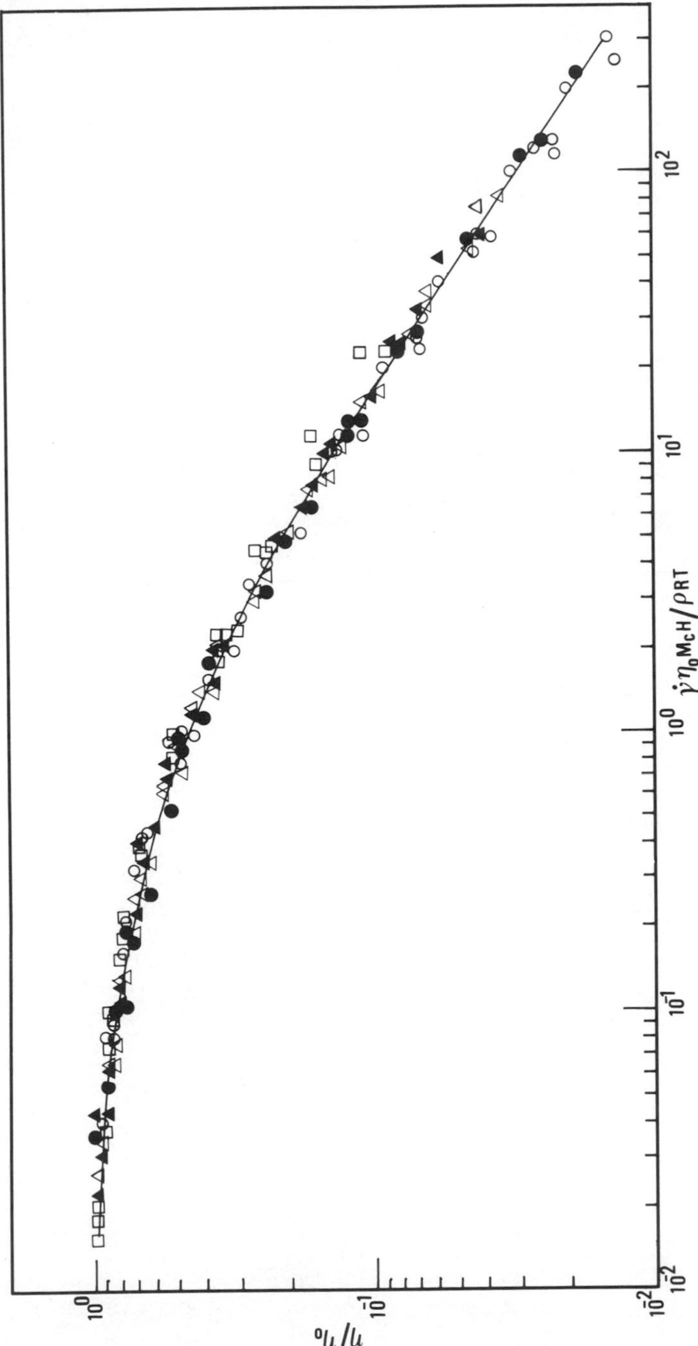

Figure 3. Master curve for PS/PMMA blends obtained by plotting (η/η_o) versus $(\dot{\gamma}\eta_o M_c H/\rho RT)$ at temperatures of 180°, 200°, 220° and 240°C respectively: (○) PS; (●) PS/PMMA = 75/25; (△) PS/PMMA = 50/50; (▲) PS/PMMA = 25/75; (□) PMMA

The results of such superpositions are shown in Figure 2 for homopolymers and three blends of PS and PMMA at four different temperatures. It is apparent from Fig. 2 that the superposition is achieved very well within the experimental error for all the blends at fixed composition.

Recently, Lyngaae-Jørgensen[19] predicted and observed that the master curves of the PS and PMMA with different molecular weight distribution superposed into a single master curve when plotted as (η/η_o) versus $(\eta_o\dot{\gamma}M_cH/\rho RT)$, in which M_c is the critical molecular weight where the slope in the Equation $\eta_o = K\overline{M}_w^a$ changes to 3.5 and H is the heterogeneity $(H = \overline{M}_w/\overline{M}_n)$.

Figure 3 shows the reduced plots of η/η_o versus $(\eta_o\dot{\gamma}M_cH/\rho RT)$ for all pure samples and the blends of PS and PMMA. It is seen from Fig. 3 that the master curves of PS, PMMA and PS-PMMA blends fall onto single master curve. The shear rate-temperature superposition principle is commonly found for homopolymers which are homogeneous and single phase in nature. In analogy to the homopolymers, the superposition of steady state viscosity data onto master curves indicate that all the blends of PS and PMMA are either compatible in the molten state under shear flow conditions or have a morphology which is independent of temperature.

REFERENCES

1. H.Van Oene, J.Colloid Interfac.Sci. 40, 448 (1972)
2. J.M. Starita, Trans.Soc.Rheol. 16, 339 (1972)
3. C.D. Han and Y.W. Kim, Trans.Soc.Rheol. 19, 245 (1975)
4. S. Danesi and R.S. Porter, Polymer 19, 448 (1978)
5. C.J. Nelson, G.N. Argeropoulos, F.C. Weissert and G.G.A. Böhm, Angew.Makromol.Chem. 60/61, 49 (1977)
6. N. Alle and J. Lyngaae-Jørgensen, Rheol.Acta (in press)
7. N. Alle and J. Lyngaae-Jørgensen, Rheol.Acta (in press)
8. M. Yakob, M.V. Tsebrenko, A.V. Yudin, and G.V. Vinogradow, Int.J.Polym.Metals 3, 99 (1974)
9. A.P. Plochocki, Pololefin Blends, in "Polymer Blends", vol. 2, D.R. Paul and S. Newman, ed., Academic Press, New York (1978)
10. N. Alle, Ph.D. Thesis, The Technical University of Denmark Copenhagen (1980)
11. B. Rabinowitsch, Z.Phys.Chem.A 145, 1 (1929)
12. B.L. Lee and J.L. White, Trans.Soc.Rheol. 18, 467 (1974)
13. G.V. Vinogradow and A.Y. Malkin, J.Polym.Sci. A 2, 2, 2357 (1966)
14. G.V. Vinogradow and A.Y. Malkin, J.Polym.Sci. A 2, 4, 135 (1966)
15. R.C. Penwell, W.W. Graessley and A. Kovacs, J.Polym.Sci., Polym.Phys.Ed. 12, 1771 (1974)
16. F. Bueche, J.Chem.Phys. 22, 603 (1954)
17. W.W. Graessley, J.Chem.Phys. 43, 2696 (1965)
18. S. Middleman, "The Flow of High Polymers", Interscience, New York (1968)
19. J. Lyngaae-Jørgensen, Dr.techn.Thesis, The Technical University of Denmark, Copenhagen (1980)

EXPERIMENTAL STUDY OF THE RHEOLOGICAL PROPERTIES OF ION CONTAINING AMORPHOUS POLYMERS

Guy Broze, Robert Jérôme, Philippe Teyssié,
Claude Marco+

Laboratory of Macromolecular Chemie and Organic Catalysis, University of Liègc, Sart-Tilman, 4000 Liège, Belgium.

+Laboratoire de Chimie Générale et de Sciences des Matériaux, Université de Mons, Faculté des Sciences, Av. Maistriau, 19, 2000 Mons, Belgique.

INTRODUCTION

During the two last decades, the interest in block and graft copolymers has been largely increasing. As the chemical blocks linked together are thermodynamically uncompatible they form heterophase systems with specific properties of each block. A typical example of such a compound is the poly(styrene-b-butadiene-b-styrene). In this block copolymer, the hight Tg sequence (poly styrene) acts as a physical thermoreversible reticulation. Such compounds recieved the name of "thermoplastic elastomers" and promoted industrial applications.

Eisenberg has shown that ionic polymers could also be heterophase materials (1); in the ionomers, it is well known that the ions are located in separate domains, the "clusters". Several papers have been published about the behaviour of these ionic polymers in relation with their morphology, but all the experimental results are far from being satisfactorily explained.

To gain a straightforward analysis of the physico-mechanical behaviour of the ionic polymers, we looked for polymers with well-defined structure and composition which can be systematically modified. For this purpose, the dicarboxylic telechelic polymers are very attractive model compounds to study the properties of ionic polymers.

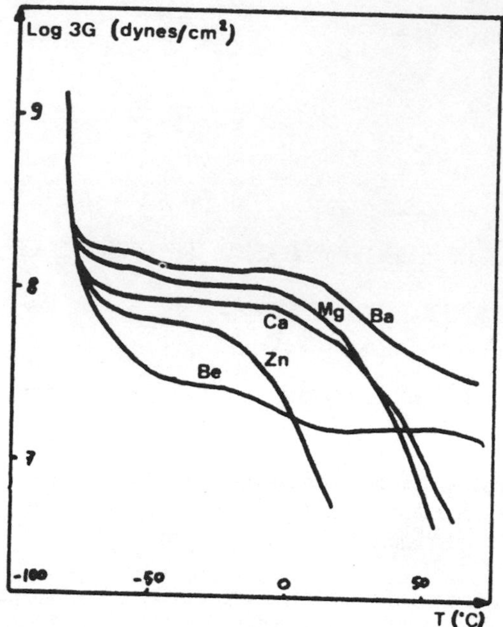

Fig. 1 PBD Salt: Shear Mod vs Temp.

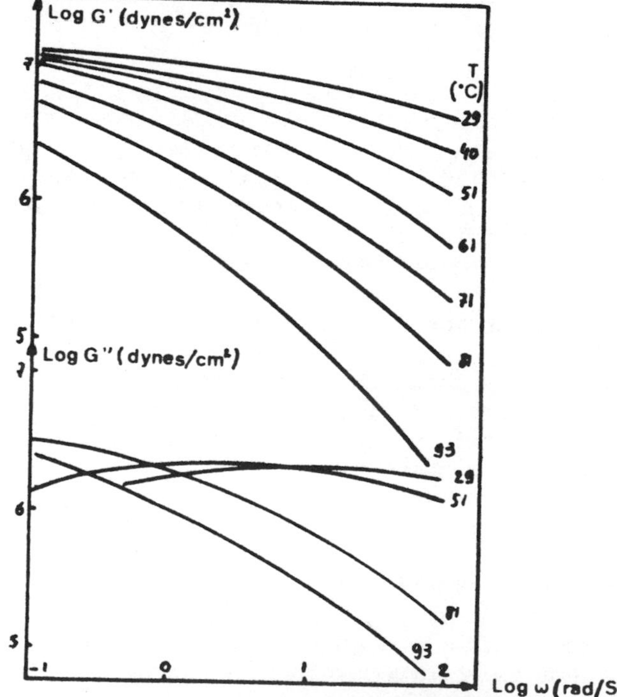

Fig. 2 PBD Mg: Dynamic Properties

CARBOXYL TERMINATED POLYBUTADIENE

In a first set of experiments, we used, like Pineri (2), α, ω dicarboxylic polybutadiene (PBD, $\bar{M}n$ = 4600, functionnality = 2) neutralised with various divalent metal ions. By thermal differential analysis, we observed the Tg of the PBD (-80°C), but also another second order transition at about room temperature. As well known, TDA is not very sensitive to second order transitions and we mesured the isochrone variations of the shear modulus with the temperature. As shown by figure 1, Tg is not affected by the presence of the ions, but there is a second transition around room temperature, that is affected by the nature of the cation. This second transition is surely in connection with the presence of the ions.

This is clearely demonstrated by the study of the complex shear modulus as a function of frequency and temperature. In the case of magnesium-neutralised polybutadiene (PBD-Mg), figure 2 shows the effects of time and temperature on the viscoelastic properties. The shift factors has been determined for each temperature and at figure 3, it may be seen the master curves for the storage and the less moduli G' and G". The relaxation spectrum H, is also given and has been calculated by the first order approximation of Ninomiya and Ferry (3). So, the presence of the metal carboxylate leeds to a new relaxation process. The variation of the shift factors with the temperature does not follow the W.L.F. equation, but well the Arrhenius equation. An activation energy of 30 kcal/mole is found,a reasonnable value for an ionic process. Moreover, the activation energy is deeply affected by the nature of the cation : as illustrated in the table 1, smaler is the cation radius then higher is the activation energy.

TABLE 1

Cation	Ionic radius (Å)	Act. energy (kcal/mole)
Mg^{2+}	0,65	30
Zn^{2+}	0,74	25
Ca^{2+}	0,99	23
Ba^{2+}	1,35	15

The PBD-salt has been investigated also by thermally stimulated depolarisation currents. The results are in perfect agreement with the mechanical experiments (figure 4). The α relaxation (Tg) is complex. There are in fact two α relaxations; the more intense one (α_1, lower temperature) is the real Tg, the weaker one (α_2, higher temperature) seems to be in relation with amorphous PBD situated near the clusters. The relative intensity

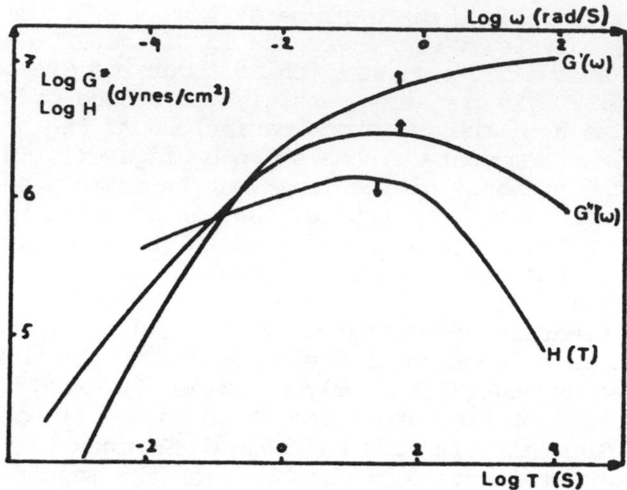

Fig. 3 PBD Mg: Master Curves (23°C)

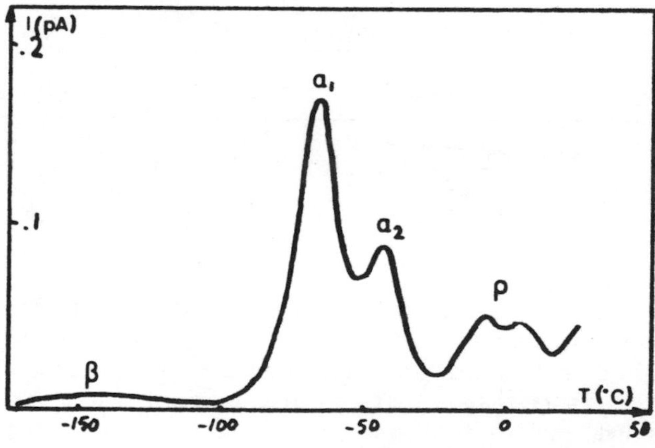

Fig. 4 PBD Mg: T.S.D.

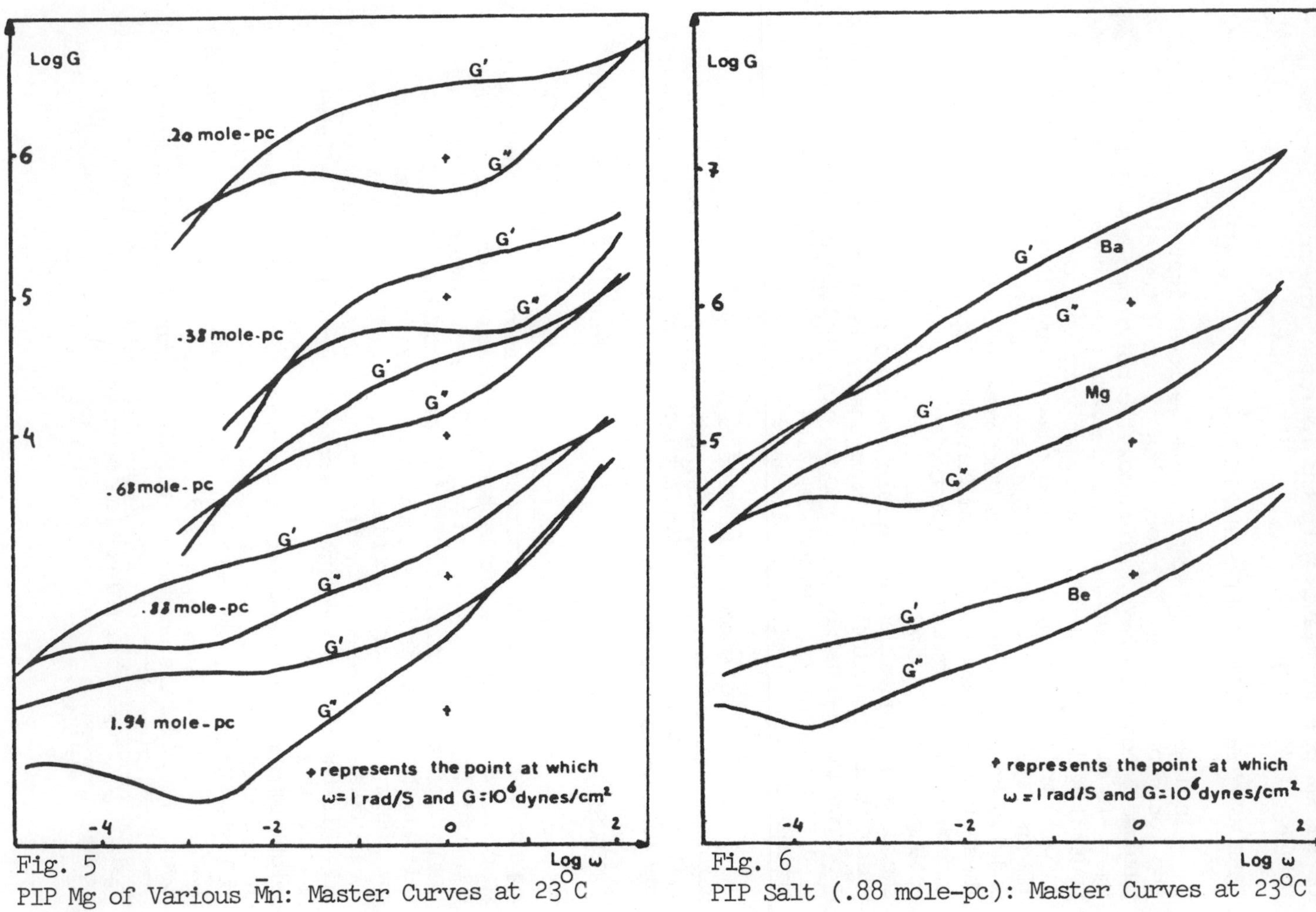

Fig. 5
PIP Mg of Various $\bar{M}n$: Master Curves at 23°C

Fig. 6
PIP Salt (.88 mole-pc): Master Curves at 23°C

of the $\alpha 2$ peak increases with the radius of the metal ion, as the cluster size surely do.

CARBOXYL TERMINATED POLYISOPRENES

In a second set of experiments we investigated the effect of the ionic content on the properties of the material. For this purpose 5 homodisperse α,ω dicarboxylic polyisoprenes (PIP, Tg = -10°C) of various molecular weights were synthetized, and neutralised with magnesium. Only at very low ion concentrations (< 0,75 mole-percent), the Tg and the rubbery plateau are noticable (figure 5). The W.L.F. equation is appropriate, but the storage modulus on the rubbery plateau increases with the ion content. These ions are associated in multiplets dispersed quite homogeneously in the amorphous matrix, and acts as thermoreversible entanglements. Above a critical concentration (0,75 mole % in the case of magnesium carboxylate), a second transition takes place at higher values on the time scale, and the time-temperature superposition works no more. So we can conclude to the appearance of a second relaxation mechanism with a different activation process associated with the clustering of the ions.

The critical concentration of clustering depends on the nature of the cation. With Be^{2+}, there are two relaxations at 0,75 mole % concentration, while there is only one transition in the case of Ba^{2+} (figure 6).

CONCLUSIONS

From these experiments, it can be concluded that the appearence of a relaxation process associated with the ionic sites is conditionned by the labile character of the ions, but also by the mobility of the chain. If the labile character of the ions is of the same order of magnitude, the formation of an ionic relaxation process will be conditionned by the characteristics of the chain. If the chain mobility is high (PBD), there will be an ion relaxation process. On the opposite (high molecular weight PIP), if the relaxation times associated with the entanglements are too long, no ionic relaxation is observed. If the molecular weight of the PIP is lowered, the relaxation times of the entanglements will be smaler, so that the ionic relaxation process could be effective. At constant chain mobility, it is the labile character of the ions that is determinant. Smaler is the cation, lower is its labile character, higher are the relaxation times associated with the ions, and higher are the probability of leeding to a specific relaxation process.

REFERENCES

(1) A. Eisenberg and M. King : Ion Containing Polymers. Academic Press, New York, San Francisco, London, 1977.
(2) A. Moudden, A.M. Levelut and M. Pineri. J. Pol. Sci., Phys., 15, 1707, 1977.
(3) K. Ninomiya and J.D. Ferry. J. Colloïd Sci., 14, 36, 1959.

ENTANGLEMENT FORMATION NEAR THE GLASS TRANSITION TEMPERATURE

H.H. Kausch, K. Jud

Chaire de Polymères
Ecole Polytechnique Fédérale
CH-1007 Lausanne

INTRODUCTION AND THEORETICAL BACKGROUND

The formation and breakdown of entanglement coupling points between molecular coils plays an important role in polymer rheology. Their effect on the viscosity of solutions and on the elastic modulus of rubbery networks is well investigated. Until recently very little was known, however, on the kinetics of entanglement formation at temperatures close to the glass transition temperature (T_g) although the practical importance of chain interpenetration and entanglement coupling for good mechanical properties of thermoplastic materials, especially at long times, higher temperatures and in the presence of active environments, is generally recognized. In the last few years, however, the self-diffusion and interpenetration of chain molecules in entangled polymer systems has attracted growing attention following the development of the reptation model of longitudinal diffusional motion by De Gennes (1) and Doi and Edwards (2). Experimental studies have been carried out meanwhile concerning the kinetics (3-5) and the mechanical effects (6) of chain penetration. In this paper the mechanism of entanglement formation and resolution is further investigated by variation of the conditions of entanglement formation and by measurement of the energy necessary for the mechanical separation of the partially interdiffused networks.

The employed experimental method (6) utilizes the fact that at temperatures above T_g the diffusion of chain segments across an interface between two blocks of compatible polymers A and B gives rise to an interpenetration of chains and to the formation of physical links between molecules from different sides of the interface. If the interpenetration is sufficiently strong the interface disappears optically and mechanically; static and dynamic forces can be transmitted across the interface. The strength of the bond established between A and B by chain interpenetration at $T>T_g$ can be quantitatively determined if one measures the material resistance R of the material in the interfacial region at ambient temperature.

For our experiments so-called compact tension (CT) specimens were used which permit the determination of the energy release rate G_{Ii} associated with the onset of breakage at the interface between the two welded blocks. According to linear elastic fracture mechanics G_{Ii}, R, and the fracture toughness K_{Ii} of our samples are related by:

$$R = G_{Ii} = K_{Ii}^2(1-\nu^2)/E \tag{1}$$

where E is Young's modulus and ν Poisson's ratio.

The fracture toughness K_{Ii} is determined from specimen cross-section WD, length a_o of a notch introduced at the interface, and fracture load P in a tensile experiment as

$$K_{Ii} = A(W/a_o)P\sqrt{a_o}/WD \tag{2}$$

with $A(W/a_o)$ being a slowly varying correction function (6). The material resistance R, which is a direct measure of the energy necessary for the deformation and breakage of the material in the fracture plane, is dominated by the energy U_p of plastic deformation (30-300 Jm^{-2}) of the interfacial region. Bond breakage (0.4 Jm^{-2}) and surface tension (0.08 Jm^{-2}) are of minor importance. As shown by earlier studies (7) with polystyrene samples of different molecular weight, the rupture strength of a sample increases with the (initially molecular weight dependent) volume concentration c of entanglement coupling points, i.e. R = R(c). In our experiments we also observe an increase of R with c which here is due to entanglement formation with increasing spatial interpenetration of the chain molecules.

EXPERIMENTAL

Experiments have been carried out using polymethylmethacrylate PMMA 7H (M_w=170'000; $M_w/M_n \simeq 2$) from Röhm GmbH., and styrene acrylonytril SAN Luran 368 R copolymer (≃25 mol % AN) from BASF. Both have the same glass transition temperature T_g=102-104°C and are completely compatible. The compact tension specimens (CT) were compression moulded from granulates. A geometry LxWxD=26x26x3 mm^3 was employed.

Cracks were introduced by sawing, they were subsequently sharpened by a razor blade. The specimens were fractured in tension. The fracture events led to smooth fracture surfaces FS. For a first series of specimens (crack healing) these surfaces FS were brought in contact again and placed in a preheated hotpress. A slight pressure of about 1 bar was applied normal to the major specimen surface. For a time t_p the specimens were kept at a temperature T_p set to be 1 to 15 K above the glass transition temperature. After this time the specimens were shock-cooled to room temperature thus freezing-in the state of network interpenetration achieved during t_p. For a second series of experiments flat surfaces PS were prepared by polishing the fracture surfaces FS. Arbitrarily selected surfaces PS were then welded together as described above.

RESULTS AND DISCUSSION

The fracture toughness K_{Ii} (initiation) has been measured on rehealed and welded CT-samples as a function of contact time t_p and temperature T_p. The fairly large scatter involved in these measurements required a large number of experiments.

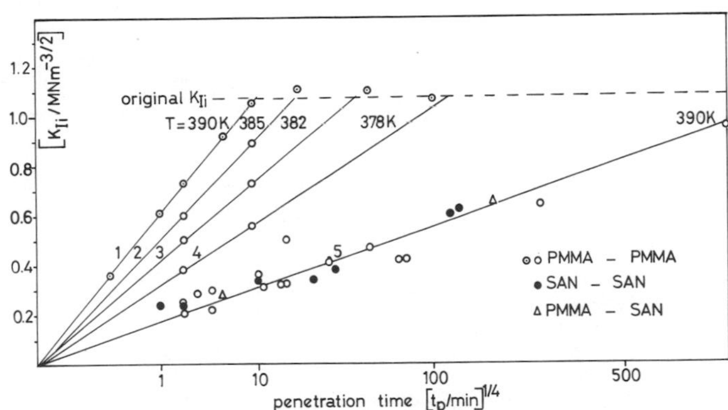

Fig. 1. Fracture toughness of the region of chain interpenetration as a function of penetration time (t_p) and penetration temperature (T).

In Fig. 1 each point on curves 1 to 4 is an average value from about 20 to 30 measurements. In 5 all measured data are indicated. In a log K_{Ii} vs. log t_p plot the data are very well represented by straight lines which turn out to have a slope of 0.25. In a K_{Ii} vs. $(t_p)^{1/4}$ plot all lines are initially straight and go through the origin (Fig. 1). The K_{Ii} values of the rehealed specimens do not exceed, however, the fracture toughness K_{Io} of the virgin material. The curves change their slope abruptly as soon as K_{Ii} (t_p) becomes equal to K_{Io}.

From Fig. 1 it can be seen that there is a time lag $t_{po}(T)$ before the broken specimens have regained the full original short time fracture toughness K_{Io}. The four values taken from curves 1-4 are very well described by an Arrhenius equation:

$$t_{po}(T) = t_{po}(T_o) \exp \left(\frac{E_a}{RT} - \frac{E_a}{RT_o}\right) \tag{3}$$

with an energy of activation E_a equal to (274±20) kJ mol^{-1}.

In order to obtain further information on the kinetics of entanglement formation one does need the relation between R and the number N of entanglement points formed per unit area of the fracture surface. We know that R is zero if no entanglement points have been formed ($T_p<T_g$ or $t_p<<1$ min) and that R does reach the value R_o of the virgin material after a penetration time $t_{po}(T)$. We call the value of N reached at this moment N_o. In the absence of any other information we interpolate R(N) in this range linearly between the two known values:

$$R(N) = N(t)\ R_o/N_o. \tag{4}$$

The number N(t) of newly formed cross-links certainly increases with the depth Δx of interpenetration of the molecules. In the following we will test the hypothesis that the longitudinal chain displacement across an interface is a diffusional mechanism. We assume that N is proportional to Δx. Under this condition one derives from the Einstein diffusion equation:

$$\frac{N^2(x|t|)}{N_o^2} = \frac{<\Delta x^2(t_p)>}{<\Delta x^2(t_{po})>} = \frac{2\,Dt_p}{2Dt_{po}} \tag{5}$$

From Eqs. 1, 4 and 5 one obtains

$$\frac{K_{Ii}(t_p)}{K_{Io}} = \frac{E(t_p)^{\frac{1}{2}}}{E(t_{po})^{\frac{1}{2}}} = \frac{t_p^{\frac{1}{4}}}{t_{po}^{\frac{1}{4}}}. \tag{6}$$

If one neglects the rather small change of the elastic modulus E with t_p one concludes from Eq. 6 that a plot K_{Ii} vs. $t_p^{\frac{1}{4}}$ should be a straight line. Exactly this is confirmed experimentally (Fig.1).

It follows from Eqs. 3 and 5 that the diffusion coefficient D(T) must also obey an Arrhenius equation with the same activation energy E_a of 274 kJ mol^{-1}. This energy of 274 kJ mol^{-1} is larger than the energies between 25 and 165 kJ mol^{-1} thus far reported in the literature for such a mechanism. This difference is probably due to the fact that up to now all diffusion studies have been carried out at temperatures consistently higher than T_g, i.e. at $T-T_g>85$ K, whereas in our experiments $T-T_g<15$ K.

We have attempted to estimate the diffusion coefficient D(T) using the reptation model (8). Based on the work of Doi and Edwardse Grassley (8) recently proposed the following relation:

$$D = \frac{G_o}{135} \left(\frac{\rho RT}{G_o}\right)^2 \left(\frac{R_e^2}{M}\right) \frac{M_c}{M^2 \eta_o(M_c)}. \quad (7)$$

The rubber plateau modulus G_o was determined to be $3 \cdot 10^6$ MPa. For the other quantities values were taken from the literature (9) namely for the density $\rho=\rho(T)$, the mean square end-to-end distance R_e^2 of a PMMA molecule of molecular weight M=170'000 ($R^2/M = 7.29 \cdot 10^{-4}$ nm^2 mol g^{-1}), the critical molecular weight for entanglement (M_c=5'900), and the zero shear viscosity for the critical molecular weight ($\eta_o = 3.78 \cdot 10^4$ Ns cm^{-2} at T=375 K). At T=378 K Eq. (7) gives a value of D equal to $1 \cdot 10^{-16}$ cm^2 s^{-1}, at T=390 K D is equal to $1.5 \cdot 10^{-15}$ cm^2 s^{-1}. Using these values of D we calculate from the Einstein diffusion equation $(<\Delta x^2(t_{po})>)^{\frac{1}{2}}$, the depth of penetration for "fully rehealed" specimens to be 15 nm. This seems to be a reasonable depth, since the end-to-end distance for PMMA of molecular weight M=170'000 is of the order of 11 nm.

Looking at the measurements on polished interfaces, PS, the difference with regard to the rehealing measurements can presumably be explained by a resistance to diffusional flow (3) which makes the diffusion constant more than two orders of magnitude lower ($D=10^{-17}$ cm^2 s^{-1} at T=390 K).

Our present investigations aim at an elucidation of the role of microfibrils in rehealing and a direct quantitative determination of the depth of penetration by infrared spectroscopy and energy dispersive X-ray analysis (EDS).

SUMMARY

Chain diffusion across polymer-polymer interfaces leads to the gradual optical disappearance of the interface and to the establishment of mechanical strength. A measure of this strength, the fracture toughness K_{Ii}, has been determined quantitatively using broken and rehealed compact tension specimens. It has been found that in polymethylmethacrylate K_{Ii} increases with the $\frac{1}{4}$ power of t_p, the time of rehealing at an elevated temperature T_p. This behavior can be explained by the reptational diffusion of chains at $T>T_g$ with a diffusion coefficient of $D(T)=7.3 \cdot 10^{21}$ exp $- E_a/RT$ and an activation energy of 274 kJ mol^{-1}.

ACKNOWLEDGEMENTS

The authors thank Drs. M. Dettenmaier and N.Q. Nguyen, EPF Lausanne, and Dr. J. Klein, The Weizmann Institute of Science, Rehovot, for their constructive comments, I. Adamou for his help with the measurements, and the Swiss National Science Foundation for its support of this work.

REFERENCES

1. P.G. De Gennes, J. Chem. Phys. 55, 572 (1971)

2. M. Doi, S.F. Edwards, J.C.S. Faraday II 74, 1789 (1978)

3. J. Klein, B.J. Briscoe, Proc. Roy. Soc. A 365, 53 (1979)

4. H. Koch, R. Kimmich, 26. IUPAC Sympos. Makro Mainz 1979, Vol. II, 1078

5. H. Sillescu, G. Zimmer, W. Wetterauer, ibid., 1082

6. K. Jud, H.H. Kausch, Polym. Bull. 1, 697 (1979)

7. J.F. Fellers, B.F. Kee, J. Appl. Polym. Sci. 18, 2355 (1974)

8. W.W. Graessley, to be published in J. Polym. Sci., Polym. Phys. Edn.

9. D.W. Van Krevelen, Properties of Polymers, Elsevier, Amsterdam (1976)

IRREVERSIBLE NETWORK DISENTANGLEMENT AND TIME-DEPENDENT FLOWS OF POLYMER MELTS

Manfred H. Wagner

Schlossbergweg 116
D-8851 Kühlenthal, W-Germany

INTRODUCTION

For small strains Lodge's rubberlike-liquid theory is a valid description of rheological behaviour of polymer melts, but at higher strains the theory fails: The phenomenon of shear thinning is not explained which is characteristic for shear flow of nearly all polymer liquids, and in elongation deviation of experimental data from predictions of Lodge's theory reflect also a flow thinning, and not a strain hardening in spite of the pronounced S-shape of the stress-strain diagrams. Comparing measured stress growth and stress relaxation data with predictions of the theory, it can be concluded that the temporary network structure of the polymer melt is destroyed increasingly with the magnitude of deformation. Hence, the number of entanglements decreases with increasing strain. Finally the irreversibility of the disentanglement process is considered.

CONSTITUTIVE EQUATION A

Lodge's rubberlike-liquid constitutive equation[1-3] is a valid description of material behaviour of polymer melts and polymer solutions in the second order limit, i.e. in the limit of small strains[4,5]. It is based on a number of assumptions[3] the most important of which is the assumption of the existence of a temporary network structure, which is not affected by the deformation and flow of the material. Formulated in space tensors, Lodge's equation is given by

$$\underline{p}(t) = -p\underline{I} + \int_{-\infty}^{t} \mathring{\mu}(t-t')\underline{C}_t^{-1}(t')dt' \quad (1),$$

where $\underline{p}(t)$ is the stress tensor at time t, $p\underline{I}$ the isotropic pressure contribution, $\underline{I}$ the unit tensor, and $\underline{C}_t^{-1}(t')$ the relative Finger strain tensor. The memory function $\mathring{\mu}(t-t')$ can be expressed in molecular terms as a function of the creation rate, N_i, and the relaxation time, τ_i, of the network strands of type i,

$$\mathring{\mu}(t-t') = \rho kT \sum_i N_i e^{-(t-t')/\tau_i} = \sum_i a_i e^{-(t-t')/\tau_i} \quad (2).$$

ρ is the density, k the Boltzmann constant, and T the absolute temperature.

For larger strains Lodge's constitutive equation fails to describe adequately the material behaviour of polymer liquids[6-9]. It appears from an analysis of material behaviour in shear and elongation[7, 10-12] that temporary junction networks disentangle with increasing deformation. This strain-dependent disentanglement can be expressed in the following statement, given below as assumption (A)[13-15], which leads to a constitutive equation adequately describing large-strain material behaviour in shear and elongation as long as the strain is non-decreasing:

(A) The disentanglement assumption.
There are two independent decay mechanisms for network strands of type i:
   (a) time-dependent relaxation with probability $1/\tau_i$, where the probability of a network strand i to survive a time difference (t-t') is equal to $\exp(-(t-t')/\tau_i)$,
   (b) disentanglement by deformation. The probability of a network strand to survive a relative deformation between the instant of creation (t') and the instant of observation (t) is independent of index i and equal to $h(I_1(t,t'),I_2(t,t'))$, where I_1 and I_2 are the first and second invariant of the relative Finger strain tensor $\underline{C}_t^{-1}(t')$, respectively.

Together with the rubberlike-liquid assumptions[3], assumption (A) leads to a modified rubberlike-liquid constitutive equation[7, 10-12] here referred to as theory A:

$$\underline{p}(t) = p\underline{I} + \int_{-\infty}^{t} \mathring{\mu}(t-t')h(I_1,I_2)\ \underline{C}_t^{-1}(t')dt' \quad (3).$$

It is interesting to note that theory A is similar to the equation of Doi and Edwards[16], which is of the form

$$\underline{p}(t) = -p\underline{I} + \int_{-\infty}^{t} \mathring{\mu}(t-t')\left[h_1(I_1,I_2)\underline{C}_t^{-1}(t') + h_2(I_1,I_2)\underline{C}_t(t')\right]dt' \qquad (4),$$

and has been developed for concentrated systems of linear polymers on the basis of the "tube model"[17]. In both cases of eqs. (3) and (4), the kernel functions are separated into a time-dependent factor, the rubberlike-liquid memory function $\mathring{\mu}(t-t')$ (eq.(2)), and a strain-dependent part, $h(I_1,I_2)$ (called the damping function[7]), and $h_1(I_1,I_2)$ and $h_2(I_1,I_2)$, respectively. This separation has been verified experimentally in shear by the step-strain experiments of Osaki[18] and Laun[10], and in elongation by the analysis of Wagner[12].

Comparison between theoretical predictions of theory A and experimental evidence has been performed mainly for a LDPE melt thoroughly investigated formerly (Melt I[19-20]). Once the linear-viscoelastic relaxation spectrum and the damping function $h_s(\gamma^2)$ for shear and $h_e(\varepsilon)$ for uniaxial elongation are supplied in the constitutive equation (3), predictions are in general accordance with available experimental data.

In Fig. 1 and 2 the strain-dependent kernel functions of the equation of Doi and Edwards are compared with experimental results as analysed by theory A in shear and elongation, where both theory A (eq.(3)) and Doi's equation (4) are reduced to

$$p_{12} = \int_{-\infty}^{t} \mathring{\mu}(t-t')h_s(\gamma^2(t,t'))\ \gamma(t,t')dt' \qquad (5),$$

and

$$p_{11} - p_{22} = \int_{-\infty}^{t} \mathring{\mu}(t-t')h_e(\varepsilon(t,t'))\ \left[e^{2\varepsilon(t,t')} - e^{-\varepsilon(t,t')}\right]dt' \qquad (6).$$

p_{12} and $p_{11}-p_{22}$ are the shear stress and the tensile stress, $\gamma(t,t')$ and $\varepsilon(t,t')$ the relative shear strain and the relative Hencky strain, and h_s and h_e the strain-dependent parts of the kernel functions in shear and uniaxial elongation, respectively. It is seen that the strain dependence of the branched polyethylene Melt I as described by theory A is inbetween Lodge's assumption (no strain dependence) and Doi and Edwards' prediction in both shear and elongation, while more linear polymers show a disentanglement behaviour closer to Doi and Edwards' theory.

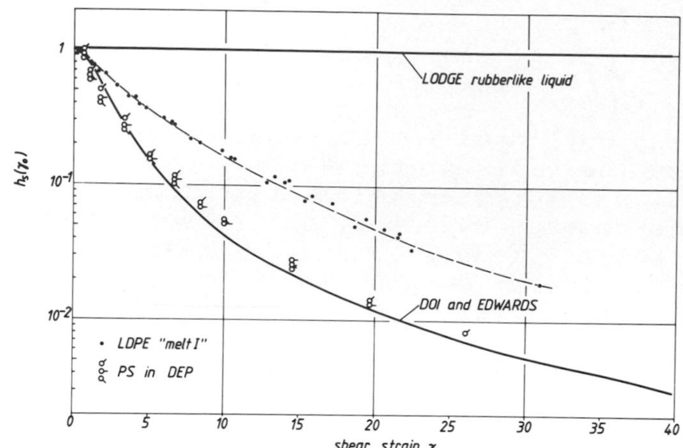

Fig. 1. Disentanglement function h_s for shear as a function of shear strain γ_0. h_s is proportional to the number of entanglements that survive a given shear deformation γ_0. Solid lines indicate Lodge's assumption (no strain dependence) and Doi and Edwards' prediction[17]. Dotted line indicates theory A as adjusted to data for Melt I[10] (full points), open points show data for a polystyrene solution[24].

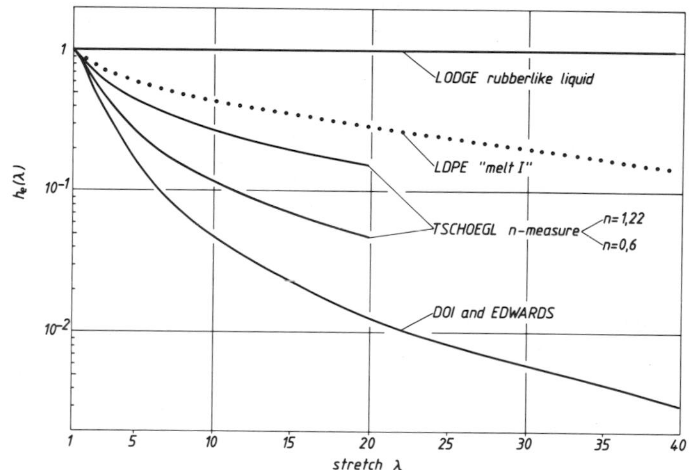

Fig. 2. Disentanglement function h_e for uniaxial extension as a function of stretch ratio $\lambda = \exp(\varepsilon)$. h_e is proportional to the number of entanglements that survive a given stretch ratio λ. Solid lines indicate Lodge's assumption, Tschoegl's data for crosslinked and uncrosslinked SBR[25,26], and Doi and Edwards' prediction[17]. Dotted line indicates data for Melt I as analysed by theory A[12].

CONSTITUTIVE EQUATION B

By means of theory A, description of rheological behaviour of a well-characterized low-density polyethylene melt (Melt I[19-21]) and a polystyrene melt[8] is excellent as long as the strain is increasing[7,9-12]. However, recovery calculations in elongation have shown that an additional assumption is necessary concerning the irreversibility of strain-dependent network disentanglement[12-15,22,23]. Since the damping function $h(I_1,I_2)$ represents the probability for a network segment to survive a given relative deformation, it is to be expected that strain-dependent disentanglement is an <u>irreversible</u> process. Therefore a further assumption (B) has been proposed[13-15]:

(B) The irreversibility assumption.
Segments lost during a non-decreasing deformation are not reformed during a decreasing deformation. The deformation dependent survival probability is therefore a <u>functional</u> of the strain with

$$\mathfrak{h}(I_1,I_2) = \underset{t' \leq t'' \leq t}{\mathrm{Min}} \left[h(I_1(t'',t'),I_2(t'',t')) \right] \quad (7),$$

where $h(I_1,I_2)$ is the survival probability for a non-decreasing deformation, and Min is a functional operator, which gives the minimum value attained by h in the time interval (t,t') for a fixed creation time t'.

Assumptions (A) and (B) give the constitutive equation here referred to as theory B,

$$\underline{p}(t) = -p\underline{I} + \int_{-\infty}^{t} \mu(t-t')\, \mathfrak{h}(t,t')\, \underline{C}_t^{-1}(t')\, dt' \quad (8),$$

where the damping functional $\mathfrak{h}(t,t')$ is defined by eq. (7).

As demonstrated in Fig. 3 and 4, theory B gives clearly a better description of recovery experiments than theory A. However, it should be mentioned that the form of $\mathfrak{h}$ was chosen because of simplicity reasons, and other choices are conceivable. The emphasis lies on the fact that $\mathfrak{h}$ should be considered as a functional of the strain. Close inspection of Doi and Edwards' theory reveals that their equation (4) is valid only for increasing deformations (increasing length of the "tube"), and that decreasing deformations, as for example in the case of elastic recovery, will result in an equation containing a functional of the strain history.

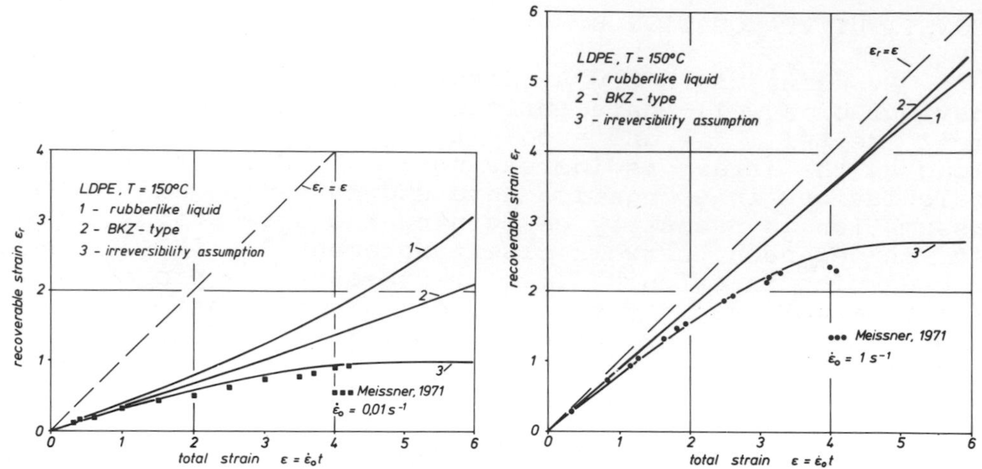

Fig. 3 and 4. Recoverable strain ε_r as a function of total Hencky strain ε for constant strain rate tests with elongation rates $\dot{\varepsilon}_o$ = 0.01s^{-1} and 1s^{-1}. Data for Melt I[19]. Dotted line indicates complete recovery. Solid lines are predictions: 1 - Lodge's rubberlike -liquid theory, eq. (1)
2 - theory A, eq. (3)
3 - theory B, eq. (8).

REFERENCES

1. Lodge, A.S., Trans. Faraday Soc. 52, 120 (1956).
2. Lodge, A.S., "Elastic Liquids", London-New York 1964.
3. Lodge, A.S., Rheol. Acta 7, 379 (1968).
4. Chang, H., and A.S. Lodge, Rheol. Acta 11, 127 (1972).
5. Wagner, M.H., Rheol. Acta 15, 133 (1976).
6. Lodge, A.S., "Body Tensor Fields in Continuum Mechanics", London-New York 1974.
7. Wagner, M.H., Rheol. Acta 15, 136 (1976).
8. Gortemaker, F.H., H. Janeschitz-Kriegl, and K.te Nijenhuis, Rheol. Acta 15, 487 (1976).
9. Laun, H.M., M.H. Wagner, and H. Janeschitz-Kriegl, Rheol. Acta 18, 615 (1979).
10. Laun, H.M., Rheol. Acta 17, 1 (1978).
11. Wagner, M.H., and H.M. Laun, Rheol. Acta 17, 138 (1978).
12. Wagner, M.H., J. Non-Newtonian Fluid Mech. 4, 39 (1978).
13. Wagner, M.H., paper presented at the IUTAM Symposium on Non-Newtonian Fluid Mechanics, Louvin-La-Neuve, 1978.
14. Wagner, M.H., and S.E. Stephenson, J. Rheol. 23, 489 (197
15. Wagner, M,H., and S.E. Stephenson, Rheol. Acta 18, 463 (1979).
16. Doi, M., and S.F. Edwards, J. Chem. Soc., Faraday Trans. II, 74, 1789-1832 (1978), 75, 38 (1979).

17. De Gennes, P.G., J.Chem. Phys. 55, 572 (1971).
18. Osaki, K., S. Ohta, M. Fukuda, and M. Kurata, J. Polym. Sci. 14, 1701 (1976).
19. Meissner, J., Rheol. Acta 10, 230 (1971).
20. Meissner, J., J. Appl. Polym. Sci. 16, 2877 (1972).
21. Meissner, J., Rheol. Acta 14, 201 (1975).
22. Wagner, M.H., Rheol. Acta 18, 33 (1979).
23. Wagner, M.H., Rheol. Acta 18, 681 (1979).
24. Fukuda, M., K. Osaki, and M. Kurata, J. Polym. Sci. 13 A, 1563 (1975).
25. Chang, W.V., R. Bloch, and N.W. Tschoegl, J. Polym. Sci. 15, 923 (1977).
26. Bloch, R., W.V. Chang, and N.W. Tschoegl, J. Rheol. 22, 1 (1978).

THE EFFECT OF PRESSURE ON MELT VISCOSITY OF HIGH IMPACT POLYSTYRENE CONTAINING ANTIMONY TRIOXIDE

F. Večerka and Z. Horák

Chemopetrol, Synthetic
Rubber Research Institute
Kralupy n/V., ČSSR

INTRODUCTION

A considerable attention has been devoted to the organic polymers modified by solid additives dispersed in polymer matrix. The additives used to improve some properties of polymer (or just only as a filler) affect, besides the properties in solid phase, the rheological behaviour of polymer melts. The influence of solid particles suspended in molten polymers on flow behaviour were studied by many authors. Most of fundamental information on the multiphase (polymeric) systems is summarized in the monography of Han /1/. However, while the effect of temperature and shear rate on rheological properties of such systems were intensively studied, the effect of pressure has not been considered for the time being.

The fact that at a certain value of the applied pressure the viscisity of polymer melts ceases to be independent of pressure is well known and fundamental studies were c arried out for an interpretation of this phenomenon /2 - 6/. Changes of viscosity with pressure are usually described by a pressure viscosity coefficient. This value, which was determined for several important polymers /2 - 6/ and copolymers /7/, seems to be convenient for the description of multiphase system, too.

There is another interesting observable phenomenon in the behaviour of system polymer - low molecular solid particles. At low concentrations of additive, under certain conditions the viscosity of blend melt decreases in comparison with the viscosity of the original polymer /8,9/. The viscosity drop is observable in the entire range of the applied pressures, however, following the anomaly in the range where $\eta = f(P)$ gives better possibility to find the minimum on the curve as a function of external factors.

This contribution deals with some changes of flow properties of three phase composite polystyrene - polybutadiene - antimony trioxide in the range, where viscosity is a function of pressure.

EXPERIMENTAL

The samples of high-impact polystyrene (HIPS) with 5,5 wt.% of butadiene elastomer and concentration $c_{Sb_2O_3}$ = = 0, 1, 5, 15 and 25 wt.% antimony trioxide were prepared by mixing of the components in the Henschel homogeniser and following plastication in an extruder. Homogenity of the samples was checked by means of the microscope. The molecular weight $\bar{M}_w$ of polystyrene component was 180 000. Antimony trioxide (of 99,95% purity) with particle size distribution from 1-7 μm and average particle size 1,6 μm was used. No aglomerates were present in the oxide.

The experiments were carried out by means of the high pressure capillary viscometer Goettfert at temperatures T=180 and 200°C. The pressure required was achieved by application of five capillaries with different ratios of length to inner diameter L/d = 5, 10, 40, 60 and 80 mm. All capillaries had diameter 1 mm. Shear stress τ_w and shear rate D_w at the capillary wall were calculated by common methods /10/. Data obtained were corrected for nonparabolic velocity profile and end effects /11/. Corrected values of shear stress τ and shear rate D were used for determination of corrected viscosity . Pressure viscosity coefficient b was calculated for given pressure P from eq. (1):

$$b = (\partial \ln \eta \ / \ \partial P)_{T=const} \qquad (1)$$

RESULTS AND DISCUSION

Viscosity dependence on the shear rate for two levels of pressure is shown in Fig. 1. As expected, the viscosity differences of the samples with different concentration of antimony trioxide are reduced with increasing

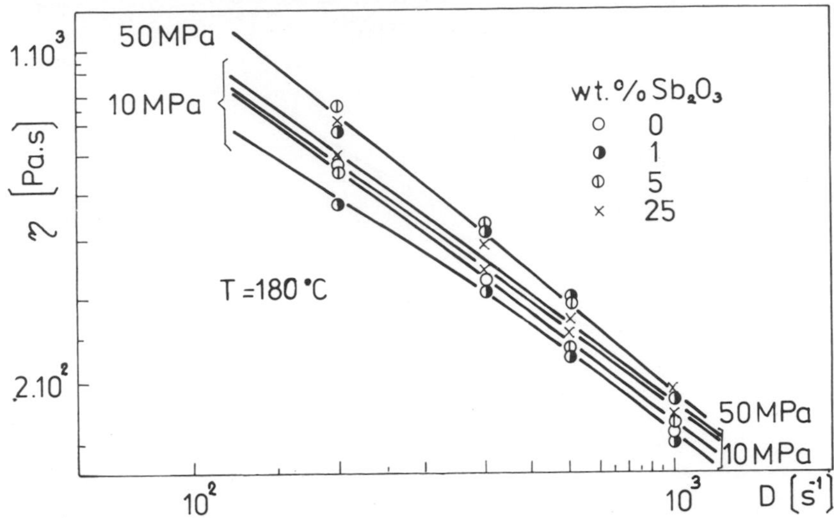

Fig. 1. Viscosity of HIPS containing different concentrations of anitimony trioxide as a function of shear for two preassures.

shear rate, what is in accordance with literature /12/. By increasing of temperature from 180 to 200°C the analogic effect was observed. For the demonstration of pressure effect on the melt viscosity the pressures 10 and 50 MPa were chosen. The first value of pressure lays in the range where $\eta \neq f(P)$, the second one in the range of strong viscosity dependence. Owing to the pressure increase from 10 to 50 MPa, the viscosity rises by about 15 - 40 % in the followed shear rate range.

The pressure effect on the melt viscosity of HIPS containing 1 wt.% of antimonity trioxide for four values of shear rate can be seen from the Fig.2. The dependence for other samples measured is an alogical. These graphs served for the calculation of pressure viscosity coefficients. There is also apparent how the range of viscosity independence of pressure is extended with shear rate.

Viscosity dependence of the composite melts on the antimony trioxide content (Fig.3) verified the fact observed formerly that an addition of comparatively small quantity of solid additive can cause a viscosity drop/8,9/.

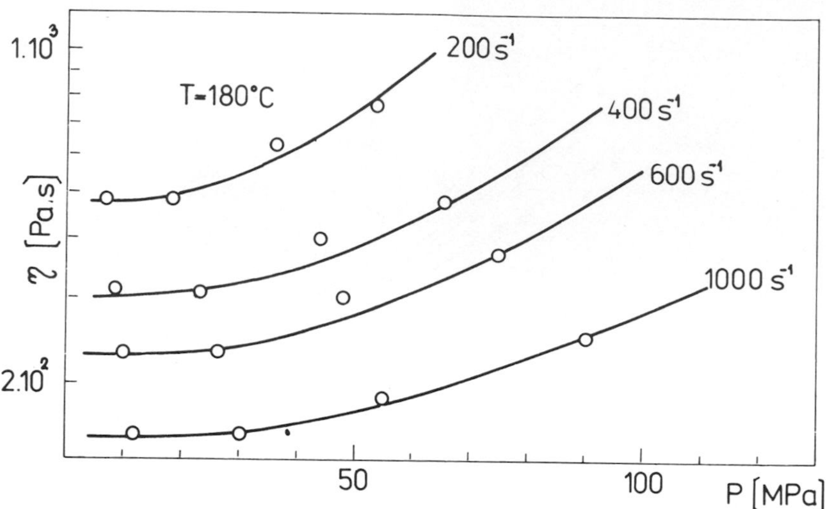

Fig. 2. Viscosity as a function of pressure at given at given shear rates.

The illustration in the Fig.3 shows the viscosity dependence at P = 4 MPa, i.e. in the range where $\eta \neq f(P)$. The minimum is reduced with temperature, shear rate and pressure, and at certain values of these external factors quite disappears. In our case the anomaly was no more observed at T = 240°C for the D = 200 s^{-1} and at T=180°C for D = 1100 s^{-1}. Further experiments to explain this phenomenon are proceeding.

Pressure viscosity coeficient is also sen sitive to the concentration of antimony trioxide in the range about 1 wt.%. In Fig.4 is illustrated such dependence for pressure P = 50 MPa. As it follows from the definition of this coeficient, this magnitude should be dependent on the same factors as viscosity. Although a strong dependence of coefficient b on shear rate and pressure for given composite is evident, no change with temperature was found.

No study of a system like this one was found in literature. However, several papers are concerned with determination of pressure viscosity coefficient of the main component of our composite - polystyrene. According to Penwell et al./2/ b = $2,9.10^{-3}$ bar^{-1} (at T = 140°C). Ramsteiner /6/ found b = $4,8.10^{-3}$ bar^{-1} (at T = 190°C and D = 0,02 s^{-1}). From the extrapolation of our data can be concluded that the coefficient b for our system will not be very different from the values mentioned above.

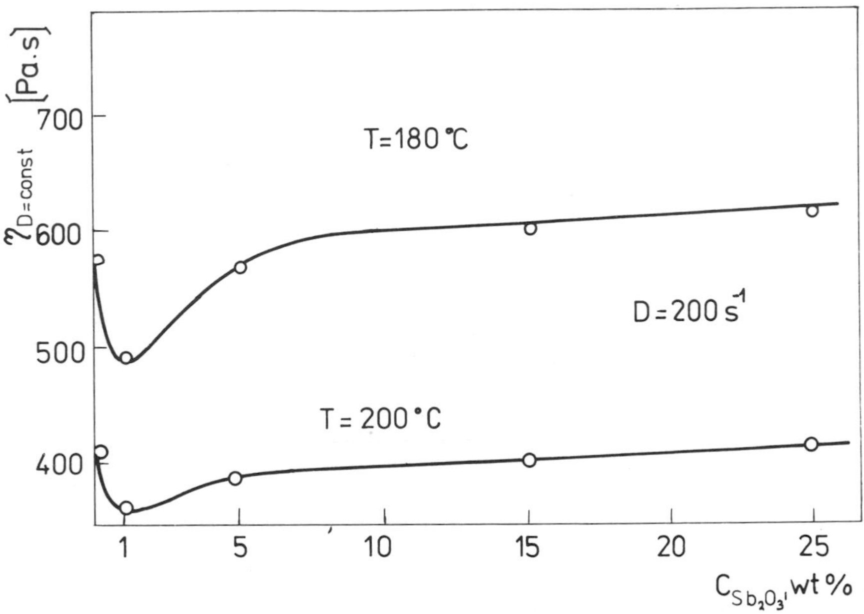

Fig. 3. The effect of antimony trioxide concentration on the viscosity of composite melt at given shear rate for two temperatures.

CONCLUSIONS

The measurements on the three phase composite system polystyrene - polybutadiene - antimony trioxide have shown:

a) the range where $\eta \neq f(P)$ extends with shear rate at given temperature
b) the viscosity differencies for samples with different content of antimony trioxide rise with decreasing shear rate, temperature and pressure (if in the range where $\eta = f(P)$).
c) the viscosity dependence on the concentration of antimony trioxide shows an anomalous drop at concentration about 1 %.
d) pressure viscosity coefficient depends on concentration of antimony trioxide strongly only at its low concen-

tration. Coefficient b strongly depends on shear rate and from the certain value of pressure it depends slightly on pressure. No difference of b on temperature under our conditions was found.

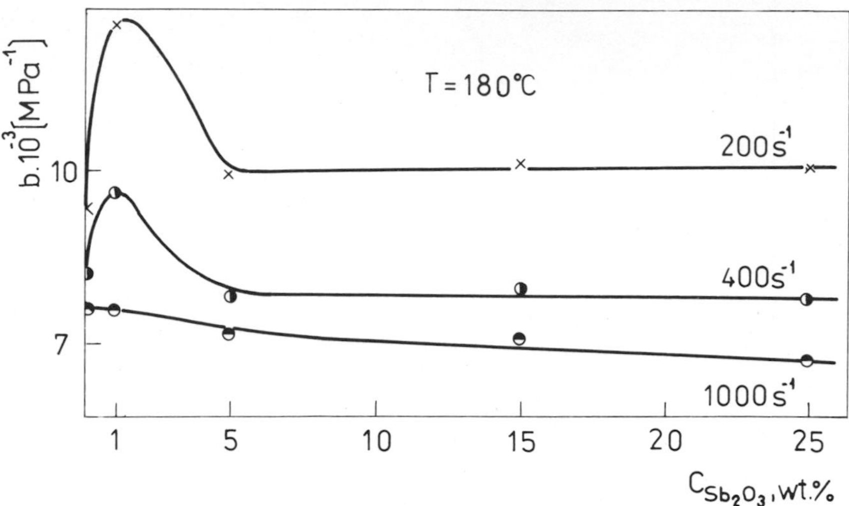

Fig. 4. Pressure viscosity coefficient as a function of antimony trioxide concentration.

REFERENCES

1. C.D.Han,Rheology in Polymer Processing,Chapter 7, Academic Press,NewYork 1976.
2. R.C.Penwell,R.S.Porter and S.Middleman, J.Polymer Sci. A2,9,731 (1971).
3. S.Y.Choi and N.Nakajima,Proceedings of the Fifth Int. Congress on Rheology,Vol.4,287,1970.
4. M.R.Kamal and H.Nyum,Rheol.Acta 12,263 (1973).
5. P.H.Goldblatt and R.S.Porter,J.Appl.Polymer Sci. 20, 1169 (1976).
6. F.Ramsteiner,Rheol.Acta 15,427 (1976).
7. F.Večerka,Z.Horák,Chem.Prům. 26,363 (1976).
8. Z.Horák,F.Večerka,Plaste u. Kautschuk 24,564 (1977).
9. F.Večerka,Z.Horák,Chem.Prům. (in press).
10. J.R.vanWazer,J.W.Lyons,K.Y.Kim,R.E.Colwell,Viscosity and Flow Measurements, Wiley,New York 1963.
11. E.A.Bagley,J.Appl.Phys. 28,624 (1957).
12. N.Minagawa and J.L.White, J.Appl.Polymer Sci.20,501 (1976).

INFLUENCE OF BLENDING UPON RHEOLOGICAL PROPERTIES OF POLYMERS IN THE MELT

Viorica Dobrescu

Central Institute of Chemistry
202 Splaiul Independentei
Bucharest, Romania

This contribution, which is part of an extensive program of investigation carried out on polymer blends, deals with the dependence of viscosity of molten PE blends of HDPE/HDPE and HDPE/LDPE types on composition and viscosities of components.

EXPERIMENTAL

The characteristics of pure components are listed in Table 1.

Blending was performed by rolling at 170°C for 10 minutes. Testing of dispersion homogeneity and reproducibility of blending operation was described elswere[1].

The flow curves were determined by an INSTRON-3211 capillary rheometer, using capillaries 0.06" in diameter, 4" long, with 90° entrance angle. The viscosity of each sample was the average of four runs, and the Rabinowich correction was applied.

Table 1. Viscosities of PE samples employed in blending at 190°C and a shear rate of 2 sec^{-1}.

Nr.	Polymer	η (P)	Nr.	Polymer	η (P)
1.	HDPE-1	110566.4	6.	HDPE-6	19114.7
2.	HDPE-2	117066.3	7.	LDPE-1	102875.9
3.	HDPE-3	104703.7	8.	LDPE-2	34762.2
4.	HDPE-4	53931.8	9.	LDPE-3	10652.7
5.	HDPE-5	15538.4	10.	LDPE-4	3763.1

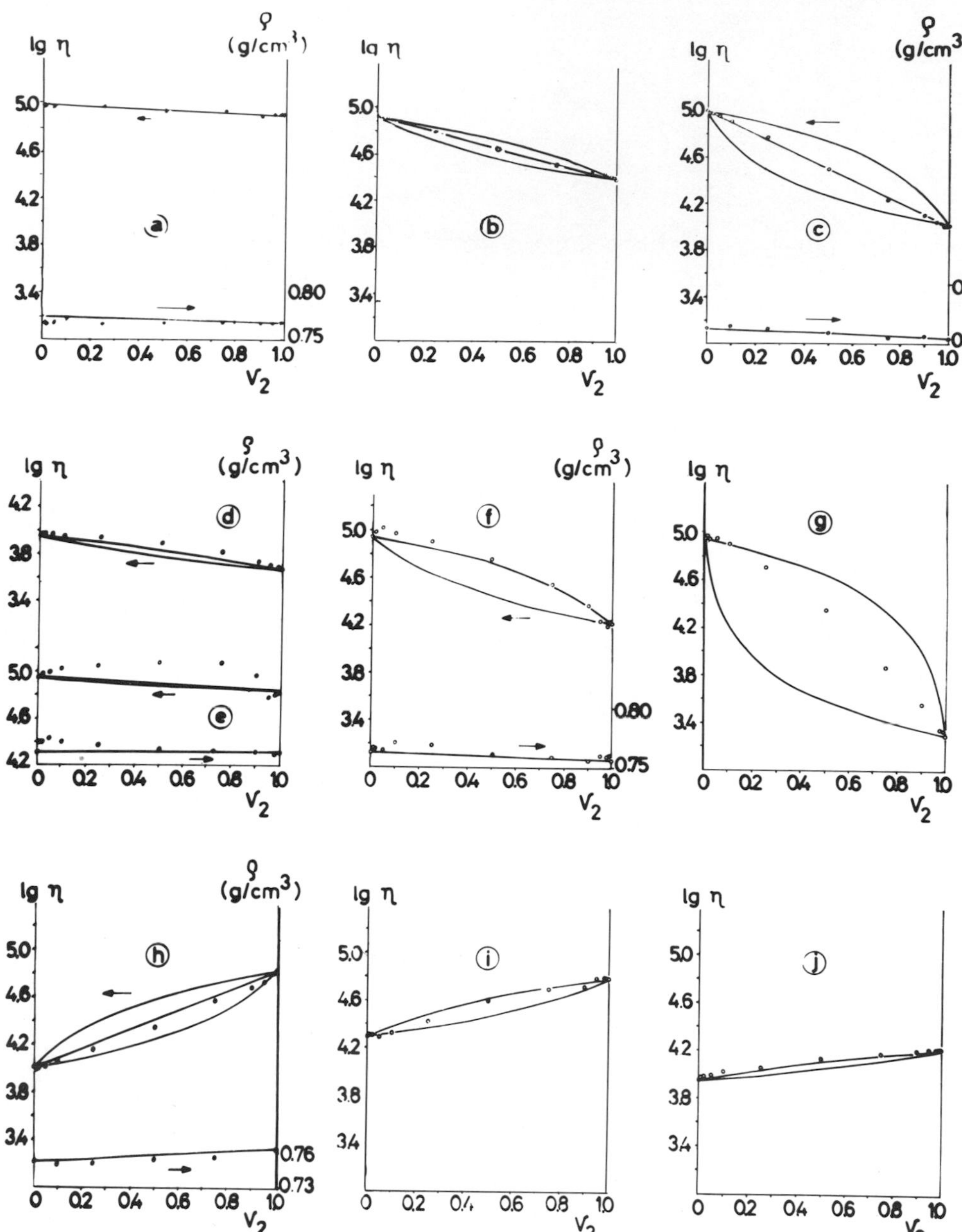

Fig. 1. Dependence of viscosity on composition ($\tau = 3.10^5$ dyne/cm^2) (a) HDPE-2/HDPE-3, 190°C; (b) HDPE-3/HDPE-4, 190°C; (c) HDPE-2/HDPE-6, 190°C; (d) HDPE-5/LDPE-3, 190°C; (e) HDPE-1/LDPE-1, 190°C; (f) HDPE-1/LDPE-2, 190°C; (g) HDPE-1/ /LDPE-4, 190°C; (h) HDPE-6/LDPE-1, 190°C; (i) HDPE-5/LDPE- -2, 150°C; (j) HDPE-5/LDPE-2, 190°C.

The experimental values for shear stress (τ) and apparent viscosity (η) have allowed the calculation of the following equation's parameters:

$$\lg \eta = A + B \lg \tau + C \lg^2 \tau + D \lg^3 \tau \qquad (1)$$

This model has proved to hold for $\eta = f(\tau)$ curves of polyolefins and their blends[2].

The density, at the temperatures of viscosity measurement, was determined using the INSTRON capillary rheometer[1] at s shear rate of 20.71 sec^{-1}.

RESULTS AND DISSCUSION

Fig. 1 presents the viscosity (at 3.10^5 dyne/cm^2) versus composition for several PE blends. The curves corresponding to limiting cases, i.e. additivity and series coupling of components, are plotted next to experimental points.

The additivity, at constant shear stress, is expressed by the equations:

$$\tau = \tau_1 = \tau_2 \qquad \eta = \eta_1 v_1 + \eta_2 v_2$$

where τ and η are the shear stress and viscosity of blend, respectively τ_1, τ_2 and η_1, η_2 the similar values for pure components and v_1, v_2 the volume fractions of components in the blend.

Additivity may be considered as mixing of the components at molecular level (i.e. compatibility) provided that intermolecular interactions among 1-1 and 1-2 type chains are identical. In those instances when blending does not cause the establishing of 1-2 interactions stronger than those among like macromolecules, or the alteration of chain packing density, the additive values are the maximum ones.

The series coupling is described by the equations:

$$\tau = \tau_1 = \tau_2 \qquad (1/\eta) = (v_1/\eta_1) + (v_2/\eta_2)$$

and fits those blends whose components retain their individuality and yield 1-2 interactions weaker than 1-1 ones. The viscosity thus calculated represent minimum values, in absence of interphase slippage[3] or other interphase phenomena.

Experimental data for PE blends show that the composition dependence of viscosity is a function of chain linearity and viscosity of components.

The viscosities of blends of linear polyethylenes are lower than the additivity ones, confirming the incompatibility of polyethylenes; in semilogarithmic coordinates the viscosity varies almost linearly with composition, according to the logarithmic mixing rule.

The case of blends of linear PE with branched ones is more intricate. The following was established:

- Addition of LDPE, of lower viscosity, to HDPE entails the rising of viscosity up to values higher than those of HDPE. For η(HDPE)/ / η(LDPE) ratios ranging within 1-7.63, the viscosity of blends are higher than the additive values throughout the whole composition range. At η (HDPE)/ η (LDPE) higher than 17.5, after an initial rise above additive values, the increase of LDPE content causes the viscosity to fall off below additive values. The maximum value of viscosity is a function of LDPE viscosity: the higher the η(HDPE)/ η(LDPE) ratio, the more the maximum shifts from predominantly branched compositions towards those of very low LDPE concentrations.

- The higher the viscosity of components, the more the additive values are exceeded by blend viscosity.

- The shape of viscosity versus composition curve does not depend on temperature but only on viscosities of components at given temperature and on their ratio.

Since this behaviour could not be accounted for in terms of formation - on blending - of some intermolecular interactions stronger than in components, it may possibly originate in packing density. The values of density for components and blends, at the temperature of viscosity measurement, have supported this hypothesis. Thus - Fig. 1 - the densities of HDPE/HDPE blends are about equal to or lower than additive values, which points out to an increase of free volume which brings about the decrease of viscosity below additive values.

The densities of HDPE/LDPE blends are higher than additive values, which suggests a closer packing of linear and branched chains in the melt. The density reaches a maximum at a composition which is not the same with that for the viscosity peak; this pleads for the existence of other effects, probably related to the different way the linear and branched chains are oriented in the shearing field.

- Addition of LDPE, of higher viscosity, to HDPE does not alter significantly the viscosity of HDPE up to about 20-25% LDPE in the blend. Higher amounts of LDPE cause the increasing of viscosity. The viscosities rest below the additive ones, and so do the densities over the whole composition range. The shape of the viscosity - composition curve is a function of component viscosities ratio.

The following equation is proposed to describe the dependence of blend viscosity on the viscosities of components and composition:

$$\lg \eta = v_1 \lg \eta_1 + \mathcal{J} v_1 v_2 + \lg \eta_2 \qquad (2)$$

where $\mathcal{J}$ stands for "packing coefficient"; $\mathcal{J} \cong 0$ for HDPE/HDPE blends and $\mathcal{J} \gtrsim 0$ for HDPE/LDPE blends, depending on component viscosities and their ratio. The coefficient $\mathcal{J}$ may be evaluated by eq. (2) using the viscosities of components, their volume fractions and the experimental viscosity of just one blend, 0.5/0.5. For HDPE/ /LDPE blends an empirical equation was developed which allows calculation of $\mathcal{J}$ only in terms of component viscosities[4].

Fig. 2 shows the curves of viscosity, calculated by eq. (2), versus composition and the experimental points; the experimental and calculated data agree fairly well. The deviations are larger at the ends of the curves. However, eq. (2) describes correctly the shape of the dependence and the composition ranges where the viscosity peak is located.

In Fig. 3 some experimental data are compared with the flow curves calculated by eq. (2) wherein the η_1 and η_2 corresponding to a given shear stress were replaced by eq. (1). It turns out that eq. (2) allows the fairly accurate calculation of flow curves of blends, the only experimental data required being the flow curves of pure components and their blend 0.5/0.5.

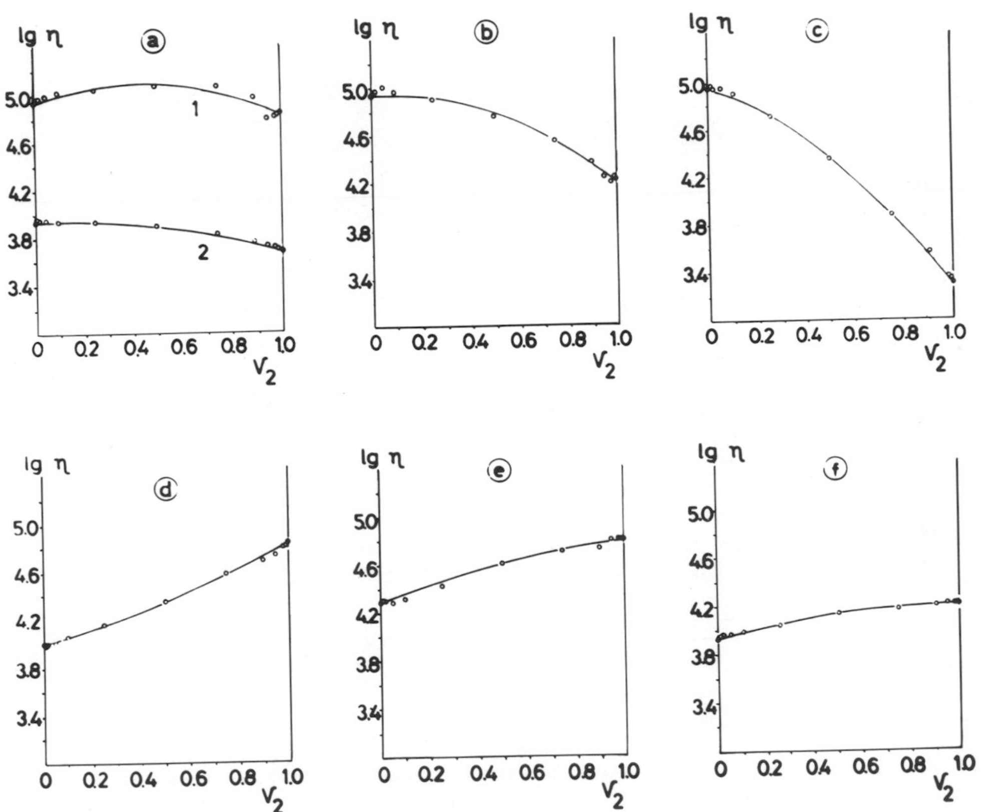

Fig. 2. Dependence of viscosity on composition ($\tau = 3.10^5$ dyne/cm^2)
(O) - experimental; (—) - calculated, eq. (2).
(a-1) HDPE-1/LDPE-1, 190°C; (a-2) HDPE-5/LDPE-3, 190°C;
(b) HDPE-1/LDPE-2, 190°C; (c) HDPE-1/LDPE-4, 190°C;
(d) HDPE-6/LDPE-1, 190°C; (e) HDPE-5/LDPE-2, 150°C;
(f) HDPE-5/LDPE-2, 190°C.

Eq. (2) have also proved to apply to HDPE blends with other polyolefines, e.g. PP (considered a branched component); in this case $\mathcal{J}$ is negative[4].

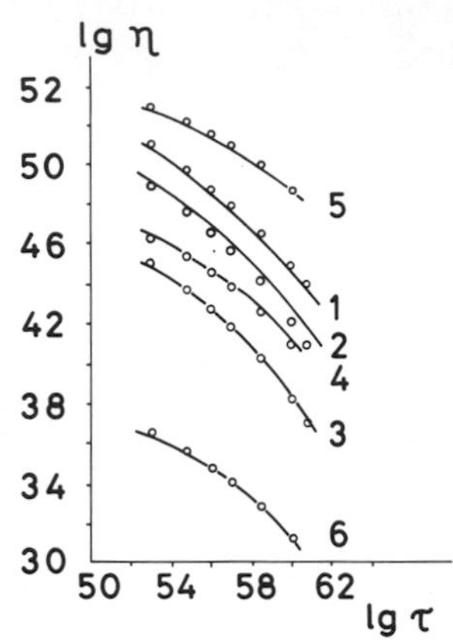

Fig. 3. Flow curves calculated using eq. (1) and (2) for different HDPE/LDPE blends. (0) - experimental points.

1. V. Dobrescu and A. Constantinescu, Rheological Properties of Polyolefine Blends, Materiale Plastice, in press.
2. V. Dobrescu and I. Antonovici, Rheological models for polymers and polymer blends in the melt, Paper presented at the 1st National Congress of Chemistry, Bucharest 11-14 sept. 1979, Revista de Chimie, in press.
3. K. Hayashida, J. Takahashi, and M. Matsui, Some Contributions to the Flow Behavior of Polymer Melts, in: "Proceedings of the Fifth International Congress on Rheology, vol. 4, pg. 525 S. Onogi, ed., Univ. of Tokyo Press, Tokyo, 1970.
4. V. Dobrescu, Rheology of Polymeric Composites. Part 8, Revista de Chimie, in press.

S U S P E N S I O N S

A PROCEDURE FOR THE DEFINITION OF A THIXOTROPIC INELASTIC MODEL

Romano Lapasin, Aldo Alessandrini, Franco Sturzi

Istituto di Chimica Applicata e Industriale

Università degli Studi, Trieste, Italy

INTRODUCTION

Different approaches can be followed to give a description of the thixotropic behaviour. Some advantages are derived from the phenomenological characterization suggested by Cheng[1], because steady and unsteady states of a material can be described by the resulting thixotropic inelastic model, with satisfactory approximation at any kinematic condition.

The model is composed of two constitutive equations, namely, the state equation

$$\tau = \tau\ (\lambda,\dot{\gamma}) \tag{1}$$

and the rate equation

$$\frac{d\lambda}{dt} = g\ (\lambda,\dot{\gamma}) \tag{2}$$

where λ is an arbitrary structural parameter[2]. Alternatively, it can be represented by the following equation

$$\frac{d\lambda}{dt} = \alpha(\tau,\dot{\gamma})\ \frac{d\dot{\gamma}}{dt} + \beta(\tau,\dot{\gamma}) \tag{3}$$

obtained by differentiating eq. (1) and substituting eq. (2). In the latter case, it is unnecessary to specify the functional dependence of α and β on τ and $\dot{\gamma}$ and the two functions can be deter- directly from experimental data on condition that a combination of constant-$\dot{\gamma}$ and constant-τ experiments is performed. Even if such a procedure is attractive, it is not suitable to be employed with the viscometers generally available, in which only the kinematic variable can be controlled.

The present work examines the problems which arise immediately after the determination of the experimental data obtained from constant-$\dot{\gamma}$ experiments according to the procedure originally suggested by Cheng[1]. Particular experimental data and/or a particular thixotropic behaviour can strictly indicate the procedure to be followed for the definition of the thixotropic model, like in the case of cement pastes which are partially thixotropic systems[3]. When the experimental data relative to the stress-time transients (at constant shear rate) and to constant-λ curve map are just sufficient or superabundant with reference to the complexity of the model, an appropriate procedure for the definition of the model must be selected. The present work reports an investigation of the problems arising from the different steps of the procedure with reference to a typical thixotropic material.

EXPERIMENTAL

The material examined is a clay/kaolin aqueous suspension, with 37.5% of solids by volume and .4% of deflocculant (50% sodium silicate, 50% sodium carbonate).

Experimental data were determined by means of a rotational viscometer Rotovisko Haake RV2 in the shear rate range 10-1200 s^{-1}. The measuring device was a bob/cup system with the narrowest gap at disposal (R_i/R_o = .96). Tests were carried out at 25$\pm$ .1°C. In accordance with the Cheng procedure the constant-λ curves relative to four different shear rates ($\dot{\gamma}$ = 150.4, 300.8, 601.6, and 1203 s^{-1}) were determined. The transients shear stress-time were registered by means of a x-t Perkin-Elmer recorder.

RESULTS AND DISCUSSION

Starting from data relative to constant-λ curve map, the first problem is the choice of an equation, suitable to describe the shear-dependent behaviour of the material in all the structural conditions examined. This aim can be achieved by examining separately each curve at λ constant. For the material examined, the equation chosen was that suggested by Carleton-Cheng-Whittaker[4]:

$$\tau = \tau_o + k\dot{\gamma}^n \tag{4}$$

In order to define a state equation linear in λ and, consequently, a simple thixotropic model of easy application, in the spirit of Cheng's approach, firstly it is necessary that n value is common to all the constant-λ curves. It is possible if we consider the slight differences between the n values resulting from the least square fitting of each constant-λ curve. First estimates of n, $\tau_{o,i}$ and k_i for all the curves can be obtained through a simultaneous fitting of all the data of constant-λ curve map. The analysis of residual distribution for each curve has shown the validity of the choice of the equation suggested by Carleton et al.

Both the parameters $\tau_{o,i}$ and k_i of the state equation vary appreciably from curve to curve and, therefore, an examination of $\tau_{o,i}$ and k_i values is necessary to test the possibility of defining a state equation linear in λ. In fact, the linearity in λ is expressed by:

$$\tau_{o,i} = \tau_{o,o} + (\tau_{o,1} - \tau_{o,o})\lambda_i \qquad (0 \leq \lambda_i \leq 1) \tag{5}$$

$$k_i = k_o + (k_1 - k_o)\lambda_i \tag{6}$$

from which

$$\lambda_i = \frac{\tau_{o,i} - \tau_{o,o}}{\tau_{o,1} - \tau_{o,o}} = \frac{k_i - k_o}{k_1 - k_o} \tag{7}$$

or

$$k_i = m\,\tau_{o,i} + q \tag{8}$$

Accordingly, it is necessary that the linear correlation index between the k_i and $\tau_{o,i}$ values derived from the simultaneous fitting of constant-λ curves is sufficiently high, and, in the case considered, it was .9356.

The values of $\tau_{o,o}$, $\tau_{o,1}$, k_o, k_1 and of λ_i of the constant-λ curves cannot be obtained by solving the system:

$$\begin{cases} \tau_{o,o} + (\tau_{o,1} - \tau_{o,o})\lambda_i = \tau_{o,i} & (i=1,2..c) \\ k_o + (k_1 - k_o)\lambda_i = k_i & (c \geq 4) \end{cases} \tag{9}$$

as it appears from an examination of the form of the system.

Since the structural level present in the material in equilibrium conditions is assumed to be a monotonic function of the shear rate $\bar{\dot{\gamma}}$, at this time it seems convenient to consider the structural parameter λ as a function of $\bar{\dot{\gamma}}$. Thus the state equation becomes:

$$\tau = \tau_{o,o} + (\tau_{o,1} - \tau_{o,o})\lambda(\bar{\dot{\gamma}}) + (k_o + (k_1 - k_o)\lambda(\bar{\dot{\gamma}}))\dot{\gamma}^n \tag{10}$$

If the functional dependence of λ on $\bar{\dot{\gamma}}$ is specified, the values of $\tau_{o,o}$, $\tau_{o,1}$, k_o, k_1 and n can be determined directly from data of the map.

The implicit function $f(\lambda,\bar{\dot{\gamma}})$ can be derived from the rate equation:

$$\frac{d\lambda}{dt} = C_1(\dot{\gamma})\ (1 - \lambda)^m - C_2(\dot{\gamma})\ \lambda^n \tag{11}$$

in equilibrium conditions, for which is

$$\frac{d\lambda}{dt} = 0, \qquad \dot{\gamma} = \bar{\dot{\gamma}} \tag{12}$$

The function $\lambda(\bar{\dot{\gamma}})$ is defined if m, n and the dependence of C_1 and C_2 on $\bar{\dot{\gamma}}$ are assumed.

In the case examined, the transients shear stress-time can be described satisfactorily by:

$$\tau = \tau_e + \frac{A_1 \exp(-Qt)}{1 - A_2 \exp(-Qt)} \tag{13}$$

where

$$Q = \sqrt{4\ C_1(\dot{\gamma})C_2(\dot{\gamma}) + C_1(\dot{\gamma})^2} \tag{14}$$

The relative residuals are lower than .025. On condition that the state equation is linear in λ, eq. (13) can be derived from the rate equation

$$\frac{d\lambda}{dt} = C_1(\dot{\gamma})(\ 1 - \lambda) - C_2(\dot{\gamma})\lambda^2 \tag{15}$$

The dependence of λ on $\bar{\dot{\gamma}}$ is defined, through $C_1(\dot{\gamma})$ and $C_2(\dot{\gamma})$, by:

$$\lambda = \frac{\sqrt{4\ C_2(\dot{\gamma})/C_1(\dot{\gamma}) + 1} - 1}{2\ C_2(\dot{\gamma})/C_1(\dot{\gamma})} \tag{16}$$

If we assume, in accordance with other Authors[4-6]:

$$C_1 = a_1 \tag{17}$$

$$C_2 = b_2\dot{\gamma} \tag{18}$$

the eq. (16) becomes:

$$Q = \sqrt{4\ a_1 b_2\ \dot{\gamma} + a_1^2} \tag{19}$$

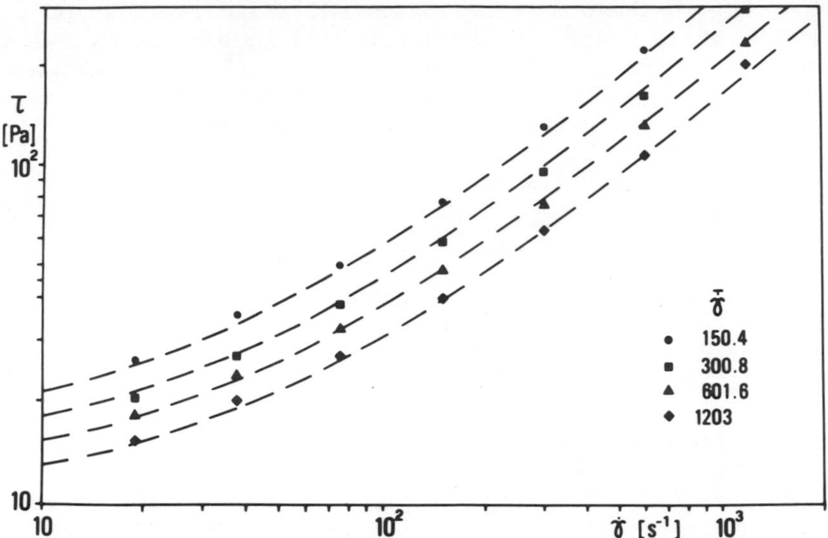

Fig. 1. Comparison between the constant-λ curves calculated from the state equation and the experimental data in a log τ-log $\dot{\gamma}$ plot.

From a least square fitting of Q vs. $\dot{\gamma}$ the values of a_1 and b_2 are derived. Thus, the rate equation and the function $\lambda(\dot{\gamma})$ are defined.

Subsequently, the parameters of the state equation are determined, by substituting $\lambda(\dot{\gamma})$ in eq. (10), through a fitting of the constant-λ curve map.

The values of the parameters of the state and the rate equations are:

$\tau_{o,o}$ = 6.604 Pa	m = 1
$\tau_{o,1}$ = 29.17 Pa	n = 2
k_o = .0881 Pa	a_1 = 3.32 10^{-3}
k_1 = 1.639 Pa	b_2 = 7.08 10^{-5}
n = .870	

Fig. 1 reports a comparison between the constant-λ curves calculated from the state equation and the experimental data.

According to the procedure suggested, the parameters of the rate and the state equations are determined in two successive steps. If a different expression of the eq. (17) and (18) does not allow the identification of the parameters of the rate equation from the fitting of Q vs. $\dot{\gamma}$, a different procedure must be followed. All the data relative to the transients and to the constant-λ curve map must be treated simultaneously.

REFERENCES

1. D. C.-H. Cheng and F. Evans, "Phenomenological characterization of the rheological behaviour of inelastic reversible thixotropic and antithixotropic fluids", Brit. J. Appl. Phys., 16: 1599 (1965).
2. D. C.-H. Cheng, "A differential form of constitutive relation for thixotropy", Rheol. Acta, 12:228 (1973).
3. R. Lapasin, V. Longo and S. Rajgelj, "A thixotropic model for cement pastes", VIIIth Int. Congr. Rheol., Napoli, (1980).
4. A. J. Carleton, D. C.-H. Cheng and W. Whittaker, "Determination of the rheological properties and start-up flow characteristics of waxy crude and fuel oils", Inst. Petr., IP 74-009 (1974).
5. F. Moore, "The rheology of ceramic slips and bodies", Trans. Brit. Ceram. Soc., 58:470 (1959).
6. B. T. Storey and E. W. Merrill, "The rheology of aqueous solutions of amylose and amylopectine with reference to molecular configurations and intermolecular association", J. Polym. Sci., 33:361 (1958).

SOME ASPECTS OF THE FLOW PROPERTIES OF HIGH-BUILD PAINTS

Romano Lapasin (°),Adriano Papo (°°),Giovanni Torriano (°)

(°) Istituto di Chimica Applicata e Industriale
Università degli Studi di Trieste
Via A.Valerio,2 - 34127 Trieste (Italy)

(°°) Istituto di Chimica
Università degli Studi di Udine
Viale Ungheria,43 - 33100 Udine (Italy)

INTRODUCTION

High-build paints are at present very popular on account of both technical and economical reasons. They are dispersions of one or more pigments and extenders in a low-molecular weight polymer solution; they contain additives and rheological agents; the volume fraction of the disperse phase usually ranges between 0.15 and 0.25.

In previous studies[1-3] Casson model[4] was found generally suitable to describe the equilibrium flow behaviour of these materials. Casson τ_o and η_∞ dependance on rheological agent content was shown. Excellent correlations were brought out between Casson τ_o and sag resistance[1-3]. For a deeper insight into sagging phenomena, non-equilibrium behaviour was investigated as well[2-3]. The master-curve procedure proved to be particularly suitable to give a compact representation of equilibrium flow properties over a wide shear interval[5].

Recently, a study was started on the rheological properties of a family of low-molecular weight chlorinated rubber, titanium dioxide high-build paints, at present widely employed in the marine field and for industrial maintenance.

A first series of investigations was carried out in order to get a full picture how the variation of a number of composition parameters affects flow-dependent technological application properties and film-forming characteristics. Tests were designed and performed mainly in view of practical applications. So paints to be tested were formulated on the basis of given technological properties (0.25 Pa.s at 10^4 s^{-1}) and performance requirements (PVC range).

This paper reports the results of a second series of investigations planned with the aim of obtaining a better understanding of flow behaviour in connection with paint composition variations. Variations were selected independently of any limitation arising from technological exigencies and service requirements and extended over a vast range.

EXPERIMENTAL

Materials

12 samples of paint were prepared. Binder was Alloprene R10, ICI, low-molecular weight chlorinated rubber, plasticized with Cloparin 45, Caffaro, chlorinated paraffin; pigment was rutile titanium dioxide RS 66,Montedison, rheological agent Thixomen, ICI, a castor oil derivative.

Samples A-G were formulated by adding different amounts of titanium dioxide to the same paint vehicle (Alloprene R10:24g/100cc; Thixomen:2.47g/100cc). The pigment volume fraction,Φ_P, was increased up to 0.26.

In samples K-O Thixomen amounts ranging from 0 to 3.93g/100cc were added to the same suspension (Φ_P=0.1373; vehicle:24g/100cc of Alloprene R10).

Paint characteristics are given in Table 1.

Apparatus

A commercial coaxial cylinder viscometer, Rotovisko-Haake RV11, and a measuring device MVI, 0.96 mm clearance,were employed. Apparent shear rate range: 7.05 - 1142 s^{-1}.

Tests were carried out at 25 $\pm$ 0.1 °C.

RESULTS AND DISCUSSION

Shear rates were corrected according to Krieger-Elrod[6].

Equilibrium data were correlated with the Casson equation in its original simple form[4]. The Matsumoto et al. correction[7] was not necessary, since the continuous phase did not display appreciable deviation from the newtonian behaviour in the shear rate range con-

Table 1. Paint Characteristics

Paint	PVC	$\Phi_P \cdot 100$	$\Phi_T \cdot 100$
A	0	0	2.47
B	5.0	1.50	2.43
C	15.4	4.96	2.34
D	25.0	8.75	2.26
E	34.9	13.76	2.14
F	46.0	19.23	2.01
G	55.0	26.00	1.84
K	37.1	13.73	0
L	36.1	13.58	1.08
M	35.0	13.44	2.15
N	34.1	13.29	3.18
O	33.5	13.19	3.93

sidered.

τ_o and η_∞ values are given in Table 2.

Excellent correlations with the Casson model were shown by all the systems, with the exception of system G (the highest titanium dioxide content) and O (the highest Thixomen content), where deviations at low shear rates were evident. Similar deviations were observed elsewhere too[1,2,8].

An interpretation of the experimental results was attempted on the basis of a model suitable to correlate the yield value τ_o and the relative viscosity at infinite shear rate, $\eta_{r,\infty}$, with the composition parameters varied, i.e. the volume fraction of pigment, Φ_P, and the volume fraction of Thixomen, Φ_T.

First of all, the validity of the expressions which correlate τ_o and $\eta_{r,\infty}$ with the solid volume fraction according to the Casson model in the form given by Asbeck[9] was verified on the basis of the following equation derived from[9]:

$$\sqrt{\tau_o} = K(\sqrt{\eta_{r,\infty}} - 1) \qquad (1)$$

Unsatisfactory results were obtained. Consequently, other equations were taken into consideration.

As regards $\eta_{r,\infty}$, five of the equations suggested in the literature were tested. They are reported below.

Table 2. Casson τ_o and η_∞ Parameters

Paint	τ_o (Pa)	η_∞ (Pa.s)
A	0	0.086
B	0	0.103
C	0	0.129
D	0.008	0.173
E	0.369	0.245
F	8.033	0.354
G	20.75	0.649
K	0.031	0.203
L	0.712	0.230
M	2.368	0.232
N	8.674	0.270
O	13.99	0.286

Landel-Moser-Bauman[10]:

$$\eta_{r,\infty} = (1 - \Phi/\Phi_m)^{-2.5} \tag{2}$$

Ham[11]:

$$\eta_{r,\infty} = (1 - \Phi/\Phi_m)^{-1} \tag{3}$$

Eilers[12]:

$$\eta_{r,\infty} = \left(1 + \frac{1.25\ \Phi}{1 - \Phi/\Phi_m}\right)^2 \tag{4}$$

Frankel-Acrivos[13]:

$$\eta_{r,\infty} = 1 + \frac{9}{8}\,\frac{(\Phi/\Phi_m)^{1/3}}{1 - (\Phi/\Phi_m)^{1/3}} \tag{5}$$

Mooney[14]:

$$\eta_{r,\infty} = \exp\left\{\frac{2.5\ \Phi}{1 - \Phi/\Phi_m}\right\} \tag{6}$$

The equations considered involve the volume fraction Φ of the disperse phase and the volume fraction Φ_m of the disperse phase at its maximum packing.

Equations containing other adjustable parameters were not taken into account.

The suitability of the equations (2) - (6) was checked by means of a least square fitting of $\eta_{r,\infty}$ versus the total nominal volume fraction of the disperse phase $\Phi = \Phi_P + \Phi_T$, assuming Φ_m as the adjustable parameter.

The best fitting was given by the Landel et al. equation with a Φ_m value of 0.494 and a mean relative deviation of 0.045. For the other equations tested the mean relative deviation was in the range of 0.20.

A more meaningful correlation could be obtained by taking into account adsorption phenomena and, accordingly, the effective total volume fraction of the disperse phase; moreover, Φ_m variations connected with variations of the pigment/Thixomen ratio should be considered as well. But the determination of Φ_m for Thixomen is difficult. On the other hand, the satisfactory results obtained with the Landel et al. equation by assuming Φ_m as constant and $\Phi = \Phi_P + \Phi_T$ indicate that the assumptions made are acceptable from a practical point of view.

For τ_o dependance on concentration and size of the disperse phase, reference is made to[15,16]. Difficulties are encountered in applying these equations, because measured values of floc size and concentration are needed together with hypotheses on the aggregation state.

On the other hand, a simple examination of data shows that yield value τ_o cannot be correlated with nominal total volume fraction (as it was the instance for $\eta_{r,\infty}$). Hence, it is suggested to consider the separate contributions of the single species of the disperse phase, that is to relate τ_o with Φ_P and Φ_T.

The equation proposed is the following:

$$\tau_o = K_P \Phi_P^n + K_T \Phi_T^n \tag{7}$$

where the functional dependance of τ_o on Φ_P and Φ_T is assumed to be the same and K_P and K_T account for the different interactions and the different mean size of the disperse phase species.

Data fitting gives: $K_P = 6.31 \cdot 10^3$ Pa, $K_T = 1.22 \cdot 10^7$ Pa, n =4.22; mean relative deviation: 0.104.

The importance of the Thixomen contribution to τ_o is clearly brought out.

REFERENCES

1. R. Lapasin, A.Papo and G. Torriano, Ind.Vernice, 30:3 (1976).
2. A. Papo and G. Torriano, J.O.C.C.A., in press.
3. I. Kikic, R. Lapasin and G. Torriano, VII International Congress

on Rheology, Proc., Gothenburg (1976), p.584.
4. N. Casson, "Rheology of Disperse Systems", C.C. Mill Ed., Pergamon Press, London (1959), p.84.
5. R. Lapasin, A. Papo and G. Torriano, XIV F.A.T.I.P.E.C. Congress, Proc., Budapest (1978), p.653.
6. I.M. Krieger and H. Elrod, J.Appl.Phys., 24:134 (1935).
7. T. Matsumoto, A. Takashima, T. Masuda and S.Onogi, Trans.Soc. Rheol., 14:617 (1970).
8. T. Matsumoto, C. Hitomi and S.Onogi, Trans.Soc.Rheol.,19:541 (1975).
9. W.K. Asbeck, Off.Digest, 33:65 (1961).
10. R.F. Landel, B.G. Moser and A. Bauman, IV International Congress on Rheology, Proc., U.S.A. (1963).
11. R.K. Ham, M.S. Thesis in Chem. Eng., University of Washington (1965).
12. H.Eilers, Kolloid Z., 97:313 (1941).
13. N.A. Frankel and A. Acrivos, Chem.Eng.Sci., 22:847 (1967).
14. M. Mooney, J.Coll.Sci., 6:162 (1951).
15. A.S. Michaels and J.C. Bolger, Ind.Eng.Chem.Fund., 1:153 (1962).
16. B.A. Firth and R.J. Hunter, J.Coll.Interface Sci., 57:248 (1976).

DETERMINATION OF THE THIXOTROPIC PROPERTY OF AN EMULSION GEL PAINT

D.C-H. Cheng and R.A. Richmond*

Warren Spring Laboratory
Stevenage, SG1 2BX, UK
(*now at ICI Mond Division, Winnington, CW8 4DJ, UK)

INTRODUCTION

The term "thixotropy" was coined in the 1920's to describe the phenomenon of isothermal reversible sol-gel transformation due to mechanical disturbance. The word took on additional meaning in the subsequent years and is now defined in a variety of glossaries (see Refs. 1 and 2). A review of the literature was made by Bauer and Collins[3] in 1967. The most recent review is by Mewis[2] in 1979. The latter lists three types of experimental methods for thixotropy:

(1) step change in shear rate or shear stress (step-shear test),
(2) consecutive linear increase and decrease in shear rate (loop test), and
(3) sinusoidal change in shear rate (oscillatory test).

All of these can be used to study the physical and chemical bases, the molecular and the particulate nature of thixotropic phenomena (the material science application). They can also be used to provide data for engineering applications, such as solution of equations of motion, mathematical modelling, empirical correlation of engineering data etc.

The results of the loop and oscillatory tests, provided the maximum shear rate and amplitude of oscillation are sufficiently large to take the sample past its yield stress, can be interpreted to give data suitable for engineering applications[4]. However the data treatment procedure is rather complicated and has not been put to serious use. On the other hand, the step shear rate test can be readily carried out and the results reduced for engineering applications. This paper gives an example of the use of this method.

The method will be supplemented by testing on the Warren Spring Laboratory Gun Rheometer, which is based on the tube geometry and designed to measure the development of yield stress of a thixotropic sample recovering at rest.

The step-shear rate and the Gun Rheometer tests are based on a "single-structured" theory of thixotropy[5-8]. This theory has been found to be valid for certain materials, such as aqueous bentonite dispersions[9]. But for other materials it is inadequate and the emulsion gel paint to be discussed below falls into this category. Although it is possible to consider a "multi--structured" theory and test complex material accordingly, the method is rather complicated[10,11]. However, the single--structured theory will be seen to be a first approximation as well as the first step in the application of the multi-structured theory. The results which will be presented on the emulsion gel paint and the interpretation according to the single-structured theory are to be appreciated in this context.

THEORY, EXPERIMENTAL AND RESULTS

Limitation of space does not allow even a resumê to be given. The reader is assumed to be familiar with Refs. 5-8. In the barest summary, the experimental measurement is concerned with the determination of the constant-structure (λ) and the constant-rate (β) curves, in a plot of shear stress (τ) vs shear rate ($\dot{\gamma}$), by means of the step-change in shear rate test. This was carried out on a specially equipped Weissenberg Rheogoniometer. The equilibrium curve (β = 0), also obtained on the Rheogoniometer, was supplemented by the use of the Ferranti-Shirley viscometer. For the development of yield stress (τ_y) at rest, the WSL Gun Rheometer (now available from Deimos Ltd, Canterbury) was used. Details of the use of these equipment and data treatment can be found in Refs. 9 and 12.

In the present paper, the material was a proprietary vinyl emulsion gel paint, manufactured by Donald MacPherson and Co. Ltd, and purchased in a local store. The results are plotted in Figs. 1-4.

DISCUSSION

The experimental techniques and data treatment methods used in this paper have been used previously on aqueous bentonite dispersions[9], showing the validity of the single-structured theory with these materials. The results on the vinyl emulsion gel paint indicate that the methods are again useful in providing quantitative data on the complex thixotropic behaviour, and as such there is not a lot to say about them. The interesting features of these results, however, are that they show deviations from predictions of the single-structured theory. This means that the behaviour of the emulsion paint is more complicated than can

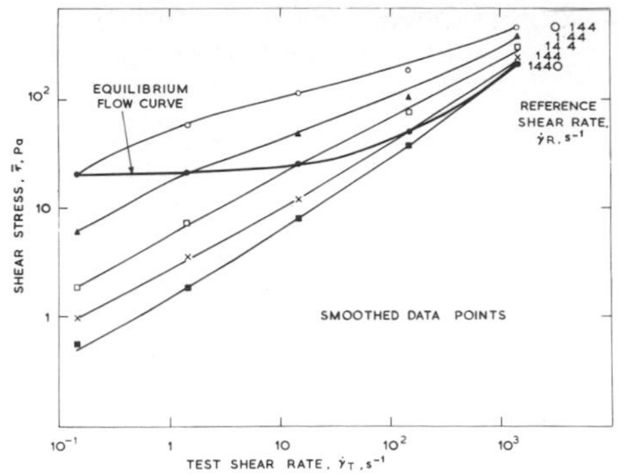

Fig. 1 Constant-Structure Curves

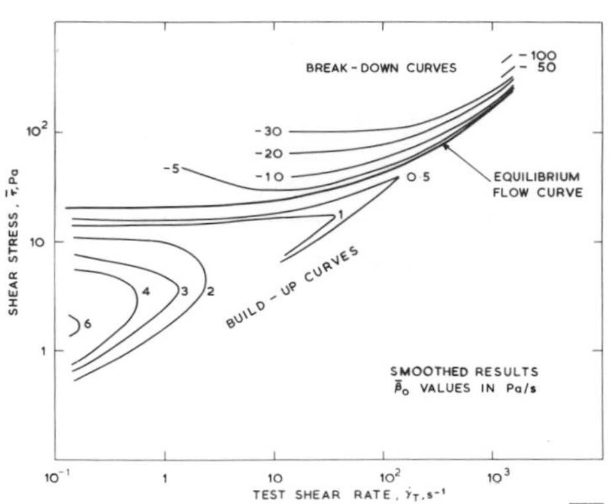

Fig. 3 Constant-β_o Curves

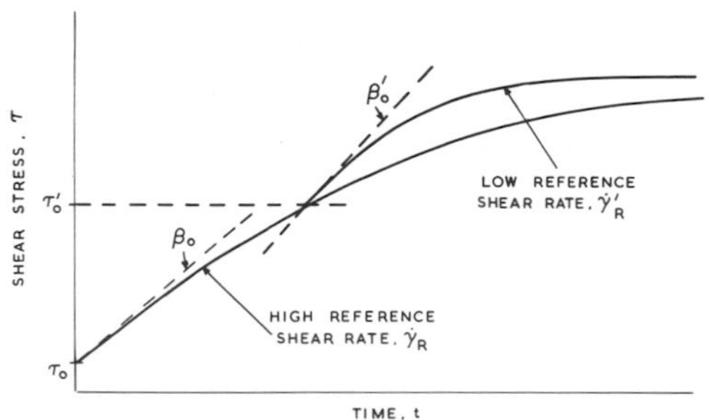

Fig. 5 Build-Up at a Given Test Shear Rate After Step-Change from Different Reference Shear Rates

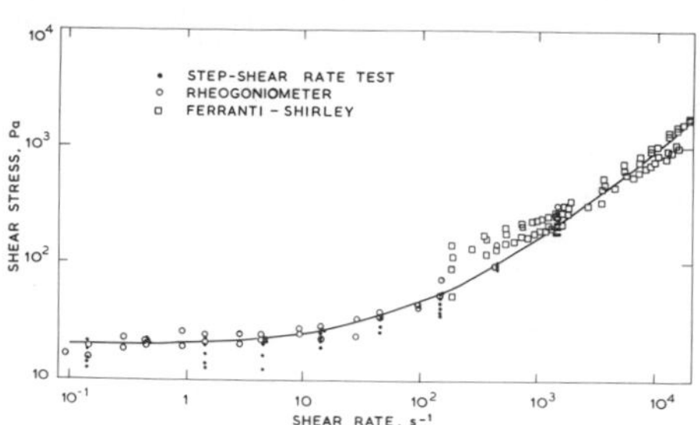

Fig. 2 Equilibrium Flow Curve

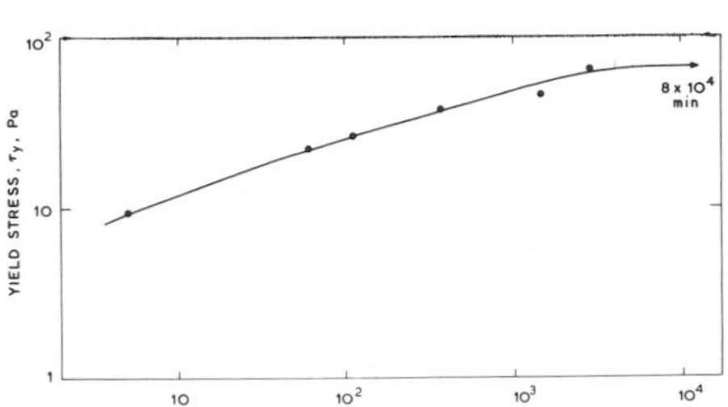

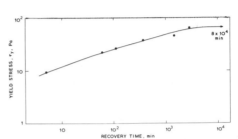

Fig. 4 Development of Yield at Rest

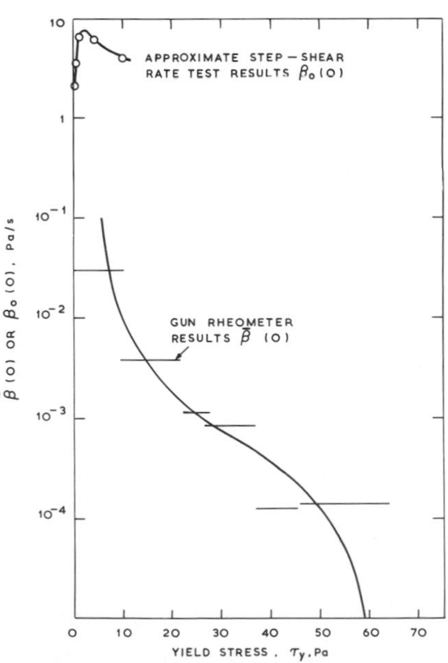

Fig. 6 Build-Up Rate β at Rest

be adequately described by the single-structured theory. In this section, the evidence is discussed. It is seen that the single-structured theory is a first approximation as well as a first step in the description of multi-structured thixotropy.

There is good agreement between the three sets of equilibrium flow curve results (Fig. 2). The Ferranti-Shirley data appeared to deviate from the Rheogoniometer data as the shear rate was reduced; this was considered to be caused by friction on the cone tip, but the overall scatter was that usually encountered in other materials of a similar nature to the paint tested. The notable feature of the constant-λ curves (Fig. 1) is that the slopes remain quite steep even at 0.1 s^{-1}, at which the equilibrium flow curve is very flat. This means that the yield stress of the broken-down structures (corresponding to high $\dot{\gamma}_R$), if existing at all, would be very low, much less than 1 Pa. Compared with the initial τ_y in the Gun Rheometer results (Fig. 4), this could mean that the initial structural state in the Gun Rheometer tests were not as broken-down as achieved in the Rheogoniometer. Alternatively there could have been significant build-up during the Gun Rheometer test. The interpretation of the low recovery time results needs further study.

According to the single-structured theory the yield stress corresponding to the fully built-up state is given by the yield stress indicated by the equilibrium flow curve. Fig. 2 indicates that this is about 20 Pa. The Gun Rheometer results (Fig. 4) shows that at rest τ_y can build-up significantly higher to more than 60 Pa. This is an example of the well-known difference between the dynamic yield stress and the static value[13] and is an indication that the single-structured theory does not hold for this emulsion gel paint.

Another indication of the inadequacy of the single-structured theory to explain the present results in total is found with the constant-β_o results for build-up (i.e. positive β_o) in Fig. 3. Instead of τ being single valued, as expected from the single-structured theory, they are double-valued at any given $\dot{\gamma}_T$. This behaviour can be traced to the recorder traces of τ vs t. Shown diagrammatically in Fig. 5, the traces are such that they cannot be made to coincide by being shifted along the t-axis. If the single-valued theory applied, one should be able to do this, and β would decrease monotonically from β_o as τ increased to reach τ_e under constant $\dot{\gamma}_T$. These results therefore show that β was not only a function of τ and $\dot{\gamma}_T$, but depended also on $\dot{\gamma}_R$. It must be concluded that this emulsion gel paint was multi-structured in its thixotropic property. It may be possible to explain some of these results by using a double-structured theory[10,11]. But the data have not yet been so assessed.

The Gun Rheometer results can in principle be further compared with the step-shear rate results, in terms of $[\tau_y, \beta(\dot{\gamma}_T{=}0)]$ for the former and $[\tau_y, \beta_o(0)]$ for the latter. If the single-structured theory applies, a single curve would be obtained when the data are plotted. $\beta(0)$ and $\beta_o(0)$ are the values at zero shear rate and were determined as follows.

Theoretically, $\beta(0)$ can be obtained by differentiating the Gun Rheometer (τ_y, t) data. Because of the paucity of the present results, an average $\overline{\beta}(0)$ was calculated from adjacent data points,

$$\overline{\beta}(0) = \frac{\tau_y(t_2) - \tau_y(t_1)}{t_2 - t_1}$$

and plotted versus τ_y in Fig. 6.

With the step-shear rate results, τ_y would be obtained by extrapolating the constant-λ curves on the $(\tau, \dot{\gamma}_T)$ plot to zero $\dot{\gamma}_T$. The corresponding $\beta_o(0)$ would be obtained by replotting β_o vs $\dot{\gamma}_T$ and extrapolating the constant-$\dot{\gamma}_R$ (equivalent to constant-λ) curves to zero $\dot{\gamma}_T$. But the present data did not allow accurate determination of τ_y and $\beta_o(0)$ to be made. Nevertheless, some approximate estimates were made, and plotted on Fig. 6. Instead of a monotonic curve as required by the single-structured theory, the $[\tau_y, \beta_o(0)]$ relationship was double-valued. It should be stressed that $\beta_o(0)$ are the initial values, not $\beta(0)$ values for $t > 0$. If it were possible to obtain the τ_y and $\beta(0)$ values for $t > 0$ in the step-shear rate experiment, a family of $[\tau_y, \beta(0)]$ curves would be obtained, one for each $\dot{\gamma}_R$ used in the tests, because of the multi-structured property of the emulsion paint. Furthermore the disposition of the equilibrium flow curve shows that these curves would tend asymptotically to zero $\beta(0)$ as τ_y approaches 20 Pa (whereas the Gun Rheometer results tended to zero $\beta(0)$ as τ_y tended to about 60 Pa.

The comparison of the Gun Rheometer and the step-shear rate results for the emulsion paint again illustrates that the single-structured theory is inadequate. But it shows how the complicated multi-structured thixotropic behaviour may begin to be measured in quantitative terms.

Further comparison of the Gun Rheometer and the equilibrium flow curve results may be made using the modified Moore model[14,15], which is a specific mathematical form of constitutive relationship. From this, numerical values may be derived for certain build-up and break-down rate constants. But there is no space in this paper to go into details.

CONCLUSION

Based on the single-structured theory of thixotropy, step-shear rate tests are devised for use with the Weissenberg Rheogoniometer to determine the thixotropic constitutive relationship and the Gun Rheometer is developed to measure the development of yield stress in a sample recovering at rest. The Ferranti-Shirley Cone-Plate Viscometer is used to extend the upper shear rate range in the equilibrium flow curve. These methods have been described previously. In this paper they are used on a vinyl emulsion gel paint, giving the results shown on Figs. 1-4. Thixotropic behaviour is a complex phenomenon and these figures express the emulsion paint property in quantitative terms.

Analysis of these results show, however, that the emulsion paint behaviour is even more complex than can be described in full by the single-structured theory, although this theory has been shown to be valid for certain materials. For the more complex multi-structured behaviour, it is shown that the experimental methods and data interpretation used in this paper gave a first approximation and provided a first step towards describing it in quantitative terms.

ACKNOWLEDGEMENT

This work was carried out under the Rheological Research Service, which was a co-operative project between the Department of Industry, acting through the Chemicals and Minerals Requirement Board, and industrial members.

REFERENCES

1. "Rheological terminology" in Rheol. Acta, 14:1098 (1975).
2. J. Mewis, J. Non-N. Fluid Mech., 6:1 (1979).
3. W.H. Bauer and E.A. Collins in "Rheology Theory and Applications", F.R. Eirich, ed, Academic Press, New York, 4:423 (1967).
4. D.C-H. Cheng, "Res. Rept. No. LR 158 (MH)", Warren Spring Laboratory, Stevenage, (1971).
5. D.C-H. Cheng and F. Evans, Brit. J. Appl. Phys., 16:1599 (1965).
6. D.C-H. Cheng, Nature, 216:1099 (1967).
7. D.C-H. Cheng, "Res. Rept. No. LR 157 (MH)", Warren Spring Laboratory, Stevenage, (1971).
8. D.C-H. Cheng, Rheol. Acta, 12:228 (1973).
9. D.C-H. Cheng, Determination of the thixotropic property of bentonite-water dispersions. Paper to Joint Meeting of British, Italian and Netherlands Societies of Rheology, Amsterdam, April 1979.
10. D.C-H. Cheng, "Res. Rept. No. LR 203 (MH)", Warren Spring Laboratory, Stevenage, (1974).
11. D.C-H. Cheng, J. Phys. D.: Appl. Phys., 7:L155 (1974).
12. D.C-H. Cheng, Nature, 245:93 (1973).
13. D.C-H. Cheng, Bull. Brit. Soc. Rheol., 21:60 (1978).
14. D.C-H. Cheng, Lecture 7 in "Postgraduate School in Rheology", Chemical Society, London, p. 45 (1973).
15. D.C-H. Cheng, Chem. Ind., to be published, (1980).

CONSTITUTIVE EQUATIONS FOR CONCENTRATED SUSPENSIONS

D.V.Boger, R.R.Huilgol[+], N.Phan-Thien[++]

Monash University
[+] Flinders University
[++] University of Newcastle

(Abstract)

Highly concentrated suspensions exhibit yield stress, shear thinning, rheological dilatancy as well as thixotropy when sheared. The above phenomena depends on the shear rate and concentration. An excellent example of a material which shows all of the above behaviour is red mud, which is a waste product in the production of alumina from bauxite. This suspension has a pH value of 11 as well.

The exeperimental results from this material have led to the development of constitutive equations to describe such suspensions. It is clear that a viscosity dependent on concentration and shear rate is required; such a theory can be obtained from suspension mechanics. To predict yield stress, an anisotropic fluid theory is essential; finally, to predict dilatancy and thixotropy, the concentration should vary with time as well as from one point in the fluid to another.

In this paper, an amalgamation of anisotropic fluid theory, concentrated suspension mechanics and granular material theory is used to describe the rheological behaviour of a concentrated suspension. The theoretical predictions are compared with experimental results from the red mud.

A MIXTURE THEORY FOR SUSPENSIONS†

S. L. Passman
J. W. Nunziato
P. B. Bailey

Sandia Laboratories‡
Albuquerque, NM 87185 U.S.A.

E. K. Walsh

University of Florida
Gainesville, FL 32611 U.S.A.

INTRODUCTION

There are a wide variety of engineering problems involving suspensions of particles in a fluid. In many cases, it is important to be able to predict the rheological properties of the suspension in the situation of interest. Classically, this problem has been addressed from the micromechanical point of view.[1] That is, the detailed behavior of an individual particle in the flow field is determined, often with the aid of kinematical simplifying assumptions. Then, assuming that the suspension is dilute so that interactions between the particles are negligible, a rheological equation of state for the suspension is developed. Inherent to this approach is the assumption that the length scale of the gross motion of the suspension is much larger than the average particle size.

In this work, we have considered a somewhat different approach to modelling the rheology of suspensions. It is physically plausible to think of the particles and the suspending fluid as a two-phase mixture. Thus, using some recently developed concepts for multiphase mixtures,[2,3] we propose a theory for a suspension involving two constituents in which we account for the discrete nature of the particles by permitting the volume fraction and the mass density of each phase to vary independently. The fact that the volume fraction is an independent variable results in an additional balance equation, not occurring in conventional continuum

†This work was supported by the U.S. Department of Energy under Contract DE-AC04-76-DP00789.

‡A United States Department of Energy Facility.

theories, which relates to the interfacial force system and surface tension. This theory is unique in that it also makes use of the constitutive equations of each phase and of the micromechanical models already developed to describe such interactions as momentum and energy transfer between the phases. Furthermore, there are no limitations as to particle concentration.

The utility of this theory can best be demonstrated by considering the solution of specific boundary-value problems. Here, we consider the case of steady isochoric shearing between two parallel plates. For constitutive equations with constant viscosity, the velocity profile is linear. However, the particle distribution is nonuniform and depends on the normal stresses and volume fractions at the boundaries. For dilute suspensions, our results indicate the particles tend to accumulate near the boundary plates. This is in general qualitative agreement with experiments on dilute suspensions of monodisperse spheres.[4] For large concentrations, the particles tend to accumulate near the center line of the flow.

BASIC EQUATIONS[2,3]

Our particular interest here is in a two-phase mixture of solid particles in a viscous fluid. Each constituent is assigned a <u>local density</u> γ_a which represents the mass of the ath constituent $(a = 1,2)$ per unit volume occupied by the constituent. The <u>partial density</u> ρ_a is related to γ_a by

$$\rho_a = \gamma_a \varphi_a \quad , \quad a = 1,2 \quad , \tag{1}$$

where φ_a, $0 \leq \varphi_a \leq 1$, is the <u>volume fraction</u> of the ath constituent. The <u>mixture density</u> ρ is given by

$$\rho = \rho_1 + \rho_2 \quad , \tag{2}$$

and, assuming the mixture is saturated, we have

$$1 = \varphi_1 + \varphi_2 \quad . \tag{3}$$

To characterize the suspension, we assume that the mixture is ideal[3] and that for each phase, the stress $\underset{\sim}{T}_a$, the surface stress $\underset{\sim}{h}_a$, the momentum interaction $\underset{\sim}{m}_a^+$, and the interfacial force interaction v_a^+ $(a = 1,2)$ depends linearly on the difference of volume fraction rates, the diffusion velocity $\underset{\sim}{u} = \underset{\sim}{v}_1 - \underset{\sim}{v}_2$, and the rate of deformation $\underset{\sim}{D}_a$

$$\underset{\sim}{D}_a = \frac{1}{2} [\text{grad } \underset{\sim}{v}_a + (\text{grad } \underset{\sim}{v}_a)^T] \quad , \tag{4}$$

where $\underset{\sim}{v}_a$ is the velocity of phase a. Further assuming that both the fluid and solid particles are <u>isotropic</u> and <u>incompressible</u>, we have

<u>Phase 1</u>:

$$\underset{\sim}{T}_1 = -p_1 \underset{\sim}{1} + 2\mu_1 \underset{\sim}{D}_1 - \underset{\sim}{h}_1 \otimes \text{grad}\, \varphi_1 \quad , \tag{5}$$

$$\underset{\sim}{h}_1 = 2\alpha_1 \, \text{grad}\, \varphi_1 \quad , \tag{6}$$

$$v_1^+ = p_1 - \lambda - F(\grave{\varphi}_1 - \grave{\varphi}_2) \quad , \tag{7*}$$

$$\underset{\sim}{m}_1^+ = \lambda \, \text{grad}\, \varphi_1 - d\, \underset{\sim}{u} \quad , \tag{8}$$

<u>Phase 2</u>:

$$\underset{\sim}{T}_2 = -p_2 \underset{\sim}{1} + 2\mu_2 \underset{\sim}{D}_2 - \underset{\sim}{h}_2 \otimes \text{grad}\, \varphi_2 \quad , \tag{9}$$

$$\underset{\sim}{h}_2 = 2\alpha_2 \, \text{grad}\, \varphi_2 \quad , \tag{10}$$

$$v_2^+ = p_2 - \lambda - F(\grave{\varphi}_2 - \grave{\varphi}_1) \quad , \tag{11}$$

$$\underset{\sim}{m}_2^+ = \lambda \, \text{grad}\, \varphi_2 + d\, \underset{\sim}{u} \quad , \tag{12}$$

where p_1 and p_2 are pressures to be evaluated from the boundary conditions and λ, the mean interface pressure, is a consequence of the saturation constraint. Here also μ_1, α_1 are functions of φ_1; μ_2, α_2 are functions of φ_2; and F and d may depend on φ_1 and φ_2.

It is important to note that if phase 1 (say, the viscous fluid) is absent, then the constitutive equation for the stress in phase 2 (the solid particles) is identical to that proposed by Goodman and Cowin[6] for flowing granular materials. The viscosity μ_2, then, is a result of particle-particle interactions which we would expect to be significant for high particle concentration. The stress $\underset{\sim}{h}_2$ serves to correct the stress for any effects of surface tension. Similar statements apply for the fluid viscosity μ_1 and the surface stress $\underset{\sim}{h}_1$.

SHEARING FLOW BETWEEN TWO PARALLEL PLATES

An idea of the physical applicability of a theory may be obtained by considering a specific boundary-value problem and solving the equations of motion in conjunction with appropriate boundary

---

*The backward prime indicates time differentation following the motion of a particle of the ath constituent.

conditions. One of the simplest cases, but still most interesting, is that where the boundaries are two parallel, infinite plates, a fixed distance apart. Deformation is caused by moving one plate at a constant speed parallel to the other, and the motion is called simple shearing.

The analysis of the simple shearing problem is straightforward. Let the plates be a distance L apart. Choose fixed Cartesian coordinates with the origin on the fixed plate, x_1 parallel to the direction of motion of the top plate, and x_2 orthogonal to the plates and pointing from the fixed plate toward the moving plate. Then, with all fields steady we have

$$\underset{\sim}{v}_1 = [v_1(x_2),0,0] \quad , \quad \varphi_1 = \varphi_1(x_2) \quad , \tag{13}$$

$$\underset{\sim}{v}_2 = [v_2(x_2),0,0] \quad , \quad \varphi_2 = \varphi_2(x_2) \quad . \tag{14}$$

The gravitational field acts in the $-x_2$ direction and is given by

$$\underset{\sim}{b} = [0,-b,0] \quad . \tag{15}$$

Substituting the constitutive equations (5)-(12) into the appropriate balance equations 2,3,5, it can be shown that the motion of each phase is isochoric. The nonvanishing components of the linear momentum equations are*

$$(\mu_1 v_1{}')' + d(v_1 - v_2) = 0 \quad , \tag{16}$$

$$(\mu_2 v_2{}')' + d(v_2 - v_1) = 0 \quad , \tag{17}$$

$$-(p_1 + 2\alpha_1(\varphi_1')^2)' + \lambda\,\varphi_1' - \gamma_1\varphi_1 b = 0 \quad , \tag{18}$$

$$-(p_2 + 2\alpha_2(\varphi_2')^2)' + \lambda\,\varphi_2' - \gamma_2\varphi_2 b = 0 \quad , \tag{19}$$

while the interfacial force balance equations yield

$$2(\alpha_1\varphi_1')' + p_1 - \lambda = 0 \quad , \tag{20}$$

$$2(\alpha_2\varphi_2')' + p_2 - \lambda = 0 \quad . \tag{21}$$

In addition to these equations, we have the saturation condition (3).

---

*Hereafter a prime on a quantity, ( )′, denotes differentiation with respect to x_2.

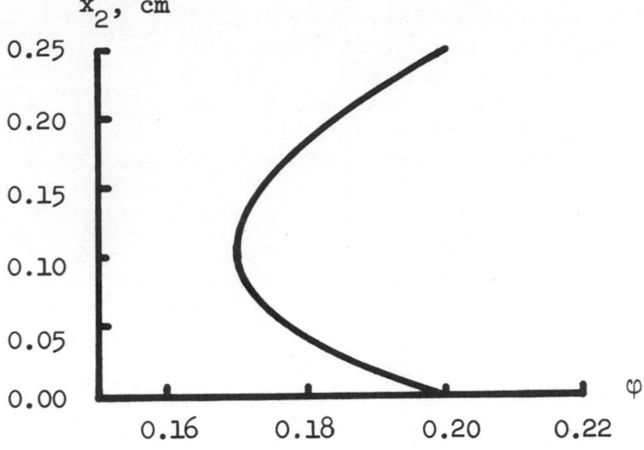

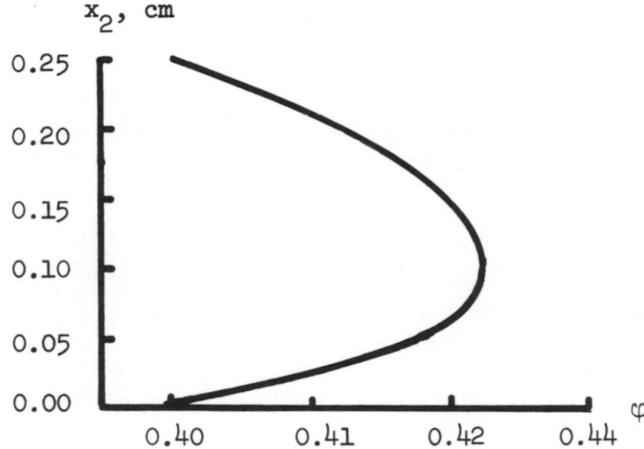

Fig. 1. Typical volume fraction profiles in simple shearing.

Assuming the constitutive functions are given, these equations (16)-(21) and (3) compose a nonlinear system of seven (7) ordinary differential equations in seven (7) unknowns. The system is partially uncoupled in the sense that (3) and (18)-(21) yield φ_1, φ_2, λ, p_1, and p_2 independent of v_1 and v_2. In turn, v_1 and v_2 depend on these quantities only through the dependence of μ_1 and μ_2 on φ_1 and φ_2.

For the present analysis, we assume the material coefficients μ_1, α_1, d, μ_2, α_2 are all constants typical of a sand-water suspension. In this case (16) and (17) can be solved analytically to give linear velocity profiles. To find the volume fraction profile $\varphi_2(x_2)$ and hence the particle distribution, we solve (18)-(21) and (3) numerically. The results of the calculations for two different boundary values of φ_1 for a plate distance L of 0.25 cm, and for vanishing normal stresses are shown in Figure 1.

For a boundary condition of $\varphi_2(0) = \varphi_2(L) = 0.2$, the volume fraction of particles φ_2 decreases toward the center of the flow. In this case, a average particle distribution between the plates is $(\varphi_2)_{ave} = 0.195$. Thus in simple shearing the particles tend to accumulate in the neighborhood of the plates. This is consistent with experimental observations on dilute suspensions of monodisperse spheres.

For a boundary condition of $\varphi_2(0) = \varphi_2(L) = 0.4$, the particle distribution has been reversed, i.e., the volume fraction φ_2 increases toward the center of the flow field. Here an average particle distribution is $(\varphi_2)_{ave} = 0.415$, thus in simple shearing the particles tend to accumulate in the vicinity of the center of the flow.

These results suggest that the particle distribution in suspensions undergoing simple shearing are sensitive to initial concentrations.

REFERENCES

1. E. J. Hinch, "The mechanics of fluid suspensions," in: "Theoretical Rheology," J. F. Hutton, J. R. A. Pearson and K. Walters, ed., Applied Science Publishers, London (1975).
2. S. L. Passman, "Mixtures of granular materials," Int. J. Engr. Sci., 15:117 (1977).
3. J. W. Nunziato and E. K. Walsh, "On ideal multiphase mixtures with chemical reactions and diffusion," Arch. Rat. Mech. Anal., forthcoming.

4. D. J. Jeffrey and A. Acrivos, "The rheological properties of suspensions of rigid particles," AIChE J., 22:417 (1976).
5. J. W. Nunziato and S. L. Passman, "Application of multiphase mixture theory to two-phase flows," forthcoming.
6. M. Goodman and S. C. Cowin, "Two problems in the gravity flow of granular materials," J. Fl. Mech., 45:321 (1971).

DISPERSION IN LAMINAR FLOWS

R.L.Powell°, S.G.Mason

McGill University, Montreal, Canada
° Washington University, St.Louis, U.S.A.

(Abstract)

The dispersion of clusters of small spherical particles (20-400 m) suspended in a liquid has been studied by subjecting them to linear two-dimensional flow fields which include pure shear as one limit pure rotation as the other with simple shear as an intermediate case. If the liquid used to form the suspension is the same as the bulk medium, dispersion proceeds in a well-defined fashion depending upon: the amount of vorticity in and the strength of the undisturbed flow, the initial radius of the cluster, the radius and volume fraction of particles. It was Found that a model based upon the assumption that the rate at which the particles leave the surface of the cluster is proportional to its surface area adequately describes the dispersion process.

When a fluid otheR than that used to produce the flow was used to make the suspension from which the clusters were formed, the interfacial tension between the two liquids qualitatively changes the dispersion process. The cluster behaved as a liquid drop having finite surface tension with a viscosity dependent upon the volume fraction of particles. The effect of changing the surface tension by the addition of surface active agents was also examined.

STRUCTURE FORMATION IN CONCENTRATED DISPERSE SYSTEMS UNDER DYNAMIC CONDITIONS

N.B. Uriev

Institute of Physical Chemistry of the USSR
Academy of Sciences, Moscow, USSR

(Abstract)

Regularities of combination of mechanical (vibration) effects and surface-active additives of various chemical nature and structure in the processes of formation and destruction of structures in concentrated disperse systems have been studied: in highly dispersed powders (such as SiO_2, $CaCO_3$, CaF_2 and others) and also in systems with a liquid dispersion medium (in aqueous dispersions of calcium aluminates and silicates, of calcium bentonite and in non-aqueous dispersions, such as graphite in vaseline oil and others).

It is shown that the formation of a saturated absorption layer of surface-active substances (for instance, octadecylamine for quartz powder, stearic acid for calcium carbonate), while substantially weakening the strength of adhesion in contacts, at the same time, as a result of a reduction in the liophobic-liophilic mosaicity of the surface or particles, drastically diminishes the scatter in the strength of these elementary atomic and coagulation contacts. There has been found an effect of mutual strengthening of the action of mechanical factors (vibration) and adsorption-active medium, characterized by a continuous growth of the relationship $D = I_o/I_1$ with an increase in the degree of destruction of the structure, i.e. of the decrease of effective viscosity from its maximum value ηV_o to the minimum viscosity of the ultimately destroyed structure ηV_{min} (I_o being specific power of mechanical (vibration) actions without the introduction of surface-active additives and I_1, in combination with such additives). A concept is introduced concerning the dynamic criterion of the action of surface-active substances, which is determined by this relationship $D_m = I_o/I_1$ with the extreme destruction of the structure $\eta V = \eta V_{min}$. The maximum value D_m is shown to be reached with the introduction into concentrated disperse systems of chemically adsorbing surface-active substances with a developed hydrocarbon chain, which form a saturated adsorption layer on the surface of the particles. The results of investigations in the regularities of combination of mechanical (vibration) actions and surface-active additives have found wide practical application in various technological processes, including the technology of blending multicomponent disperse systems; in heterogeneous mass transfer processes used in chemical engineering and carried out in a fluidized bed; in the technology of producing various kinds of dispersed materials rich in solid phase.

VISCOELASTIC PROPERTIES OF HIGHLY FILLED LIQUID

POLYESTER COMPOSITIONS

Ya. Ivanov, R. Kotsilkova and Y. Simeonov

Bulgarian Academy of Sciences, Central Laboratory
of Physico-Chemical Mechanics., Acad. G. Bontchev Str.
IV Bl., Sofia 1113, Bulgaria.

I. INTRODUCTION

The rheological behavior of concentrated suspensions in polymers has recently been a subject of much interest. This interest is due to the increasing use of composite materials. Knowledge of their extremely complex and diverse properties is helpful in determining correctly the proportion of mixing. By means of rheological characteristics it is furthermore possible to regulate existing properties as well as to create new structural, mechanical or other desired properties. This report deals with the investigation of rheological properties of concentrated suspensions consisting of highly filled liquid polymer compositions.

II-1. MATERIALS

Polyester resin "Vinalkid-550 PE" was used as a suspending medium (viscosity 1.2 Pa.s, density 1.12 g cm^{-1}). The fillers used were mineral powders of the diabaz stone. Its surface area determined by means of Blayn's method was 400 m^2g^{-1}. The particle size distribution determined by granulomatry: 82% < 10 μ; 12% between 10-20 μ; 4.4% between 20-30 μ; 1.2% between 30-40 μ; some particles up to d = 50-80 μ are present. The particles were nearly spherical.

II-2. APPARATUS AND METHODS

The viscoelastic properties of highly filled compositions were investigated using a Weissenberg Rheogoniometer (R-18) with parallel plates (Radius 2.5 cm, gap 0.33 cm). Steady flow properties were also measured using a coaxial cylinder rheometer (Rheotest 2) and Vibrorheometer VR-74. Studies have been performed at temperatures ranging from 293 to 333°K and different total solid concentration varying from 10 to 90% in volume.

In order to increase the accuracy of the measurements, especially in the low frequency range, we used a special electronic filtering system eliminating noise in output signals of the tranducers.

III. EXPERIMENTAL RESULTS

III-1. Concentration effects on the rheological properties

Highly filled suspensions of polyester resin become tixotropic several hours after their preparation. Fig. 1 shows the viscosity in steady shear as a function of time for various volume concentrations of filler. According to P.A. Rehbinder [1] their structures are due to interactions between filler particles and the thin layers surrounding them. Differences in the tixotropic structure reflect themselves in differences in the flow behavior of the polyester compositions. Fig. 2 gives the flow curves in reduced coordinates* for the polyester resin and suspensions of 30, 50, 60, 62 and 65 % volume fraction. For several values of filler concentration the disperse systems are viscoplastic (curves 2 and 3). In these cases the systems exhibit a yield stress (τ_o) and maximal distruction of structure occurs at low shear rates. If the filler concentration is larger than the critical value, an anomaly occurs in the flow curves (5, 6 and 6). The latter is shown by the fast decrease of shear stress when the limit resistance of the structure is reached (points A_1, A_2 and A_3) [3,4].

We assume that for these high concentrations of filler all resin ends up in the thin adsorbed layers. At concentrations larger than 62% the number of the direct contacts between the particles rises rapidly. The structure formed in the suspension is not homogenous and when the limit resistance is reached, a snowball effect in the destruction of the structure takes place.

---

* For the presentation of temperature dependence on viscosity, the method of reduced variables was used (time-temperature equivalence superposition of Williams-Landel-Ferry). In the calculation of shift factor a_T a reducing temperature of T_o = 293°K was chosen. The curves were plotted as $\lg \tau\, T_o/T$ vs $\lg \dot{\gamma}\, a_T$ [2]. The presence of the filler changes the reduced temperature to 279°K.

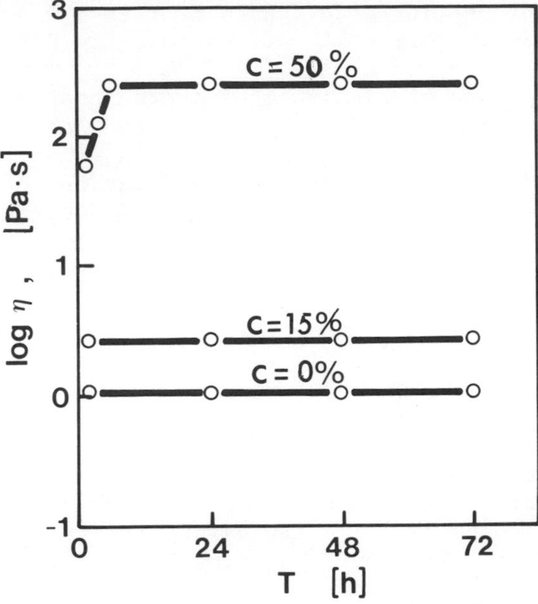

Fig. 1 - Viscosity as a function of time of polyester compositions with various volume fractions at room temeprature (293°K).

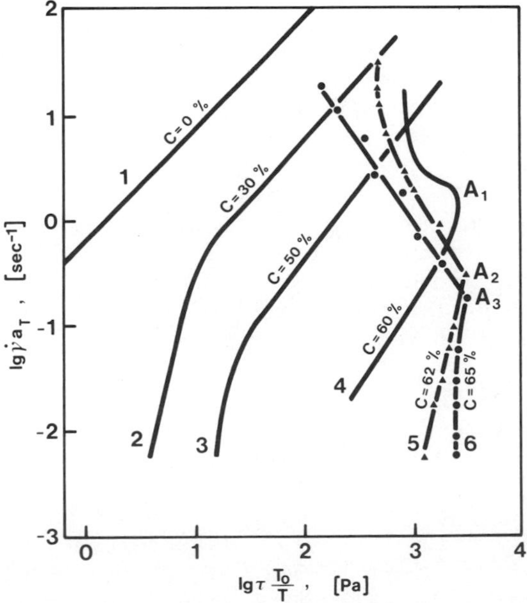

Fig. 2 - Flow curves for the filled polyester compositions in reduced coordinates.

III-2. Dynamic properties as a function of the concentration, frequency and temperature

The investigation of highly concentrated suspensions subjected to dynamic deformations (due to small amplitude ascillatory shear) indicates that for a filler concentration 60%, the response to deformation is linear. Fig. 3 shows the relation between the real parts of the dynamic shear modulus G', the dynamic viscosity η' and the maximum deformation $\dot{\gamma}_{max}$ for different concentrations, temperatures and frequencies. The calculations of $\dot{\gamma}_{max}$ were done by means of the equation [5]:

$$\dot{\gamma}_{max} = \frac{1}{A} (\theta^2 + \phi^2 - 2\phi\theta \cos \zeta)^{1/2}$$

where A is an instrument shear rate constant. θ and ϕ are the amplitudes of input and output signals. ζ - phase angle.

The experimental data show that the linear part of the curves shorten to a marked degree with increasing filler concentration and frequency. The temperature influence in the range from 293 to 333°K is very small. For suspensions with 62% filler only the nonlinear part of the curves was observed. G' and η' depend on $\dot{\gamma}_{max}$. In this case we assume that the number of direct contacts between the particles increases significantly. This is a result of insufficient wetting. As a result the structure formed has an insufficient elasticity and hence is very sensitive to deformation [6]. This is the reason why the dynamic properties of these disperse systems were measured with amplitude as small as possible.

Fig. 4 shows the frequency dependence of the dynamic properties. It shows a plateau in the curves G' = f(ω) at small frequencies. By analogy to gel-like systems and three dimensional structural solutions, G' in the region of the plateau is called the "pseudoequilibrium" modulus. Increasing the filler's concentration causes the rise in the values of G' and η'. The transitive point for the pseudoequilibrium modulus G' moves to smaller values of ω. Changing the concentration from 30 to 60% corresponds to a change of one order magnitude in G' and η'.

The existence of the plateau for G' confirms the assumption made above for the three dimensional structure of our systems. It also confirms the experimental investigations on the yield stress in these systems.

The pseudoequilibrium modulus G' also depends on the filler concentration. This is shown in Fig. 5. The slope of the curve indicates the degree of the structure formed in the system. The marked increase of this slope beyond 50% volume concentration of filler shows that a perceptible thickening of the structure occurs.

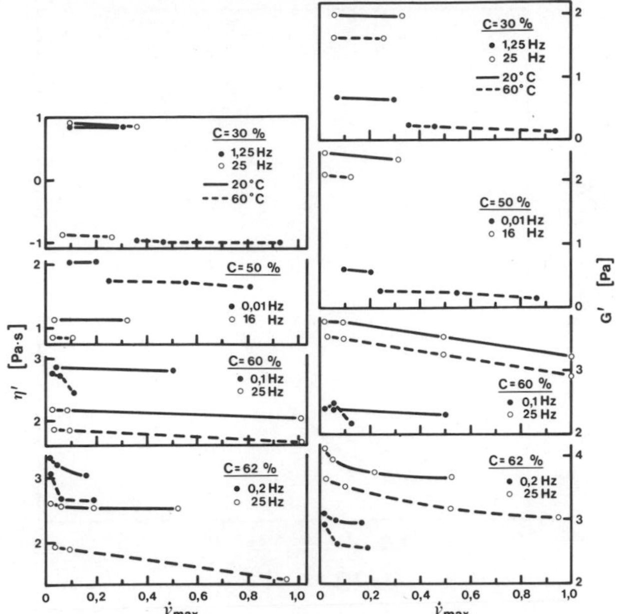

Fig. 3 - G' and η' as a function of concentration, frequency and temperature.

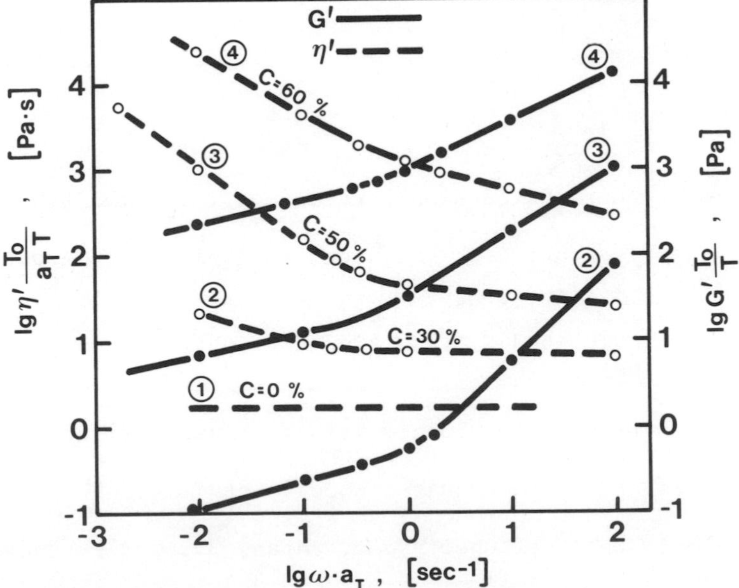

Fig. 4 - G' and η' as a function of the frequency in reduced coordinates.

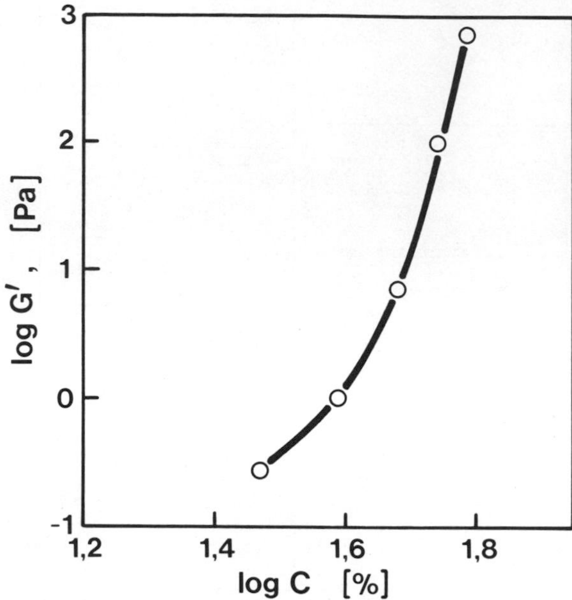

Fig. 5 - Double logarithmic plot of the modulus G' vs the filler concentration.

REFERENCES

1. P.A. Rehbinder, Discussion Faraday Soc., 18, 151 (1954).

2. L.A. Faytelson and E.E. Yacobson, Mechanika polymerov, 6, 105-1086 (1977).

3. M.P. Volarovich and N.I. Malkin, Eng. phys. jurnal, 10, 6, 804-812 (1966).

4. G.V. Vinogradov, Rotationnie pribory, Moskow (1968).

5. J. Harris and R. Maheshuaru, Rheol. Acta, 14, 457-465 (1975).

6. M. Takano, Bull. of Chem. Soc., Japan, 37, 78-79 (1964).

AN EXPERIMENTAL STUDY OF THE DYNAMIC VISCOSITY OF EMULSIONS WITH SMALL DROPLETS IN THE kHZ REGION

M. Oosterbroek, J.S. Lopulissa, and J. Mellema

Department of Applied Physics
Twente University of Technology
Enschede, The Netherlands

INTRODUCTION

In the fifties the theoretical work of Rouse[1] and Zimm[2] on the linear viscoelastic properties of diluted polymer solutions stimulated the experimental investigation of such solutions. Although Fröhlich and Sack[3] already in 1946 and Oldroyd in 1953[4] and 1955[5] predicted viscoelastic properties of dispersions and emulsions respectively, little experimental work in this field has been done ever since. The aim of our investigation is to demonstrate the existence of linear viscoelasticity of emulsions caused by drop deformation. This paper will show some preliminary results.

MECHANICAL MODELS FOR EMULSIONS

Perhaps the most simple model (I) of an emulsion is a system in which monodisperse spherical droplets with radius a of some Newtonian fluid with viscosity η_1 are immersed in another Newtonian fluid with viscosity η. The interfaces between these liquids are infinitely thin and can just be characterized with a constant interfacial tension γ. The dynamic viscosity $\eta^* = \eta' - i\eta''$ can be easily calculated if only the influence of the first order of the volume concentration ϕ of the droplets is considered[4]. It has one relaxation time τ, approximately given by:

$$\tau = (16\eta + 19\eta_1)(3\eta + 2\eta_1)a/\{40(\eta + \eta_1)\gamma\}. \qquad (1)$$

In a more sophisticated model (II) of an emulsion the infinitely thin surface of a droplet is not only characterized with an interfacial tension, but it is conceived as a two-dimensional rheological body with linear viscoelastic properties: a surface dilatational modulus κ and viscosity σ and a surface shear modulus μ and viscosity ζ. This conception leads to two relaxation times, τ_1 and τ_2, for the dynamic viscosity. The viscosity has been calculated by Oldroyd[5] in the first order of concentration. If the times are separated a few orders of magnitude, three different cases are possible. They are approximately given by:

$$\kappa \gg \mu, \gamma$$

$$\tau_1 = (32\eta + 23\eta_1 + 16\,\zeta\, a^{-1})a/\{8(3\gamma + 2\mu)\},$$

$$\tau_2 = Q/\{2\kappa(32\eta + 23\eta_1 + 16\,\zeta\, a^{-1})\}; \qquad (2)$$

$$\mu \gg \kappa, \gamma$$

$$\tau_1 = (12\eta + 13\eta_1 + 8\,\sigma\, a^{-1})a/\{8(\gamma + \kappa)\},$$

$$\tau_2 = Q/\{4\mu(12\eta + 13\eta_1 + 8\,\sigma\, a^{-1})\}; \qquad (3)$$

$$\gamma \gg \mu, \kappa$$

$$\tau_1 = (5\eta + 5\eta_1 + 6\,\sigma\, a^{-1} + 4\,\zeta\, a^{-1})a/\{2(3\kappa + 2\mu)\},$$

$$\tau_2 = Q/\{8\gamma(5\eta + 5\eta_1 + 6\,\sigma\, a^{-1} + 4\,\zeta\, a^{-1})\}, \qquad (4)$$

where

$$Q = (16\eta + 19\eta_1)(3\eta + 2\eta_1)a + 46\,\sigma\,\eta_1 + 32\,\sigma\,\zeta\, a^{-1} + $$
$$+ 64\,\sigma\,\eta + 25\,\zeta\,\eta_1 + 48\,\zeta\,\eta.$$

For real emulsions, in particular microemulsions, the droplet surface can presumably not be considered as infinitely thin[6].

Another simple model of an emulsion (III) is one in which the interface has finite thickness d and consists of an isotropic linear viscoelastic fluid. Both the inner and outer surface of the interface have interfacial tensions, γ_1 and γ_2, respectively. The calculation of the complex viscosity η^* for this model at low droplet concentration is a variant of the work of Sakanishi and Takano[7]. If the interface consists of a Newtonian fluid and $d/a \ll 1$, the two resulting relaxation times are separated a few orders of magnitude and can be given approximately by:

$$\tau_1 = 5(a/d)^2(\eta + \eta_1)a\Big/\{24(1/\gamma_1 + 1/\gamma_2)\},$$

$$\tau_2 = (16\eta + 19\eta_1)(3\eta + 2\eta_1)a/\{40(\eta + \eta_1)(\gamma_1 + \gamma_2)\}. \qquad (4)$$

For most instruments which can measure η'' accuracy problems arise when one tries to measure the effects predicted by these three models at low concentration. In practice we investigated emulsions with volume concentration up to 27%. So the concentration effect on the models given above has been considered as well. To estimate this influence we incorporated the models of the emulsion droplet in the cell model of Simha[8].

SAMPLES

In our investigation we used two different kinds of O/W emulsions, both with small droplets. Firstly we made non-ionic emulsions of benzene emulsified with Tween 20 (polyoxyethylene (20 mol) sorbitan monolaurate) in water. This system is comparable with the system made by Matsumoto and Sherman[9]. The solution benzene+water +Tween 20 was heated at 65 degrees centigrade and then shaken at that temperature. Next the solution was slowly cooled down to room-temperature.

Another kind of emulsion we investigated were the non-ionic microemulsions. These systems form spontaneoulsy and are investigated after equilibrium is reached. As an example we will show one consisting of cyclohexane emulsified with DPE (polyoxyethylene 8.5 mol) dodecyl phenylether) in water. The phase diagrams have been studied by Shinoda and Kunieda[10].

MEASUREMENT OF THE DYNAMIC VISCOSITY

The last years a new apparatus for the measurement of the dynamic viscosity in the kHz region has been developed in our laboratory[11]. It consists of a nickel tube which has been magnetized circumferentially. A transmitting coil which produces an alternating axial magnetic field forces the tube in a torsional vibration. This is a magnetostrictive effect. A receiving coil detects the movement of the tube by means of the reverse effect. Comparison of characteristic quantities of the resonance curves of the tube surrounded by air and liquid gives in principle the shear impedance $Z_L = R_L + i X_L$ of the fluid. The relation between Z_L and η^* at angular frequency ω and for a liquid with density ρ is given by:

$$\eta' = 2 R_L X_L / (\omega\rho), \quad \eta'' = (R_L^2 - X_L^2)/(\omega\rho). \tag{5}$$

RESULTS AND DISCUSSION

In fig. 1 the measurements of a system A consisting of 18.5% benzene, 76.7% bidestilled water, and 4.8% Tween 20 (V/V), are shown. In fig. 2 the effects of a system B, consisting of 17.0% benzene, 78.0% bidestilled water, and 5.0% Tween 20 (V/V), can be seen. Both systems were measured at 20 degrees centigrade. A microemulsion indicated with C consists of 27% cyclohexane, 70% bidestilled water and 3% DPE (V/V). The dynamic viscosity of C can be found in fig. 3.

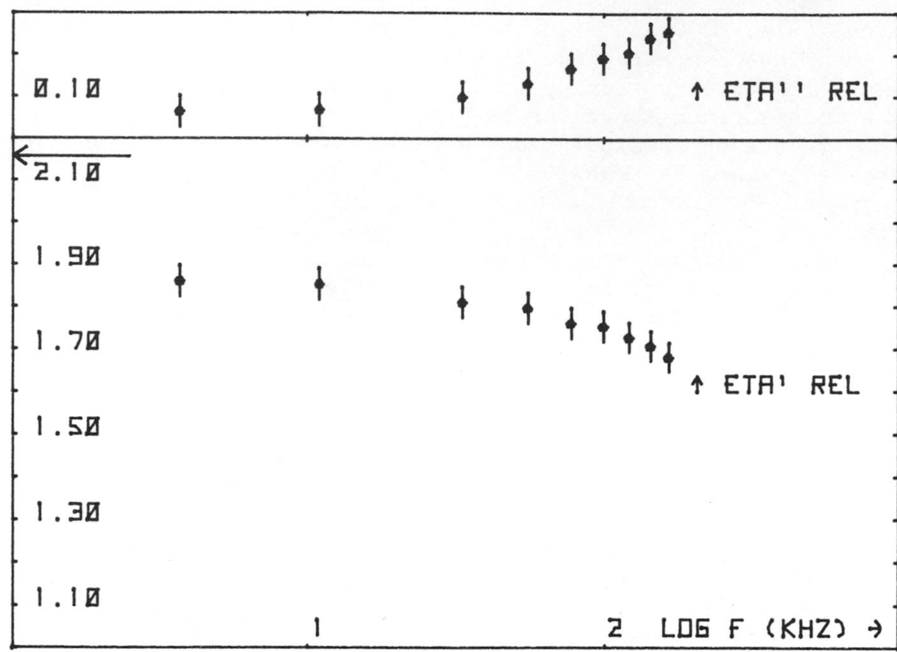

Fig. 1. The relative dynamic viscosity of 18.5% benzene, 76.7% bidestilled water, and 4.8% Tween 20 (V/V) at 20° C. System A. The arrow indicates the steady-state viscosity.

It was measured at 24 degrees centigrade. The arrows in the figures indicate the steady-state viscosities measured with a Contraves, LS30 (rates of shear between 1 s^{-1} - 100 s^{-1}). Due to the measuring procedure η' and η" are divided by the viscosity of a Newtonian calibration fluid and the quotients are indicated by η'REL

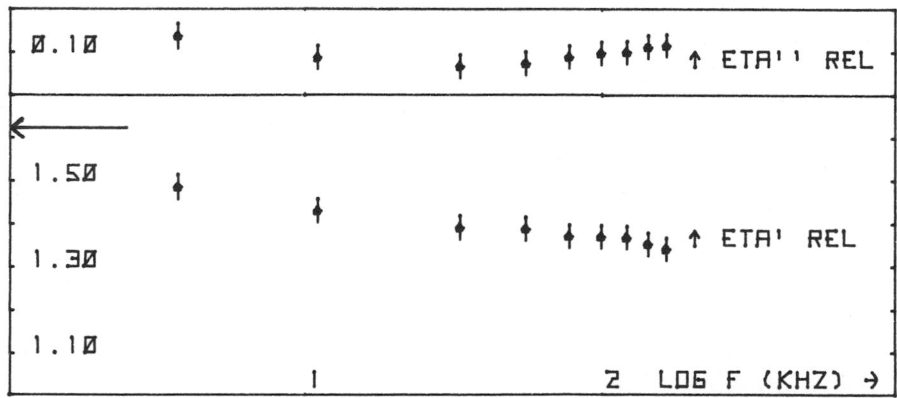

Fig. 2. The relative dynamic viscosity of 17.0% benzene, 78.0% bidestilled water, and 5.0% Tween 20 (V/V) at 20° C. System B. The arrow indicates the steady-state viscosity.

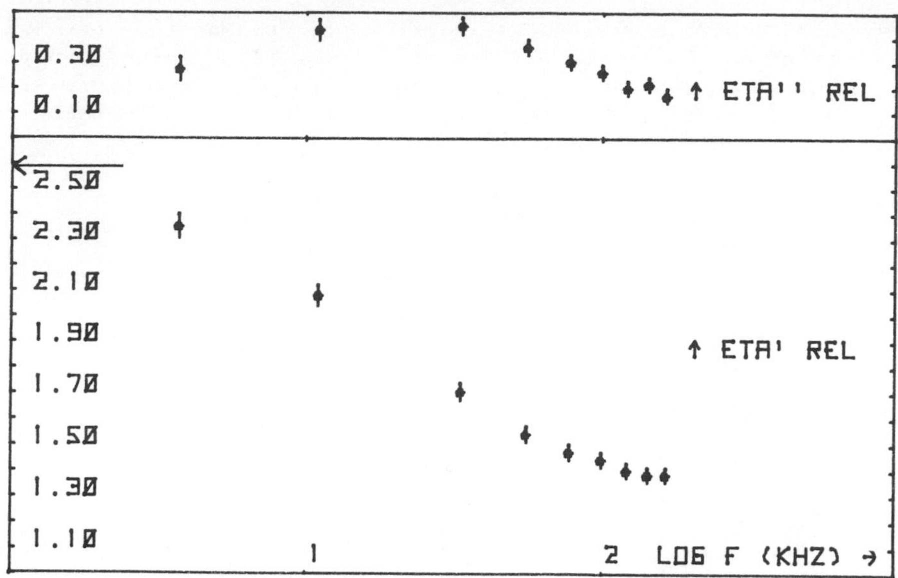

Fig. 3. The relative viscosity of 27% cyclohexane, 70% bidestilled water and 3% DPE (V/V) at 24° C. System C. The arrow indicates the steady-state viscosity.

and η''REL. To give an idea of the accuracy error bars are added. For system A an estimation of the droplet radius was derived by means of ultracentrifugation. The mean radius was about 50 nm. With the same method a sample of the continuous phase could be derived. Next it was analyzed. A solution of the derived composition was used for the dilution of system A to about 15, 10, and 5% (V/V) benzene. The steady-state viscosity appeared to be almost equal to values known for hard spheres of the same concentration. Both model II and model III allow this. A good fit is only possible when two or more relaxation times are involved. With model II and one set of parameters the four series of measurements could be fitted. The best fit appeared to be with $\mu \gg \gamma, \kappa$ and the following parameter values: $\mu = .53\ \mathrm{mNm^{-1}}$, $\gamma + \kappa = .71\ \mathrm{\mu Nm^{-1}}$, $\zeta = .22\ \mathrm{nNsm^{-1}}$ and $\sigma =$ $= 5.1\ \mathrm{pNsm^{-1}}$. However, the fit ($\gamma \gg \kappa, \mu$), with $\gamma = .5\ \mathrm{mN/m}$, $\mu + \frac{3}{2}\kappa = .47\ \mathrm{\mu Nm^{-1}}$, $\zeta = 36\ \mathrm{pNsm^{-1}}$ and $\sigma = 0$ was also quite good. A fit with $\kappa \gg \gamma, \mu$ was very bad. At this moment we cannot distinguish between these sets of parameters. Obviously the form of η' and η'' curves does not convincingly show the existence of two relaxation times. Therefore we also show the effects of system B. In fig. 2 the measurements suggest clearly two relaxation times which are closer together than for system A. Model III has not been used yet for a fit. System C shows clearly a maximum in η'' and an inflection point in η', whereas systems A and B do not reveal the complete effect of one relaxation time.

The results are not yet analyzed completely in terms of the models given. But even this will not be sufficient, because this reference system might be too simple. We are studying real emulsions and purely mechanical models may be inadequate to explain the effects[12]. We should also carefully examine if gel formation or electroviscoelastic effects do play a role.

REFERENCES

1. P.E. Rouse, Jr., J. Chem. Phys. 21: 1272 (1953)
2. B.H. Zimm, J. Chem. Phys. 24: 269 (1956)
3. H. Fröhlich and R. Sack, Proc. Roy. Soc. A, 185: 415 (1946)
4. J.G. Oldroyd, Proc. Roy. Soc. A, 218: 122 (1953)
5. J.G. Oldroyd, Proc. Roy. Soc. A, 232: 567 (1955)
6. Microemulsion, Theory and Practice, L.M. Prince, editor, Academic Press, N.Y., 1977
7. A. Sakanishi and Y. Takano, Jap. J. Appl. Phys. 13: 882 (1974)
8. R. Simha, J. Appl. Phys. 23: 1020 (1952)
9. S. Matsumoto and P. Sherman, J. Coll. Interf. Sc. 30: 525 (1969)
10. See ref. 6, p. 57
11. H.A. Waterman, M. Oosterbroek, G.J. Beukema, and E.G. Altena, Rheol. Acta 18: 585 (1979)
12. M. van den Tempel, J. Non-Newt. Fluid Mech. 2: 205 (1977)

MICRORHEOLOGY OF SUSPENSIONS: OSCILLATIONS IN VISCOSITY OF SHEARED SUSPENSIONS OF UNIFORM RIGID RODS

Ya. Ivanov[1], Z. Priel and S.G. Mason

Department of Chemistry
McGill University
Montreal, Canada, H3A 2A7

ABSTRACT

Previously developed theory predicts that when a dilute suspension of monodisperse axisymmetric rigid particles is subjected to a simple shear flow, assuming no particle interaction and negligible Brownian motion, the orientation distribution function undergoes undamped oscillation of frequency twice that a particle rotation about the vorticity axis. Thus, the instantaneous rheological properties of the suspension, such as the intrinsic viscosity which is obtained as averages with respect of orientation, should also oscillate with time. Experiments on model suspensions confirm these oscillations which, however, turn out to be damped, presumably as a result of: 1) a small spread in particle shape; 2) particle interactions; 3) rotary Brownian motion; and 4) nonuniformity of shear flow.

In all cases the rotating particles came to an equilibrium distribution of orientations and of the associated rheological properties.

Experiments on the viscosity oscillation due to the distribution of orientations in suspensions of nearly monodisperse rigid rods subjected to steady shear flow are discribed. The significance of the particle concentration, particle axis ratio, initial particle orientation and shear rate are shown and discussed.

[1]On leave from Bulgarian Academy of Science, Sofia, Bulgaria.

RHEOLOGICAL CHARACTERIZATION OF PASTE-LIKE DISPERSE SYSTEMS

Dimiter Hadjistamov

CIBA-GEIGY AG

4002 Basle, Switzerland

The paste-like disperse systems, referred to as formulations in this paper, can be characterized according to flow behaviour $\tau = f(D)$, as well as to time-related behaviour $\tau = f(t)$. The flow curves will be approximated in the logarithmic representation $\lg \tau = f(\lg D)$ through straight line sections (1). Depending on the graphic shape of the flow curves, four different types of flow behaviour are distinquished: type A - newtonian, type B - structural viscous, type C - pseudoplastic and type D - plastic flow behaviour (see Fig. 1) (1). The four types of flow behaviour can be charac-

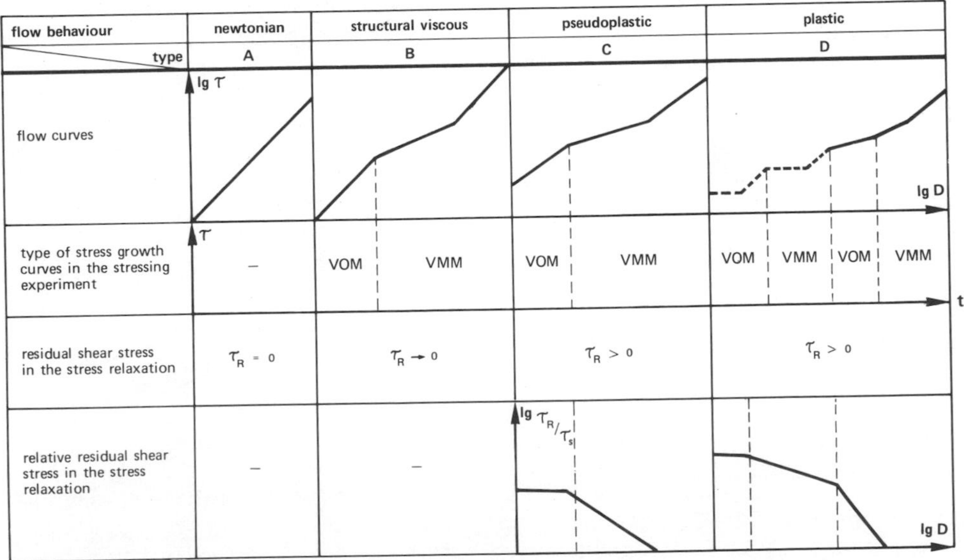

Fig. 1 Relationship between flow and time-related behaviour of paste-like disperse systems

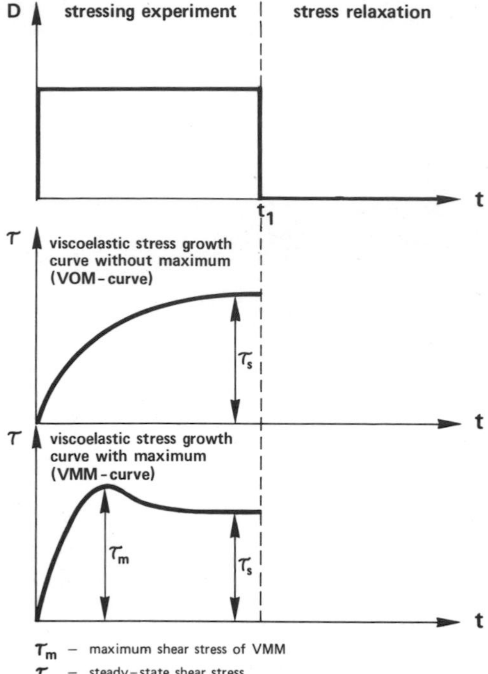

Fig. 2 Types of stress growth curves in the stressing experiment

terized according to the slope n of the straight line sections.

The newtonian liquids have one straight line with n = 1. The formulations with structural viscous flow behaviour have 3 straight line sections with $n_1 = n_3 = 1$ (so-called 1. and 2. newtonian region) and $n_2 < 1$. The formulations with pseudoplastic flow behaviour have also 3 straight line sections with $n_1 < 1$, $0 \leqslant n_2 < n_1$ and $n_3 > n_2$. The main feature of the formulations with plastic flow behaviour is the existence of one or more yield stress sections with $n \rightarrow 0$. The sections between the yield stress sections will be called transition sections. After the last yield stress sections, the slope of the straight line sections increases from one section to the next.

The time-related behaviour was examined in the stressing experiment (stress growth experiment) with subsequent stress relaxation. In principle, 2 stress growth curves can be distinquished in the stressing experiment - viscoelastic stress growth curve without maximum (VOM-curve) and with maximum (VMM-curve) (see Fig. 2) (2). The four types of flow behaviour show differences with regard to stress growth (upon inception of steady shear flow) and stress relaxation (after cessation of steady shear flow) (see Fig. 1) (2). The time-related behaviour of a system with a certain flow behaviour depends on the shear rate at which the stressing experiment is carried out. A formulation exhibits similar time-related behaviour only at shear rates within a given straight line section. This behaviour changes if the stressing experiment is carried out at shear rates from another straight line section of the flow curve.

We cannot speak of time-related behaviour in newtonian liquids, because the shear stress rises abruptly in the stressing experiment and returns abruptly to 0 on relaxation. In practice, the abrupt change in shear stress is somewhat delayed. The reasons are, among others, D = const. which does not occur immediately, inertia of the measurement device etc.

Formulations with structural viscous behaviour exhibit in the stressing experiment, at shear rates within the straight line, a viscoelastic stress growth curve without maximum, a VOM-curve. In stressing experiment with shear rates within the 2. or 3. straight line sections, the shear stress reaches equilibrium after passing through a maximum, i.e. viscoelastic stress growth curve is obtained with a maximum, a VMM-curve. The shear stress declines towards the zero point on relaxation - no residual shear stress is observed.

Formulations with pseudoplastic flow behaviour exhibit in the stressing experiment, at shear rates within the 1. straight line, a VOM-curve. At shear rates within the 2. or 3. straight line sections, a VMM-curve is observed. In the stressing experiment at shear rates within the 2. straight line section the shear stress m to rise again after passing through a maximum. This shear stress increase of some formulations is caused by a dynamic strengthening of the thixotropic agent structure. A certain residual shear stress is obtained on relaxation. The relative residual shear stress (the ratio of absolute residual shear stress value to steady state shear stress) changes insignificantly in the region of the 1. straight line section. Obviously, a structural strengthening occurs in this area, capable of balancing the increased shear effect. Only after this area begins a rapid decrease of the relative residual shear stress, or destruction of the thixotropic agent structure with shear rate. The relative residual shear stress remains smaller than 10^{-2} in the region of the 3. straight line section.

The formulations with plastic flow behaviour exhibit a rather complicated time-related behaviour. Let us consider a formulation with 3 yield stress sections (see Fig. 1). In the stressing experiment with shear rates in the 1. yield stress and transition section, a VOM-curve is observed. At shear rate in the 2. yield stress and transition section, a VMM-curve is observed. In the stressing experiment with shear rates within the 3. yield stress section, a VOM-curve is again obtained. If the stressing experiment is carried out with still higher shear rates, once more a VMM-curve is obtained. Following the stressing experiment with shear rates from the 1. yield stress section, surprisingly high relative residual shear stress values were found on relaxation (approximately 0.75). On relaxation following the stressing experiment at higher shear rates, the relative residual shear stress declines gradually. Rapid drop of the relative residual shear stress occurs only after the stressing experiment with shear rates in the 3. yield stress section.

By thixotropy and rheopexy, we mean the time-related flow behaviour. The viscosity (or τ, because D = const.) decreases (thixotropy) or increases (rheopexy) toward a final value as a result of steady shear flow. Because during this steady shear flow, the shear rate is kept constant, we speak of a stressing experiment. It is true that viscosity decreases at thixotropy, but first of all the viscosity increases rapidly (stress growth curve) and only after passing a maximum (VMM-curve) it begins to decrease. We will also observe in the presence of thixotropy a VMM-curve and respectively in the presence of rheopexy - a VOM-curve.

It is assumed according to definition that, if thixotropy or rheopexy is present, the viscosity will rise (thixotropy) or decrease (rheopexy) following cessation of the steady shear flow and a rest period. With cessation of steady shear flow, begins the stress relaxation which continues during the rest period. Because the viscosity cannot be measured at rest, the formulation has to be submitted again to steady shear flow (stressing experiment), following the rest interval.

From the above it follows that thixotropy or rheopexy can be measured in practice by conducting first a stressing experiment (D = const) followed by a stress relaxation (D = 0) and a rest period and then a subsequent stressing experiment.

The already described time-related behaviour (stressing experiment with following stress relaxation) can be used to elucidate the concepts thixotropy and rheopexy. If a VMM-curve is obtained during the stressing experiment, then thixotropy can be expected because viscosity decreases after τ_m (see Fig. 2) with time. Correspondingly, rheopexy is to be expected with a VOM-curve. However, these two curve types are observed with all formulations having structural viscous, pseudoplastic and plastic flow behaviour. The corresponding stress growth curve type depends on the shear rate at which the stressing experiment is carried out. Consequently, it cannot be maintained that a formulation is thixotropic or rheopexic. In actual fact, the formulations exhibit thixotropic or rheopexic properties in certain shear rate regions.

For the purpose of studying the concepts thixotropy and rheopexy more closely, formulations having structural viscous, pseudolastic and plastic flow behaviour were submitted to a stressing experiment followed by a stress relaxation and rest period and then again to a stressing experiment.

For the sake of simplicity we shall examine the shear stress values rather than the viscosities (the viscosity is proportional to the shear stress, because D = const). A comparison is made between the stationary (steady state) shear stress τ_s values from the 1. and 2. stressing experiment.

With formulations having structural viscous flow behaviour (see Fig. 3) either a VOM- or a VMM-curve is obtained in the 2. stressing experiment at shear rates from the 1. straight line section (see Fig. 4A). The stationary shear stress is even after rest periods of different duration either the same as, or higher than that, of the 1. stressing experiment. In the stressing experiment with shear rates in the 2. or 3. straight line section, the stationary shear stress is either the same as, or lower than that, of the 1. stressing experiment (see Fig. 4B).

Formulations with pseudoplastic flow behaviour (see Fig. 5) yield in the 2. stressing experiment with shear rates of the 1. straight line section a VMM-curve with stationary shear stress higher than the 1. stressing experiment (see Fig. 6A). In many instances a renewed shear stress increase can occur after the maximum, i.e. a dynamic structural strengthening. It is also noteworthy that the residual shear stress exhibits higher values after the 2. stress-

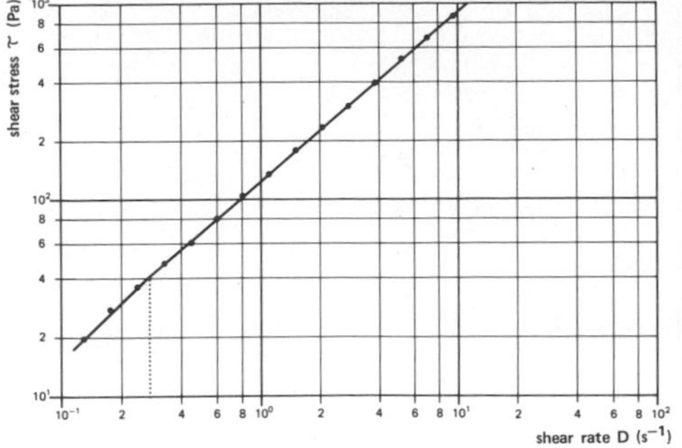

Fig. 3 Flow curve of a formulation with structural viscous flow behaviour
(XB 3008A, op. 14/79; Rheomat–30, MS–KP2)

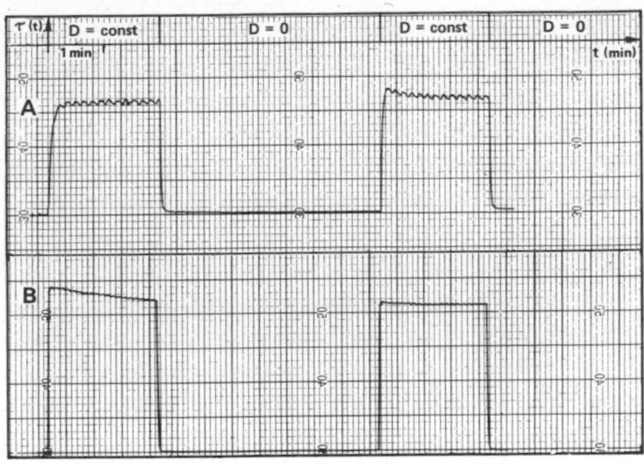

Fig. 4 Stressing experiment (D = const) – stress relaxation (D = 0; t = 4 min) – stressing experiment (D = const) – stress relaxation (D = 0)
(XB 3008A, op. 14/79; Rheomat–30, MS–KP2)
A → D = 0.13 s^{-1}, τ(t) = 1.25 · scale division (Pa)
B → D = 2.80 s^{-1}, τ(t) = 12.50 · scale division (Pa)

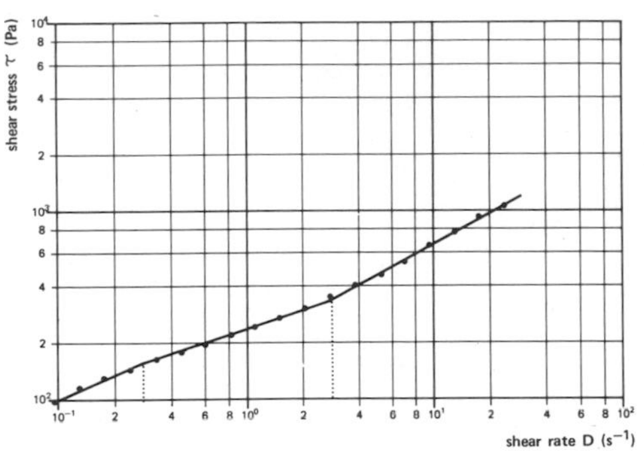

Fig. 5 Flow curve of a formulation with pseudoplastic flow behaviour
(Exp. 802, 2nd day after produce; Rheomat–30, MS–KP2)

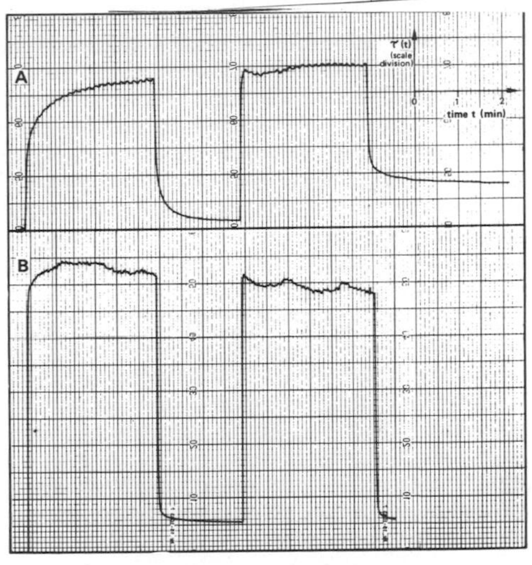

Fig. 6 Stressing experiment (D = const) – stress relaxation (D = 0; t = 2 min) – stressing experiment (D = const) – stress relaxation (D = 0)
(Exp. 802; Rheomat–30, MS–KP2)
A → D = 0.13 s^{-1}, τ(t) = 5.0 · scale division (Pa)
B → D = 1.51 s^{-1}, τ(t) = 12.5 · scale division (Pa)

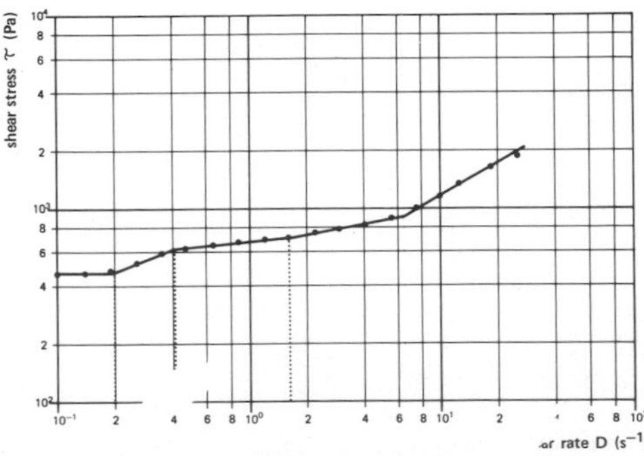

Fig. 7 Flow curve of a formulation with plastic tiow behaviour
(Exp. 534, 6 months after produce; Rheomat–30, MS–KP4)

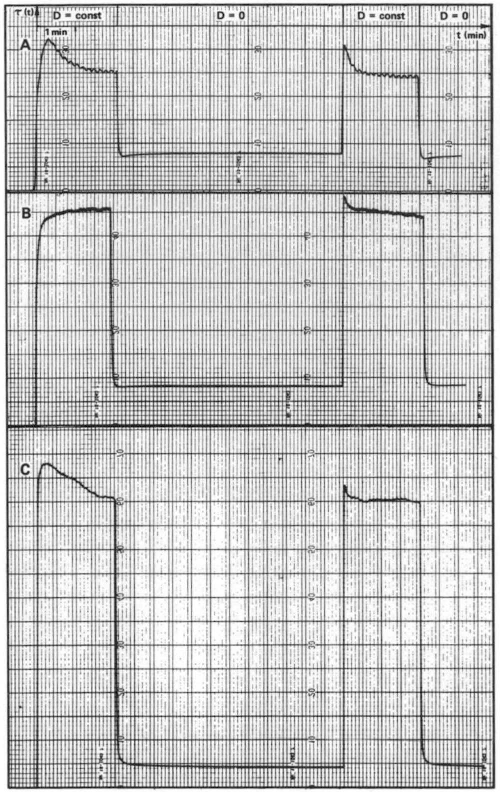

Fig. 8 Stressing experiment (D = const) – stress relaxation (D = 0; t = 6 min) – stressing experiment (D = const) – stress relaxation (D = 0)
(Exp. 534; Rheomat–30, MS–KP2)
$\tau(t)$ = 12.5 · scale division (Pa)
A → D = 0.13 s^{-1}, B → D = 0.443 s^{-1}, C → D = 2.80 s^{-1}

ing experiment which, again points to structural strengthening. In the stressing experiment with shear rates of the 2. or 3. straight line section, the stationary shear stress remains either the same as, or is lower (or sometimes higher) than, in the 1. stressing experiment (see Fig. 6B).

For formulations with plastic flow behaviour (see Fig. 7) it can be assumed that the stationary shear stress in the 2. stressing experiment remains about the same as in the 1. stressing experiment (see Fig. 8).

It is now established that in shear rate regions where rheopexy is to be expected (1. straight line section of formulations with structural viscous and pseudoplastic flow behaviour and in the 2. yield stress section (see Fig. 7) of formulations with plastic flow behaviour) no decrease of the stationary shear stress occurs after rest periods of different duration. Similarly, in shear rate regions where thixotropy is to be expected (2. and 3. straight line sections of formulations with structural viscous and pseudoplastic flow behaviour and with formulations having plastic flow behaviour at shear rates higher or lower than the 2. yield stress section) no increase (in most cases) of the stationary shear stress occurs after the rest period.

In this case it is not possible to speak of rheopexy and perhaps of thixotropy in accordance with the usual definition, when dealing with paste-like disperse systems. It seems more appropriate to refer to shear time-increasing or - decreasing properties in considering the time related behaviour of paste-like disperse systems (at D = const). Hence, the formulations can exhibit shear time-increasing as well as shear time-decreasing properties. The display of these properties depends on the shear rate at which the stressing experiment is conducted.

REFERENCES

1. Hadjistamov, D., Degen, K., Rheol. Acta 18:168 (1979).
2. Hadjistamov, D., Rheol. Acta, in press

RHEOLOGY OF AGEING OF COAL-OIL SUSPENSIONS

S. N. Bhattacharya and L. Barro

Department of Chemical Engineering
Royal Melbourne Institute of Technology, Melbourne
Australia

INTRODUCTION

There are abundant quantities of low grade coal around the world like lignite or Victorian brown coal which could be efficiently utilised for burning in boilers as coal-fuel oil mixture or in the production of liquid fuels. In either of these energy conversion processes one of the important operations is the handling of coal-oil slurries. A knowledge of the handling characteristics of these slurries is therefore required in slurry preparation as well as in mixing, bulk tanker transport, pumping through short and long distance pipe lines, heating, cooling, etc.

Most of the work described in literature on the rheology of coal-oil suspensions [1-4] is simply concerned with the determination of the equilibrium flow diagram for bituminous coal-mineral oil suspensions. While these works have been very useful, they have not taken into account the many abnormal rheological characteristics exhibited by coal-oil suspensions. Moreover there has been no work reported on brown coal-oil rheology except by these authors [5, 6]. Flow characteristics of brown coal-oil slurries is quite complex and many peculiar types of flow behaviour have been observed with these slurries. One of the characteristics is the ageing of the slurry which can significantly alter the rheological property with time. This behaviour which is dependent on time is however, quite different from the usual time dependent behaviour exhibited by many suspensions.

Many materials like gel paints or suspensions of bentonite, sodium or calcium montmorillonite in water exhibit thixotropic behaviour where the structure of the material breaks down quite readily under shear. The structure of these materials has been found to recover with time once the applied shear is removed. Alternatively for materials like gypsum suspension or starch paste, the applied shear helps to build a more rigid structure which again reverts back to its original form once the shear is removed. This phenomenon of build up and break down of structure for these materials is purely physical and is by and large reversible in nature. With certain coal-oil suspensions, however, the behaviour appears to be quite different. These materials may develop some structure quite irreversibly with time. Such build up of structure results in an increase in shear stress on storage for any applied shear rate, and in many instances the flow behaviour of the material may change from a Newtonian to a non-Newtonian type.

For a concentrated suspension of these materials the increase in apparent viscosity with time on storage may be very significant and on certain cases the suspension may turn almost to a solid state, exhibiting enormous amounts of yield stress.

It is the objective of this paper to describe the change in the flow characteristics of some brown coal-oil suspensions on storage. The experimental study of the variation of yield stress and apparent viscosity with storage time will be presented in this paper. The effect of solid concentration on the rheological characteristics and the transition from Newtonian to non-Newtonian flow behaviour of the suspensions on storage will also be demonstrated.

EXPERIMENTAL

Coal-oil suspensions may be of settling type if the particle size is large or the oil is not very viscous. Amongst the various rheometers which are available for characterising these suspensions Capillary rheometer is certainly highly suitable if appropriate correction for end effect and wall effect are incorporated.

A Capillary rheometer has been developed in this department for the rheological study of suspensions of different types at various temperatures and pressures. The main body of the rheometer consists of a pressure vessel fitted with a magnetic stirrer. Capillary tubes of various diameters and lengths may be connected at the horizontal outlet close to the bottom of the pressure vessel through which discharge occurs under pressure. The absolute pressure in the pressure vessel is measured by means of one of a number of Budenberg pressure gauges appropriate for the pressure range. The differential pressure across the capillary test section is measured by means of a differential pressure gauge. The whole apparatus is immersed in a constant temperature bath to maintain the entire contents of the pressure vessel and the capillary tube at the same constant temperature. Further details of this apparatus are presented elsewhere [5].

Brown coal supplied by the State Electricity Commission of Victoria, Australia, from Morwell opencut has been used in this work. The coal was dried by evaporative means until the moisture content was reduced to 10 percent (ω/ω). The ultimate analysis of coal on dry basis was carbon, 67.6 percent; hydrogen, 4.9 percent; oxygen, 23.9 percent; sulphur 0.6 percent and other inorganic materials, 3.0 percent. The material was pulverised and the coal particles used here were a close cut between 0.085 mm and 0.06 mm giving an average linear diameter of 0.0725 mm. The objective of taking a close cut was to eliminate particle size as a variable in this study. The oil used was a coal derived oil (KC 220) supplied by Koppers Australia Pty Ltd, Mayfield, N.S.W. The boiling point range of the oil was 234°-$319^{\circ}C$ with 80 percent of the oil boiling within the range of 234°-$265^{\circ}C$. The specific gravity was 1.058 at $20^{\circ}C$ and the viscosity was 6.4 mPaS at $20^{\circ}C$. This oil is considered as a suitable starting material for coal liquefaction. The suspension was prepared by slowly adding coal to oil which was stirred continuously.

RESULTS

Experimental study in Capillary rheometer is liable to error due to end effect and wall effect. Study carried out with tubes having different length to diameter ratios (L/D) indicated that for L/D ratios higher than about 200 the end effect was negligible. For all data reported in this work L/D ratio was above 300. Coal-oil suspension has been observed to exhibit significant wall effect. Wall effect was found to occur even for flow through the 5.5 mm tube having a tube diameter to coal particle diameter ratio (D/d) of 76. This is because coal particles in oil suspensions have been observed to agglomerate resulting in an effective particle size which was much larger than the average particle size of 0.0725 mm used to prepare the suspension. All experimental data were corrected for wall effects by the method outlined elsewhere [7].

Ageing Characteristics

One of the important rheological characteristics of brown coal-oil suspensions is that the flow behaviour changes on storage. Generally the apparent viscosity of a suspension has been found to increase with time. This is considered to be due to physico-chemical interactions between coal particles and oil. Flow characteristics of a 35 percent and a 38 percent coal-oil suspension on storage at $20^{\circ}C$ has been reported in this section.

Flow Behaviour of 35 Percent Suspension

A plot of shear stress (τ_{ω}) versus apparent shear rate (8V/D)

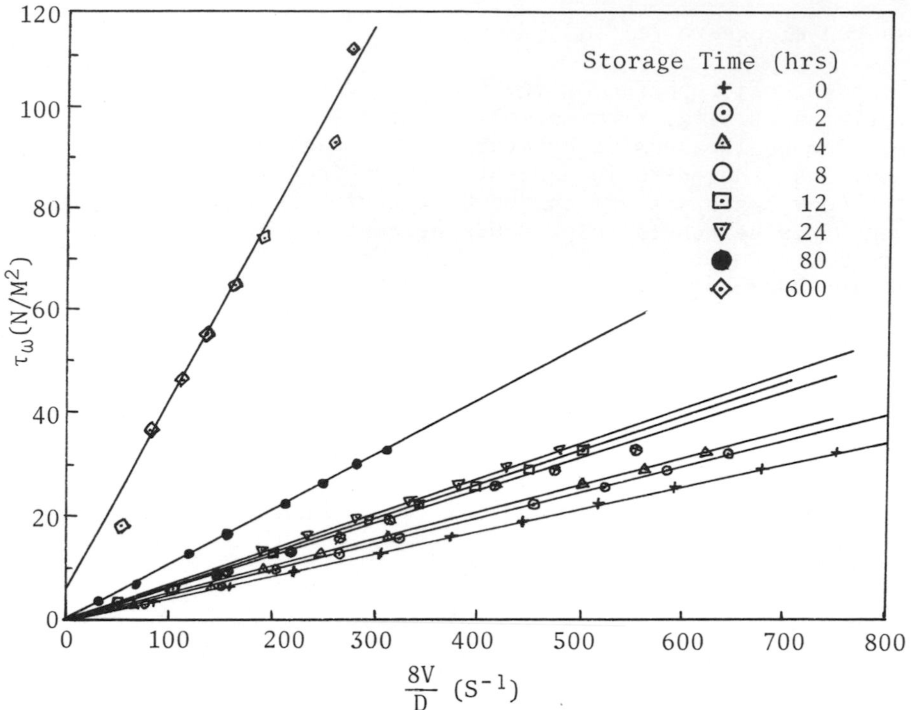

FIGURE 1: Flow Diagrams Showing Ageing of 35% Suspension

for a 35 percent suspension is shown in Figure 1. The rheograms were obtained after storing the suspension at 20°C for different storage times varying from 0 hours to 600 hours. For any value of 8V/D, τ_ω was found to steadily increase with storage time. The suspension was observed to behave as Newtonian liquid for all runs taken until 80 hours of storage. The data for 600 hours storage indicates that the flow behaviour was non-Newtonian when a yield stress developed. The conclusion here is that there are physico-chemical interactions between coal particles and the suspending medium on storage. Part of the oil is absorbed within the porous structure of the coal, providing an increased contact between coal particles. The result is the slow build up of a structure of the material which accounts for increase in viscosity with time and subsequent change from Newtonian to non-Newtonian behaviour.

The viscosity values $\{\tau_\omega/(8V/D)\}$ calculated from Figure 1 is plotted as a function of storage time and represented by curve A in Figure 2. The rate of increase in viscosity was quite high in the initial period of 24 hours, the gradient being low beyond that region. It is therefore believed that the build up of structure of the suspension in the initial 24 hour period is significent. The viscosity was found to increase even beyond 600 hours of storage, eventually becoming asymptotic at infinite time.

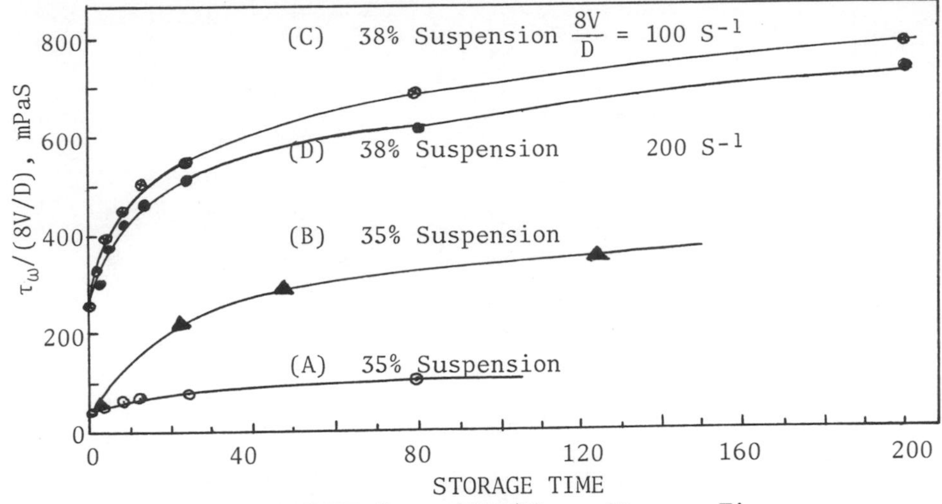

FIGURE 2: $\tau_\omega/(8V/D)$ vs Storage Time

In Figure 2 the viscosity-storage time relationship for another 35 percent suspension is shown by curve B. In this case no run was taken between 0 hours and 24 hours indicating that the material was undisturbed for the initial 24 hours resulting in considerable build up of structure and increase in viscosity. For rheological studies of materials like coal-oil suspensions, shear history is found to be a very important factor.

Viscosity - storage time relationship given by curve A could be expressed by means of an exponential relation of the form

$$\eta = a + b(1 - \bar{e}^{ct}) \qquad \text{.... (1)}$$

where η = viscosity, in mPaS; a, b, c = constants,
t = storage time, hr

Values of 'a', 'b' and 'c' have been found to be 45, 480 and 0.0025 respectively.

Curve B could not be expressed by a simple relation like equation (1). However, it could be expressed by a combination of two exponential relations. No mathematical expression is provided for this curve in this paper.

Flow Behaviour of a 38 Percent Suspension

Flow diagrams of a 38 percent suspension for various storage time are provided in Figure 3. This suspension was considerably thicker than the 35 percent suspension and it behaved as non-Newtonian liquid exhibiting yield stress almost from the time of its preparation. For the first 24 hours the increase in yield stress

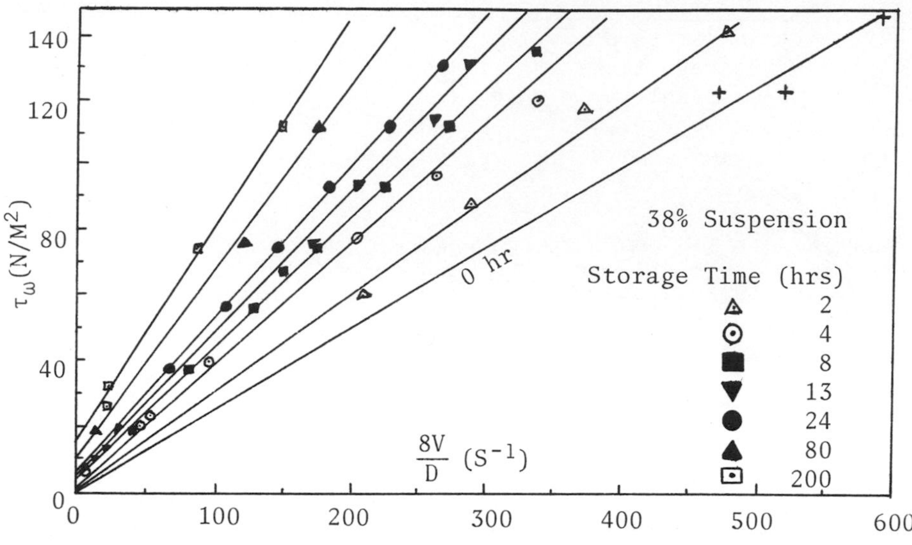

FIGURE 3: τ_ω vs 8V/D for 38% Suspension

with time was significant and the increase was observed to continue even at 200 hours of storage. The effective viscosity $\{\tau_\omega/(8V/D)\}$ calculated at 8V/D values of 100 and 200 sec^{-1} are shown as a function of storage time by curves C and D in Figure 2. The behaviour was similar to the curves A and B and an approximate relation of the form.

$$\tau_\omega/(8V/D) = 250 + b\ (1 + e^{-0.0053t}) \quad .... (2)$$

similar to equation (1) was obtained for curves C and D. Values of b for curves C and D were 950 and 890 respectively.

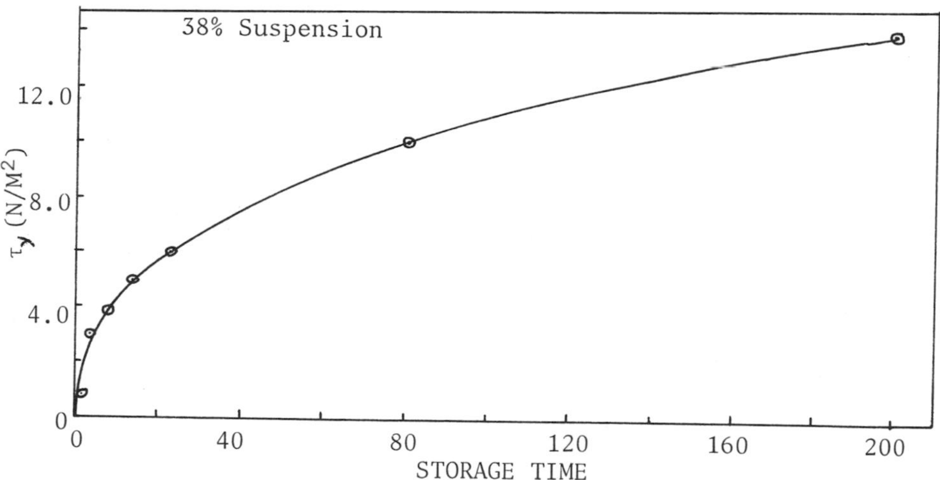

FIGURE 4: Yield Stress vs Storage Time

The yield stress estimated from Figure 3 was plotted as a function of storage time in Figure 4. The yield stress appeared to attain a constant value after long storage time. Yield stress, $\tau_y (N/M^2)$ could be expressed as a function of storage time, t (hr) by

$$\tau_y = 18\ (1 - \bar{e}^{0.0085t}) \qquad .... (3)$$

CONCLUSION

Brown coal-oil suspension has been found to alter its flow characteristics on storage. 35 percent coal-oil suspension was found to increase in viscosity with storage time and the flow characteristics changed from a Newtonian to a non-Newtonian behaviour on long storage.

A 38 percent suspension was observed to possess much higher viscosity than the 35 percent one and the flow behaviour of the former was non-Newtonian exhibiting yield stress from almost the time when the suspension was first prepared. The effective viscosity and yield stress were observed to increase with storage time until the curves became asymptotic at infinite storage time.

ACKNOWLEDGEMENT

The authors wish to thank Victorian Brown Coal Council for providing financial support in the form of a post-graduate scholarship to one of the authors.

REFERENCES

1. Berkowitz, G. Coal in pipelining, Oil in Canada, 25 (27407) (1961)

2. Moreland, C., Viscosity of Suspensions of Coal in Mineral Oil, Can. Jr. of Chem. Engg'., 24-28, (1963).

3. Okutani, T., Yokoyama, S. and Maekawa, Y., Viscosity of Coal Paste under Pressure. Govt.Ind.Div.Lab, Hokkaido, Japan, (1979).

4. Thomas, M. G. and Traeger, R. K., Am. Soc. Div. of Fuel Chemistry, 24, 3, (1979).

5. Barro, L. and Bhattacharya, S. N., Rheological Study of Brown Coal-oil Suspensions, First National Conf. on Rheo., Melbourne, May 30 - June 1, 83, (1979).

6. Bhattacharya, S. N. and Barro, L., Paper to be presented at the Aust. Inst. of Min. and Met. Conf., New Zealand, May (1980).

7. Skelland, A.H.P., Determination of flow properties in non-Newtonian flow and heat transfer, John Wiley & Sons, N.Y., (1967).

RHEOLOGICAL PROPERTIES OF ALLOPHANE COLLOID

Rokuro Yasutomi
Faculty of Agriculture
Ibaraki University
Ami-machi, Ibaraki

Manabu Senō
Inst.of Industrial Sci.
Tokyo University
Roppongi, Minato, Tokyo

INTRODUCTION

There are many volcanic soils containing allophane in the countries of volcanic soil like Italy, Newzealand, and Japan. The most particular rheological property of allophane is thixotropic, and it happens often in earth work that remolding of soil by machine makes difficult conditions to continue work. The shearing strength of allophane soil goes down to a half or more and soil erosion increases when it gets wet or dry excessively. These mechanical properties result from the surface forces between soil particles.

This paper treats the thixotropic behaviour of allophane colloid obtained from natural volcanic soil of Kanto Plain (Japan).

MATERIALS

The size of primary particles is reported to be less than 10 nm in diameter and their shape is spherical, sometimes thin thread.[1] The particles flocculate together to make honeycomb structure, in which a rather large amount of water is confined by the aid of surface adsorption. This is one of the reasons for the high water-holding capacity of allophane soil.

The chemical composition of allophane is known to be $nSiO_2 \cdot Al_2O_3 \cdot mH_2O$,[2] which is similar to that of Kaolinite. In many cases, it contains a small amount of ferric oxide (sesquioxide) and Fe is partially replaced by Al. Allophane has a very high adsorption capacity for phosphate and a high cation-exchange capacity, which is 10-60 meq/100g in the pH region from 3 to 8. Isoelectric point is ca 6.5

Preparation of Samples

A suspension of allophane soil was deflocculated by adding a small amount of hydrochloric acid and 1 ml of a 30% hydrogenperoxide solution per 100 g of allophane was added to decompose organic matters. Then, the suspension was set in an ultrasonic bath for 15-30 min. A solution of NaCl, KCl, LiCl or $CaCl_2$ was added to prepare corresponding cation-form samples. Besides, urea was added to Na-form soil to obtain a well-dispersed sample by breaking hydrogen bonds. Deferated samples were also prepared. The sample suspensions were set in a centrifuge of 2000 G for 1 hr to remove particles of diameter larger than 100 nm. The supernatant suspension after centrifugation was collected and dialysed in cellulose tubes against the controlled pH solution.

Characterization of Samples

(1) DTA test: A typical DTA curve shows a sharp endothermic peak at ca 130°C owing to evaporation of crystalline water and a weak peak due to gibsite at ca 300°C. This is consistent with that of allophane.

(2) X-ray analysis: X-ray diffraction analyses show no existence of crystalline materials even after tratment with KCl or $CaCl_2$.

(3) Specific surface area: The measured values are dependent of the methods of sample treatments and measurement; about 211.4 m^2/g (average) by BET method and ca 300 m^2/g by the H_2O vapour method.

(4) Cation-exchange capacity (CEC): The CEC values are 43.0 for NH^+-form, 32.5 for Na^+-form, 26.0 for K^+-form, and 21.5 meq/100 g for Ca^{2+}-form sample. The sedimentationvolume is 2.52 cm^3/g.

EXPERIMENTAL

The thixotropic behaviour is considered to come from rearangement of particles dispersed in the suspension. Then, in order to obtain some informations about interparticle interactions, the compression test was carried out and the distance and forces between particles are examined in relation to viscoelastic properties.

Compression Test

The measurements of forces between particles were made by a compression tester.[3] This is based on the principle that the total pressure is balanced by the interstitial and repulsive forces between particles; this reasoning is common to that for drained consolidation process of soil. When the shape of particles is regular, spherical or plate, the average distance between particle can be estimated by the formula; $H_0 = 2V/S \cdot m$, where S is specific surface area, m is sample weight, and V is volume of water drained by the pressure applied. The calculation was done by using the value of S by the vapor pressure method.

Rheological Measurements

A Weissenberg rheogoniometer is used for measurement of dynamic viscosity and rigidity.[4] The flow curve is drawn by a MacMichael viscometer. The apparatus is convertible for both the measurements; therefore, the dynamic and the static measurements could be carried out successively on the same sample in the same vessel. The temperature is controlled by circulation of thermostatted water. The period for dynamic measurements is 5 to 150 sec, and the amplitude is adjusted to 0.14 in most cases to get a linear responce. But the linearity may be violated at higher concentrations.

RESULTS

The thixotropic properties are affected by the ionic forms of particles, additives and temperature. The concentration here used is about 1 or 2 % in weight and the suspensions are thixotropic enough to take a gel state after setting for several hours. The behaviour is very sensitive to temperature.

Viscoelastic Behaviours

The limiting viscosity number is very large and the Einstein equation was not applied even at the concentration less than o.5 %. Bingham yield value θ increases exponentially with an increasing concentration ϕ, that is, $\theta = 1 + K\exp(\phi - \phi_0)$, where ϕ_0 is the concentration at which the yield value appears and K is a constant. The viscosity depends pH and temperature, and above all treatments with urea. The area enveloped by flow curve change scarecely according to the ionic form of samples, such as Na^+, K^+, Li^+ and Ca^{2+} exchanged allophanes. These suspensions are coagulative and show no hysteresis loops in a pH range 6-7. At lower pH's the area of loop and the yield value increase. Addition of urea affects remarkably the behaviour to strengthen thixotropic properties.

The viscosity increses with a rise in temperature. For the urea-treated samples, the behaviour is extremely sensitive to temperature. The yield value increases and the plastic viscosity decreases with an increasing temperature, as shown in Fig. 1. These trends are observed for K^+ and other ionic form samples except for Ca^{2+} form sample. The dynamic viscoelasticity behaves similarly with temperature. A typical result is shown in Fig. 2.

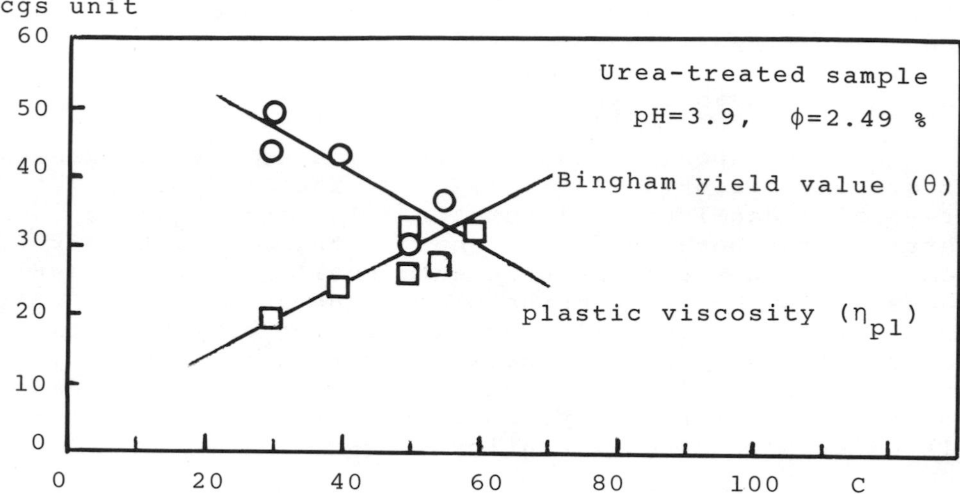

Fig. 1. Relationship between plastic viscosity (η_{pl}), Bingham yield value (γ) and temperature.

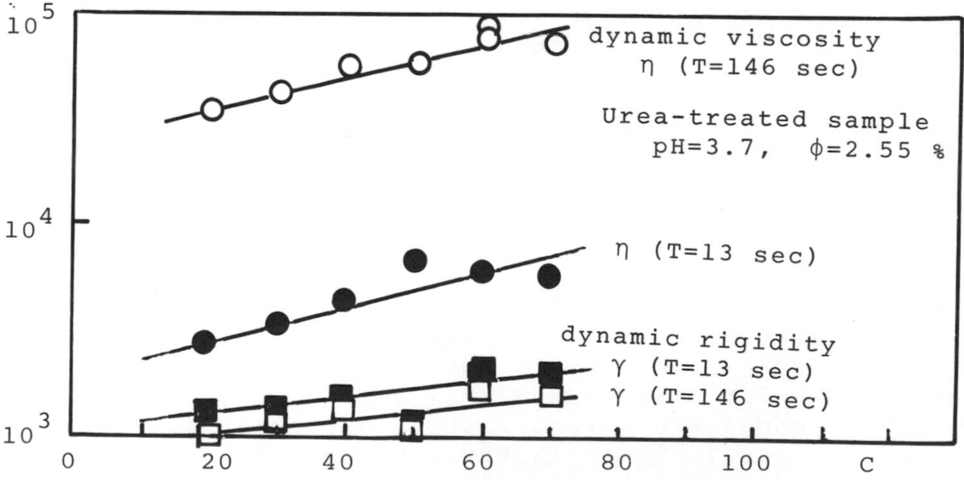

Fig. 2. Dynamic viscoelasity vs temperature.

Compression Test

The pressure of compression was increased from 1 to 100 atm. The distance between particles during course of compression was estimated from the water content at the top pressure as a standard. For the irregular arrangement of particles may give a serious effect to the result, the preliminary compression up to the top pressure is essential to obtain a reproducible result. The results show that the repulsive forces between suspended particle increases with an decreasing pH.

DISCUSSION

The allophane susensions exhibit very remarkable thixotropic behaviour sensitive to temperature even at very low concentrations less than 1 %. The suspension treated with urea shows particularly high sensitive behaviours. With an increasing temperature, the allophan particles take a well-dispersed state and the extent of structure formation proceeds. A trend of tightning the structure results a decreasing fluidity, as Fig. 2 shows. The dynamic rigidity and viscosity increase as the temperature rises. On the other hand, the result in Fig. 1 shows that the yield value increases and the plastic viscosity decreases with an increasing temperature. This suggests that the larger forces are necessary for breakdown of the well-dispersed structure, but the broken dispersion system exhibits the lower viscosity. These behaviours would be correlated with the pH change, as shown in Fig. 3.

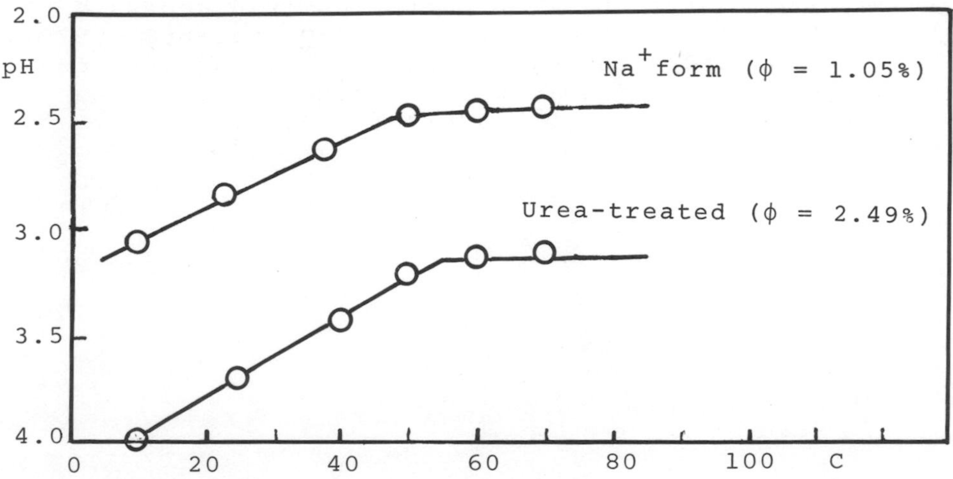

Fig. 3. pH change of allophanes by temperature.

It is considered that water is responsible for the structure formation of allophane suspensions. As the temperature rises, the dissociation of surface functional groups proceeds, as Fig. 3 shows. The desorption of adsorbed substances on the particle surfaces also proceed. The increase in surface charges aids the particle deflocculation and strengthen the electrostatic interaction with water. The primary hydration layer around the hydrophilic particles is formed firmly at these temperatures and the hydration shell being commom to both the particles serves a binding medium between the particles.

SUMMARY

Very finely divided particles of allophane are dispersed to form a remarkably thixotropic suspension. The thixotropic behaviours are partly described by the static and the dynamic rheological measurements. The extent of thixotropy is large for the suspension consisting of particles trated by urea and increases as the temperature rises. These behaviours are originated from the network structure formed by well-dispersed particles with highly developed hydration shell.

REFERENCES

1) Y. Kitagawa, "Unit particles of allophane", Amer. Mineral., 52, 690-708 (1967).
2) K. Wada, "Structural schem of soil allophane", Amer. Mineral., 52, May-June (1967).
3) L. Barclay, A. Harrington and R. H. Ottewill, "The measurement of forces between particles in disperse systems", Kolloid-Z. u. Z. Polymere, 250, 655-666 (1972).
4) R. Yasutomi and S. Sudo, "A method of measuring some physical properties of soil with a forced oscillation viscometer", Soil Sci., 105, 336-341 (1967).

STEADY SHEAR AND VISCOELASTIC PROPERTIES OF STERICALLY STABILISED NONAQUEOUS DISPERSIONS

T.H.Milkie, M.L.Hair, M.D.Croucher

Xerox Research Centre of Canada, Mississauga,
Ontario, Canada

(Abstract)

Steady shear and viscoelastic measurements have been carried out on a dispersion of PMMA particles suspended in n-hexadecane as a function of particle concentration and temperature.

The PMMA particles were prepared using a nonaqueous dispersion polymerisation in the presence of poly(dimethyl-siloxane-g-methyl methacrylate). The particles produced were submicron in size and sterically stabilised by the poly(dimethyl-siloxane) chains which were terminally attached to the particle surface and soluble in the dispersion medium.

The shear rate dependent viscosity of the dispersion, $\eta(\dot{\gamma})$, was found to be Newtonian at low particle concentrations, becoming progressively non-Newtonian as particle-particle interactions increased. It is thought that the steady shear rheological behavior is profoundly influenced by the interactional strength of the steric barrier, which varies with temperature, and also with the Brownian motion of the particles. It has also been observed that there is a discontinuity in the $\eta(\dot{\gamma})$ curve. This discontinuous increase in the viscosity at a critical shear rate is thought to be due to the onset of a disordered flow pattern within the dispersion.

Measurement of the viscoelastic properties of the dispersion, as a function of particle concentration and temperature, also indicates that the elastic response of the material is dominated by the interaction of the steric barrier around the particles. This has important implications for the thermodynamic theories of steric stabilisation which will also be discussed briefly.

RHEOLOGY OF SUSPENSIONS

S.G. Mason

Department of Chemistry, McGill University
Montreal, H3A 2A7, Canada

The Editors regret that poor health made it impossible for prof. Mason to prepare the typescript of this review lecture in time for inclusion in the Proceedings of the Congress.

A TWO-FLUID MODEL FOR HIGHLY CONCENTRATED SUSPENSION FLOW THROUGH NARROW TUBES AND SLITS: VELOCITY PROFILES, APPARENT FLUIDITY AND WALL LAYER THICKNESS

Daniel Quemada, Jacques Dufax et Pierre Mills

Laboratoire de Biorhéologie, Université Paris VII
Paris, France

INTRODUCTION

Viscometric flows of highly concentrated suspensions in narrow slits and pipes exhibit a two phase structure if the particle size cannot be considered as infinitesimally small in comparison with the transverse dimension of the vessel. In pipes this leads to a well-known [1,2] annular structure with a particle rich axial core surrounded by a particle depleted wall layer. The former results in a blunted velocity profile and in a reduced mean concentration in the tube, compared to the feed reservoir concentration [1]. The latter works as a lubricant layer which lowers the apparent viscosity, hence,which decreases extensively the viscous energy loss, even if this layer is very thin, as it was usually observed with highly concentrated suspensions.

The aim of this work is to use a non-newtonian viscosity equation, both shear rate and concentration dependent, previously proposed [3b,3c], in order (i) to predict blunted velocity profiles and apparent fluidity values, (ii) to get a wall layer analysis by comparison of these theoretical results with laser velocimeter and pressure flow rate measurements in slits.

RHEOLOGICAL MODEL

Minimization of the viscous energy dissipation within a two phase flow led [3a] to a relative viscosity η_r as a function of volume concentration of particles, ϕ. Using the (relative) fluidity $\psi = \eta_r^{-1} = \eta_F/\eta$, where η is the viscosity of the whole suspension

----

See in this issue a companion paper by J. DUFAUX et al.

and η_F the viscosity of the suspending fluid, the viscosity equation becomes:

$$\psi = (1 - \tfrac{1}{2}k\phi)^2 \tag{1}$$

where k is an effective intrinsic viscosity, closely related to the actual structure of the system. Hence k shows a shear dependence $k = k(\dot{\gamma})$ which takes the following form [3b]:

$$k = k_\infty + \frac{k_0 - k_\infty}{1 + g^p} \tag{2}$$

where $g = \tau\dot{\gamma}$ is a reduced shear rate, which involves a mean relaxation time of the structure, τ (or, alternatively, a critical shear rate $\dot{\gamma}_c = \tau^{-1}$),

k_0 and k_∞ are the limiting values of k as $g \to 0$ and $g \to \infty$, respectively,

p is an empirical constant, the value of which was found close to $p = \frac{1}{2}$ for many systems.

It may be noticed that in the present model, the value k_0 is related ($k_0 = 2/\phi_m$) to the packing concentration ϕ_m, which gives a zero shear fluidity $\psi_0 = \psi(g = 0)$ equal to zero. The Casson equation, which is recovered [3c] at "high" shear rates $g \gg 1$, is also recovered, as $\phi \to \phi_m$. Putting $\psi_\infty = \psi(g \to \infty)$ one finds in this limit:

$$\eta_\infty = \frac{\eta_F}{\psi_\infty} \quad \text{and} \quad \sigma_Y = \frac{\eta_\infty}{\tau} \tag{3}$$

as the Casson viscosity and the Casson yield stress, respectively.

VELOCITY PROFILES

Dimensionless variables for laminar flow through a slit $-h \leq y \leq h$ (or a pipe $0 \leq r \leq R$), with time τ and length $\lambda = h$ (or R), are used:

$$\rho = \frac{y}{h} \left(\text{or } \frac{r}{R}\right); \quad u = \frac{\tau}{\lambda} v \; ; \quad g = \tau\dot{\gamma} \; ; \quad \sigma_p = \frac{\tau\sigma_W}{\eta_F}$$

where v, $\dot{\gamma}$ and σ_W are the velocity, the shear rate and the wall shear stress, respectively. The calculation of velocities involves the reduced shear rate g as a function of ρ since, as it results from momentum equation, the shear stress distribution is:

$$\frac{\sigma}{\sigma_W} = \frac{y}{h} \left(\text{or } \frac{r}{R}\right) \text{ with } \sigma_W = Ph \; (\text{or } \tfrac{1}{2} PR) \tag{4}$$

P being the pressure gradient. Eq. (1) and (2) can be expressed as a shear rate-shear stress relationship in the following (dimensionless) form:

$$g = \tau\dot{\gamma} = \frac{1}{4}\left[A\rho^{\frac{1}{2}} - 1 + \left((A\rho^{\frac{1}{2}} - 1)^2 + 4B\rho^{\frac{1}{2}}\right)^{\frac{1}{2}}\right]^2 \tag{5}$$

where $A=(\sigma_p\psi_\infty)^{\frac{1}{2}}$ and $B=(\sigma_p\psi_0)^{\frac{1}{2}}$.

The two phase structure is defined by $0 \leqslant \rho \leqslant \beta$, for the axial core, and $\beta \leqslant \rho \leqslant 1$ for the peripheral layer. Assuming newtonian behaviour for the latter (w subscript) and shear thinning behaviour, defined by (1) and (2), for the former (s subscript), leads to the following dimensionless velocities:

$$u_w(\rho) = \tfrac{1}{2}\sigma_p\, \psi_w(1-\rho^2) \qquad (\beta \leqslant \rho \leqslant 1) \qquad (6)$$

$$u_s(\rho) - u_s(\beta) = \int_\rho^\beta g_s(\rho')\,d\rho' \qquad (0 \leqslant \rho \leqslant \beta) \qquad (7)$$

accounting for boundary conditions $u_w(1) = 0$ and $u_s(\beta) = u_w(\beta) = \frac{1}{2}\sigma_p\psi_w(1 - \beta^2)$ because of the no-slip condition at the wall, $\rho = 1$, and the velocity continuity at $\rho = \beta$. Eq (6) and (7) are not sufficient for velocity determination since the value of β is required.

PRESSURE-FLOW RATE RELATIONSHIP. APPARENT FLUIDITY

Dimensionless flow rates in the wall layer, Q_w, and in the axial core, Q_s can be written in terms of reduced apparent shear rates $g_{ai} = \tau Q_i/\lambda S$, S being the cross sectional area of the vessel. This gives the relations:

$$g_{aw} = \tfrac{1}{2}\sigma_p\psi_w F(\beta); \quad g_{as} = \tfrac{1}{2}\sigma_p\psi_w G(\beta) + \int_0^\beta g_s(\rho)\rho^m d\rho \qquad (8)$$

both valid for slits (m = 1) and pipes (m = 2), with

$$F(\beta) = \frac{m\ \beta^{m+2} - (m+2)\,\beta^m + 2}{m+2} \quad \text{and} \quad G(\beta) = \beta^m(1-\beta^2)$$

The apparent relative fluidity ψ_a is related to the dimensionless total flow rate as:

$$g_a = \tau(Q_w + Q_s) / \lambda S = \sigma_p\psi_a/(m+2)$$

which leads to the fluidity equation,

$$\psi_a = (1-\beta^{m+2})\ \psi_w + \beta^{m+2}\ \psi_s \qquad (9)$$

where the core fluidity ψ_s is defined as:

$$\psi_s = \frac{m + 2}{\sigma_p \beta^{m+2}} \int_0^\beta g_s(\rho)\rho^m\, d\rho \qquad (10)$$

WALL LAYER THICKNESS

As axial core and wall layer concentrations, ϕ_s and ϕ_w respectively,are not known, neither velocity profile nor fluidity equation can be fitted on data.

In order to get the value of β , i.e. the wall layer thickness $\delta = (1-\beta)\lambda$, an additional equation is found writting the mass balance equation for the suspended phase:

$$(Q_s + Q_w)\ \phi_a = Q_s\ \phi_s + Q_w\ \phi_w$$

where ϕ_a is the (constant) feed reservoir concentration. At high ϕ_a values, one may assume $\phi_w = 0$, as resulting from geometrical exclusion of particles near the wall (known as Vand effect). From (8), ϕ_s is then defined by:

$$\phi_s = \phi_a \left(1 - \frac{F(\beta)}{\psi_a}\right)^{-1} \qquad (11)$$

Given the values of feed concentration and those of rheological variables of the whole fluid, (which can be independently, obtained from Couette viscometry), calculation of the actual β value needs one more equation. For that, many works [4/5] used the ratio of the mean channel concentration ϕ_t to the feed concentration ϕ_a, $K = \phi_t/\phi_a = \beta^m\phi_s/\phi_a$. In pipes, K was found dependent on ϕ_a and tube radius R and very slightly on σ_w.[5] Nevertheless, in moderately narrow slits or tubes, with λ about 20 to 30 times the particle "radius," K is very close to unity and its evaluation requires a verry high (and hardly reached) accuracy. In the present work it was thought better to get the needed relation from data of apparent fluidity versus wall shear stress. This was performed from pressure flow rate measurements [6] using blood flows throught a narrow slit with different thickness, e = 2h. Fig 1a displays variation of ψ_a plotted as a function of σ_w for blood flow in a slit (e = 350 µm) with $\phi_a = .57$. Adjusting the value of β until agreement between experimental and theoretical values of ψ_a is obtained, gives the wall (plasma) layer thickness δ. In some extent, this procedure resembles the ones that THOMAS[4] and CHARM and KURLAND [7] proposed.

RESULTS AND DISCUSSION

Fig 1b gives the corresponding variations of calculated δ vs. σ_w . It shows that, as σ_w increases from zero, δ increases and reaches a maximum (about 2 µm) for σ_w near $\sigma_{w0} = 6$ dynes/cm^2, then decreases and seems to tend towards zero at very high σ_w (limiting value ψ_∞ was used as $\sigma_w \to \infty$). It was observed that axial core concentration ϕ_s always takes a value very close to that which corresponds to infinite value of zero shear viscosity. Neverthless, ϕ_s reaches a slight maximum near σ_{0w} . Very similar results were obtained at different feed concentration and slit thickness (e=350µm, $\phi_a = 0.35, 0.40, 0.66$; e = 200 µm, $\phi_a = 0.35, 0.50$).

Although the present work concerns suspension flow both through slits and pipes using slit data, strictly limits our conclusions on wall layer properties to narrow slits. Hovever, in moderately narrow pipes, with β very close to unity, it can be expected that wall curvature would not have any dramatic effect and that no essential differences could be found between pipes and slits. Therefore, our conclusions can be compared with various available pipe data. Indeed, similar variations of δ were observed yet, especially its decrease as ϕ_a increases and as e decreases at least at high σ_w.

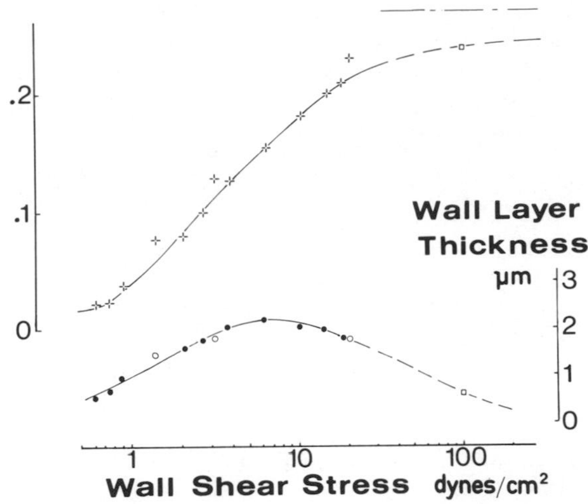

Figure 1(a) Apparent fluidity ψ_a and (b) Wall layer thickness δ vs.wall shear stress σ_w.Normal Human Blood Flow(ϕ_a=.57) through a slit (350µm × 1.1cm×10cm).T=23°C. Rheological variables of the same blood sample (from Couette Viscometry):k_0=3.24, k_∞=1.68, τ =.135sec.(Points O: from fitted curve on ψ_a data; Points □: from theory).

What seems to confirm the present model is the recovering of the maximum on the curve δ vs.σ_w, recently observed [8] with normal blood (ϕ_a = .25) in 200 µm diameter tube, by microphotography under dark fied illumination. The maximum in δ was found for values of σ_w between 1.5 and 5.0 dynes/cm^2, which satisfactorily agrees with our results. Moreover, the order of magnitude of δ , for instance, about 1.5 - 2 μm at ϕ_a = .57, in also in a good agreement with values generally accepted [4,7] ,accounting for Red Blood Cell deformability in wall exclusion effects [9].

Furthermore, one may stress that using narrow slits facilitated low shear measurements, since it gives, at a same pressure gradient a flow rate higher than the one observed in a pipe of equal size (2R=e). The lower scattering of the present data, in comparison with similar data on pipes (see[7] for instance) is believed to originate from this difference.

Finally , after value of β has been obtained from ψ_a data for a given value of σ_w, the corresponding theoretical velocity profile can be calculated from (6) and (7) and then it can be compared to the experimental one measured using a Laser-Doppler Velocimeter. Results of this comparison are given in the companion paper by J. Dufaux et al.

References

1 R.FAHRAEUS, The suspension Stability of Blood, Physiol.Rev. 9: 241-274 (1929).

2 A.D.MAUDE and R.L.WHITMORE, Theory of the Blood Flow in Narrow Tubes, J.Appl.Physiol. 12:105-113 (1958).

3 D.QUEMADA, Rheology of Concentrated Disperse Systems, Rheol. Acta (a) 16:82-94(1977),(b)17:632-642 (1978), (c) 17:643-653 (1978).

4 H.W.THOMAS, The Wall Effect in Capillary Instruments, Biorheology 1:41-56 (1962).

5 J.H.BARBEE and G.R.COKELET, The Fahraeus Effect, Microvasc. Res. 3:6-16 (1971).

6 J.DUFAUX, B.GRINBAUM, D.QUEMADA and P.MILLS, Indirect Determination of Plasma Layer Thickness in Plane Capillaries. Proceed. of the European Symposium "Hemorheology and Diseases",17-19 Nov.1979, Nancy (France),Douin,Paris (in press).

7 S.E.CHARM, G.S.KURLAND and S.L. BROWN, The Influence of Radial Distribution and Marginal Plasma Layer on the Flow of Red Cell Suspensions, Biorheology 5:15-43 (1968).

8 T.DEVENDRAN and H.SCHMID-SCHONBEIN, Axial Concentration in Narrow Tube Flow for Various RBC Suspensions as Function of Wall shear stress. Pflügers Arch.355:R20(1975).

9 R.L.WHITMORE, Rheology of the Circulation, Pergamon Press, Oxford (1968).

SUSPENSIONS FLOW DESCRIBED BY MEANS OF A MICROPOLAR FLUID THEORY AND APPARENT VISCOSITY OF AGGREGABLE PARTICLES SUSPENSION IN A COUETTE FLOW

P. Mills[+], J.M. Rubi[o], D. Quemada[+]

[+] Laboratoire de Biorhéologie, Université PARIS VII, Paris, France [o] Departemento de Termologia, Universidad de Barcelona, Barcelona, España

INTRODUCTION

In this paper suspension flow described by means of a micropolar fluid theory is studied. We will assume that the shear stress is weak enough to neglect the influence of a particle free marginal zone. The apparent viscosity of aggregable suspensions is calculated in a Couette flow. The result is compared to human blood viscometric data.

SIMPLE SHEAR FLOW BETWEEN TWO PLATES

The Field Equations

Let us consider a suspension of non deformable particles flowing between two parallel plates distant from h. The first plate is fixed and the second one moves with a constant velocity V_0. We put the flow lines along x and the y axis is normal to the plates (y=0 corresponds to the fixed plate). In the following equations c will be the volume concentration and τ'_W the wall shear stress. In two dimensions the micropolar equations governing the flow are [1]:

- balance of linear momentum:

$$\frac{1}{2}\frac{dv_x}{dy} = -\Omega_z = \frac{1}{2(\eta_1'+\eta_2')}\left[\tau_W' - 2\eta_2'\,\omega_z\right] \qquad (i,1)$$

- balance of first moment stresses:

$$\beta_1'\frac{d^2\omega_z}{dy^2} = 4\eta_2'\,(\omega_z - \Omega_z) \qquad (i,2)$$

- no marginal layer condition:

$$\frac{dc}{dy} = 0 \qquad (i,3)$$

The set of equations (i) governs the velocity field, the microrotation ω_z being considered as an independant variable. In equations (i), η_1' shear viscosity, η_2' is rotation viscosity, β_1' a concentration dependant parameter (when $c \to 0$, $\eta_2' \to 0$, $\beta_1' \to 0$). Physical significance of β_1' parameter will be given in section II.

For low concentration, Brenner [2] calculated η_1' et η_2' for spheres:

$$\eta_1' = \eta_0 \frac{1 + 1,5c}{1 - c} \quad , \quad \eta_2' = \eta_0 \frac{1,5c}{1 - c}$$

where η_0 is the suspending medium viscosity.

Introducing the dimensionless quantities:

$$\eta_1 = \frac{\eta_1'}{\eta_0} \quad , \quad \eta_2 = \frac{\eta_2'}{\eta_0} \quad , \quad \beta_1 = \frac{\beta_1'}{4\eta_0 h^2} \quad , \quad \alpha = \frac{\eta_2}{\eta_1}$$

$$Y = \frac{y}{h} \; , \; \Omega = \frac{h}{V_0} \Omega_z \; , \; \omega = \frac{h}{V_0} \omega_z \; , \; V = \frac{V_x}{V_0} \; , \; \Gamma = \frac{h}{V_0}$$

$$\tau_w = \frac{\tau'_w}{\eta_0 \Gamma} \quad , \quad K^2 = \frac{\eta_1}{\beta_1} \frac{\alpha}{1 + \alpha}$$

Equations (i) become:

$$\frac{dV}{dY} = \frac{1}{\eta_1 + \eta_2} [\tau_w - 2\eta_2 \omega] \qquad (ii,1')$$

$$\frac{d^2\omega}{dY^2} - K^2 \omega = \frac{K^2 \tau_w}{2\eta_1} \qquad (ii,2')$$

$$\frac{dc}{dY} = 0 \qquad (ii,3')$$

Particle Microrotation

The general solution of (2') is:

$$\omega = c_1 e^{KY} + c_2 e^{-KY} - \frac{\tau_w}{2\eta_1}$$

The microrotation profile depends on the boundary conditions imposed on ω at the walls. We set this condition in a similar way to the Condiff and Dalher's one [3]:

$$\omega(1) = s\,\Omega_0(1) \; , \; \Omega_0(1) = \frac{1}{2\eta_1}$$

We must point out that s is generally related to the concentration. When c is decreasing to wards zero, it is logical to assume s to tend to the limit calculated for an individual particle near a fixed solid boundary [4]:

$$\frac{\omega(1)}{\Omega(1)} = 1 - A(\frac{d}{a})^3$$

where a is the particle radius and d the distance of its center from the wall:

A = 5/16 for spheres

A = 0 for elongated particles

i.d. $s < 1$ and $s = 0$ respectively. Since

$$\omega(0) = 0$$

if $s = 1$, $\omega(Y) = \Omega(Y) \quad \forall \quad Y$

Kline [5] pointed out that, for Poiseuille flow, if particle spin magnitude $|\omega|$ exceeds local fluid vorticity $|\Omega|$ at one point of the flow field then $|\omega| \geqslant |\Omega|$ throughout the flow. So it is difficult to assume $s > 1$, for constant transport coefficients. The problem remains open if a concentration discontinuity exists near the walls (for example in two fluids models) as it would be at higher shear rate than those considered in the present study.

Apparent Viscosity of Aggregated Particles Suspension

If particles built up elongated aggregates we can assume $\omega = 0$ for $Y = 0$ and $Y = 1$.

The solution of (1',ii) is:

$$V(Y) = \frac{1}{\eta_1 + \eta_2} [\tau_W Y - 2\eta_2 \int_o^Y \omega(t)\,dt]$$

$$\int_o^Y \omega(t)\,dt = \frac{\tau_W}{2\eta_1} [\Psi(K) + \Psi(-K) - Y]$$

where $\Psi(K) = \frac{1}{2K\,shK} (1 - e^{-K})(e^{KY} - 1)$

Let V(1) be the velocity of the moving plane for Y = 1:

$$V(1) = 1 = \frac{\tau_W}{\eta_1 + \eta_2} [1 - \frac{\eta_2}{\eta_1} (\frac{2(ch\ K - 1)}{K\ sh\ K} - 1)]$$

The relationship between the velocity of the moving plane and the wall stress is thus obtained; and thus the relative apparent viscosity η_a :

$$\eta_a = \eta_1 (1 + \alpha) \quad [1 - \alpha (\frac{2(ch\ K - 1)}{K\ sh\ K} - 1)]^{-1} \tag{5}$$

PHYSICAL SIGNIFICANCE OF THE K PARAMETER

It can be shown by arguments involving time dependant flows [5] that the parameter β_1' can be expressed as a fonction of the gyration radius k of the microstructure:

$$\beta_1' = \eta_1' \ k^2 \qquad , \ K^2 = \frac{4\alpha}{1+\alpha} \ \frac{h^2}{k^2}$$

we shall distinguish between suspensions of particles without attraction and suspensions of aggregable particles.

a) For the first ones we can define a shear rate independant length related to the dimension of individual particles at low density and related of the dimension of large clusters for density above a critical value c_0 (by analogy with percolation arguments [6]). Cluster formations can increase dramatically the length k. Then we can predict an increase of the apparent viscosity in an apparatus where the cluster size is of the same order of magnitude as h (a similar effect is described in [6]).

b) For aggregable particles, aggregate formation is both concentration and shear rate dependant. In a Couette viscometer human red blood cells (RBC) built a rouleaux network which is able to bridge the narrow instrument gap. We have experimentally established [7] a relation between the number of RBC, m, in a rouleau and the apparent shear rate Γ in a Couette viscometer:

$$m = m_0 \left(\frac{\Gamma_C}{\Gamma} \right)^{1/2}$$

We interpret Γ_C as the shear rate required for breaking the three dimensional network built by aggregated red blood cells. We propose the relation for the gyration radius k:

$$k = h \left(\frac{\Gamma_C}{\Gamma}\right)^{1/2}$$

and $K^2 = G(c)\dot{\Gamma}$ where $G(c) = \dfrac{4\ \alpha}{(1+\alpha)\Gamma_C}$

We have compared the theoretical expression of viscosity to experimental results, for example [8], for normal blood H = 45% and with our experimental results H = 37% (Figure 1). The numerical value of η_1 was given in [8] for rigidified blood cells.

Our theoretical model agrees in a very satisfactory way with the experimental results for shear rate such as $0 < \dot{\Gamma} < 1\ s^{-1}$; beyond, the micropolar model cannot be used since the anisotropic deformation of the cells becomes an important parameter in the expression of apparent viscosity.

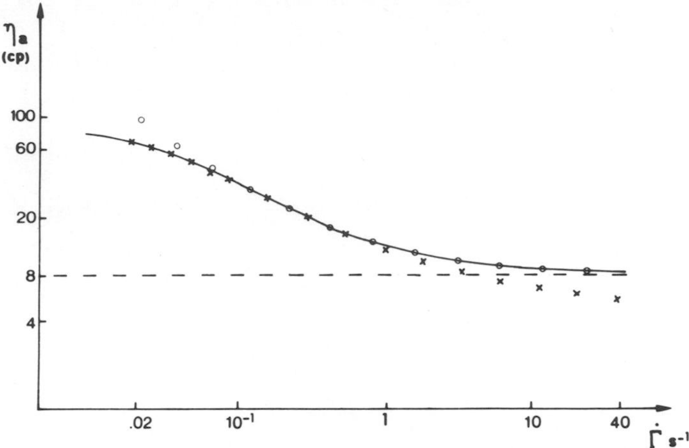

Fig.1: Apparent viscosity in Couette viscometer for normal human blood H = 37% : theoretical curve (G(c)=25s); x experimental points; o Casson's viscosity

CONCLUSION

We may conclude that, the theoretical model developped here can not only account for numerous experimental results concerning blood viscometry in a satisfactory way but it can also lead to further interesting developments since it enable us to relate a macroscopic quantity, the apparent viscosity in a Couette viscometer, to parameters attached to the microscopic structure of the suspension. In capillaries, one can assume such a rheological equation holds, relating transport coefficient as functions of local concentration and shear rate. Then changes in of velocity and rotation fields are observed.

References

1 A.C. ERINGEN, Theory of micropolar fluids, J.Math.Mech., 16, p.1 (1966).

2 H.BRENNER, Ann.Rev.Fluid Mech., 2: 137-76 (1970).

3 CONDIFF and J.S. DALHER, Phys.Fluids, 7, 842 (1964).

4 H.L.GOLDSMITH and S.G.MASON: The microrheology of dispersions. F.R.Eirich 4 p.152-53. New-York & London: Academic 522 pp(1967).

5 K.A.KLINE, Trans.Soc.Rheol. 19:1, 139-145 (1975).

6 P.G.de GENNES, Journ. de Phys. 40:783-787 (1979).

7 P.MILLS et al., Jour. Maladies Cardiovasculaires, Paris, 4 , pp 91-94 (1979).

8 S.CHIEN and al., Science, 157, 827 (1967).

ON BOUNDARY CONDITIONS FOR MICROPOLAR FLUIDS

J.M. Rubi°, P. Mills+, D. Quemada+, J. Casas-Vázquez°

°Depto. Termología - Universidad Autónoma de Barcelona, Bellaterra, Barcelona, Spain
+L.B.H.P.-Université Paris VII - 75221 Paris, France

1. INTRODUCTION

The use of additional degrees of freedom to explain the behaviour of micropolar systems, namely, spin, microinertia, antisymmetric pressure tensor and spin flux, makes the problem of choosing boundary conditions more acute. For micropolar fluids in contact with a wall, diverse intuitive boundary conditions have been proposed by several authors [1-3]. Some of these boundary conditions are incompatible and others cannot be applied to any micropolar fluid. In our opinion the difficulty lies in the fact that intuitive arguments are not sufficient to understand the behaviour of micropolar fluids near a boundary and therefore it is necessary to make use of another type of approach.

Using a method developed by Bedeaux-Albano-Mazur (BAM)[4], we deal in this paper with the boundary condition problem of a system consisting in two immiscible micropolar fluids in contact. From this method, boundary conditions for the dissipative fluxes are obtained and boundary conditions for the spin are outlined. Since micropolar fluids seem able to describe flow suspensions, the result is extended to such systems.

2. BRIEF RESUME OF BAM METHOD

We can summarize the fundamental steps of BAM method in the following way:

1) The total system is described by means of a function $f(\underline{r},t)$ defined as

$$f(\underline{r},t) > 0 \qquad \text{fluid I}$$
$$f(\underline{r},t) = 0 \qquad \text{interface} \tag{1}$$
$$f(\underline{r},t) < 0 \qquad \text{fluid II}$$

2) Heaviside functions in both fluids $\theta^{\pm}$ and a delta function at the interface δ^{s} are introduced. All the variables appearing in the set of balance equations are split as

$$\Psi = \Psi^{+}\theta^{+} + \Psi^{s}\delta^{s} + \Psi^{-}\theta^{-} \tag{2}$$

where Ψ denotes a generic variable and the superscripts +, - and s stand respectively for the variables that become to I, II and to the interface.

3) Inserting (2) in the balance equations and making use of some properties of $\theta^{\pm}$ and δ^{s} one recovers the balance equations for both fluids and for the interface.

4) In the framework of the local equilibrium hypothesis one gets the entropy balance equation on the interface, which compared with the entropy balance equation on the interface obtained from 3) give rise to the entropy flux and to the entropy production. The linear constitutive equations obtained from the entropy production are the phenomenological equations at the interface and the boundary conditions for both fluids.

3. BOUNDARY CONDITIONS FOR MICROPOLAR FLUIDS

Micropolar fluids are governed by the balance equations of mass, microinertia momentum, moment of momentum and internal energy which can be written respectively as[5]

$$\partial\rho/\partial t = -\nabla\cdot(\rho\underline{v}) \tag{3}$$
$$\partial\rho j/\partial t = -\nabla\cdot(\rho j\underline{v}) \tag{4}$$
$$\partial\rho\underline{v}/\partial t = -\nabla\cdot(\underline{\underline{P}} + \rho\underline{v}\underline{v}) \tag{5}$$
$$\partial\rho j\underline{\omega}/\partial t = -\nabla\cdot(\underline{\underline{Q}} + \rho j\underline{v}\underline{\omega}) - 2\underline{P}^{a} \tag{6}$$
$$\partial u_{v}/\partial t = -\nabla\cdot(\underline{q} + u_{v}\underline{v}) - \tilde{\underline{\underline{P}}}:\nabla\underline{v} + 2\underline{P}^{a}\cdot\underline{\omega} - \tilde{\underline{\underline{Q}}}:\nabla\underline{\omega} \tag{7}$$

Here ρ is the mass density, j the microinertia, $\underline{v}$ the velocity, $\underline{\omega}$ the spin, $\underline{\underline{P}}$ the pressure tensor, $\underline{\underline{Q}}$ the spin flux, $\underline{q}$ the heat flux, u_v the internal energy per unit volume and $\underline{P}^{a}$ the axial vector related to the antisymmetric part of the pressure tensor. A dot denotes inner product and the tilde stands for the transposition. Following the procedure indicated in 3) and considering that there is no mass on the interface, it is possible to get[6] the balance equations for the momentum, internal angular momentum and internal energy on the interface

$$\nabla\cdot\underline{\underline{P}}^s + \underline{n}\cdot(\underline{\underline{P}}^+ - \underline{\underline{P}}^-) = 0 \quad (8)$$

$$\nabla\cdot\underline{\underline{Q}}^s + \underline{n}\,(\underline{\underline{Q}}^+ - \underline{\underline{Q}}^-) = -\,2\underline{P}^{as} \quad (9)$$

$$\partial u_v^s/\partial t + \nabla\cdot(\underline{q}^s + u_v^s\underline{v}^s) = -\,\tilde{\underline{\underline{P}}}^s : \nabla\underline{v}^s + 2\underline{P}^{as}\cdot\underline{\omega}^s - \tilde{\underline{\underline{Q}}}^s : \nabla\underline{\omega}^s -$$
$$-\,\underline{n}\cdot(\underline{q}^+ - \underline{q}^-) - \underline{n}\cdot\underline{\underline{P}}^+\cdot(\underline{v}^+ - \underline{v}^s) - \underline{n}\cdot\underline{\underline{P}}^-\cdot(\underline{v}^- - \underline{v}^s) - \underline{n}\cdot\underline{\underline{Q}}^+\cdot(\underline{\omega}^+ - \underline{\omega}^s) -$$
$$-\,\underline{n}\cdot\underline{\underline{Q}}^-\cdot(\underline{\omega}^- - \underline{\omega}^s) \quad (10)$$

and the conditions

$$\underline{n}\cdot\underline{\underline{P}}^s = \underline{n}\cdot\underline{\underline{Q}}^s = \underline{n}\cdot\underline{q}^s = 0 \quad (11)$$

In (8)-(11) $\underline{n}$ is the unit vector pointing from II to I. Assuming the local equilibrium hypothesis to be valid, we postulate a Gibbs equation on the interface

$$d^s s_v^s/dt = (1/T^s)\,d^s u_v^s/dt \quad (12)$$

where s_v^s is the surface entropy per unit volume and T^s the surface temperature. Moreover, the material derivative at the interface is defined as $d^s/dt = \partial/\partial t + \underline{v}^s\cdot\nabla$. Making use of the substantial balance of internal energy obtained from (10), equation (12) leads to

$$\partial s_v^s/\partial t + \nabla\cdot(\underline{J}_s^s + s_v^s\underline{v}) = \sigma^s \quad (13)$$

$\underline{J}_s^s$ and σ^s being the entropy flux and the entropy production on the interface respectively. Now if we compare (13) with the entropy balance equation obtained according to 3) we get the form of the entropy flux and entropy production at the interface given respectively by

$$\underline{J}_s^s = \underline{q}^s/T^s \quad (14)$$

$$\sigma^s = \underline{q}\cdot\nabla(1/T^s) - (1/T^s)\{\tilde{\underline{\underline{P}}}^{vs}:\nabla\underline{v}^s + \tilde{\underline{\underline{Q}}}^s : \nabla\underline{\omega}^s - 2\underline{P}^{as}\cdot\underline{\omega}^s\} +$$
$$+\,\underline{n}\cdot(\underline{q}^+ + \underline{q}^-)(1/T^+ - 1/T^-)/2 + \underline{n}\cdot(\underline{q}^- - \underline{q}^+)\left[1/T^s -\right.$$
$$\left.-\,(1/T^+ + 1/T^-)/2\right] - (1/T^s)\{\underline{n}\cdot(\underline{\underline{P}}^+ + \underline{\underline{P}}^-)\cdot(\underline{v}^+ - \underline{v}^-)/2 +$$
$$+\,\underline{n}\cdot(\underline{\underline{Q}}^+ + \underline{\underline{Q}}^-)\cdot(\underline{\omega}^+ - \underline{\omega}^-)/2 + \underline{n}\cdot(\underline{\underline{Q}}^- - \underline{\underline{Q}}^+)\cdot\left[\underline{\omega}^s - (\underline{\omega}^+ + \underline{\omega}^-)/2\right]\} \quad (15)$$

where $\underline{\underline{P}}^{vs}$ is the viscous pressure tensor at the interface.

To obtain this last expression we have also made use of the Euler equation $u^s_v = T^s s^s_v - p^s$, where p^s is the hydrostatic surface press ure, and of some simple transformations of the fluxes and forces[6]. Since surface quantities do not vary in the normal direction it follows $\underline{n}\cdot\nabla\underline{v}^s = \underline{n}\nabla\underline{\omega}^s = 0$. Taking into account this result and after straightforward manipulations we arrive to the boundary conditions

$$\underline{n}\cdot(\underline{q}^+ + \underline{q}^-) = L_1\nabla\cdot\underline{v}^s + L_2(1/T^+ - 1/T^-) + L_3\left[1/T^s - (1/T^+ + 1/T^-)/2\right] \quad (16)$$

$$\underline{n}\cdot(\underline{q}^- - \underline{q}^+) = L_4\nabla\cdot\underline{v} \quad + L_5(1/T^+ - 1/T^-) + L_6\left[1/T^s - (1/T^+ \quad 1/T^-)/2\right] \quad (17)$$

$$\underline{n}\cdot(\underline{\underline{P}}^+ - \underline{\underline{P}}^-)\cdot(1-\underline{\underline{nn}}) = L_7(\underline{v}^+ - \underline{v}^-) + L_8\,\nabla(1/T^s) \quad (18)$$

$$\underline{n}\cdot(\underline{\underline{Q}}^+ + \underline{\underline{Q}}^-) = L_9(\underline{\omega}^+ - \underline{\omega}^-) + L_{10}(\underline{\underline{nn}}\cdot\mathrm{rot}\underline{v}^s - 2\underline{\omega}^s) + L_{11}\left[\underline{\omega}^s - - (\underline{\omega}^+ + \underline{\omega}^-)/2\right] \quad (19)$$

$$\underline{n}\cdot(\underline{\underline{Q}}^- - \underline{\underline{Q}}^+) - L_{12}\left[\underline{\omega}^s - (\underline{\omega}^+ + \underline{\omega}^-)/2\right] + L_{13}(\underline{\underline{nn}}\cdot\mathrm{rot}\underline{v}^s - 2\underline{\omega}^s) + + L_{14}(\underline{\omega}^+ + \underline{\omega}^-) \quad (20)$$

where the L_i's are phenomenological coefficients obeying Onsager-Casimir reciprocity relations. If in (9) we make use of (20) together with the constitutive equation for $\underline{\underline{P}}^{as}$ obtained from (15) we get a relation among $\underline{\omega}^s$, $\underline{\omega}^\pm$ and T^s. When $L_{10}=L_{14}/2$, the rotational viscosity at the interface η^s_r is equal to L_{13}/T^s and if then we allow for the De Groot-Mazur approach ($\underline{\underline{Q}}\sim 0$), we obtain the boundary condition $\underline{\omega}^s = (\underline{\omega}^+ + \underline{\omega}^-)/2$ which is equivalent to the boundary condition for the spin field selected in [4], namely $\underline{v}^s = (\underline{v}^+ + \underline{v}^-)/2$. In any other case, more complicated bounday conditions for the spin field can be obtained.

4. APPLICATION TO DILUTE SUSPENSIONS

Polar fluid theories are able to describe the behaviour of dilute suspensions of neutrally buoyant particles[7-9]. Proceeding as in ref.8 we explain the motion of the particles by means of a diffusion equation

$$\rho\dot{c} + \nabla\cdot\underline{J} = 0 \quad (21)$$

where ρ is the density, c the particle's concentration and $\underline{J}$ the diffusion flux. If we proceed as indicated in the last section we get from (21) the condition

$$\underline{J}^+\cdot\underline{n} = \underline{J}^-\cdot\underline{n} \quad (22)$$

which indicates that there is no transfer of particles across the interface.

Neglecting thermal effects we obtain now the following entropy production at the interface

$$\sigma^s = -(1/T^s)\{\tilde{\underline{\underline{P}}}^{vs} : \nabla \underline{v}^s + \tilde{\underline{\underline{Q}}}^s : \nabla \underline{\omega}^s - 2\underline{P}^{as} \cdot \underline{\omega}^s\} - (1/T^s)\{\underline{n} \cdot (\underline{\underline{P}}^+ + \underline{\underline{P}}^-) \cdot$$

$$\cdot (\underline{v}^+ - \underline{v}^-)/2 + \underline{n} \cdot (\underline{\underline{Q}}^+ + \underline{\underline{Q}}^-) \cdot (\underline{\omega}^+ - \underline{\omega}^-)/2 + \underline{n} \cdot (\underline{\underline{Q}}^- - \underline{\underline{Q}}^+) \cdot$$

$$\cdot [\underline{\omega}^s - (\underline{\omega}^+ + \underline{\omega}^-)/2] + \underline{J}^+ \cdot \underline{n}\ (\mu^- - \mu^+)/T \tag{23}$$

where $\mu^{\pm}$ stands for the chemical potential of both fluids. From this equation one obtains the boundary condition for the diffusion flux

$$\underline{J}^{\pm} \cdot \underline{n} = \alpha_1 \nabla \cdot \underline{v}^s + \alpha_2 (\mu^+ - \mu^-) \tag{24}$$

the α_i's being phenomenological coefficients and the boundary conditions for the remainder dissipative fluxes similar to (18)-(20). All the phenomenological coefficients that appears in (18)-(20) and (24) are function of c. Since the polar behaviour of the fluids is due to the existence of the particles, when $c^{\pm} \to 0$, $L_i \to 0$ (i=9,.., 14) and $\alpha_i \to 0$. Physical situations in which the particle's concentration in one of the fluids vanishes can be observed, for instance, when the particles into a pipe migrate to central positions (Segré-Silbeberg effect) . Boundary conditions (18)-(20) and (24) are able to be applied to this case. Moreover, it is possible to consider a more general situation in which the existence of particles at the surface may be taken into account. In this situation (22) would not be valid and the set of boundary conditions would be increased by relations among $\underline{J}^+$ and $\underline{J}^-$ likely to (16) and (17).

REFERENCES

1. D.W. Condiff and J.S. Dahler, Fluid mechanical aspects of antisymmetric stress, Phys. Fluids, 7 : 842 (1964).
2. E.L. Aero, A.J. Bulygin and E.V. Kuvshinskii, Asymmetric hydromechanics, J. Appl. Math. Mech., 29 : 333 (1965).
3. S.C. Cowin and C.J. Pennington, On the steady rotational motion of polar fluids, Rheol. Acta, 9 : 307 (1970).
4. D. Bedeaux, A.M. Albano and P. Mazur, Boundary conditions and non-equilibrium thermodynamics, Physica, 82A : 438 (1975).
5. J.M. Rubí and J. Casas-Vázquez, Thermodynamical aspects of micropolar fluids. A non-linear approach, J. Non-Equilib. Thermodyn., (in press).
6. J.M. Rubí and J. Casas-Vázquez, Boundary conditions for systems with spin, (to be published).
7. K.A. Kline and S.J. Allen, A thermodynamical theory of fluid suspensions, Phys. Fluids, 14 : 1863 (1971).

8. A.S. Popel, S.A. Reginer and P.I. Usick, A continuum model of blood flow, Biorheology, 11 : 427 (1974).
9. C.K. Kang and A.C. Eringen, The effect of microstructure on the rheological properties of blood, Bull. of Math. Biol., 38 : 135 (1976).

RHEOLOGY OF DISPERSE SYSTEMS - INFLUENCE OF NaCl ON VISCOUS PROPERTIES OF AQUEOUS BENTONITE SUSPENSIONS

Mario Ippolito , Carmine Sabatino

Istituto di Idraulica - Università di Napoli

80125 NAPLES, ITALY

INTRODUCTION

Clay suspensions of engineering interest are obtained through dispersion in water of clay minerals, whose particle dimensions are of the order of microns. Viscous properties of these suspensions are strongly influenced by electric charge distribution on the solid-liquid interface, whose extensions is very large. To distribution and nature of electric charge the stability against separation is also connected.

Experimental investigations on drag reduction due to viscosity reducing agents are of technical interest in hydraulic transport of suspensions in pipes, often supplied by pumping stations. The particular origin of viscous behaviour suggests to study the effect of changing ionic composition of disperding phase and, subsequently, distribution of electric charges on solid-liquid interface, by adding variable amounts of salts.

The experience shows that salt concentration influences stability and viscosity of suspension, but the available data are not sufficient to take useful indications for optimal design of transport processes

This work is devoted to ascertain the influence of NaCl on viscosity of aqueous bentonite suspensions, whose rheological properties, previously investigated by capillary viscometer, have been presented in a previous note (Ippolito 1972).

EXPERIMENTAL EQUIPMENT

The experimental equipment consists of a capillary viscometer whose tank is supplied in closed circuit, for preventing separation through a continuous movement up to bottom. Regulation of supply is able to take constant head during experiment. The viscometric tube (stainless steel, 3mm diameter, 1.5m lenght) is horizontally positioned, with inlet section sharply connected to the wall of the tank; the end section is connected to a short tube of larger diameter; the efflux to the open atmosphere happens throught a nozzle whose converging section prevents separation of solid particles, mantaining stability of viscometric flow. During experiments the composition of suspension is invariable. The experimental equipment was placed in a thermoregulated room, at 20°C. Heat produced in the closed circuit was taken off by a heat exchanger. Pressure difference applied to viscometric tube is measured by differential manometers. Discharge is taken by direct method.

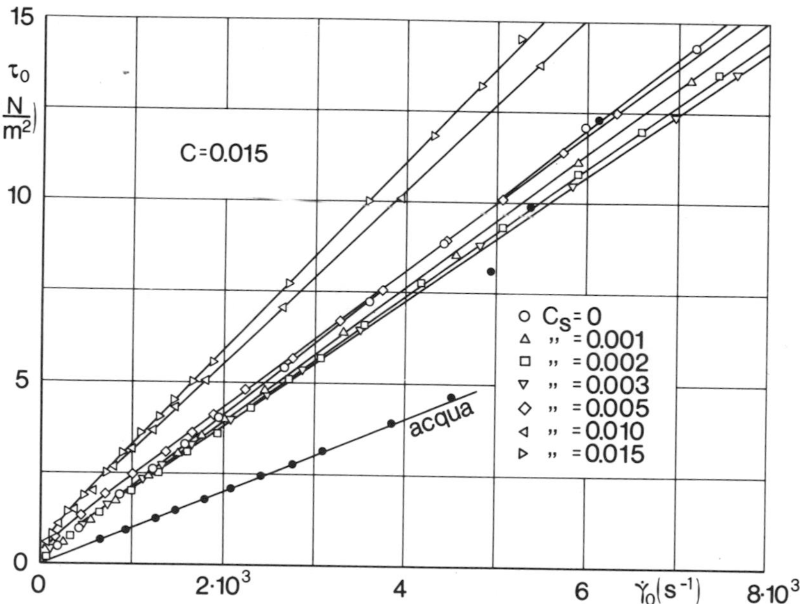

Fig.1 - Rheological curves $\tau_o(\dot{\gamma}_o)$; C = 0.015; $0 \leq C_s \leq 0.015$

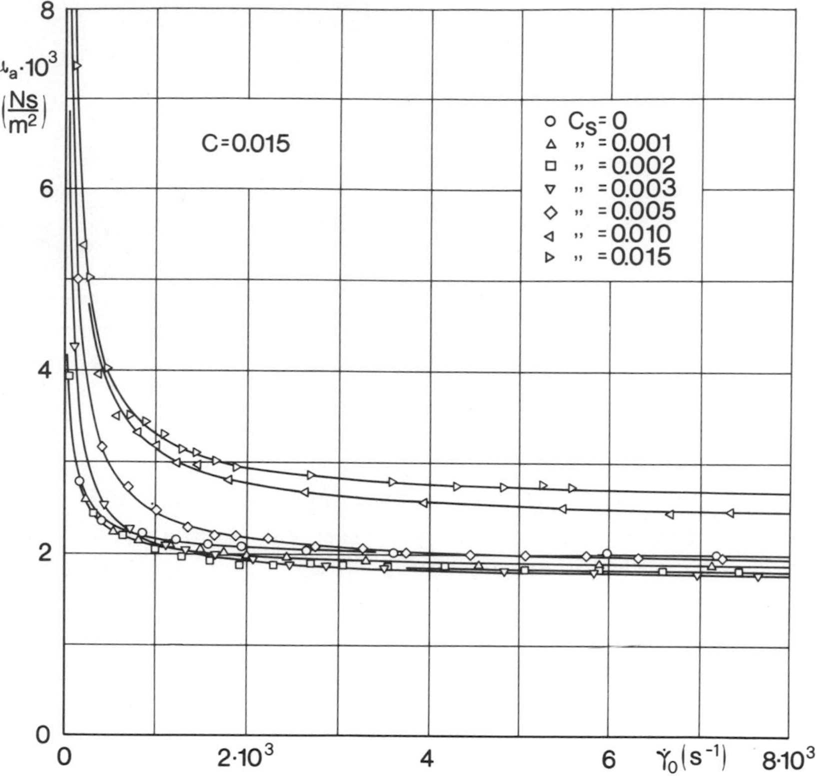

Fig.2 - Apparent viscosity μ_a $(\dot{\gamma}_o)$; C = 0.015; $0 \leq C_s \leq 0.015$

Preliminary investigations, in water flow, show that movement up to bottom in the tank of viscometer does not influence appreciably the flow in viscometric tube.

PROCEDURE AND RESULTS

The experimental work concerns the influence of NaCl on rheological behaviour of aqueous suspensions of bentonite supergel, substantially composed by montmorillonite, whose solid particles are less than 2 microns (about 84%), less than 5 microns (about 98%). The influence of NaCl is investigated for two different weight concentrations of bentonite (C=0.015 and C=0.089), that represent dilute and concentrated suspensions. The data are treated to determine experimental correlation between shear stress and shear rate at the wall.

Shear stress τ_o is obtained through pressure difference applied to viscosimetric tube, previous introduction of end correction $2.2 \ \rho V^2/2$, taking into account the kinetic energy and inlet phenomena for sharp edge, such that :

$$\tau_o = \frac{D}{4L} \left(\Delta p - 2.2 \ \rho \frac{V^2}{2}\right) \tag{1}$$

Shear rate $\dot{\gamma}_o$ is calculated by :

$$\dot{\gamma}_o = \left(\frac{dv}{dy}\right)_o = \frac{3 + 1/n'}{4} \ \frac{32 \ Q}{\pi \ D^3} \tag{2}$$

where n' is taken by experimental data fitted by power law :

$$\tau_o = k' \left(\frac{32 \ Q}{\pi \ D^3}\right)^{n'} \tag{3}$$

In fig.1 the experimental values $(\dot{\gamma}_o ; \tau_o)$ obtained for equeous suspensions of bentonite at weight concentration $C = 0.015$, and at weight concentration of additive NaCl $0 \leq C_s \leq 0.015$ are

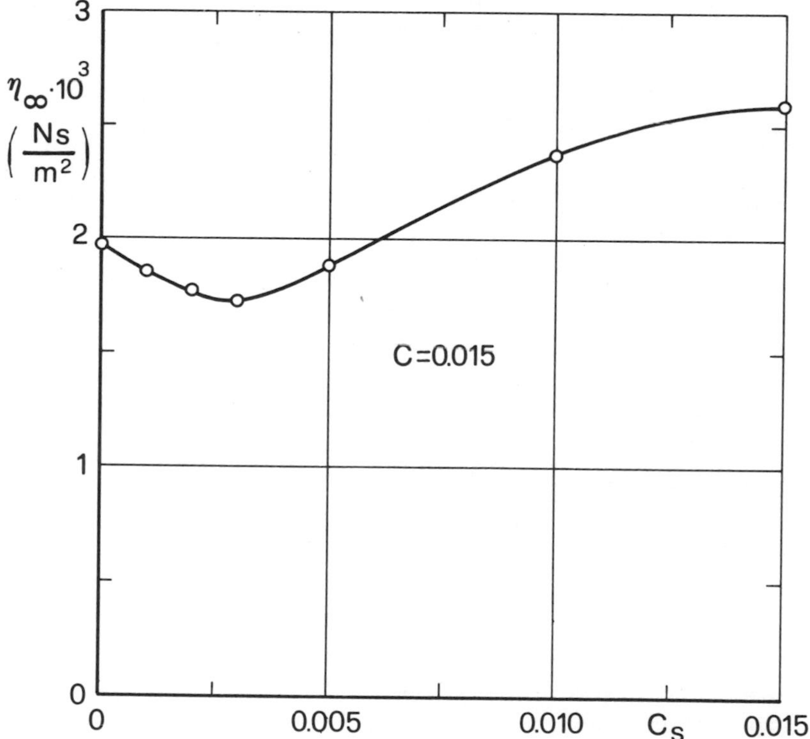

Fig.3 - Asymptotic viscosity η_∞ versus C_s; $C = 0.015$

reported, together with the data obtained with water, whose experimental points lie on a straight line defined by the water viscosity at 20°C $\mu = 1.01 \times 10^{-3}$ N s/m^2, up to $\dot{\gamma}_o = 4000$ s^{-1} ; for higher values of $\dot{\gamma}_o$ the flow becomes turbulent.

Rheological curves of suspensions are correlated following the model proposed by Briant :

$$\tau = \eta_\infty \dot{\gamma} \left(1 + \frac{\tau_\infty}{m \eta_\infty \dot{\gamma}}\right)^m \tag{4}$$

where m, τ_∞ and η_∞ are rheological parameters which take into account the tendency to a Bingham behaviour, that approaches to experimental data in large ranges of shear rate.

With increasing NaCl concentration, rheological curves goes through a minimum at $C_s = 0.003$ and then for $C_s = 0.005$ the τ_o vs $\dot{\gamma}_o$ curve returns near to that of the unfilled suspension but presents higher τ_o in the range of small values of $\dot{\gamma}_o$. Further increasing of C_s causes a large increase of τ_o .

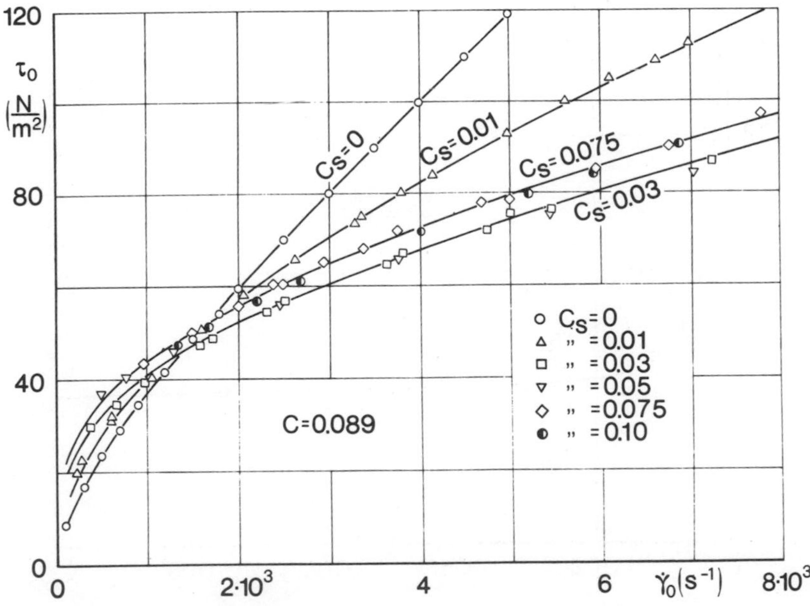

Fig.4 - Rheological curves $\tau_o(\dot{\gamma}_o)$; $C = 0.089$; $0 \leq C_s \leq 0.10$

In fig.2 apparent viscosity $\mu_a = \tau_o/\dot{\gamma}_o$ is reported versus shear rate $\dot{\gamma}_o$. The experimental points are correlated by μ_a vs. $\dot{\gamma}_o$ curves obtained through the model of Briant. In the range of small values of shear rate, appreciable effect of salt addition is observed for $C_s \geq 0.003$. At increasing shear rate, suspensions show viscosity quickly decreasing to values very close to η_∞, that represent asymptotic viscosity. In the range of higher values of $\dot{\gamma}_o$, addition of NaCl reduces viscosity μ_a for $C_s \leq 0.003$; at further increasing of C_s, rapidly increasing values of μ_a are shown.

The asymptotic viscosity η_∞ versus C_s, (shown in fig. 3) at first decreases, reaching a minimum for $C_s = 0.003$ and, at further increasing of C_s, η_∞ increases.

Fig.4 concerns experimental data ($\dot{\gamma}_o$; τ_o) obtained for concentrated suspension (C = 0.089) with addition of NaCl $0 \leq C_s \leq 0.10$. Analytical correlation of experimental points by the model of Briant

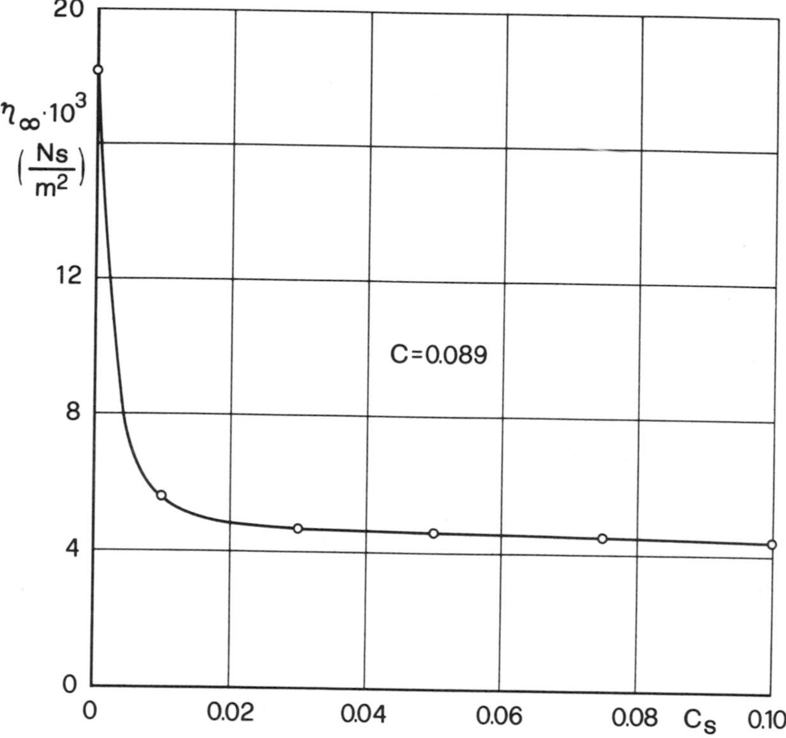

Fig.5 - Asymptotic viscosity η_∞ versus C_s; C = 0.089

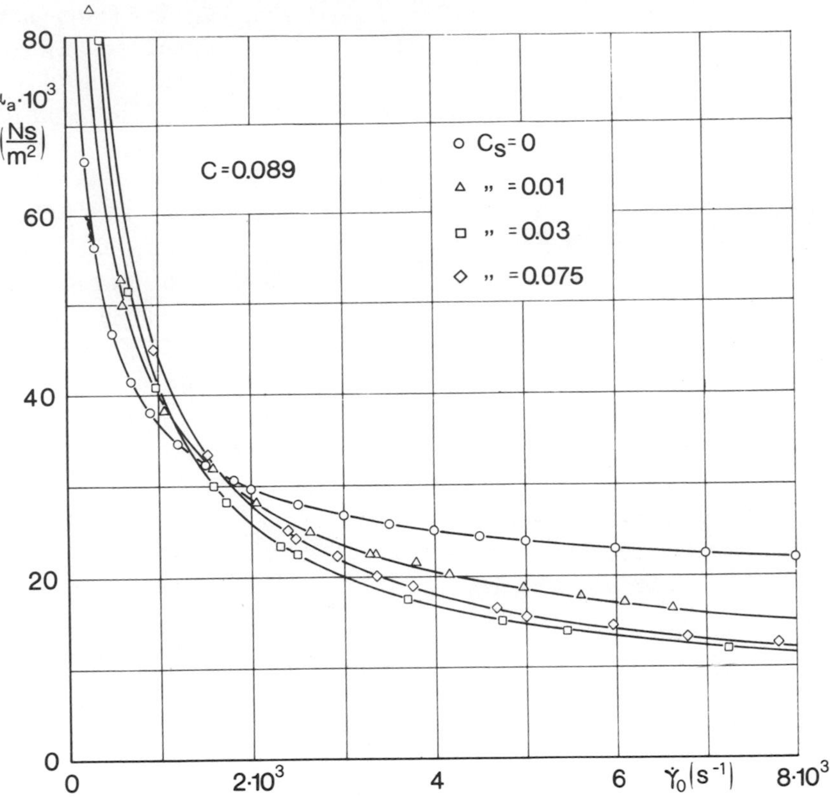

Fig.6 - Apparent viscosity $\mu_a(\dot{\gamma}_o)$; $C = 0.089$; $0 \leq C_s \leq 0.10$

is shown for $C_s = 0$; 0.01; 0.03; 0.075; experimental points concerning $C_s = 0.05$ and 0.10 are very near to that of $C_s = 0.03$ and $C_s = 0.075$ respectively.

The pseudoplastic behaviour of suspension, at increasing C_s, gradually approachs a near plastic behaviour, described by higher values of yield stress; the plastic viscosity decreases for $C_s \leq 0.03$, and then remains substantially constant at increasing NaCl concentration. The strongest reduction of plastic viscosity η_∞ happens in the range $C_s \leq 0.01$ (see fig.5).

Fig.6 shows experimental values of apparent viscosity μ_a versus shear rate $\dot{\gamma}_o$, for $C_s = 0$; 0.01; 0.03; 0.075. The full lines are obtained following the model of Briant.

Apparent viscosity μ_a increases with increasing C_s

in the range of small values of $\dot{\gamma}_o$, in accordance with previously shown results.

In the range of higher values of $\dot{\gamma}_o$ a gradual decrease of μ_a is shown. However their values are far from η_∞.

CONCLUSIONS

Experimental data show the influence of NaCl at different concentrations on the viscous properties of a dilute and concentrated suspensions of bentonite in water.

For dilute suspensions significant η_∞ data are obtained. In accordance with previous results (v. Olphen, 1963), viscosity η_∞ decreases at first with increasing NaCl concentration C_s , reaching a minimum value; at further increasing C_s , it is observed an increase of viscosity, that becomes greater than viscosity of suspension without additive. Maximum reduction of viscosity is about 15%.

For concentrated suspension apparent viscosity μ_a in the range of experimental data is too far from asymptotic viscosity for taking significant results on the basis of values of η_∞. Nevertheless it is important to observe that viscosity η_∞ strongly decreases (about 75%) in the range $C_s \leq 0.03$, while remains substantially invariable at further increasing of C_s up to 0.10. Different influence of increasing concentration of NaCl on apparent viscosity μ_a is related to strongly pseudoplastic behaviour of suspension, evermore approaching a plastic behaviour of the Bingham type, with increasing yield stress.

REFERENCES

Ippolito M. , 1972, Proprietà reologiche di sospensioni non newtoniane, in: "Atti del XIII Convegno di Idraulica e Costruzioni Idrauliche",Milano.

v.Olphen H. , 1963, "An Introduction to Clay Colloid Chemistry", J.Wiley and S., New York.

ANALYSIS AND ESTIMATION OF THE YIELD STRESS OF DISPERSIONS

Elizabeth R. Lang and ChoKyun Rha

Food Materials Science and Fabrication Laboratory
Department of Nutrition and Food Science
Massachusetts Institute of Technology
Cambridge, Massachusetts

INTRODUCTION

Yield stress affects mass transport, presents processing constraints during production and is responsible for functional performance of final products. Yield stress is of particular importance in suspensions and dispersions such as paints, printing, inks, catsup, salad dressings, cements, and clay slurries. The yield stress dominates the apparent viscosity at low shear rates, and is directly responsible for the coating characteristics, shape retention, and solvent holding capability of the dispersions. Yield stress also presents processing limitations in the form of material losses as residues remaining adhered to pipes and vessels, increased pumping requirements to initiate flow, and decreased overall rate or effects of processing. Yield stress remains a particularly important rheological property of suspensions and dispersions as it is a common and useful parameter for quality control, product identity, and for evaluating functional performance.

The criteria for the onset of yield stress have not yet been established and the accuracy of the experimental methods for measuring yield stresses has not been thoroughly evaluated. This paper discusses the yield stress in terms of structural parameters and empirical measurements.

DEPENDENCE OF YIELD STRESS ON DISPERSION PARAMETERS

When the pair interparticle interactions are present in sufficient magnitude and frequency a network of particles is formed, thus, particles in the dispersion no longer behave independently under applied shear stresses. The yield stress is the force per unit area necessary to overcome such interparticle interactions, and its magnitude is determined by the overall strength of the interparticle network. The structural parameters affecting the yield stress, can be summarized as shown in Table I.

For rigid spheres, Hunter and coworkers showed (Firth and Hunter, 1976) that the yield stress increases with the inverse of the particle radius (thus with the ratio of surface area to volume) and with the square of the volume concentration. As the axial ratio of the particles increases (i.e. the contact area to volume ratio increases) the volume concentration at which the particles are first able to interact with one another decreases, thus the yield stress increases. The type of packing determines the number of nearest neighbors to each particle, thus the interparticle association per unit area.

YIELD STRESS FOR AN IDEAL SUSPENSION

A yield stress equation can be derived, analagous to the case of perfect rupture in solids, assuming that flow occurs as a result of simultaneous rupture of the interactions normal to the shear plane. As a result of interparticle interactions, the dispersion has taken on order which is characteristic of a solid.

$$\tau_o = (F \times N)A^{-1} \quad (1)$$

Table 1. Factors Affecting Yield Stress

I. STRENGTH OF PAIR INTERPARTICLE INTERACTION

- A. Nature of Operative Force
- B. Interparticle Distance
  1. volume concentration of particles
  2. net interparticle potential
  3. type of packing
- C. Contact Area/Volume of Particle
  1. particle size
  2. particle shape
  3. surface roughness of particle

II. NUMBER OF PAIR INTERACTIONS PER UNIT AREA

- A. Volume Concentration of Particles
- B. Type of Packing Order of Particles
  1. particle shape
  2. size distribution of particles
  3. packing density

Where τ_o is the yield stress, F is the force of one interparticle association, and N is the number of associations perpendicular to the shear plane with cross sectional area A. Equation (1) assumes, that interactions normal to the shear plane are simultaneously separated, that the dispersion is homogenous and isotropic, and that energy is dissipated only by separating pairs of interacting particles. The yield stress of the particle network is a function of the strength of the pair interaction (F), as well as the number of pair interactions (N/A).

INTERPARTICLE INTERACTIONS

In dispersions, different types of forces, Van der Waals forces, hydrogen bonds, hydrophobic interactions, liquid bridges, shared counterions, or electrostatic repulsions, may act concurrently to contribute to the interparticle interactions. Some of the forces typically encountered in dispersions and their relative

Table 2. Interparticle Interactions

FORCE	MECHANISM	ENERGY (Kcal/mole)	DISTANCE Å
ATTRACTIVE			
Van der Waals	Mutual induction of dipole moments in electrically apolar groups.	2-3	3-5
Hydrophobic "Bond"	Association of non-polar molecules in aqueous solution to exclude water and reduce free energy of the system.	3-5	1-5
Hydrogen Bond	Hydrogen shared between electronegative atoms, often O or N.	5-8	2-3
Liquid Bridge	Reduction in free energy by minimizing interfacial area between two phases.	3-10	2-5
Shared Counterion	Multivalent ion shared by two ions of the opposite charge.	~30	2-3
REPULSIVE			
Electrostatic	Increase in free energy as diffuse double layers, formed due to surface charge and areas of countercharge, overlap and deform.	9-12	10-100
Steric Hindrance	Increase in free energy as macromolecules in the absorbed layer are re-arranged and compressed.	10-13	~50

strength and interactive distances are listed in Table 2 (adapted from Verwey and Overbeeck 1948, Gray 1973, Lykema 1968, Friberg 1972, Overbeek 1972, Neville and Hunter 1974). In any dispersion, the net interparticle force is determined by the sum of all forces present. Combined attractive and repulsive forces give an interparticle potential of the type shown in Figure 1. At an interparticle distance or interactive radius, r', the interactions become significant. In addition there exists one or more prefered distances which correspond to the minima in the actual interparticle potential.

The interactive radius will determine the critical minimun concentration necessary for onset of yield stress because the volume fraction of the particles must be at least such that the interactive domains of the particle overlap. The interparticle force corresponding to the distance as determined by the volume concentration and packing order of the particles determines the value of F in equation (1). The required interparticle potential (Figure 1) resulting in a particle network, in the case of a dispersion of rigid uniform spheres with radius 0.1 μm, must have a secondary minimum greater than 50 kT or an energy barrier to the primary minimum less than 10-20 kT. (Based on flocculation energies calculated according to DLVO theory, Verwey and Overbeck 1948). Both magnitude and operative range vary with the nature of the interparticle forces (Table 1).

Thus, the distance at which a particular force dominates depends

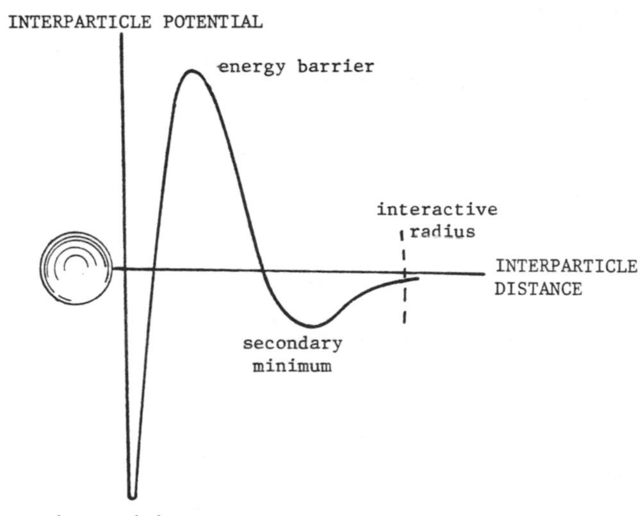

Figure 1. The net interparticle potential as a function of interparticle distance, showing the primary and secondary minima, the energy barrier to the primary minimum, and the interactive radius, r'.

on the nature of the force. It is interesting to note that some of the interactions fall to zero at a specific distance, as in the case of a liquid bridge which forms only if the interparticle distance is less than the particle radius, or in the case of steric repulsion between absorbed polymers on the particle surface which can act over a distance no longer than the absorbed film thickness.

NUMBER OF INTERACTIONS PER AREA

In the presence of the interparticle interactions and as a result of their size, shape and volume fraction, the particles will adopt a packing order which minimizes the overall energy of the system. As volume fraction of particles increases, packing often shifts from random to more ordered arrangements. The type of packing order determines the number of interactions per unit area (N/A), while the packing density and volume fraction control the interparticle distance. For example, if simple cubic packing is assumed, equation (1) will become (Bache 1976)(r" is the distance between particle centers)

$$\tau_o = F[4(r'')^2]^{-1} \qquad (2)$$

ONSET OF YIELD STRESS

As the particle size and shape change, the required conditions for onset of yield stress shift, and scaling parameters allow evaluation of the requirements. The particles must be large enough so that Brownian motion will not be appreciable and sufficiently small that the sedimentation is negligible. At a fixed volume concentration, as the particle size increases, so does the force required for onset of yield stress (Lang and Rha 1980). The ratio of contact area to volume is the appropriate scaling factor. The required minimum interparticle force established the critical volume concentration, i.e. the volume concentration must be at least that to enable the particles to maintain the distance corresponding to the required force.

MEASUREMENT OF YIELD STRESS

The most commonly used yield stress is the extrapolated Bingham yield stress. There is an important fundamental difference between this empirical yield stress and the yield stress defined in equation (1). The extrapolated yield stress is mearly a measure of the energy dissapated by means other than viscous flow in the upper Newtonian region. Dimensional analysis shows that the relative importance of interparticle interactions decreases at high shear rates (Wagstaff and Chaffey 1977) and it has been shown (Firth and Hunter 1976) that phenomena other than the rupture of the pair interparticle interactions dominate the extrapolated Bingham yield stress value. An extrapolated yield stress indicates the state of a dispersion at the shear rate from which the extrapolation is made. An extrapolated yield stress is not a measure of the stress to initiate flow because at high shear rates pair interactions no longer account for the energy dissapated.

Several other methods for determining yield stress have been defined. The yield stress determined in creep is the largest stress under which complete elastic recovery is exhibited. Equation (1) does not include time dependent terms, however, in actuality yield may occur over a range of stresses. Interparticle "bonds" will stretch, break and reform in response to an applied stress, causing viscous flow in combination with elastic deformation. Each of these processes have characteristic times. Since the "bonds" usually are composed of different types of forces, the dispersion will posses a range of characteristic relaxation times. The empirical yield measurement represents the net result of the structural breakdown and is dependent upon the time frame of the experiment and the applied load. Once an applied stress exceeds the yield value, the structure breaks down and flow occurs. However, if the stress is applied over relatively longer periods of time, relaxation and orientation will occur causing flow at lower applied stresses. It has indeed been demonstrated thay many dispersions have inherent viscoelastic properties as a result of their interparticle interactions (Van de Ven and Hunter 1979).

CONCLUSION

The yield stress is an important rheological property and it results from structural elements, thus material properties, of the dispersion. However, the inherent time dependency of interactions makes the magnitude of the creep yield stress dependent upon the time scale of the experiment. The importance of factors which dissipate stress shift with shear rate, making the extrapolated yield stress dependent upon the shear rate of the extrapolation.

REFERENCES

Bache, H.H., 1976, Tensile strength of cohesive powders, "Proceedings of the VII International Congress on Rheology", 606.

Firth, B.A., 1976, Flow properties of coagulated colloidal suspensions II, J. Colloid Interface Sci., 57(2):257.

Firth, B.A., and Hunter, R.J., 1976, Flow properties of coagulated colloidal suspensions III, J. Colloid Interface Sci., 57(2):266.

Friberg, S., 1972, "Food Emulsions", Academic Press, London.

Gray, H.B., 1973, "The Chemical Bond", W.A. Benjamin, Menlo Park.

Lang, E.R., and Rha, C.K., 1980, Yield stress of hydrocolloid dispersions, J. Texture Stud., in press.

Neville, P.C., and Hunter, R.J., 1974, The flow behavior of sterically stabilized latexes, J. Colloid Interface Sci., 49(2):204.

Overbeek, J.Th.G., 1972, "Colloid and Surface Chemistry II", MIT Press, Boston.

Van de Ven, and Hunter, R.J., 1979, Viscoelastic properties of coagulated suspensions, J. Colloid and Interface Sci., 68(1):135.

Verwey, J.W.E., and Overbeek, J.Th.G., 1948, "Theory of the Stability of lyophobic colloids", Elsevier, Amsterdam.

Wagstaff, I., and Chaffey, C.E., 1977, Shear Thinning and Thickening Rheology I and II, J. Colloid Interface Sci., 59(1):53.

ACKNOWLEDGEMENT

This study was partially supported by NIH Biomedical Research Support Grant No. 87804 from the Office of the Provost, M.I.T.

Index entries employ article rather than page numbers. The two-letter abbreviations refer to the different sections. See page vii to discover in which volume a given section is located.

AUTHOR INDEX

Index entries employ article rather than page numbers. The two-letter abbreviations refer to the different sections. See page vii to discover in which volume a given section is located.

SUBJECT INDEX